THE OXFORD HANDBOOK OF GENERALITY
IN MATHEMATICS AND THE SCIENCES

The Oxford Handbook of Generality in Mathematics and the Sciences

Karine Chemla

Renaud Chorlay

and

David Rabouin

Laboratoire SPHERE UMR 7219 (ex-REHSEIS),
Université Paris 7—CNRS, Paris, France

OXFORD
UNIVERSITY PRESS

OXFORD

UNIVERSITY PRESS

Great Clarendon Street, Oxford, OX2 6DP,
United Kingdom

Oxford University Press is a department of the University of Oxford.
It furthers the University's objective of excellence in research, scholarship,
and education by publishing worldwide. Oxford is a registered trade mark of
Oxford University Press in the UK and in certain other countries

© Oxford University Press 2016

The moral rights of the authors have been asserted

Impression: 1

Published in the United States of America by Oxford University Press
198 Madison Avenue, New York, NY 10016, United States of America

British Library Cataloguing in Publication Data
Data available

Library of Congress Control Number: 2016943983

ISBN 978–0–19–877726–7

Printed and bound by
CPI Group (UK) Ltd, Croydon, CR0 4YY

"A Matthias, ces espaces ouverts à l'exploration"

Contents

List of Contributors

Jean-Gaël Barbara Institut de Biologie, laboratoire Neuroscience Paris Seine-IBPS (CNRS UMR8246/Inserm U1130/UMPC UMCR18) and Laboratoire Sphere, UMR 7219 (ex-REHSEIS), Université Paris 7—CNRS, 75205 Paris Cedex 13, France.

Evelyne Barbin Laboratory of Mathematics Jean Leray UMR 6629, University of Nantes—CNRS, Nantes, France.

Jacqueline Boniface Maître de conférences (retired), Philosophy Department, University of Nice.

Frédéric Brechenmacher LinX-SHS, École Polytechnique, Université Paris-Saclay, 91128, Palaiseau Cedex, France.

Yves Cambefort Laboratoire SPHERE UMR 7219 (ex-REHSEIS), Université Paris 7—CNRS, 75205 Paris Cedex 13, France.

Karine Chemla Laboratoire SPHERE UMR 7219 (ex-REHSEIS), Université Paris 7—CNRS, 75205 Paris Cedex 13, France.

Renaud Chorlay Laboratoire SPHERE UMR 7219 (ex-REHSEIS), Université Paris 7— CNRS, 75205 Paris Cedex 13, France.

Olivier Darrigol Laboratoire SPHERE UMR 7219 (ex-REHSEIS), Université Paris 7— CNRS, 75205 Paris Cedex 13, France.

Evelyn Fox Keller Emeritus, MIT, USA.

Emily R. Grosholz Center for Fundamental Theory, Institute for Gravitation and the Cosmos, Pennsylvania State University, USA.

Frédéric Jaëck Laboratoire SPHERE UMR 7219 (ex-REHSEIS), Université Paris 7—CNRS, 75205 Paris Cedex 13, France.

Eberhard Knobloch Berlin- Brandenburg Academy of Sciences and Humanities, Jaegerstrasse 22/ 23, 10117 Berlin, Germany; and Berlin University of Technology, Strasse des 17. Juni 135, 10623 Berlin, Germany.

Igor Ly Maître de Conférences en Philosophie, laboratoire Ceperc UMR 7304, Aix-Marseille Université—CNRS, France.

David Rabouin Laboratoire SPHERE UMR 7219 (ex-REHSEIS), Université Paris 7—CNRS, 75205 Paris Cedex 13, France.

Anne Robadey Laboratoire SPHERE UMR 7219 (ex-REHSEIS), Université Paris 7—CNRS, 75205 Paris Cedex 13, France.

Tatiana Roque Instituto de Matemática, Universidade Federal do Rio de Janeiro, Rio de Janeiro, Brazil.

Stéphane Schmitt Laboratoire SPHERE UMR 7219 (ex-REHSEIS), Université Paris 7— CNRS, 75205 Paris Cedex 13, France.

THE OXFORD HANDBOOK OF GENERALITY
IN MATHEMATICS AND THE SCIENCES

1

Prologue: generality as a component of an epistemological culture

KARINE CHEMLA, RENAUD CHORLAY, AND DAVID RABOUIN

Generic, general, universal.

Uniform, unified.

"For almost all," "except for a set of measure zero," particular, special, exceptional, pathologic.

Principle, law, general method, ad hoc solution.

Model, example, case, paradigm, prototype.

All these adjectives, terms, and expressions have been used, and sometimes shaped, by actors in the context of scientific activity. However, they do not occur uniformly, independently of the setting. This statement holds true diachronically. It also holds true synchronically: at the same time period, different mathematical milieus, for instance, show collective use of different terms related to the general.[1]

This simple remark takes us to the core issue of this book. It aims to show *how*, in given contexts, actors have valued generality and *how* they worked with specific types of "general" entities, procedures, and arguments. Actors, we claim, have *shaped* these various types of generality. Depending on factors in the context in which they are or were operating to be elucidated, actors have introduced specific terminologies to distinguish

[1] This book is the outcome of a collective work that took place between 2004 and 2009 in the context of the research group of CNRS and University Paris Diderot at the time called REHSEIS. In the meantime, REHSEIS has merged with another research group to constitute a larger entity, newly named SPHERE. The collective work developed in a seminar that was organized by Karine Chemla, Renaud Chorlay, David Rabouin, and Anne Robadey. It allowed us to explore multiple facets of generality. We are happy to thank all the participants and contributors for the insights they gave us, as well as Rebekah Arana, who helped us with the polishing of some of the articles. Karine Chemla was able to benefit from the generous hospitality of Professor Lorraine Daston and the Max Planck Institut für Wissenschaftsgeschichte as well as the unflinching support of the librarians in Berlin during summer 2014 to work on the completion of the book, and in particular its prologue. Our thanks extend to Richard Kennedy for his contribution to the preparation of the final version of this prologue.

between different levels or forms of generality, and have designed means to work with them, or to work in relation to them. Actors have in some cases discussed which virtues they attached to the general, why it was essential to their project, and in relation to which other values they prized it.[2]

This book aims to inquire into this diversity.[3] It intends to highlight how "the general" does not assume any *a priori* meaning, which would be valid across contexts, at a given time, or even in a single discipline. Nor, in our view, would a history of scientific progress as reflected by the achievement of an ever higher level of generality be faithful to our sources. On the contrary, the goal of this book is to reveal how actors worked out what the meaningful types of generality were for them, *in relation to* their project, and the issues they chose to deal with. If such a view holds true, as we claim, it implies that evidence exists of different ways of *understanding* the general in different contexts. Accordingly, it suggests a nonlinear pattern for a history of generality. The book does not claim to offer such a history. However, it intends to open a space for such a historical approach to generality to become possible.

The claim we make about the various ways of working out and practicing generality is a facet of a more general thesis, which draws on the assumption that the scientific cultures in the context of which actors have operated do not fall ready-made from the skies. The thesis holds that actors have shaped these cultures, and reshaped them constantly, *in relation to* the scientific work they have carried out.[4] These cultures bring into play types

---

[2] Philosophers, especially those in a pragmatist tradition like Hilary Putnam, have insisted on the importance of taking values into account when dealing with scientific practice. See, for instance, Putnam (2002: 30*ff*). In philosophy of science, in the last decades, "epistemic values" have been at the center of a certain attention. See, for instance, Kuhn (1977), Laudan (1984), and Carrier (2012). In conclusion, we will situate our project with respect to this developing field of study. Let us simply note for now that simplicity, beauty and other values have often been mentioned as examples in this context. However, it seems to us that generality has not often been discussed as a value, except, as we suggest in conclusion, indirectly. This is one of the main claims of the book: that it should be addressed as an epistemic value.

[3] Hagner and Laubichler (2006) is the outcome of a similar project. Outlining some differences between the two projects enables us to be more precise on the specificity of our own project, and also on how the two endeavors complement each other. The book edited by Hagner and Laubichler addresses the issue of generality essentially as a concern that gained momentum, for instance in the second half of the nineteenth century in Germany, in view of the increasing specialization of scientific activity, and the related emphasis placed on detail and accuracy. Here, the main antonyms of the "general" are the "special" or the "partial." Was it possible, some actors worried, to maintain a global approach within a discipline? At an even higher level, they asked: Could one maintain a global and reflexive outlook on science? Was there a discipline able to represent the general level for all other disciplines? How, more generally, could one act in favor of the unity of science despite ever finer disciplinary differentiations? Typical in this respect are the congresses of the Gesellschaft Deutscher Naturforscher und Ärtzte around 1900, described by Ziche (2006), whose actual organization institutionalized the concern by opposing "general" lectures—in a sense of general that Ziche discussed—and specialized sections. Correlatively, the approach to generality most represented in Hagner and Laubichler (2006), which focuses mainly on the nineteenth century and the first half of the twentieth century, is at the level of disciplines or beyond. By contrast, in our book, we place ourselves at the micro-level of scientific activity, in settings much smaller than those of disciplines. As a rule we also examine generality closer to scientific *practice*, aiming to uncover variety behind what has often been assumed to take a rather obvious and uniform meaning. Accordingly, we pay close attention to differences between the constellations of terms linked to generality that vary across settings and reflect the categories of the specific actors. In addition, Hagner and Laubichler (2006) consider disciplines ranging from physics to history and philosophy, whereas we focus on mathematics, physics, and the life sciences.

[4] This is one of the main theses put forward and discussed in Chemla and Fox Keller (2016).

of texts and inscriptions, instruments, and other material entities. They also bear on ways of engaging with these entities. Further, they include shaping of the social organization of scientific activity as well as epistemological factors.[5] This book focuses in the first place on one such epistemological factor: generality. We do not want only to understand *how* it was conceived and implemented in different contexts, but we are also interested in how actors related this facet of their activity to other facets. It is our hope that our project can inspire similar efforts that will allow us to better understand how epistemological factors are key components of scientific cultures.

The book could have included case studies ranging from antiquity to the modern times, and occurring in any part of the planet.[6] To understand diversity in a more restricted environment, and thus be able to perceive relationships between contexts with respect to generality, we have instead chosen to concentrate on mainly, even if not exclusively, early modern and modern Europe. However, we have found it useful, for the sake of our reflection, to explore our set of issues in the context of various disciplines. The reader will thus find chapters in this volume devoted to mathematics, physics, and the life sciences.

1.1   Actors' historiography of generality and their meditations upon its value

Some actors have shown interest in the history of generality, and some have developed a reflection on it. They sometimes manifested their awareness of differences in how generality had been handled and practiced in the past. As Frédéric Brechenmacher highlights, in a chapter devoted to a controversy on generality to which we return below, for some actors like Leopold Kronecker (1823–1891), in line with a certain way of thinking about history in the nineteenth century, progress in mathematics meant, in particular, progress in the achievement of ever higher levels of generality. This points to one way of perceiving differences in the past. Actors also regularly manifested their awareness that other practitioners at their time did not approach generality in the same way as they did—they sometimes added "in the right way." This is one facet of Kronecker's dispute with Camille Jordan (1838–1922). Kronecker interprets his disagreement with Jordan in this respect as the result of a history of progress in the forms of reasoning. In his outline

---

[5] Different concepts of "scientific cultures" that could be useful for historians, philosophers, or anthropologists of science were put forward in the last decades. Knorr-Cetina (1999) introduces the concept of "epistemic cultures" based on ethnographic case studies carried out in laboratories working on high-energy physics and molecular biology in the twentieth century. Fox Keller (2002) develops her ideas about "epistemological cultures" through case studies related to the history of biological development. In that publication, she highlights in particular the epistemological factors that differ across scientific cultures and account for the problems that sometimes occur in interdisciplinary exchanges. In particular, she analyzes how in work on the same biological problems, physicists and biologists differ on what counts as an "explanation." See also Chapter 17 by Fox Keller in this book. Chemla (2009) outlines a concept of "mathematical culture," inspired by these previous publications. The introduction to Chemla et al. (2016) discusses the relationship between these concepts and others.

[6] Indeed, one of the editors has intensively published on the valuing of generality attested in mathematical writings from ancient China (see, for instance, Chemla, 2003, 2005).

of the history of general reasoning, Kronecker attributes to mathematicians of the past (and to Jordan) a practice of reasoning that was flawed—Hawkins (1977: 122) referred to this practice as "generic reasoning." In Kronecker's eyes, the Berlin mathematician Karl Weierstrass (1815–1897) had shaped a new practice, from which essential epistemic benefits derived. Kronecker's views on generality and its history, Brechenmacher argues, have left their imprint on the historiography of mathematics, which embraced them. An interpretation of the history of generality as a linear unfolding thus has roots in the past that can be uncovered. As a result of observers adopting actors' historical accounts in this respect, views like Kronecker's have overshadowed the different conceptions of other actors at the time. Historiography, Brechenmacher adds, has perhaps failed to recognize diversity in this respect.

The same conclusion derives from Olivier Darrigol's description of James Clerk Maxwell's (1831–1879) practice of generality. Historiography, Darrigol emphasizes, has adopted Pierre Duhem's (1861–1916) assessment of this practice, which was formulated from the viewpoint of another conception of how generality should be achieved. As a result, Darrigol suggests, we lack an account of how Maxwell fulfilled the ideal of generality he set himself.

These remarks highlight two benefits that derive from our approach. Gaining a better understanding of the diversity of practices of generality promises to yield tools for revisiting, in a critical way, parts of the historiography of science. Historiography has in some way made the past appear uniform, in relation to some actors' representations of generality and its history. The book will help restore some diversity and yield source material for thinking about generality in a broader perspective. Moreover, the episode on which Brechenmacher concentrates shows that the different types of general reasoning practitioners brought into play are correlated with different types of general mathematical features on which they concentrated: invariants, for Weierstrass and Kronecker, and the reduction into simpler elements, for Jordan. In addition to a history of the modes of general reasoning, we thus see the emergence of a history of the various ways of shaping generality in the subject matter itself.

1.1.1 What are the issues at stake in generality? Epistemic and epistemological values

The case studies evoked above already suggest our sources testify to motley practices of generality. Accordingly, actors have offered dissonant historical accounts of generality. These different historiographies naturally constitute an important resource for us and this is in fact where our book begins its inquiry.

Its second chapter is devoted to Michel Chasles's (1793–1880) *Aperçu historique sur l'origine et le développement des méthodes en géométrie*,[7] completed in 1837. The book

---

[7] The full title is: *Aperçu historique sur l'origine et le développement des méthodes en géométrie, particulièrement de celles qui se rapportent à la géométrie moderne, suivi d'un mémoire de géométrie sur deux principes généraux de la science : la dualité et l'homographie.*

appeared in a specific context. Chasles was researching on geometry in the framework of a mathematical culture that took shape in France at the end of the eighteenth century and the first half of the nineteenth century in engineering schools, and more specifically in relation to the Ecole Polytechnique. These geometers, who collectively shaped projective geometry, were all obsessed by the question of the relationship between analytic and "purely geometrical" approaches to geometry. In particular, they acknowledged the immense power of analytical approaches in solving geometrical problems, but could not understand why pure geometry was comparatively so weak when dealing with the same questions. For them, generality appeared to be one of the main assets of analytical approaches. Our geometers thus collectively began a reflection on the *sources* of generality in analysis, and the *means* to be developed to equip pure geometry with forms of generality able to compete with those of the rival approach. Their reflection did not develop only from within their mathematical practice, but they also approached the question from a historical perspective. Following predecessors and colleagues like Carnot and Poncelet, Chasles conducted a mathematical and historical reflection on the history of geometry from the viewpoint of generality.

In Chapter 2, Chemla relies on Chasles's reflection to capture an actor's perspective on the various facets of generality in geometry. Indeed, his book provides an amazing source of material for the examination of how a geometer perceives generality in his field, how he understands the shaping of means to achieve generality throughout history, and how he contributes to shaping further means of introducing new forms of generality. In his reflections, generality appears to be multi-faceted. What is essential for us is that it takes different forms from those we see in the context of the Kronecker–Jordan controversy.

In line with his project, Chasles surveys historical transformations of generality in geometry in two historical traditions (the analytic and the geometric approaches), his intention being to deploy another type of generality. For Chasles, ancient Greek geometry seriously lacked generality in many senses of the term. Chasles identifies a turn in this respect in about the sixteenth century.

His historical account, Chemla shows, first identifies a history of geometrical *objects*, which shapes objects with increasing generality. Chasles attributes to Girard Desargues (1591–1661) a key role in this regard. Further, in Chasles's view, the change in geometrical objects Desargues brought about was correlated with the development of means to *transfer* properties between objects formerly perceived as unrelated, or to put these properties in relation with each other. For Chasles, these operations represent another facet of generality. More generally, he underlines the introduction, in the seventeenth century, of *methods*, which established *connections* between objects and also between properties.

Chasles likewise emphasizes the emergence of *uniformity* in the treatment of problems dealing with different objects, as one important aspect of the shaping of generality in geometry, associating it in particular with René Descartes (1596–1650). From the perspective of generality, Chasles thus also sketches a history of the changing means of proving in geometry. He offers in particular a history of reflections on the actual scope of the conclusions that could be derived from a proof.

Like Kronecker's, Chasles's history is a history of progress, but it is not linear. He appreciates how in the context of different approaches, different groups of actors operating at the same time shaped different means of achieving different types of generality. Generality was explored in different directions, and each of these lines of inquiry, he shows, contributed to the history of the shaping of generality in geometry.

Chemla further highlights how specific features of Chasles's treatment of the history of geometry can be correlated in various ways with his own contributions to geometry. Let us outline only one of these facets, which will prove important for the argument of the book. Chasles's meditations on general objects in geometry, the uniformity of their treatment and the possible scope of the conclusion of a proof, Chemla suggests, inspired his reformulation of a principle, made explicit by Poncelet in 1822, under the name of "the principle of continuity."

Poncelet's principle had the aim of enabling practitioners to claim a conclusion for a general geometrical object, after having completed a proof of the proposition on a narrower set of objects. First, the principle derived from a reflection on what general objects were in geometry. It was based on a new conception, for which objects were no longer figures, but configurations of geometrical elements (lines, curves, planes, etc.) that could present different ranges of general states. Secondly, Poncelet considered that a conclusion, obtained through reasoning based on one general configuration, could be asserted about any other general configuration, as long as the latter configuration derived from the original through continuous deformation (hence the name given to the principle). In particular, when some relations between elements of the configuration had become "ideal," the conclusion, Poncelet suggested, still applied. The principle clearly contributed to the shaping of a form of generality in geometry. It stated how conclusions about general objects could be derived from a proof, despite the fact that it was not valid for all their states. For Poncelet, the principle simply accommodated in geometry an assumption that was implicitly used in analytical reasoning about geometrical figures and from which this type of reasoning derived its generality. We return to this formulation of the principle below.

The principle embodies but one facet of generality in this context. On a higher level, the reflection on generality that developed within projective geometry is characterized by the emergence of similar "principles," and also by the amount of work carried out in discussing and reformulating them. Chasles ponders this fact more widely. In this specific case, his reflections will precisely lead him to offer a new analysis for the "principle of continuity," and a new formulation, which he will then call "the principle of contingent relationships."

His analysis suggests another conception of a general object in geometry. His key concept in this respect is that of "figures," for which one can distinguish between different "general circumstances of construction." For such objects, Chasles distinguishes between permanent and contingent features, and his principle suggests another way of determining, or interpreting, the generality of the conclusion deriving from a given proof. In his view, the conclusion of a proof that has been carried out on one set of "general circumstances of construction" of a figure can be stated for the figure in any other "general circumstances of construction." Chasles further suggests, on this basis, a new practice of proof: he invites geometers to develop proofs using only permanent properties, which will apply uniformly to the figure in any set of "general circumstances of construction."

Accounting for this reflection on, and reformulation of, principles belongs to a history of projective geometry that does not only deal with results and theories, but also includes how actors produced, and discussed, the means of achieving types of generality in their practice. In this case, it is particularly important to highlight the history of the principle, since, as we show below, the related reflection on generality, and the concepts the reflection produced, actually inspired similar developments in other mathematical domains.[8] Focusing on generality and actors' reflections on it thus opens a new page in the history of science, by showing that not only concepts, methods, and results circulate, but also actors' reflections on generality.

Chasles's historical considerations are an opportunity for him to discuss the virtues he attaches to the most general statements. In his view, they are also the most simple to formulate, the easiest to prove, the most widely true, as well as being the most fruitful. By the latter term, he means these statements yield, through almost no proof at all, the other true statements in the theory to which they belong. It comes as no surprise that the identification of these statements—the "source" of all the others, whose existence he assumes for all theories—is the goal he sets for his practice. Chemla suggests this goal also echoes the aim pursued in the analytical organization of knowledge at the time.[9]

To approach the issues at stake here in more abstract terms, we suggest the introduction of a distinction, whose usefulness will be further illustrated in what follows. For Chasles, like for Kronecker, we claim generality is an "epistemic value," in that its pursuit, like that, say, of "coherence" in scientific practice, would be conducive to truth. Indeed for actors like Kronecker, as we show in greater detail below, the *types* of general statement and uniform approach he advocated guide us to true knowledge.[10] But for Chasles, generality is not merely an "epistemic value." It is *also*, we emphasize, an "epistemological value,"

---

[8] See Section 1.3.4.

[9] Lagrange (1799 (Thermidor, An VII): 280) gives a general description of this goal when he deals with the analytical treatment of spherical trigonometry.

[10] The reader should be warned that the expression "epistemic value" is used elsewhere with a markedly different meaning, for instance in the title of the book by Haddock, Millar, and Pritchard (2009). In that book, the expression refers to what makes the value of knowledge, as opposed to true belief, for instance. Knowledge is thus considered given a priori to the analyst, and what is at stake is its *evaluation* by contrast to other forms of beliefs. In our book, in general the term "value" refers to values like simplicity or elegance, which actors prize and put into play in the production of knowledge. What is at stake is how this valuing is correlated with the knowledge produced, and how these values can take different meanings. In our case, the expression "epistemic value" refers to a distinction we establish between values, the second pole of the distinction being "epistemological value" (see below). For a discussion about epistemic values in scientific practice, see Putnam (2002: 30–3). Like Putnam in his book, we distinguish between a distinction and a dichotomy. We are *not* claiming that we establish a dichotomy between values: generality is a perfect example of values that are found in both sides of the distinction. Carrier (2012) also discusses the parts played by "epistemic values" in the production of scientific knowledge. He identifies two main roles. For him, "epistemic values" contribute to the selection of the goals chosen and the evaluation of their significance. Moreover, "epistemic values" are values conducive to truth. Carrier does mention generality as one factor that enables us to capture variety in the goals actors choose as significant (p. 240). This book is predicated on the assumption that one can go further in the analysis of generality as a value. The *ways* in which the values on which Carrier focuses in his analysis are conducive to truth differ from how in our view actors perceive generality sometimes performs the same task. This calls for an analysis of the modalities according to which various types of epistemic values fulfill this function. This issue goes beyond the framework of our book. However, it appears as a promising avenue for a future general inquiry into epistemic values.

in the sense that its pursuit according to some criteria yields a kind of knowledge suited to *how* we want to know. We can obtain different forms of knowledge in geometry, that is, different types of formulations for theorems and different kinds of proof. Chasles has expectations regarding the satisfactory form of knowledge we should aim for. The link that, as we just described, Chasles establishes between generality, on the one hand, simplicity and fruitfulness as he envisages them, on the other, relates precisely to the latter concern.[11]

In the same way that actors form different historical accounts of generality, they also perceive the forms taken by generality and its virtues in different ways. Chapter 3, by Eberhard Knobloch, illustrates the fact by focusing on Leibniz (1646–1716), a particularly rich case for a discussion of generality in this respect. What is interesting for us is that the virtues Leibniz lends to generality will *at the same time* relate to Chasles's account and yet present important differences.

As we have seen, for Chasles a key manifestation of generality in geometry derived from a credo regarding the structure of knowledge in any theory. By contrast, the assumption from which for Leibniz generality derives is a theologico-philosophical principle, which holds there is a "universal harmony" in the world. This harmony is brought to light through the reduction of the variety of things to the highest order possible—this is another form generality can take. Mathematics can be used to highlight order. Order can also be disclosed within mathematics. This goal assigned to mathematics and mathematical endeavor thus grants generality its value. To use the terminology introduced above, it is in this regard an epistemic value. Knobloch's chapter focuses on the latter dimension: the disclosure of order within mathematics. The possibility and the meaning of the general are thus postulated, and identifying the general in Leibniz's view will allow us to perceive beauty in mathematics. It will take the form of "divine theorems," "laws," "methods," and so on.

A first type of manifestation of the general is the production of theorems that "link together the most dispersed things," or are common to all formulas in a given context. This appears as a first way of reducing a given variety and showing order in it. Although the theme of connecting truths was also present in Chasles's account, Leibniz attaches different virtues to this reduction, which brings to light in which regards generality is also an epistemological value for Leibniz. For him, such theorems are "excellent summaries of human understanding," constituting "abridgements," which ease memorization and the work of thought. With reductions of this type, the practitioner is saved the labor of repeating similar treatments in situations that are shown to be related.

For Leibniz, the general also has the property of being simple, however, in his case, prominently in relation to the fact that it is *concise*: all irrelevant details have been

---

[11] Values of this type are also those that are at play when actors choose between competing theories that equally account for facts. Simplicity, beauty, and other values have been evoked in this respect. We suggest that these values can be epistemic or epistemological, depending on how actors justify their use. They have been discussed mainly in the philosophy of physics and in relation to theory choice. The case in mathematics that we discuss here indicates that these values are used at different levels (or scales) and play different parts. This also awaits further description.

eliminated. In this manner, Leibniz suggests, fruitfulness is implemented, and this value takes meanings close to those we outlined above. However, in line with Leibnizian reflections on the characteristic, it comes as no surprise that the meditation takes a specific course. For him—a fact that played no role in Chasles's reflections—adequate notations yield conciseness. They thus play a key role in the exhibition of the general. Once notations paint the "intimate nature of things," they disclose the universal in them, and reasoning becomes a computation. The reduction that adequate notation carries out also allows again the expression to "be easily retained" and to "the labor of thinking" to be "diminished." As Leibniz shows, a reduction of this type also provides help in carrying out further reductions of the first type. They are both tools in the service of the art of invention.

Finally, for Leibniz as for Chasles, the general is the subject of an unending quest, the assumption being that there can always be higher levels of generality, yielding even more powerful resources. In this quest, a field appears as fundamental for Leibniz, namely, combinatorics. The key position of this field in relation to generality can be correlated to a specific mode of expression of the general: "laws" of formation. Laws of this kind highlight patterns in general expressions, formulas, or tables, allowing practitioners to produce these entities, without memorizing them, and thus dispense with them. The fact that these laws represent higher forms of generality echoes the fundamental position Leibniz grants to combinatorics in mathematics.

In conclusion, we see clearly that the meaning of the general, the values attached to it, as well as the forms it can take vary significantly from one context to another, in close relation to actors' scientific activity.

How do actors understand the way(s) in which the general can be established and the way(s) in which the extension of its validity may be captured? How do they consider it can be worked with appropriately? These are the issues to which the next chapters turn, while displaying yet other meanings attached to, and other forms taken by, the general in other contexts.

1.1.2 Actors' reflections about generality

Actors do not always formulate their reflections on generality explicitly, nor do they always make explicit which options they take in this respect. Yet, aspects of their reflections and choices on this issue can very often be gathered from clues we find in their writings. In some cases, background knowledge about actors and their immediate context, especially the scholarly culture in the context of which they operated, can complement the sources that come down from them and help us describe how they understood generality and how they worked with it. However, this is not always the case.

As usual, ancient history provides the most critical examples in this respect, where we have isolated documents, with no meta-level statements that might reveal actors' reflections on generality and their practice of it. How can we, in such cases, interpret the clues and describe our actors' take on generality? The book illustrates the problem, and examines this issue, with an example taken precisely from ancient history. This constitutes its main incursion into earlier time periods. In fortunate cases, we can find other documents that, although they may have been written centuries apart from the sources under consideration, present connections with them regarding the issue of generality and, further, make reflections on this topic more explicit. Even though this information must clearly be used with discernment, it yields precious evidence to interpret the clues our sources contain on actors' understanding of, and practice with, generality.

Chinese mathematical texts from antiquity offer an example of a situation of this kind. For example, the classic *The Nine Chapters on Mathematical Procedures* (completed ca. first century CE) gives many clues that generality was a major epistemological value for its authors. And yet, at first sight, the book includes no comments on this fact. Nor can we find, strictly speaking, any contemporary document that would fill this gap in our documentation. However, we are fortunate enough to have another classic, probably completed a century earlier, *The Gnomon of the Zhou [Dynasty]*, as well as commentaries from the third and the seventh centuries on *The Nine Chapters*, handed down with that classic. These documents all yield essential information enabling us to interpret features of the approach to generality in the context of *The Nine Chapters*, as well as in its commentaries (Chemla, 2003, 2005). Greek mathematical texts of antiquity confront us with a similar situation.

Clearly, Euclid's *Elements* also reflects careful consideration, as well as practice, of generality.[12] However, no immediately related evidence can allow us to describe Euclid's cogitations other than clues in his text. In this case, Aristotle's detailed discussions on generality, which additionally often evokes mathematics as a key example, provide crucial information for the interpretation of these clues. Chapter 4, by David Rabouin, is devoted to this case study. Rabouin considers features of Euclid's approach to generality more specifically in the example of the theory of ratios.

As is well known, the issue of generality is essential to Aristotle's discussion of science, since for him the principal characteristic feature of scientific knowledge is that it is knowledge of the general. In this context, Aristotle frames the issue in a very specific way. He considers that the general is attached to "genres," and that it formulates essential attributes of entities falling under a "genre," which derive from a "demonstration" holding for all of them. This specific approach to generality nicely correlates with the structure of Euclid's *Elements*. Indeed, the *Elements* can be decomposed into two parts, one dealing with geometrical objects and the other with numbers. Moreover, proportions are defined for these two domains of entities in two different ways, and similar properties of proportions are proved using wholly different characteristics of objects in each case.

At first sight, the correlation between the clues given by the mathematical text about the treatment of the general and the theory of "genres" expounded by the philosophical text thus appears to be obvious. However, Euclid's *Elements* also contains a puzzling passage, in which proportions on magnitudes and proportions on numbers are brought into relation. This point seems to challenge the interpretation of the general in Euclid's *Elements* as conforming to what Aristotle describes. It further raises problems for the interpretation of the *Elements stricto sensu*.

---

[12] *Pace* Chasles, from whose perspective Greek texts of antiquity were deficient in this regard.

It is interesting to note how the difficulty in interpretation is in line with a problem in understanding how Euclid deals with the general. In fact, Rabouin highlights that this difficulty is *itself* correlated with vexing issues in the interpretation of Aristotle's problematic statements on the general. Accordingly, the solution Rabouin offers questions received views on the general in Aristotle's theory. Widening the documentary basis on which this question has been approached, Rabouin offers a modified account of how Aristotle understands the general in science and how he prescribes it should be pursued. It is noteworthy that this modified account enables us to make sense of the features of the Euclidean text that were perplexing. Perplexing, Rabouin stresses, only for the modern reader, since we have no evidence that ancient commentators took issue with these features of Euclid's *Elements*. Perhaps, Rabouin suggests, false expectations with respect to generality in the ancient texts have created challenges for modern readers.

This conclusion is worthy of greater consideration by historians dealing with values such as generality. Different understanding and practices of a value in different contexts demand methodological prudence in order to avoid reshaping the past on the basis of our expectations. We have already met with similar concerns above.

This ancient discussion touches the question of the *entity* to which general properties can be legitimately attached. It also brings into focus the issue of which *procedures* can legitimately be used to establish a property of that kind. It sets the frame for a practice of the general. Closer to us, we still find actors, like the mathematician Henri Poincaré (1854–1912), who wrote many texts discussing the meaning of general statements and ruminating on their legitimacy. Igor Ly's analysis of these reflections, in Chapter 5, shows that the issues Poincaré addresses in this respect in his philosophical writings are wholly different from the issues outlined above. For Poincaré, as for Aristotle, there is no science without the general. However, Poincaré is interested in understanding the different meanings of generality in different disciplines, and comparing the practices of generalization in these different contexts. His aim is, in particular, to characterize the meaning of generality in mathematics and to understand the part played by mathematics in practices of generalization in other disciplines.

To begin with, what does it mean, Poincaré asks, to speak of "all the integers", for instance? And, how should we interpret a mathematical statement asserting that a property is shared by the elements of this infinite collection? Here, as elsewhere, Poincaré's discussions about generality involve the infinite. A key point in Poincaré's answer to these questions consists, Ly stresses, in his interpretation of the infinite: for him, it is never "given," but is endlessly in construction, referring in fact to the potential infinity of a sequence of operations. Accordingly, generality is thus not the *result* of, but rather the *operation* of generalization itself. With respect to integers, for instance, Poincaré suggests their collection is conceived through the "power of the mind" to repeat the addition of 1 indefinitely. This power of the mind, of which we have the intuition, is, for Poincaré, what gives meaning to such expressions as "all the integers." It also allows us to shape and grasp the mathematical continuum and other mathematical concepts. The same "power of the mind" is brought into play in mathematical induction, which for Poincaré is *not* a logical, but a purely mathematical type of reasoning. The reason for this is that induction requires an indefinite combination of the same or of similar acts, which in Poincaré's view characterizes mathematical generalization, by comparison to generalization in other scientific fields. In fact induction provides mathematics not only with an essential tool for generalization, but also for shaping mathematical concepts. These arguments help explain why in Poincaré's reflections, generality, generalization, and the infinite are always associated with each other.

Poincaré is also interested in generalization in physics and especially in the part played by mathematics in these generalizations. As Ly shows, Poincaré develops careful and specific analyses with respect to both mathematical physics and experimental physics. In the latter case, Poincaré offers an extremely subtle theory of the part played by mathematics in inductions carried out on the basis of experimental measurements. Here again, his conclusion leads Poincaré to oppose this type of induction and a generalization that operates by means of the extension of the domain of a predicate.

In all these cases, the scientist Poincaré discusses philosophical questions as an observer of scientific activity. He takes as his starting point the practices of generalization in mathematical, or experimental, physics. And he develops quite specific interpretations of what generality means and how generalization functions in different contexts, thereby justifying the legitimacy of these operations in mathematics and in physics.

Quite interestingly, the mathematician Poincaré is also essential in another respect. We have abundant evidence that in his work he proceeds in a specific way with respect to generality, constantly keeping an awareness of the generality of the situations he is dealing with. Even though, to the best of our knowledge, he remains silent about this, there is ample evidence of the fact in the statements he uses and the structure of his writings. This takes us to another range of issues that the book addresses.

1.2   Statements and concepts: the formulation of the general

We have seen so far the variety of reflections actors developed with respect to the general. These reflections partly overlap and partly diverge. They clearly constitute a precious asset for our project of a historical study of generality. However, how did actors express and state the general? How did they write it down? Part II of the book examines this question, emphasizing actors' roles in shaping concepts and statements to grasp, and express, the general. On this issue, Poincaré will provide us with a magnificent example, which highlights a key phenomenon for our history: the historicity of the statements actors used to formulate the general.

1.2.1   Developing new kinds of statement

In Chapter 6, Anne Robadey provides evidence documenting the circumstances in which a practitioner introduced a new type of general statement. The statement in question asserts that a proposition holds true for "almost all" the objects considered, where the meaning of "almost all" is *quantified precisely* using mathematical tools. Robadey notices statements of this kind are also characterized by the fact that no attempt is made to individually identify the "exceptions," the existence of which is referred to.

The practitioner to whom Robadey ascribes the invention of this type of general statements is none other than Poincaré himself. She is even able to situate the moment of this invention between early December 1889 and 5 January 1890. The historical background for it is clear. In 1888, Poincaré competed for a prize offered by King Oscar II of Sweden. Having been awarded the prize, he submitted the text of the *memoir* for publication. At the end of 1889, he became aware of a mistake in the proof of the main theorem and withdrew his publication. Within a month, he struggled to fix the problem and, on 5 January 1890, he was able to send the new version of the *memoir*, which was published. The introduction of the new type of statement was part of the resources on which Poincaré drew to formulate and establish a new result in place of the erroneous one. Robadey can document subsequent episodes more finely in the shaping of the statement under consideration.

In her chapter she emphasizes several points that are essential for our purpose. First, in his formulation of the new type of statement, Poincaré uses the word "exceptional" to refer to the cases for which the general property does not hold. He recurrently emphasizes that in these statements, this word, which is taken from ordinary language, is a technical term, making clear the mathematical meaning he has ascribed to it. Interestingly enough, Robadey notes, this fact contrasts with Poincaré's way of using another technical expression: "the most general polynomials of their degree." Clearly, the latter expression relates to yet another specific type of statement of the general. For us, its sense requires some explanation. However, Poincaré seems to take the technical meaning of that other expression for granted, never stressing it, nor even defining it. This suggests the conclusion that the latter technical expression was in common use in the mathematical culture for the members of which Poincaré writes. This thus gives us hints that in given contexts, actors use shared sets of specific technical expressions of the general. By contrast, when Poincaré introduces a new type of formulation of the general, he feels compelled to warn his readers of the technical dimension of its terms.

Second, Poincaré relies on knowledge in probability theory to ascribe a meaning to the term "exceptional." In brief, he defines "exceptional" as that which arrives with a probability equal to zero. Moreover, in the second version of the *memoir* where he uses these concepts, he makes clear how he suggests defining such a probability. This, Robadey emphasizes, is in fact the first piece of evidence we have of Poincaré's reflections on probabilities. In fact, when he brings probabilities into play to fix the flaw discovered in the first version of his *memoir*, Poincaré does not use knowledge on the topic that would be readily available. Robadey is able to show how the new version evidences Poincaré's research work on probabilities that remained otherwise unpublished at the time. What is more, Robadey suggests the critical situation of having to correct his mistake prompted Poincaré's personal work in this field, which he would later revisit. In fact, he puts his research on the topic into play to give meaning to the new type of statement he introduces. This fact highlights a key conclusion: shaping a new statement of this kind is not only a matter of formulation but also requires mathematical knowledge, which in this case was developed for this purpose. Actually, it is precisely this facet of the new statement—the definition from probability it brings into play—that underwent transformations in Poincaré's successive statements of his main new theorem between 1890 and 1891.

Third, through a remarkably fine analysis of a corpus of texts, Robadey establishes that the introduction of the new type of statement is correlated with other facets of Poincaré's mathematical work to correct his result. Poincaré modifies the meaning of the key concepts at stake in his *memoir*—like that of stability. He restructures the organization of his text, redefining the goal to be achieved. The way in which he chooses what will become the new essential result of stability is correlated with the possibility of stating something strong, which is general enough to be meaningful, even though it does not hold universally.

Finally, the nature of the statement, which points to exceptional cases without identifying precisely what they are, is correlated with the type of proof Poincaré develops, which, as Robadey stresses, is non-constructive. The new statement is altogether produced in the context of this gigantic reshaping and is what makes it possible. Again, the cultural artifacts actors produce in the context of their activity cannot be dissociated from the questions they address and the goals they set themselves. The modes of stating generality illustrate this more general assertion.

To recapitulate, the type of statement we are interested in was introduced as a resource to shape a new approach in line with the result that needed to be established. Although Robadey accurately establishes the circumstances for its introduction, we still lack a precise historical account of how other practitioners picked it up, and began using it. However, the fact is, as we shall see below (see Section 1.2.3), that statements of this kind circulated widely, being reworked and extended in various types of context, in which they served as inspiration to actors structuring collective research programs.

Such case studies are essential for our purpose, since they highlight a phenomenon that, to the best of our knowledge, has been overlooked. Statements of forms of generality have their history, which is worth addressing, and their production is as important a part of scientific work as is the production of new concepts and results. Once they have become part of the tools adopted within a given scientific culture, they yield key resources for the practice. But there is more. Robadey shows the approach to mathematical situations that statements of this kind disclose is not exceptional in Poincaré's mathematics. To the contrary, it fits with a systematic attitude toward his research topics, to which many of his writings testify. One could refer to it as a *style* of dealing with generality. Indeed, Poincaré's writings are full of explicit remarks he systematically adds to his reasoning, assessing degrees of generality of a situation under consideration. Robadey highlights the rich terminology omnipresent under his pen: "general," "particular," "exceptional," "most general," and so on. Moreover, she identifies three types of resource Poincaré puts into play to quantify a degree of generality. In addition to probability, she notes he regularly counts the number of arbitrary constants that the expression of a solution to a problem involves, as an assessment of the size of the set of solutions thereby found. He also uses insights from Georg Cantor's (1845–1918) research, using concepts that would soon become essential in topology (dense set, perfect set, and so on).[13]

---

[13] We return to these concepts in Sections 1.2.3 and 1.3.2.

The same kind of view on mathematical situations can also be perceived in how Poincaré writes down his exploration of a problem, when the treatment requires cases to be distinguished. The article by Robadey (2015) is devoted to how Poincaré proceeds when he presents his reasoning in the form of an enumeration of cases, as he often does. She shows that Poincaré *systematically* lists cases in an order of decreasing generality, opposing a general case to particular cases, whose generality *relative* to each other must also have been assessed, since they are listed accordingly. Incidental remarks throughout texts of this type reveal Poincaré's awareness of relative degrees of generality among different types of particular cases.

This type of assessment of generality can be documented from Poincaré's early writings on differential equations, in 1878 and 1879. Robadey establishes that, for this case, this feature distinguishes Poincaré from his predecessors. It constitutes the basis for the development of a new type of reasoning using the relative degrees of generality of particular cases. In fact, the related hierarchy of cases, presented in the form of an enumeration, is a key resource on which Poincaré draws to develop his new global approach to differential equations. This new approach depends on his ability to focus on what is both tractable and essential in the mathematical situation, as determined by the assessment of degrees of generality.

This remark allows Robadey (2015) to characterize the nature of the general reasoning that Poincaré carries out. It is by no means a kind of "generic reasoning" of the type Hawkins (1977) showed that Weierstrass had criticized. In fact, Robadey suggests that under the label "generic reasoning" Hawkins might have put together types of reasoning that differ substantially, and she outlines the beginning of a typology which is most interesting for the project of our book. In this setting, Poincaré's reasoning appears to be a general reasoning carried out within a framework defined by a quantification of the generality of the cases left aside, by comparison to the size of the collection of cases dealt with. In conclusion, we see a strong connection between some early works by Poincaré and his *memoir* evoked above: in the latter case, the specific concern about generality takes the shape of a statement, whereas, in the former, it takes the shape of the structure of a text. Moreover, Robadey emphasizes, the project to shape an enumeration in this way and the criteria put into play to do so are *not* stated discursively: *only* the text of the enumeration and incidental remarks on it reveal this part of Poincaré's work. The writing of the general is not only located in concepts and statements. It can also take textual forms, whose interpretation becomes more difficult for historians. In fact, elsewhere Robadey sheds light on another phenomenon of exactly the same kind, when she shows how a *memoir* by Poincaré is not in fact devoted to the topic it apparently deals with. Indeed, she establishes the topic under consideration is a *paradigm* in the context of which Poincaré chooses to present a general method.[14]

Frédéric Jaëck addresses a similar issue in the next chapter (Chapter 7) of the book, which he devotes to the introduction of what is for us today an "abstract mathematical

---

[14] Robadey (2004) further endeavors to account for why Poincaré chose to write his memoir in this way. Again, the interpretation of such texts is challenging, in relation to the fact that they express the general using a textual form.

structure." Incidentally, by contrast to the previous cases, here one would be tempted to recognize forms of generality related to abstraction. We have seen so far that these were clearly not the only types of generality. We will now see that the matter is more subtle. In the publication of his PhD thesis, in 1922, Stefan Banach (1892–1945) introduces a system of "axioms," in which we could be tempted to identify what today is called "Banach spaces." Part of these axioms relate to the fact that Banach spaces are vector spaces. The other part introduces a norm and the property of completeness. However, in contrast to what previous historians of mathematics have claimed, Jaëck suggests that in this early paper, Banach does *not* introduce the object now given his name. In his view, the axioms play a different role in 1922 from the one they play in the *Théorie des opérations linéaires* which Banach published ten years later. They do not have the same *generality*, and this remark leads him to distinguish two stages in the process of emergence of Banach spaces. Jaëck suggests introducing a distinction between *forms* of generality, which yields a different periodization in the history of science. Again, the distinction between these forms depends crucially on suggestions regarding how historians should interpret their sources.

More precisely, in 1922, Banach's aim in introducing these axioms is to identify key properties, shared by different collections of functions (a list of which is provided at the beginning of his article, and to each of which a specific norm can be attached). The introduction of these properties is tied to a specific organization that Banach intends to give to mathematical knowledge in the 1922 article. Banach's aim was to deal with integral equations. For this, in a first part, he wants to establish certain theorems that hold true for various collections of functions. His ambition is to establish these theorems once and for all, solely by using in his proof the axioms brought to light. By proving that the various collections of functions, each with a specific norm attached, satisfy the axioms, he can apply the theorems to them. The key part played by the axioms in the 1922 publication is thus to allow Banach to make *general proofs* for theorems. As a result, the generality of the axioms is *bounded* by the list of collections of functions.

All the theorems proved in relation to the axioms are used, in the second part of the article, to deal with the integral equations. Hence the consequences of the axioms are only considered in relation to a preassigned goal. In this sense, there is no study of the "Banach space" object as such, in contrast to the 1932 book. Jaëck captures this latter feature of the 1922 article by stating that it does not manifest "reflexivity" with respect to these axioms. These remarks define a first type of generality that Jaëck identifies in Banach's writings.

In the 1932 book, the same axioms have an entirely different meaning, which can be captured in the structure of the text. To begin with, in 1932, the axioms allow Banach to introduce a general object, which will later be called "Banach spaces." The book studies some properties of this object as such, without attaching the axioms to a closed list of collections of functions. In Jaëck's terms, the generality of the axioms is now "open." Moreover, the results obtained about the objects manifest "reflexivity," in that they betray an interest in their properties, rather than an intention of deriving specific applications. This is the point where generality is achieved by means of abstraction. "Reflexivity" and "openness" are the two criteria that lead Jaëck to identify, in the 1932 book, a second kind of generality. Indeed, between 1922 and 1932, the meaning of the term "general" has changed. From a textual viewpoint, the 1922 and 1932 publications present the same axioms, which in the present day we associate with "Banach spaces." However, the meaning and status of these axioms, and as a result, the kind of generality they embody differ. The same conclusion holds true for the various parts into which the 1922 axioms can be grouped. We can find the axioms of vector spaces in Peano's *Calcolo geometrico* and Riesz introduces the key properties of norms in 1916. However, the meaning and status of these statements, their generality differ from what we find in Banach's successive publications.

Jaëck raises an essential problem for history of science. How can historians determine the nature and degree of the generality of statements, like axioms, they read in their sources? Jaëck illustrates the key fact that the generality of a statement derives not only from the words composing it, but also from the *context* in which it is used and the *way* in which it is used in this context. In line with this concern, Jaëck pays specific attention to the organization of the various texts he takes into account, bringing to light that the organization with which knowledge is presented, and the deductive structure are, in this case, essential ingredients in determining the kind of generality that its statements assert. Such a method proves here to be a useful tool for conceptual history. In a sense, the process Jaëck describes is of the type the philosopher Jean Cavaillès (1903–1944) referred to as "thematization" in mathematics.[15] From this standpoint, the micro-historical analysis Jaëck develops here suggests more historical work can be carried out to observe in greater detail how precisely thematizations occurred.

But this is far from the end of the story. First, with respect to the 1932 book: its organization in fact bears witness to a more global change of perspective. Its successive chapters are devoted to various structures that are defined by a subset of the axioms introduced in 1922 (groups, general vector spaces, normed spaces, etc.). The intention, in 1922, of crafting general proofs paved the way to the introduction, in 1932, not only of a general object, but of *different* general objects of the same kind. These objects all embody the second kind of generality. Moreover, these general objects are connected with each other in a scale of decreasing generality. Secondly, now, with respect to the 1922 article: Jaëck remarks that, when dealing with this document, historians have previously focused their attention on the axioms in relation to the question of the origin of "Banach spaces." However, the article gives a prominent role to linear operations. These are also general objects Banach considers in his PhD thesis. Further, even though most theorems about them are clearly motivated by the intention of solving functional equations, some theorems seem to indicate that Banach also considers them for their own sake. Perhaps, Jaëck suggests, in the 1922 article, one can detect some reflexivity vis-à-vis operations and thus the constitution of a general object of the second type. Perhaps can we perceive here the beginning of what later would be called the theory of operators. Thematization processes might be sometimes intertwined.

---

[15] See Cavaillès (1938: 177–8).

1.2.2   A diachronic approach: continuity and reinterpretation

In Chapters 6 and 7, we have considered how specific actors introduce ways of *stating* something general. We have examined the case of expressions using a specific type of concept or statements of a technical kind. We have also advocated why we *have* to consider other types of general formulations that used more macroscopic textual features. Once these types of general entities and statements are introduced, how do other actors appropriate and rework them? In other terms, what is the historicity of these entities and statements? This is the specific issue examined in Part II.2, in which we adopt a diachronic perspective.

The first case study in this regard deals with a topic that is somewhat paradigmatic in the context of our project. It is an investigation of the concept of "genre" in natural history. Indeed, grouping living objects into species and genera can be taken as one of the paradigmatic activities, if not *the* very activity, embodying a search for generality in the sciences. In Chapter 8, Yves Cambefort analyzes the constitution of these "genres" in the long term. What the examination of this type of activity shows, which is of wider interest for our theme, is that the practices leading to the constitution of these groups in different contexts and at different time periods—practices of generality, we might say, or practices of genre making—present essential differences, despite significant continuity in the groups shaped and in the general terms used to refer to them. To highlight this point, Cambefort's chapter sketches, mainly for zoology, how a similar ambition of, and a quest for, generality have been pursued with entirely different methods, and diverging interpretations of the groups constituted.

For Plato and Aristotle, Cambefort argues, the introduction of genres and species for animals did not aim to *classify*, but rather to *differentiate* between living objects. This is coherent with what we have seen above for Aristotle, in relation to mathematics. This approach to "genres" and "species" echoes their "downward" practice of differentiation: Plato, like Aristotle, started from genera (in their sense), and introduced criteria of differentiation within these genera, to distinguish between species. Despite this similarity in the procedures of grouping, Cambefort stresses, their practices of dealing with *differentia* were not the same. However, neither the constitution of a classification nor that of a terminology were related concerns for them. Only later, did these artifacts become explicit goals for naturalists.

For example, Aristotle's concept of genre had no absolute value, each genus being understood in relation to the species into which it was divided. Any genus could, in fact, be considered a species with respect to another, higher cluster, which then was considered a genus. A key change occurs in this respect in the context of seventeenth-century botany, before it was adopted for animals.[16] Cambefort suggests *classification* then becomes a central concern. Correlatively, the *identification* of natural objects becomes a key task.

---

[16] Note the circulation of a practice of generality from one context to another. We return to this issue subsequently.

In this new environment, genres become more decisively attached to the classificatory activity, and they are interpreted as a level in a more general arborescent scheme. The latter feature illustrates how, depending on the context, the shaping of general entities can be associated with different spectra of scholarly operations. Further, at the time genres were considered to be absolute, occurring all at the *same* level (or rank) in the classification of living beings. In other words, the same "general entities" became less logical and, by contrast, more loaded with meanings related to the natural world.

In the subsequent century, we note another shift in the shaping of genres, in relation to changes in the ways of carrying out these activities. This is what can be gathered from the observation of Carl Linnaeus' (1707–1778) specific practice of identification and classification. While Linnaeus adopts some features and practices for genres and species deriving from seventeenth-century botany, he nevertheless suggests a new interpretation for genres, taking them to be entities created by God and hence natural facts the naturalist must discover. With this particular interpretation, the sort of naturalistic operations put into play to establish these groups—our general entities—undergo a transformation. The *discovery* of these genres and species becomes the purpose of the naturalist's classificatory activities, now practiced in an upward way. The effort bears on identifying key features that enable the naturalist to recognize genres. Moreover, the system of names Linnaeus suggests is tightly related to this practice of grouping, since it aims to help practitioners situate genres and species as groups that present natural divisions with one another, as reflected in the terminology. Seen from a higher perspective, these names can be considered as symbolic tools shaped to facilitate the circulation within the system of general entities. Again, we see how in each context specific practices relate to ways of making genres, which are interpreted in different ways.

The contrast between Linnaeus and Georges-Louis Leclerc de Buffon (1707–1788) illustrates clearly how other assumptions about the natural world can lead to an entirely different practice with "general entities." As Cambefort emphasizes, Buffon's assumptions are in sharp opposition with Linnaeus'. For Buffon the natural world is a continuum—and in his eyes, Linnaeus' "systematic" approach is creating arbitrary and artificial classifications. Accordingly, for Buffon, classification is not a primary activity, and interestingly for us genres—the shaping of these "general entities"—are not prominent in the naturalists' work. We return to his approach to generality subsequently.

Nineteenth-century zoology inherited the organization of activities around classification and the creation of genres from Linnaeus. However, most practitioners gave a different interpretation to genres, highlighting their *conventional* meaning. Accordingly, they also placed emphasis on *practical* considerations attached to the definition of genres, an issue that remains meaningful to the present day. Notably, the size of the groups created can make significant differences in the practice, which sheds light on the key part played by the genres and species in the operations of situating given entities in the classification. We thus see how an activity with the aim of creating genres and species can interpret, and accordingly shape, them in entirely different ways.

With the publication of Charles Darwin's (1809–1882) *Origin of Species* (1859), new ideas were introduced in the life sciences, which brought about yet another mutation in the interpretation of genres. Genres were thus again maintained as meaningful "general entities." However, by means of a genealogical reading, the trees of earlier classifications became interpreted as temporal patterns, in which the upper-level nodes—for instance, the genres—represented common ancestors for lower-level nodes—for instance, species. Consequently this new approach led to changes in the scholarly practices, through which these "general entities" were understood and thus shaped. Noteworthy is the fact, discussed by Cambefort, that in the life sciences in the present day, different groups of practitioners of systematics have developed different practices for the making of "general entities," owing to their diverging way of striking a balance between several criteria. Notably, phylogenetic considerations do not represent the only meaningful criterion for all practitioners. This evidences *contention* among diverging collective ways of shaping genres. The contenders are characterized by different ways of inheriting from past scholarship and different scientific practices of cluster making. However, if "genres" are contested, the making of such types of "general entities" remains a shared goal. Through his diachronic outline on how this task was fulfilled in different contexts, Cambefort thus highlights the wealth of factors that are likely to enter into the shaping of general entities.

Similar conclusions derive from the case study Stéphane Schmitt presents in Chapter 9 on the concept of "homology"—which embodies another way of looking for the general in the life sciences. Broadly speaking, this search attempts to identify parts in the *structures* of different species of living organisms that present a *similarity*, independent of their function. The interest in bringing to light such "homologies" has been correlated with a persisting working assumption that has taken different forms in history. This assumption basically asserts that there are a limited number of organizational plans on which all living beings are built. Aristotle had already held an assumption of this kind. For him, there even existed a single such plan. Naturalists like Etienne Geoffroy Saint-Hilaire (1772–1844) also adopted the idea of a single plan of organization for all living organisms. Other naturalists opted for the slightly different assumption that a small number of such plans existed. For instance, Georges Cuvier (1769–1832) held he could establish the existence of four basic schemes. These assumptions clearly relate to a search for generality in the life sciences. They have guided work on living organisms, leading practitioners to focus on how parts of different organisms correspond to one another.

The key thesis for which Schmitt's chapter argues is that this type of search for generality has changed meaning, and even content, throughout history, in relation to the changing contexts within which it was carried out in the life sciences. However, the basic idea persisted, and even the fundamental practices of naturalists survived key changes in the theoretical framework.

In pre-transformist comparative anatomy, Schmitt notes, scholars looked for homologies in a formal way, without attempting to interpret the results. A search of this kind can be identified in the work of Renaissance naturalist Pierre Belon (1517–1564). In order to express homologies, Belon designed specific kinds of diagrams, with which he displayed, for instance, the similarity between a human being's skeleton and that of a bird (see Figure 9.1, in Chapter 9). This practice illustrates the invention of a way of writing down the general. The argument the diagrams aimed to make led to changes in some features in the description and the drawing of skeletons in unusual ways, so that the similarity appeared more clearly. In other words, here, the search for generality is connected with a work practice, and it relates to modes of both organizing data and observing. However, in Belon's works, the remarks about similarity are not developed systematically, the comparison remaining local and bearing on a sample of only a few skeletons.

In the eighteenth century, the search for resemblances between parts of different organisms (the arm of the man and the wing of the bird, for instance) became a significant issue for practitioners. This increased interest in generality seems to have been promoted by a faith inspired by how the successes of Newtonian physics had brought order and unity into the conception of the physical world. Practitioners of the life sciences also had the aim of finding order in the living world. Like systematics, which we examined above, comparative anatomy developed in new and more methodical ways within this context. These developments illustrate how ambitions of generality circulate between fields. However, they took various forms in different life sciences: Systematics and comparative anatomy developed, so to speak, in opposition to each other.[17] This is where we return to Buffon.

Comparisons seeking to establish homologies were carried out on a large scale, and they showed what Buffon emphasized as the anatomical similarity of all living organisms. This led him to put forward for the first time a global *interpretation* of the *meaning* of this manifestation of generality. For Buffon, the global similarity derived from the fact that there had been a single "primitive and general design," on the basis of which all living beings had been created by variation. Comparisons could show at the same time the *variations* in the application of the same design and also the "hidden *resemblance*," which highlighted how the "Divine being" created. Clearly Buffon's general entities differ strikingly from Linnaeus'. Likewise, Buffon's program was an attempt to bring generality to the description of nature, but in a different way. In fact, Buffon pursued generality on different levels, which are in turn addressed in the general chapters distributed among the various volumes of the *Histoire Naturelle*.

Descriptions of animals were a key tool to achieving generality, and it had to be carried out in a specific way. This remark leads to a general conclusion, which we have already emphasized above: bringing to light a type of generality requires specific practices. Let us dwell here for a few lines on description in this context, as another practice of generality. Arbitrariness, Buffon emphasized, should be avoided as much as the excessive accumulation of unorganized information. Buffon's collaborator, Louis Jean-Marie Daubenton (1716–1799), was commissioned to write the morphological and anatomical description of animals, from book III of the *Histoire Naturelle* onward, that is, from 1753. Generality appears to have been a key epistemological value inspiring his work. In a methodological chapter devoted to how description should be carried out, Daubenton makes clear how the method of complete description he advertises has "universal value." He also emphasizes how comparison is, in his view, key to the description.

Instead of piling up facts without any hierarchy between them, he prescribes that description should rather put forward constant properties. *Terms* should be chosen to designate parts in such a way that the same term could designate parts of distinct animals

---

[17] Here and in what follows, in addition to Schmitt's chapter in this volume, we rely on Schmitt (2010: 16–17, 44–54).

that correspond to one other. In Daubenton's words, one should proscribe "particular terms," designating the "same thing" with different names, and promote instead "simple" and "universal" denominations. Clearly, the choice of terms is correlated to a systematic practice of *comparison* between animals.

The *order* of the description should also follow constant principles, in order to allow comparison: first describe the whole, and then the parts; first the external characteristics, then the internal organs, and only lastly the skeleton. In the first part of such a description, the position chosen for the animal to be described is systematically the same, both for the *text* and for the *illustration*. This practice went against previous standards, but was later adopted in zoology.

The inclusion of the anatomical part in the description was also meant to counter the practices of both those who merely observed and those who worked on systematics, on the grounds that they only remained at the surface of things. In Daubenton's hands, anatomy was also conducted in a comparative way, inaugurating what soon after came to be known as "comparative anatomy." To allow this comparison, the *text of the description* had to strictly follow the same structure. It also had to focus on elements essential to comparison, leaving aside the elements that brought nothing to the discussion of similarities and dissimilarities. In this sense, the structure of the description was thus dictated by the program of comparison and the search for generality.

Specific animals were chosen as points of reference for the description of others.[18] This feature of the practice relates to the *order* adopted in the *Histoire Naturelle* to present animals. It also relates to specificities in the text of the descriptions: in his work, Daubenton identifies stronger similarities between some groups of animals as opposed to others. These divisions are made manifest by the fact that only the first animal in the division is described, the description of the others being abridged and turned into tables of numbers, thereby instituting an animal as "model." These characteristics of the practice of description illustrate another way of taking the general into account. This form of generality is correlatively materialized in the overall structure of the text. This textual expression of a form of generality evokes what Jaëck describes with respect to Banach.

More broadly, Schmitt shows, the idea of generality captured by the concept of homology between organisms was widespread in the second half of the eighteenth century and beyond, each author subscribing to a specific idea regarding its *nature*. Evolution did *not* derive from such developments, but as soon as the idea of evolution was adopted, it inspired an interpretation of the *meaning* of this form of generality, in terms of descent. Similarities were thus enrolled as arguments in favor of evolution, which conversely offered a totally new perspective on these similarities (their distribution and their meaning). Archetypes became "common ancestors,"—a change in the underlying meaning. However, Schmitt emphasizes the work of bringing to light similarities—the *practices* of looking for the general—did not fundamentally change.

---

[18] For this and other features of the practice of description and their relation to the purpose of comparing, see Schmitt (2010: 53–57, 59–60, 66–68).

We recognize a pattern similar to that sketched out above, with respect to systematics. Similarly, in the first place, geneticists attempted to get rid of homology, with the idea it was a superficial link between organisms. However, homology resurfaced at a genetic level. Unexpectedly, the same type of search for kinds of generality continued, even though in this case the reference of the term "homology" underwent a radical mutation.

1.2.3   Circulation between epistemological cultures

The two case studies examined in Section 1.2.2 both deal with long-term continuities in ways of approaching the general in the life sciences. In both cases, a single term (genre or homology) was used in the long term and embodies a type of generality being pursued. How a concept of this type, introduced in a given cultural context, is appropriated into another requires a finer-grain analysis. This is the issue addressed in Part II.3, in Chapter 10, in which Tatiana Roque concentrates on concepts of "genericity" that were introduced into the study of dynamical systems and were constantly reshaped throughout the second half of the twentieth century.

Concepts of genericity are closely related to a type of general statement discussed above, the emergence of which, in mathematics, Robadey's chapter documents. We recall that a statement of this kind typically asserts that a property P holds true for all objects in a domain D, except for a group of objects that can be "neglected." It is thus not universally true, but true with a generality, the assessment of which requires mathematical work. Likewise, Roque studies a case in which the description and classification of *all* objects seems to be out of reach. Consequently, mathematicians settle for a classification of *almost all* objects. In fact, with this case, we return to Henri Poincaré, who is a key figure in our history.

Concepts of "genericity," Roque argues, were brought into the theory of dynamical systems in *relation* to a research strategy adopted by a collective of mathematicians. In this respect, her chapter echoes Schmitt's chapter, which describes the unfolding of a research program drawing on a shared hypothesis regarding general features living beings have in common.

Roque argues the strategy adopted in the study of dynamical systems relied on an essential *a priori* decision, that of *not* considering *single* systems in and of themselves, but *sets* of systems collectively. On this basis, the strategy consisted mainly of two key ideas. First, actors aimed to identify a collection of dynamical systems "large enough" to allow them to approximate, as closely as one might want, *any* dynamical system by one belonging to the collection. This defines a form of generality actors refer to as "genericity." It requires introducing a notion of "closeness" between systems. This notion was shaped using techniques similar to those discussed in Jaëck's chapter. Secondly, actors aimed to choose this collection of dynamical systems in such a way that it proved amenable to description. In this case, it meant that they were driven by the hope of possibly giving a classification to the systems in this collection in such a way that equivalent systems—in a sense of equivalence to be defined—would belong to a same class, and salient features would allow characterization of systems in any class.

As Roque shows, the *overall strategy*, which relied in a crucial fashion on a type of generality and a method of bringing this generality into play, was in fact initially inspired by another mathematical domain, singularity theory. In that other domain, René Thom (1923–2002) had followed the same strategy and introduced the term "generic" to refer to a collection of tractable geometric objects. The term "generic," and the type of meaning it referred to, were not appropriated alone. We see in fact the importation of a notion of generality, in relation to a strategy, from one context into the other. From this remark derives an important conclusion: Seen from the perspective of the shaping of general entities and research strategies to use them, scientific cultures do not appear as closed cells. They are in conversation with each other and regularly appropriate not only ideas and results, but also concepts and ways of working.

In the case of "genericity," Roque emphasizes the importance of the places (mainly Bures-sur-Yvette, Princeton, and Rio) in which, and the personal connections through which, the transfer was able to take place. Interestingly enough, this was not the first migration of the term "generic," since Thom had borrowed the term, and the idea, from algebraic geometry, thanks to discussions with Claude Chevalley.[19]

While terms referring to forms of generality migrated from one domain to another, in fact their actual meanings were reshaped in each case to suit the new context of their use. Subsequent research on dynamical systems testifies to exactly the same phenomenon. Roque documents the stability of the epistemic tactics outlined above in the history of the theory of dynamical systems, from Poincaré to Smale. However, she shows how research along these lines regularly highlighted problems of two kinds.

First, actors hoped to establish the "genericity" of a collection of dynamical systems they were concentrating on, but failed repeatedly in their attempts. Situations of this kind led researchers to attempt to define the collection of reference in another way that would still be appropriate for the two tasks for which the collection was meant to serve. In other terms, they strove to redefine the reference of the term "generic." These situations also led actors to attempt to identify which general phenomena had been overlooked in the shaping of the former collection. This inquiry focused frequently on "prototypes," illustrating phenomena that had mistakenly been neglected: these objects illustrate other kinds of entities meaningful for a search for the general that actors introduce in some contexts. In relation to the work done to redefine the collection of dynamical systems, the salient features on which to concentrate also underwent transformations.

Second, actors frequently felt the need to rethink the concept of "genericity" they were using, in their attempt to define the representativeness of the collection of objects—or, alternatively, to define the negligible aspects of those outside the collection—in ways that could be better suited to the difficulty they were meeting. In other terms, they strove to reshape the concept with which to capture the general. Particularly important in this respect is the fact that in two different collectives, different sets of mathematical tools were explored. In the network that took shape around Bures-sur-Yvette, Princeton, and Rio, "genericity" was approached as it had been in singularity theory, namely, with topological

---

[19] On the history of concepts of genericity in algebraic geometry, see Schappacher (2010).

tools. However, in the Soviet Union, around Andreï Kolmogorov (1903–1987), probability theory, and accordingly measure theory, was favored to capture a similar idea in the study of dynamical systems. These tools were in line with those Robadey shows Poincaré used in his study of problems in celestial mechanics. This different approach soon circulated westward, and it proved to offer new opportunities, at a moment when the topological approach to representativeness seemed to meet with intractable problems.

Thus with evolution in research, and the changing range of phenomena being concentrated on, we see actors constantly reshaping the concept of generality with which they work, and its reference, in relation to the problems dealt with, what the research has shown about these problems, and the types of mathematical knowledge used in the definition of the concept. Further, the case studied by Roque illustrates a key fact: different collectives of actors sometimes shape generality in the same domain in different ways and with different tools. This is a phenomenon to which we return below.

1.3 Practices of generality

The previous sections have given insight into actors' own reflections on generality as well as their shaping of concepts, statements, and forms of texts to express the general, refer to it, or work with it. Some of these tools, they inherit, rethink, or even reshape. Others, they simply invent in relation to the challenges they meet, or the project they set themselves. Poincaré's enumerations and Daubenton's descriptions are excellent illustrations of actors shaping of types of writing in this respect. Our analysis showed that these concepts, statements and forms of text are in most cases *related* to specific practices. Poincaré's enumerations derive from how he deals with cases and how he intends to use them. Daubenton's descriptions relate to how he organizes his naturalistic practice. Concepts of genericity are also meaningful in relation to a specific type of research program that relies on them. These examples show how the making of practices is a dimension of actors' work, inseparable from other dimensions of their work. It illustrates more widely, we claim, how they shape ways of carrying out scientific activity.

Part III of the book brings practices linked to generality into focus, to examine them in the context of the scholarly cultures in which they can be observed. In addition to analyzing how actors dealt with the general using specific practices, we are interested in how they shaped these practices in relation to the issues they selected as meaningful, and especially how their results present correlations with the practices used.

1.3.1 Scientists at work

We have previously seen Leibniz, as a practitioner of philosophy and mathematics among other things, developing a reflection on the virtues attached to generality in mathematics and the forms the general could take. We have also seen how his specific understanding of generality was related to key facets of his philosophy. The first chapter (Chapter 11) of Part III, by Emily Grosholz, now focuses on how Leibniz's reflection meshes with his own practice in mathematics. For this, Grosholz concentrates on the specific type of analysis that Leibniz shaped, and she does so by considering one kind of mathematical object to which Leibniz devoted a great deal of effort: curves.

The contrast Leibniz draws between his own approach and Descartes's analysis is telling. For Leibniz, Grosholz emphasizes, Descartes artificially restricted himself to a range of curves (the so-called "geometric curves," to which we return below) to achieve universality for his method. This is not the first time we see an actor criticizing another actor's practice of generality as artificially imposing boundaries on facts. We also return to these criticisms below. It is only in this context, Leibniz stresses, that Descartes's analysis could proceed by a systematic and uniform reduction of problems. By contrast, Leibniz's practice of generality develops in the context of specific situations, treated as *paradigms*, and aims to find patterns in them, or relating to them, that make them intelligible. Connecting a curve to sequences of numbers, to other kinds of geometrical figures, including other curves, as well as to mechanical problems, all give ways of *understanding* the curve and highlighting its hybrid character. Taken separately or in conjunction, these patterns offer resources for solving problems about the curve. Further, they frequently appear to be shared with other problems and other mathematical objects, thereby yielding means of establishing bridges between the curve and other curves, or between problems. Conversely, establishing connections between curves or between problems, and thereby studying them in analogy with each other, allows that such patterns circulate and have an extended fruitfulness.

The general is thus approached, and dealt with, in the context of a particular—a particular that is treated in a general way. In this way of proceeding, the generality of the patterns highlighted is established by progressive extension. Patterns shared across contexts shed light on links that relate the different objects at the level of their *intimate nature*. This is *how* a search for generality of this type meshes with gaining *understanding* in all related contexts simultaneously. Clearly, a procedure of this type ascribes no a priori limit to the connections that can be built. Accordingly, Grosholz can assert that analysis, as practiced by Leibniz, lends itself to generalization. Generality is in this way explored not by abstraction, but by the analysis of the conditions of intelligibility of the paradigm, conducted in a never ending process. This description of Leibniz's practice of generality powerfully evokes features of Poincaré's practice outlined above, as well as a practice evidenced in ancient Chinese mathematical texts (Chemla, 2003). In all these cases, *understanding* appears to be the crux of the matter. Perhaps however, a finer-grained analysis of these practices would reveal differences in the choice of paradigms on which to focus, and the ways of using them. Leibniz appears to focus on what Grosholz calls "canonical objects." Characterized by their simplicity, these objects become more meaningful with time, their canonicity being thus shaped through history.

Several features of Leibniz's practice of generality present an interesting parallel with a practice identified in a completely different setting, which Darrigol analyzes in Chapter 12: Physicist James Clerk Maxwell's (1831–1879) use of analogy and, one could say, of "canonical" models. In this case too, generality is not an observers' category. In 1856, Maxwell explicitly stated that a way of proceeding which he had opted for in his practice of physics aimed to "attain generality and precision," while "avoid[ing] the dangers arising from a premature theory professing to explain the cause of the phenomena." The statement is striking, since it makes explicit how an actor clearly identifies several epistemological factors that are meaningful for him (generality, precision, avoiding the dangers arising from a premature theory). It further explains how Maxwell's choice for a given practice strikes a balance between these three requirements.

Interestingly enough, as we mentioned at the beginning of this prologue, half a century later Pierre Duhem perceived a practice of physics like Maxwell's as having given up "generality and rigor." In opposition to a historiography of physics that has adopted Duhem's view, Darrigol sets himself the goal of interpreting Maxwell's own statement about his way of achieving generality. Noteworthy is Darrigol's remark that whereas, for Duhem, abstraction was essential to the practice of generality, it was not for Maxwell. This clearly captures a key difference between the two practices, which also makes sense in other contexts.

To begin with, Darrigol shows how in 1856 Maxwell describes his shaping of a practice in physics as inspired by a fruitful analogy established by William Thomson (Lord Kelvin) (1824–1907). Noticing a parallel between the search for thermal equilibrium and that of an electrical potential, which derives from structurally identical differential equations, Thompson had drawn conclusions based on a physical interpretation in the context of the former and stated, without further proof, the same conclusion in the latter. While at first sight Maxwell repeats a similar theoretical gesture as Thomson, the replacement of heat by a geometrical approach to fluid flows is not innocuous and goes along with a key feature of his practice of generality.

Indeed, Maxwell substitutes Thomson's "analogy" with the establishment of an ideal mechanical model, which he then relates to three different physical situations. The ideal model is now what will embody the generality sought. In these three contexts, elements of the model are put in correspondence with concepts derived from experiments. Further, the three sets of physical phenomena can thereby mathematically be treated conjointly. This property of the *dispositif* illustrates one feature of Maxwell's practice of generality. Interestingly enough, for Maxwell, staying only to the mathematical level would prevent the establishment of "connections" between the different situations—a benefit in terms of generality which thus in his view derives from his practice.

Darrigol then highlights how Maxwell's practice of generality developed gradually in the subsequent years, gaining additional facets. The historian's approach thus discloses a historicity in an actor's practice of generality. In fact, the same remark applies to Jaëck's discussion of Banach: his practice of generality changes in line with the change of meaning and status of the axioms he introduces. This general issue points out a most promising future research program. To return to the specific case of Maxwell, to a local use of the mechanical model as a tool to inquire further into various physical situations, he adds a global model capturing the mechanical nature of the whole range of phenomena dealt with in different domains. The fruitfulness of the "generality" of the practice is manifest through the unification between theories Maxwell achieves in this way. However, his awareness that many different mechanisms could be responsible for the "mechanical connections" uncovered leads him to add yet another facet to his practice of generality, when he attempts to capture the "general structure" common to all these models.

The way in which Maxwell conducted this search led him to move from an assumption of an underlying mechanical model to the general requirement that all these fields' fundamental equations had to have a Lagrangian structure and involve mechanical variables of a generalized type. His multi-faceted search for generality thus led Maxwell to the identification of a highly general principle, inspired by one of his earlier methods of mechanical modeling. A key conclusion emerges from the case study: it brings to light a correlation between Maxwell's practice of generality, shaped to fulfill specific epistemological constraints, and his eventual introduction of a general *principle*, which incorporates features of this practice.[20]

1.3.2   A diachronic approach: continuities and contrasts

Practices of generality sometimes present a form of diachronic stability, despite the fact that they migrate from one context into a wholly different one. This is the key issue addressed by Jean-Gaël Barbara in Chapter 13, in which he focuses on a recurring practice of shaping general objects in the life sciences. The practice under consideration, he suggests, is characterized by how it identifies objects through the converging approaches of several disciplines. Xavier Bichat (1771–1802) is the first practitioner Barbara examines in greater detail from this perspective, and more specifically Bichat's practice in what he called "*anatomie générale.*"

Bichat inherited from various practices of generality before him. To begin with, he inherited from (early) eighteenth century attempts to combine anatomical and physiological approaches. These attempts manifest a transformation in the relationship between these two domains of inquiry. Prior to this, physiological studies had been based on anatomy. Anatomy highlighted general facts, on which practitioners relied to discuss the causes of the functions of the organs. In the eighteenth century, physiological discussions freed themselves from anatomy, and the actors likewise set out to identify general principles that might characterize living things as such. These reflections introduced the idea of, and quest for, "general physiological facts." As a result, instead of having generality defined only by anatomical considerations, two different ranges of general facts could be brought together, and contrasted with each other, in the investigation. In that way, physiology was *combined* to anatomy, and did not only *derive* from it. Bichat followed such a trend and Albrecht von Haller's (1708–1777) investigations in particular. However, the specificity of Bichat's approach lay in a systematic attempt to correlate general physiological facts with general anatomical facts. As a consequence, where for example, Haller had only seen one kind of tissue in a particular study, his cross-disciplinary approach allowed Bichat to subdivide it into three types, each type referring to a similarity in function *and* in pathological transformation. Tissues as "general objects" were born at the convergence between the two domains of inquiry.

---

[20] This book does not systematically inquire into the reflections on, and practices with, principles in physics and beyond. This is, however, an important topic for a systematic study of generality. On this question, see, for instance, Seth (2006).

Bichat also had the aim of achieving generality through his practice of observation. He expected that many repeated dissections would shape "clear and general ideas" through relying on the senses. Additionally, Bichat actively sought analogies between different observations. Barbara thus concludes Bichat combined two types of generality: the general as that which derives from similar observations; and the general as what occurs in different parts of an organism and is identified through the combination of features brought to light by different ranges of issues. Tissues are one such example.

Finally, Bichat's valuing of generality has an immediate context. For the purpose of teaching anatomy, Bichat's master, Pierre-Joseph Desault (1738–1795), dissatisfied with the state of anatomical knowledge, which required large bodies of facts to be memorized, aimed to reorganize anatomical knowledge into chapters starting with general facts. Interestingly, teaching appears here as an activity, in the context of which the value of generality plays a key part. Moreover, we meet again with the correlation, encountered above, between valuing generality and aiming to alleviate the burden of memorization. Accordingly, Desault called for an "analytical surgical anatomy" in order to simplify and rationalize knowledge. Here, the term "analytical" likewise echoes a type of approach tightly linked with generality. We have seen above that Chasles likewise attempted to emulate it in geometry, for all the epistemological virtues he attached to it. We will soon return to this "analytical" type of approach in mathematics.

Bichat published Desault's lectures after the latter's death. Like Desault, after whom he taught anatomy in Paris, Bichat also promoted the quest for generality as an organizing principle in teaching. However, he took this task of importing analytical treatments into anatomy one step further, since he adopted the ideal in his own investigations. The transfer of generality practices from the activity of teaching to that of inquiry is noteworthy here. Bichat modeled his practice of inquiry on two complementary ideals, which he put into play using elementary practices of generality encountered above: looking for the *right language* that could provide an analytical tool, and looking for the *elementary components* into which to decompose reality and then to recompose it.

Tissues were precisely, in his view, the "elements" with which to carry out decomposition and recomposition. Bichat's project to classify them is in line with attempts at classification in natural history we have evoked above. In this context, Bichat follows those who value the use of a "natural method" and assumes the types of tissue identified are "real objects." In fact, it is important for us to note that the specific practice of generality for which Bichat opts, that is, approaching tissues from the perspective of different fields, appears precisely to be what *grounds* his conviction that these objects are real. In this respect, generality also constitutes for him what we have called an epistemic value.

Barbara suggests the practice of generality thereby defined was later appropriated by other scientists. He establishes his claim, by examining another practitioner's approach to general anatomy: Louis-Antoine Ranvier (1835–1922). It is to be emphasized that Ranvier focused on a level of inquiry different from Bichat's, the microscopic scale, and he begins with another general hypothesis: the generality of the cell. At stake for him was to discover general structures, in the form of parts of cells and various types of cell. To achieve this goal, like Bichat, he also combined the different perspectives that anatomy and physiology yielded on the same situation. For him, a combination of approaches of that kind likewise ensured that the objects identified were "real."

The name that Joseph-Louis Renaut (1844–1917) attributed to the practice in question, which he described explicitly, captures its essence in an interesting way: "principle of converging methods." In both cases, the key idea of the practice is the same. Ranvier's references to Bichat seem to indicate, Barbara suggests, that Ranvier actually perceives his practice of defining general biological objects as being inspired by that of his predecessor in the same field. By contrast, another case study, presented by Renaud Chorlay in Chapter 14, shows actors shaping practices of generality in opposition to that of their predecessors, which they reject.

We have evoked above how the practitioners who established projective geometry derived inspiration for their work from a careful examination of how "analytical" approaches brought into geometry types of generality that earlier geometrical approaches failed to achieve. The practice of generality in "analysis" that, among other geometers, Poncelet and then Chasles had observed in their reflections is precisely one of the three main practices on which Chorlay focuses. Investigating a classical corpus—that of the *foundations* of mathematical analysis in the nineteenth century—from a non-classical perspective—that of issues of generality—, Chorlay aims to capture key features of the scientific work bearing on generality, and related features, in what he refers to as three "epistemic configurations." His strategy is to contrast how practitioners in these three contexts approached, and worked with, the notion of "function," which had become the central notion in analysis from the mid-eighteenth century onward. The historical question of the foundations of analysis is usually investigated in terms of rigor, arithmetization, or set-theoretic thinking. Chorlay endeavors to show that questions of generality provide fresh and relevant interpretation frames.

For Joseph Louis de Lagrange (1736–1813), whose *Théorie des fonctions analytiques* (1797) illustrates the earliest practice examined by Chorlay, the introduction to the notion of function assigns no strict boundary to the object. However, essential in Lagrange's approach is a *principle* of representation of a function by a form of development holding true *uniformly* for all the functions dealt with. Faithful to his practice of analytical treatment, in a sense already encountered above, Lagrange derives from this principle the whole "calculus of functions." The way in which the development "holds true" for a function in particular also characterizes Lagrange's work with generality. It holds true *with full generality* at the level of the *form*, granted that when concrete values are given to variables, the representation *sometimes* fails to have any meaning. Actors refer to this way of dealing with the general as deriving from the "generality of algebra." Chorlay adds a description of how Lagrange captures *singular cases* by means of carefully designed *examples*, which are the simplest cases able to exemplify the phenomena.

It is this practice of generality, from which geometers like Poncelet and Chasles drew inspiration. As Chorlay emphasizes, it is also *this* practice that in Analysis Augustin Louis Cauchy (1789–1857) criticized as mere "induction," and against which he established a new practice of generality. Interestingly enough, as Chemla's chapter recalls, in 1820 Cauchy wrote a negative report on a memoir in which Poncelet introduced the "principle of continuity" into geometry, to emulate how analysis achieved generality. In this report, Cauchy addressed *exactly* the same criticism to Poncelet's principle. From the perspective of generality, we thus see how practices circulate from one culture to another. Cauchy's criticisms show how an actor perceives this circulation quite clearly. This remark sheds new light on Cauchy's criticism of Poncelet's principle. We also see how new practices are explicitly designed in opposition to earlier ways of handling generality. Whereas in the earlier context, uniformity was highly valued, in the later context, generality was redefined in relation to the valuing of rigor.

Cauchy's criticism paved the way for the shaping of the second "epistemic configuration" Chorlay examines. In this second context, the notion of function does not have strict boundaries either. However, new types of statement appear, explicitly formulating conditions on the functions as well as on their variables for a proposition asserted to be true. They derive from a new form of proof in analysis, which examines the conditions required for each step in the proof to be valid. Chorlay thus highlights a correlation between the conduct of proof and the form of general statements formulated. These statements set limits to the extension of the group of *functions* for which the proposition as established holds true. Moreover, for given functions, these statements also have the aim of defining the class of *values* of the variables, for which a proposition can be asserted. This type of inquiry will be instrumental in the later development of set theory. The practice of generality in this second context thus appears to be closely related to other features of the "epistemic configuration." Chorlay further emphasizes how in this context, a new use of *examples* emerges: singular functions are shaped as *counterexamples*, used to explore the limits of validity of propositions. This new practice would be of tremendous importance for analysis in subsequent decades.

The third "epistemic configuration" Chorlay analyzes presents several features of wider relevance for our inquiry. In this context, the approach to the notion of function has been entirely renewed. However, in line with the systematic exploration of the conditions of validity of statements, a new feature in the practice of generality has appeared: the classification of functions into classes. Accordingly, the statement of a theorem makes clear for which class it can be asserted. Further, late-nineteenth-century actors like Borel developed an interest in comparing the relative generality of a class with respect to a larger class in which it is contained—what Chorlay calls "embedded generality." Chorlay emphasizes how various types of mathematical means are put into play, and even shaped, to carry out this new task in a precise way. Depending on the purpose, the means chosen to assess the generality will differ. Moreover, Chorlay shows how in analysis new practices of proof emerge, which made use of these assessments of generality to conduct a reasoning. We have already seen how Robadey's chapter documented the emergence of a form of statement and proof of this kind in Poincaré's work. Chorlay's case study thus shows how mathematical work was carried out to develop further means to achieve similar ends. Finally, this type of reasoning pinpointed by Chorlay will precisely be, as we have seen above, an essential ingredient in the deployment of the study of "generic" cases, examined by Roque. Seen from the viewpoint of generality, history of science thus displays circulations, between contexts, not only of concepts and statements, but also of reasoning and other practices relying on a type of generality, or aiming to achieve a generality of a certain kind. It also evidences explicit disagreements between actors. They are topics essential for our inquiry.

1.3.3   A synchronic approach: controversies

Several case studies already mentioned evoke actors' criticisms of predecessors' practices of generality and their ensuing adherence to another, possibly new, practice (Leibniz criticizing Descartes, Cauchy criticizing Lagrange, and Duhem criticizing Maxwell). These episodes provide extremely useful evidence for an approach like ours, which aims to identify how actors shaped modes of expression and practices of generality in different contexts. Likewise, disagreements and debates on these matters are very interesting for the differences among cultures of scientific practice they reveal in this respect. We have already evoked a disagreement of this kind, between Linnaeus and Buffon. Their conflicting approaches to the general yielded quite different scientific outcomes, both meaningful from the viewpoint of present-day biology. This part of the book examines more closely three other disputes of that kind, the first of which takes us back to seventeenth-century geometry.

In his survey of the history of geometry from the viewpoint of the value of generality, Chasles stresses the seventeenth century as a turning point. One of the key facts he brings forth is the introduction at that time of a specific type of general procedure, namely, a "method," which allowed practitioners to deal with different objects *uniformly*, and accordingly *connect* the objects as well as the propositions established about them. Barbin's chapter (Chapter 15) takes a closer look at an episode that clearly illustrates Chasles's thesis, while shedding interesting light on differences between actors in this respect.

The episode occurred in the 1630s, and it is related to the problem of finding tangents to curves. What matters most for us here is that it gave rise to a controversy, in which the main protagonists were led to make their views explicit on what a general method should be and how it should be shaped. We can thus observe actors' shaping of their practice in the making. The main protagonists, Descartes and Pierre de Fermat (early seventeenth century–1665), both clearly valued general procedures, whose power extended beyond the treatment of single cases. They each designed a general method for the general problem, but they did so in different ways, echoing the fact they used different categories to describe them.

Descartes referred to his method for finding tangents as "universal." Accordingly, he explicitly defined the range of curves to which it applied: the curves he called geometrical, in relation to the fact that an algebraic equation could be attached to them. Descartes' method relies precisely on the equation and follows a uniform procedure to exhibit the tangent. It could thus be used equally for all these curves, but only for them: the framework was fixed in advance—we have evoked Leibniz's criticism of what he perceives as artificial in this procedure. Interestingly enough, although Desargues had a completely different approach to curves, he shared key features of Descartes's practice of generality. As Chasles emphasizes, Desargues also devised *uniform* ways of defining conical sections and accordingly developed uniform reasoning that could establish related properties of different curves in exactly the same way. He too described his approach as "universal."

Despite differences between Descartes and Desargues in the approach to curves and the way of dealing with them, Barbin shows that Desargues expressed his support to Descartes in the controversy. The conception and the practice of the general is what the two practitioners share in this case.

Fermat's way of proceeding to approach the tangent problem stands in contrast to this practice of generality. His own practice displays a general method in the context of a specific problem. We have already encountered procedures of this type several times above. Yet Fermat's practice is specific: his presentation of the general method in context establishes connections *at a higher level* between the method and another general method, associated to his name: that of *de maximis and minimis*. Further, Fermat's practice of generality consists in unfolding the potentialities of the general method, through extension and adaptation to different cases. Fermat does not set limits to the group of curves to which the method can apply. As a result, its power extends beyond the set of curves to which Descartes had from the beginning limited the scope of his method. In correlation with this "open" feature—to use the term introduced by Jaëck, which proves relevant in this context too—, Fermat does not consider it to be a priority to highlight the foundations of his method. The controversy with Descartes will compel Fermat, and also Roberval who sides with him, to formulate explicitly their views on what grounds the generality of a method. In conclusion, the episode nicely illustrates how actors shape general objects and general modes of proceeding in different ways. What is more, these different practices of generality mesh with, for example, different approaches to curves and different ways of working with them, and different working techniques to deal with equations. The type of generality pursued is also correlated with the emphasis on other epistemological values and goals. Whereas for Descartes, the uniformity of the procedure matters, for Fermat, the achievement of an ever broader or higher generality appears more meaningful. We have chosen to refer to such ways of doing mathematics, characterized by features of this kind, as "epistemological cultures." This example illustrates how, even when actors operate at the same time, on the same objects and the same problems, the epistemological cultures, in the context of which they are active, differ. Accordingly, in each context, the value of generality displays different forms.

A similar conclusion emerges from the account Frédéric Brechenmacher gives in Chapter 16 of the Jordan–Kronecker dispute in the 1870s, which we evoked in the introduction to this prologue. The conflict breaks out as a priority dispute. What is important for us is that in this context, the actors perceive that part of the dissension relates to how they practice generality. This leads them to make explicit how they believe generality should be pursued. As Brechenmacher makes clear, Jordan in Paris and Weierstrass and Kronecker in Berlin have developed different approaches to a subject (what we understand today as the various types of reduction of matrices). They grasp that their results relate to each other, since they address problems deriving from the same tradition. However, they struggle to understand fully the relationship between their results. We will focus only on what the dispute tells us about our main topic, that is, the competing practices of generality at play in mathematics at the time and the distinct epistemological values actors associate to them.

As we recalled in the introduction to this prologue, Kronecker's formulation of how he understands the difference between the two practices of generality can be interpreted as a rejection of an old practice—that of a "generic reasoning"—in favor of another one, introduced by Weierstrass a few years before, and prized for its rigor. Kronecker criticizes Jordan for a reasoning that only aims to solve a problem in general, and does not care about exceptions where the reasoning fails to apply. This was in a nutshell the criticism Cauchy, before Weierstrass, formulated against an earlier way of proceeding in mathematics. The historiography of mathematics has strongly emphasized these episodes as testifying to the development of rigor throughout the nineteenth century. For Kronecker, Jordan's approach is thus not truly general. However, it would be mistaken to believe that Kronecker's goal in this case is limited to achieving a greater certainty. Incidentally, if such were the case, the demand for generality would only be a superficial requirement and would not touch the substance of the matter.

In fact, Kronecker's demand of a full generality can be interpreted *only* if we associate it with the other value that gives meaning to it, that is, *uniformity* of the reasoning. Indeed, the fully general reasoning Kronecker expects does not only deal with all cases, but it also deals with them *uniformly*. In his view, the singular cases, for which a reasoning fails, are a precious indication that the practitioner's understanding has not yet reached the crux of the matter. They point to "the real difficulties of the study," and their dissolution, which is carried out only when the "true generality" has been achieved, is a criterion indicating that one has obtained a deeper understanding of the subject and discovered "the wealth of new viewpoints and phenomena which lie in its depths." The generality envisioned is an epistemic value: it appears to be a guide toward the essential features of a situation. The butt of Kronecker's criticism is thus not merely rigor. Kronecker's practice, like Weierstrass', also requires that problems be solved with effective means of computation. The urge to develop such means leads them to opt for an approach in terms of arithmetic *invariants*, that is, one figure of generality in mathematical terms. Accordingly, Kronecker criticizes Jordan's approach for its lack of effectiveness.

Jordan, for his part, manifests another perception of generality. The key other value that he correlatively prizes is simplicity. In his eyes, the Berlinese's computations are hard to understand and lack the simplicity of his approach. Accordingly, Jordan develops a mode of reduction of the objects involved into simpler pieces. In addition, these pieces are of the same kind as those analyzed, which embodies another figure of generality in mathematical terms. For Jordan, an approach of this kind allows the practitioner to "see" what is happening. When he defines his own general approach, simplicity and the possibility of understanding in similar terms appear to be guiding values. Further, from his perspective one virtue of this approach by reduction is that it highlights the relationship between problems that were understood as different and yields related solutions to them: generality in this context also takes the form of unifying a wider set of problems. From Kronecker's perspective, although the *existence* of the reduction is established, Jordan's approach to the problem is flawed, since it makes the actual *exhibiting* of this reduction impossible for theoretical reasons.

Brechenmacher's analysis thus highlights that in the two situations generality belongs to different complexes of values, and it is correlatively understood in different terms. These features can be further related to the fact that actors favor different types of procedures, choose different foci for mathematical research, and in the end obtain different kinds of results.

The two studies of disputes examined so far took place within the framework of the *same* discipline. Even when actors work on the same object (curve), or the same problems (which we interpret as matrix reduction), practices and ideals, goals, and values differ. In this context, the ways of understanding and practicing generality also differ. The same conclusion derives from Evelyn Fox Keller's analysis in Chapter 17, which is devoted to practices of generality in the context of two *different* disciplines, namely, physics and biology. More precisely, her study begins by examining a case in which actors belonging to these two fields diverge in their appreciation of what a general treatment of a biological problem should be.

The episode takes place in 1934. Physicist Nicolas Rashevsky (1899–1972) presents a piece of research to biologists, in which he attempts to derive features of the phenomenon of cell division from a model in which he assumes some physical forces being applied to an idealized, and simplified, cell. For him, the cell that is the basis of his work is a model. It is a tractable simplification of the object. Rashevsky thereby puts into play a common practice in physics, which aims to capture the mechanism accounting for a phenomenon. In the context of a practice of this kind, the (re-)production of a given phenomenon, using a few factors that might be at play in a situation, is perceived to provide an explanation of the phenomenon. Such a result suggests a distinction between factors that seem to matter and those that are irrelevant with respect to the phenomenon under considera- tion. Moreover, the simplicity of the model is itself perceived as an argument in favor of the possible generality of the mechanism. With these few elements, Fox Keller sketches features of the epistemological culture in which Rashevsky usually works. They include practices of research and epistemological factors, in this case, ideals of understanding and values. For the biologists who hear Rashevsky, his ideal cell is not interpreted as a model, but as a cell. As a result, in their views, the import of his results is completely different. For some, what Rashevsky talks about simply does not refer to any living organism: this cell does not exist. For others, his results are fine, but have no generality, their validity being restricted to the special case dealt with. Since the cell fails to take into account the fine details of the general cell, there appears to them to be no way in which the result can be generalized. Its relevance is minimal, if not insignificant. Through her study of the episode, Fox Keller captures the diverging expectations entertained in the two contexts with respect to generality, and she suggests this divergence partly accounts for misunderstandings that develop on the two sides of the disciplinary boundary.

These observations lead Fox Keller to concentrate on the *subject matters* dealt with in the context of the two disciplines. By contrast to the phenomena on which physicists concentrate, taken to be the products of logical and physical necessity, the properties of biological organisms (the objects biologists study) are never static, shaped by the inherently contingent nature of evolution. Clearly, this key difference between the subject matters implies that generality cannot present the same features in the two contexts. Fox Keller asks, what then are the forms generality can take in the life sciences, if one takes this key feature into account? She draws on resources provided by the history of science to offer some quite innovative suggestions. Her reflection thus illustrates the resources an inquiry like that presented in this book could offer for practicing scientists.

What is more, the conclusion Fox Keller derives from her observations has a validity that extends far beyond the case study on which she focuses in Chapter 17. Indeed, she stresses that, if the epistemological cultures on which she concentrated present notable differences, this is due in particular to the specificities of the subject matters they deal with. This yields support to, and also accounts for, the thesis we have repeatedly emphasized: collectives of actors shape their ways of doing research in *relation* to the questions they select as being meaningful. Fox Keller draws our attention to the fact that, in this process, the subject matter with which they struggle does play a key part, not least for us in contributing to the determination of forms of generality that are meaningful.

1.3.4   Circulation between epistemological cultures

We have set our inquiry into the value of generality in the context of epistemological cultures, emphasizing how depending on the context, different ways of shaping generality, interpreting this value, and working with it have been devised. However, the various case studies have regularly evidenced that these scholarly cultures are not worlds closed to one another. We have seen resources introduced in the context of one appropriated by another. This is a conclusion holding true more generally. We have seen how it is also valid with respect to an epistemic and epistemological value like generality.

In the final chapter of Part II (Chapter 10), we have examined a case where a concept, that of "genericity," was borrowed from one context to be adapted and used in another. Moreover, the concept was not adopted alone. It was used in relation to a collective research program whose broad outline was similar to the strategy followed in the former context. Likewise, the final chapter of the book (Chapter 18) highlights a striking case of appropriation of a practice of generality, shaped in a given epistemological culture, into a new culture. This case again displays the porosity of these cultures with respect to one another, precisely for the epistemological factors that are the focus of our book. The circulation in question had remained so far unnoticed. It was brought to light in the context of our collective research. Its significance illustrates the benefits that can be derived from the systematic study of a value like generality.

The case in question is Ernst Kummer's (1810–1893) introduction of the notion of ideal numbers into higher arithmetic, which represented a turn in the history of the concept of number as well as in the history of number theory. In this last chapter, Jacqueline Boniface describes the context in arithmetic, in which this innovation took place. Kummer's work followed in the path opened by Carl Friedrich Gauss (1777–1855), when the latter introduced into ordinary arithmetic the entities now called "Gaussian integers" (namely, a type of imaginary number). In 1811, Gauss had justified the introduction into analysis of imaginary magnitudes, by considerations of generality: for him, they had the virtue of bringing into the field a general and uniform validity for truths. In 1825, he further advocated the admission into higher arithmetic of "Gaussian integers" for the generality and simplicity they allowed him to introduce to the theory.

Kummer follows this direction, when, as Boniface explains, he forms the project of shaping a new form of arithmetic for "complex numbers" (with a specific meaning he gives to the expression), in analogy with ordinary arithmetic. His key idea is a hypothesis related to generality, since he assumes that the fundamental theorem of arithmetic, which asserts the unique decomposition of an integer into prime factors, should hold in that other domain.

Kummer introduces ideal numbers, in addition to imaginary numbers, as the entities necessary to ensure the uniform validity of that fundamental theorem for the class of "complex numbers" he studies. The introduction is thus premised on the idea that "complex numbers" ought to present the same properties as integers in ordinary arithmetic. The failure of some "complex numbers" to satisfy this fundamental theorem is felt as an "anomaly" to be eliminated. Ideal factors are thus introduced to give the fundamental theorem a full generality.

To capture the ideal factors, Kummer proceeds through identifying the adequate properties of the usual factors that could hold for the ideal factors. Noteworthy is the fact that Kummer, precisely like Chasles in relation to his "principle of contingent relationships," distinguishes here between permanent and contingent properties of numbers. He suggests that the properties that do not hold *uniformly* should be discarded in favor of those that are *permanently* valid for all numbers, ideal or not. This remark points to a close parallel between Kummer's reflections and what, as we have seen above, occurred in the context of projective geometry, in fact only a few years earlier. This observation leads us to notice that the term "ideal," which Kummer chose to use to refer to the new entities, also evokes projective geometry: Poncelet had introduced the concept of "ideal" into geometry in relation to his "principle of continuity." This principle, as we mentioned above, was introduced precisely to guarantee that the purely geometrical treatment of geometrical configurations has the *same generality* as the analytical treatment. It was this principle that Chasles reformulated using his "principle of contingent relationships" and the related concepts. Is this mere coincidence? As Boniface mentions, in the publication in which Kummer introduces his ideal numbers, he *explicitly* compares them to geometrical ideals, referring to ideas developed in the context of projective geometry.

If we observe the correlation between the two domains more closely, we see that Kummer borrows the term "ideal" from Poncelet. However, Kummer's interpretation of the related elements follows Chasles's approach and analysis, as formulated in his "principle of contingent relationships" discussed in Chapter 2 of this book. Chasles did not want to adopt the terminology of "ideal elements." It thus appears that Kummer somehow makes a synthesis between various means, shaped to introduce generality in the context of projective geometry. What circulated between the two contexts was a *dispositif* for mathematical work, associated with a *hypothesis* on the nature of ideality as well as a conception of *proof*, all deriving from an emphasis placed on generality. The philosophical analysis of some principles, carried out by the practitioners themselves, yielded a diagnosis regarding the means of bringing generality to a field. This diagnosis allowed the importation, into other domains of mathematics, not of *results*, but of *practices* linked to generality. The conclusion indicates how the influence of projective geometry on subsequent mathematics needs to be approached in a broader perspective and especially in relation to the reflection on the value of generality which it promoted.

We see how valuing generality leads to introducing homogeneity through the introduction of, in the first place, new relations, and, later on, of new elements. And we see that the techniques for doing so circulated from one domain of mathematics into another, before becoming a "general method" beyond the boundaries of any domain, like the "method of ideal elements" devised by David Hilbert (1862–1943).

1.4 Conclusion

The exploration of generality carried out in this book could certainly have been broader. Indeed, many chapters in the history of science that might look essential for an inquiry like ours were left aside.[21] Accordingly, many practices of generality were hardly evoked. In particular, practices of generality that were given pride of place elsewhere were evoked only tangentially here. We think, for instance, of the many studies that have analyzed the use of laws, cases, or models, in scientific activity. As we explained at the outset, we did not aim at exhaustiveness.

We have placed our collective study of generality under the auspices of a more global project that aims to understand the part played by values in scientific practice and knowledge. With this term, we did not mean economic value, or value defined in terms of usefulness, or even ethical values, even though these other values are certainly also important topics of research.[22] Instead, our project aimed at contributing to the effort of making sense of *how* actors opt for *ways of knowing* and shape them, and *which difference* it makes for the knowledge thereby produced.

This facet of scientific activity has appeared as significant in the last decades and prominently so when historians and philosophers like Thomas Kuhn (1922–1996) have attempted to account for the choice among theories in view of the underdetermination of theories by evidence.[23] Values then appeared useful as tools that could help us to understand what it meant that there can be "alternative roads to knowledge,"[24] or why dissensions over knowledge sometimes could not be solved.[25] As a contribution to this emerging field of research, we have chosen to illustrate how a value—in our case, generality, which from early on has seemed to be inseparable from scientific activity—could be explored in relatively great detail in a historical and epistemological fashion.[26] In contrast to Kuhn, however, the scale at which we have worked was not that of a whole theory, but that of a smaller scale of scientific practice.

---

[21] The reader will find additional studies in the book edited by Hagner and Laubichler (2006), which complements ours in many respects. Mathematics is not dealt with in this other volume, whereas more weight is put on human and social sciences.

[22] As is clearly illustrated by Putnam (2002), a reflection about any type of value is likely to yield insight for the study of other types.

[23] Kuhn (1977). Note that by bringing "the scope" of a theory into focus, Kuhn touches a value that has relationship with generality as discussed in this book.

[24] Carrier (2012: 242).

[25] Laudan (1984).

[26] In this respect, our project is part of another research program, which is thriving again and in many ways, after decades of quasi dormancy, namely, historical epistemology.

This exploration has suggested that in different contexts the valuing of generality had been related with different goals. In relation to this observation, we have suggested distinguishing between epistemic values and epistemological values. We have also emphasized the variety of ways of stating the general, and the variety of general entities shaped. We have further brought into focus the fact that various ways of pursuing generality could lead to different types of knowledge—to mention but one example, knowing in terms of invariants as opposed to knowing in terms of elementary building blocks. In brief, the meanings that generality has taken even in the same domain, and even for a single analyst, are multi-faceted.

Probably, generality has thereby appeared to be an even more complex value than it is usually assumed to be. Two facts are important in this respect. Generality does not assume a uniform meaning across all the "scientific community" at a given time.[27] Nor does it simply vary from individual to individual. We have rather suggested that the meanings and practices of generality would be better studied as collective facts, shaped in different ways in the context of different local epistemological cultures. In this way, we can account for shared specificities with respect to its pursuit, *as well as* for controversies that were fought, at least partly, in its name—like those between Fermat/Leibniz and Descartes, or Kronecker/Weierstrass and Jordan, and between physicists and biologists working on embryonic development. Yet, this conclusion does not mean that the cultures identified were impervious to each other in this respect.

On the contrary, distinguishing between local contexts allows us to highlight that not only results and concepts, but also ways of shaping general entities, ways of thinking about generality, ways of achieving the general, and research programs based on a way of approaching generality, circulate from one culture to another. With, for instance, Roque's case study, we have seen examples of such circulation occurring diachronically. With the case of number theory and Kummer, as studied by Boniface, we have also seen circulation occurring synchronically. Finally, with Barbara's case study, both synchronic and diachronic forms of circulation have been evidenced.

One could formulate the same remark in a different way. We have highlighted the scientific work that is involved in the making of general entities and general practices, and in their forms of interpretation. This is one way of illustrating the general claim that actors shape scientific cultures, and put knowledge into play to this end. The results of this work circulate in exactly the same fashion as the results of other facets of scientific work. As a consequence, the history of the shaping and practice of generality appears to have been non-linear, showing patterns of differentiation between scientific cultures, as well as circulation among them, and synthesis.[28] These are the unexpected lessons that

---

[27] Daston and Galison (2007) can be read as another historical and epistemological study of an epistemic value, namely objectivity. They themselves place their inquiry in the context of an exploration of "epistemic virtues." This option leads them to bring into focus the history of the making of the scientific practitioner. A key difference between their endeavor and ours is that they are interested in a historical account of the shaping of modern notions of objectivity, which they take to be linear, whereas we emphasize the diversity of ways of understanding and practicing generality in the context of different *contemporary* epistemological cultures.

[28] In this respect, our assumptions and our conclusions differ from those presented in Daston and Galison (2007).

derive from considering, like Chasles, but with a wider focus, the history of science from the perspective of generality.

Another fact of tremendous importance has also appeared throughout our study. We have noticed recurring practices of generality in contexts that at first sight seem to have been far removed from each other. This is the case, for instance, of the choice of working with, and exploring, the general in the context of a paradigm, which we have emphasized can be evidenced in ancient China, in Leibniz's practice as well as in Poincaré's.[29] This remark suggests that there could be basic modes of working with the general, whose identification remains a task for the future. It strikes us that a study like that presented in this book is an indispensable basis for such a research program to be possibly developed.

Similarly, the various case studies have shown that generality was often valued *in combination with* other values. In different contexts, we have observed different constellations of values. Yet some recurring associations have emerged, like the simultaneous valuing of generality and simplicity. Again, this remark calls for a systematic inquiry into the key reasons for the recurrence of similar constellations of values. Here too, historical and epistemological fieldwork about values like simplicity, rigor, or fruitfulness is a prerequisite for such an inquiry to become genuinely possible.[30] This is a research program whose development we call for, and to which we hope this book will contribute.

....

REFERENCES

Carrier, M. (2012) 'Historical epistemology: on the diversity and change of epistemic values in science', *Berichte zur Wissenschaftsgeschichte* **35**: 239–51.
Cavaillès, J. (1938) *Méthode axiomatique et formalisme. Essai sur le problème du fondement des mathématiques*. Paris: Librairie Scientifique Hermann et Cie.
Chemla, K. (2003) 'Generality above abstraction: the general expressed in terms of the paradigmatic in mathematics in ancient China', *Science in context* **16**: 413–58.
Chemla, K. (2005) 'Geometrical figures and generality in ancient China and beyond: Liu Hui and Zhao Shuang, Plato and Thabit ibn Qurra', *Science in context* **18**: 123–66.
Chemla, K. (2009) 'Mathématiques et culture. Une approche appuyée sur les sources chinoises les plus anciennes', in *La mathématique. 1. Les lieux et les temps*, eds. C. Bartocci and P. Odifreddi. Paris Editions du CNRS, 103–52.
Chemla, K., and Fox Keller, E. (eds.) (2016) *Cultures without culturalism in the making of scientific knowledge*. Duke University Press.
Daston, L., and Galison, P. (2007) *Objectivity*. New York: Zone Books.
Fox Keller, E. (2002) *Making sense of life: explaining biological development with models, metaphors, and machines*. Cambridge, Mass.: Harvard University Press.

[29] Aby Warburg's practice of the detail in his historical work, described in Grafton (2006), appears as another example thereof.

[30] We have begun research on the value of simplicity along the same lines as those followed in our study of generality. This was, for instance the aim of the workshop "Simplicity as an Epistemological Value in Scientific Practice", organized in 2009 by Karine Chemla and Evelyn Fox Keller in the context of the Institute for Advanced Study, Paris (http://www.paris-iea.fr/en/events/simplicity-as-an-epistemological-value-in-scientific-practice-79, accessed May 16, 2016).

Grafton, A. (2006) 'Auf den Spuren des Allgemeinen in der Geschichte: Der wilde Gott des Aby Warburg', in *Der Hochsitz des Wissens. Das Allgemeine als wissenschaftlicher Wert*, eds. M. Hagner and M. D. Laubichler. Zürich-Berlin: Diaphanes, 73–95.

Haddock, A., Millar, A., and Pritchard, D. (eds.) (2009) *Epistemic value*. Oxford: Oxford University Press.

Hagner, M., and Laubichler, M. D. (eds.) (2006) *Der Hochsitz des Wissens. Das Allgemeine als wissenschaftlicher Wert*. Zürich-Berlin: Diaphanes.

Hawkins, T. (1977) 'Weierstrass and the theory of matrices', *Archive for history of exact sciences* 17: 119–63.

Knorr-Cetina, K. (1999) *Epistemic cultures: how the sciences make knowledge*. Cambridge, Mass.: Harvard University Press.

Kuhn, T. S. (1977) 'Objectivity, value judgment, and theory choice', in *The essential tension. Selected studies in scientific tradition and change*, ed. T. S. Kuhn. Chicago: University of Chicago Press, 320–39.

Lagrange, J.-L. (1799 (Thermidor, An VII)) 'Solutions de quelques problèmes relatifs aux triangles sphériques avec une analyse complète de ces triangles', *Journal de l'Ecole Polytechnique* VI: 270–96.

Laudan, L. (1984) *Science and values. The aims of science and their role in scientific debate*. Pittsburgh Series in Philosophy and History of Science. Berkeley: University of California Press.

Putnam, H. (2002) *The collapse of the fact/value dichotomy and other essays*. Cambridge, Mass.: Harvard University Press.

Robadey, A. (2004) 'Exploration d'un mode d'écriture de la généralité: l'article de Poincaré sur les lignes géodésiques des surfaces convexes (1905)', *Revue d'histoire des mathématiques* 10: 257–318.

Robadey, A. (2015) 'A work on the degree of generality revealed in the organization of lists: Poincaré's classification of singular points of differential equations', in *Texts, textual acts and the history of science*, eds. K. Chemla and J. Virbel. Dordrecht: Springer, 385–419.

Schappacher, N. (2010) 'Rewriting points', in *Proceedings of the International Congress of Mathematicians. Hyderabad 2010*, ed. Rajendra Bhatia. Hyderabad, India: Hindustan Book Agency, 3258–91.

Schmitt, S. (ed.) (2010) *Georges Louis Leclerc de Buffon, Œuvres complètes, Vol. IV*. Paris: Honoré Champion.

Seth, S. (2006) 'Allgemeine Physik? Max Planck und die Gemeinschaft der theoretischen Physik, 1906–1914', in *Der Hochsitz des Wissens. Das Allgemeine als wissenschaftlicher Wert*, eds. M. Hagner and M. D. Laubichler. Zürich-Berlin: Diaphanes, 151–84.

Ziche, P. (2006) '"Wissen" und "hohe Gedanken". Allgemeinheit und die Metareflexion des Wissenschaftssystems im 19. Jahrhundert', in *Der Hochsitz des Wissens. Das Allgemeine als wissenschaftlicher Wert*, eds. M. Hagner and M. D. Laubichler. Zürich-Berlin: Diaphanes, 129–50.

Part I

The meaning and value of generality

Section I.1

Epistemic and epistemological values

2

The value of generality in Michel Chasles's historiography of geometry

KARINE CHEMLA

2.1 Introduction: contexts and goals of the *Aperçu historique*

In 1837, Michel Chasles (1793–1880), a former student of the Ecole Polytechnique and an already well-known geometer, published his *Aperçu historique sur l'origine et le développement des méthodes en géométrie, particulièrement de celles qui se rapportent à la géométrie moderne, suivi d'un mémoire de géométrie sur deux principes généraux de la science : la dualité et l'homographie.* [1] In the book, the issue of generality in geometry, which already comes to the surface in the title, was central. Chasles's introduction made this clear right at the outset in presenting the project of the book.

[1] Chasles (1837). The title can be translated as *General historical survey of the origin and development of methods in geometry, in particular of those that relate to modern geometry, followed by a memoir of geometry on two general principles of that science, that is, duality and homography.* Two further editions of the book appeared in the nineteenth century. The third edition (Chasles, 1889), prepared posthumously by Catalan in 1889, is the most widely available. Contrary to what the title page claims, there are differences between the first and the third editions. My quotations and references rely on the first edition. I will also indicate differences between the editions, when they are relevant for my purpose. Each chapter of the book is divided into numbered sections, and I will also refer to the section numbers. Koppelman (1971) and Raina (1999) provide elements on Chasles's biography. Kötter (1892) yields an evaluation of Chasles's geometrical contributions, including the *Aperçu historique*. The first version of this chapter was written in early 2008, after presentation in the seminar we organized on the topic as well as in other venues. It is a pleasure to thank my different audiences as well as the research group that took part in the project of the value of generality and prepared this book for their remarks and suggestions. I would like to thank in particular Renaud Chorlay, Massimo Galuzzi, Emily Grosholz, Ken Manders, and David Rabouin for criticisms that helped me improve this chapter. Needless to say, the remaining flaws are my sole responsibility. This chapter was completed while I benefited from the support of a Chinese Academy of Sciences Visiting Professorship for Senior Foreign Scientists 外国专家特聘研究员, grant number 2009S1-34 (Beijing, 2010). John Mumma has most generously offered his help to complete the elaboration of this paper. I have pleasure here in expressing my heartfelt gratitude to him.

The Oxford Handbook of Generality in Mathematics and the Sciences. First Edition. Karine Chemla, Renaud Chorlay and David Rabouin. © Oxford University Press 2016. Publishing in 2016 by Oxford University Press.

At first sight, the reader may be tempted to believe that the purpose of the *Aperçu historique* was mainly historical: Chasles aimed at giving a sketch of the development of ideas in geometry. Yet, as is well known, Chasles added notes to the main text, some historical and others presenting new results.[2] In fact, in the book, mathematics and history were combined as two kinds of tools useful in achieving a single aim. Chasles's introduction formulated this aim, while explaining why he added notes with new mathematical results. He wrote:

> These (notes) would not seem essential, if one envisaged *only the historical aim* of our work. However, through relating the course of Geometry and presenting the state of its recent discoveries and doctrines, we *mainly* had in view to show, by means of some examples, that the *main characteristic* of these doctrines is to *bring to all parts of the science of extension* a *new ease* and the *means to achieve a kind of generalization*, so far unknown, of *all geometrical truths*. This feature had also been the characteristic feature of analysis when it was applied to Geometry. We shall thus *conclude* from our general survey that the *powerful resources* that Geometry has acquired in the last thirty years or so *can be compared*, in several respects, with the *analytical methods*, with which this science *can now compete*, without being disadvantaged, in quite a large range of questions.
>
> This idea will be reproduced, shall we say justified!, in many places of this writing, since it constitutes its *origin* and it *never ceased to guide the lengthy research* that the *historical part*, the *Notes and* the two *memoirs* that compose this book have required.[3]

The mathematical notes bear witness to the fact that, for Chasles, the book had a goal beyond its historical import. The historical and the mathematical developments it included derived from the same intellectual effort and were geared toward the same aim: bringing to light how the recent advances in geometry allowed the strictly geometrical methods in this domain to rival the analytical approach to geometry, which had dominated the field in the previous two centuries. The "resources" recently "acquired" by geometry, as we shall see shortly, referred to new developments that led to the emergence of projective geometry and were associated, among others, with the names of Monge (1746–1818), Carnot (1753–1823), and Poncelet (1788–1867). Further, Chasles was also referring to his own contributions in the notes and the *Mémoire*, introduced by a sixth and final chapter entitled: "Object of the memoir." In other words, the *Aperçu historique* constituted a plea in favor of the new geometry, based on history and mathematics. Moreover, right in the introduction, Chasles identifies a key reason for this status that, according to him, had been gained by geometry: the new doctrines provided various means to "generalize," whereas generality had long been deemed the privilege of "analysis." Generality, and the related property of simplicity, are *Leitmotive* in the whole book.[4]

This theme was by no means new. Monge, Carnot, Poncelet, and other geometers of the early nineteenth century in France were all puzzled by the same question: from

---

[2] Chasles (1837: 2). The genesis of the book accounts for this mixture.
[3] Cf. Chasles (1837: 2). My emphasis.
[4] See, for instance, Chasles (1837: 47, Section 42).

Descartes's works onward, analytical methods brought about important developments in geometry. Why did purely geometrical methods not yield the same results? These geometers constantly pondered the question. Their diagnosis identified the generality that analytical methods provided as the key property that made the difference.[5] Their reflections thus mainly bore on the means with which to introduce generality in geometry proper. Projective geometry emerged as a result of this effort. Moreover, these geometers turned more and more often to history to find resources for their investigation, as is amply illustrated by the introduction to the *Traité des propriétés projectives des figures*, the 1822 book in which Poncelet first gave shape to the new geometry.[6] Chasles's book hence reflects a practice of mathematics, shared by a milieu, in which a study of history meshed with mathematical work and a meditation on methods.[7]

Chasles's *Aperçu historique* may be considered as a synthesis and development of these reflections, in which the historical analysis of geometry from the point of view of generality received its fullest treatment. Moreover, Chasles introduced in the book new mathematical ideas related to these questions. These facts explain why the *Aperçu historique* is an interesting document to guide a reflection on the value of generality in geometry, which is the aim of this chapter: in relation to its purpose, the book develops a view on geometry and its history from precisely the perspective of generality. As a result, it provides source material to think about *how* generality is achieved in geometry, which means can be used to this end and which benefits might be derived from it. In Section 2.2, I analyze Chasles's interpretation of the history of geometry from the viewpoint of generality. While doing so, I shall stress the correlations between his reading of history and his own mathematical work. In Section 2.3, I examine Chasles's own mathematical practice in relation to the value of generality. The conclusion will offer reflections on the value of generality in general, and also on the history of projective geometry from this perspective.

My goal in this chapter is not to deal with the topic of generality in geometry exhaustively. Nor is it to analyze in a critical way how Chasles selected and analyzed his historical source material. Rather, I limit myself to discussing, on the basis of his own account of the history of geometry as well as his presentation of mathematical results, some aspects of the problem of generality in geometry. At the same time my purpose is also historical. Chasles's book attests to how a philosophical and historical

[5] On Monge and Poncelet, compare Taton (1951b: 79–100), Belhoste and Taton (1992: 277–99), and Belhoste (1998), on Carnot and Poncelet, see Chemla (1998), and on Carnot, Poncelet, and Chasles, see Nabonnand (2011). Taton (1964) contains interesting remarks on generality in analytical geometry at the time.

[6] Poncelet (1822). Brianchon (1783–1864) and Carnot were some of the mathematicians who provided insight in the history of mathematics in relation to their research in geometry. For Brianchon, compare Chemla (1990: 530). As far as Carnot is concerned, compare, for instance the "Discours préliminaire" of his *Géométrie de position* (Carnot, 1803: I–XXXVIII).

[7] In particular, it must be stressed that Chasles did not practice history for its own sake. The development of his historical work is mainly motivated by the meditation on generality and the methods of the new geometry. This point is demonstrated by his explanation for not dealing with certain topics (Chasles, 1837: 93, Section 34). In particular, in these pages, Chasles stresses that his *Aperçu historique* focuses on the "successive formation of *methods*" (my emphasis) in geometry. Elsewhere, he makes clear that his emphasis remains on what he calls "geometry of forms and situations" (Chasles, 1837: 143).

reflection on generality, conducted from within mathematics, steered key developments in geometry. In fact, my analysis reveals how various actors of projective geometry in the first decades of the nineteenth century did not share the same ideas with respect to generality. Despite the lengthy treatment offered here, I am fully aware that the book would deserve a much longer analysis for its wealth of insight into the question of generality in geometry.

2.2   Chasles's historical analysis of geometry

The historical part of the *Aperçu historique* may be roughly summarized in three main theses as follows:

First, in ancient geometry, in a phase that lasted until roughly the fifteenth century, Chasles claims there were neither general conceptions, nor general methods.

Second, some generality was progressively introduced from the seventeenth century onward, and Chasles examines the means by which this change was brought about. He identifies here two main channels: the introduction of *geometrical* means, which constituted until the end of the eighteenth century a less influential tradition, and that of *analytical* means, in a tradition stemming from Descartes' *Géométrie*, which soon became dominant due to the generality and the fecundity it made possible.

Third, the modern geometry that steadily developed from the end of the eighteenth century onward constituted, in Chasles's view, a turn. It shaped new purely geometrical methods to deal with questions in geometry and yet reach levels of generality similar to what had been achieved using analysis. Chasles scrutinizes *how* methods in pure geometry recovered means to achieve generality. He further explained how he contributed to them and what for him was at stake in pursuing generality as a value.

Let us examine his analysis of each of these phases in greater detail.

2.2.1   A diagnosis about the limits of ancient geometry

In his critical investigation into ancient geometry, Chasles highlights several respects in which it lacked generality. Each of them is approached negatively and contrasted with different practices shaped from the seventeenth century onward. In other terms, these aspects are approached anachronistically, in a retrospective way. Listing them thus highlights benefits to be derived from generality in Chasles's conception. Let us examine some of them, albeit in an order different from his.

First, Chasles emphasizes how the Ancients gave, as different, propositions that in fact, from a modern point of view, amounted to the same statement. Several variants of this diagnosis can be found in his first chapter.

For instance, propositions that, in Greek geometrical books, appeared as unrelated could be shown to be "simple consequences" or "simple transformations" of a single general theorem. Such is Chasles's diagnosis, when he comments on lemmas given by Pappus in relation to the *De locis planis* by Apollonius—I delete all unnecessary details:

… One can also relate them to *one and the same proposition*, which expresses a *general* property of four points arbitrarily taken on a straight line […Chasles inserts here a reference to the theorem meant]

In that way, propositions 123 and 124, which express a relation between four points arbitrarily taken on a line and a fifth point determined through a certain condition, are *easy consequences* of this theorem.

Propositions 125 and 126 express a relation between four points arbitrarily taken on a straight line and one easily recognizes that this relation is nothing but a *very simple transformation* of the *same theorem*.

… It is remarkable enough that these four propositions (i.e., propositions 119–22, note by KC), which *look so different* from the others and *seem* to have *no relationship with them*, are also *consequences* of the *same theorem*…" (Chasles, 1837: 42–3, Section 36, my emphasis).

Several points are worth noticing here.

In this case, Chasles emphasizes that coming to know the relevant general theorem allows one to *easily* derive from it several propositions that Pappus presented as *distinct* and even *unrelated*. The generality of the theorem is here correlated to the number of different propositions deriving from it, that is, to its fruitfulness. We shall see that, in his own contributions, Chasles looks precisely for theorems that are the most general in the same sense, that is, those that by mere transformations—the word meaning here "reformulations"— can be changed into several propositions formerly thought of as being different. In Section 2.3, by providing one such example from Chasles's work, we shall analyze the tools that can be put into play to achieve this end, and we shall come back to the notion of "transformation."

Moreover, in his comment on Pappus, Chasles puts a proposition being a (logical) "consequence" of a general theorem on a par with a proposition being a "transformation" of the general theorem. One can be more specific here. The single proposition to which Chasles relates all of Pappus's lemmas has the shape of expressing "a general property of four points arbitrarily taken on a straight line." This analysis provides Chasles with a tool with which to identify how propositions relate to this general one. With this tool, some propositions appear to carry out the same task and, through further examination, Chasles identifies that they amount to "a very simple transformation of the same theorem." With the same tool, other propositions can be understood as being "easy consequences." Whether propositions are derived from transformation or as a consequence, Chasles insists on the *ease* with which one obtains them: a simple consequence appears to be a mere change of form. This is the first occurrence of the value of "simplicity" in relation to that of generality. In fact, both epistemological values will prove to have, in Chasles's conception, deep connections.

Finally, Chasles links relating distinct propositions to a single theorem and bringing them into relation with each other, highlighting a connection and a degree of similarity between them. This way of reading ancient sources and analyzing them already betrays distinctive features of Chasles's approach to generality in geometry.

In other examples Chasles examines how propositions that look different in ancient Greek books turn out, according to his analysis, merely to be various *particular cases* of a single theorem. For instance, after having introduced how in his *Conics*, Apollonius approached these curves and quoted proposition 37 of Book 3, Chasles notices:

> The 23 first propositions of book 4 relate to harmonic division of straight lines drawn in the plane of a conic. They are, for the most part, various *cases* of the general theorem we just stated (Chasles, 1837: 19–20, Section 12, my emphasis).

Chasles formulates the same conclusion regarding 43 propositions grouped by Pappus in the seventh book of his *Mathematical collection*:

> At first sight, one does not perceive the *true meaning* of these numerous propositions, nor does one see the *relationships* that can *tie them all to the same question*, and *reading* in this state *is painful*. However, with some attention, one recognizes that they all relate to the theory of the involution of six points, which was created by Desargues and which became of great use in recent Geometry. These propositions are not properties of the most general relation of involution, the one holding between six points (it even seems that the Ancients did not know the transformations of this general relation), but they are properties of several relations which today can be considered as *particular cases of this general relation* (Chasles, 1837: 39–40, Section 34, my emphasis).[8]

We come back later to the topic of involution. Let us examine what Chasles's analysis reveals here about ancient geometry as well as his own conceptions of geometry. As in the previous case, Chasles notices that, seen from another perspective, propositions presented in Greek sources as unconnected appear in fact linked to each other: They express "properties of relations" that are "*particular cases* of a general relation" holding of six points. We can grasp here *how* the relation of "being more general" leads in Chasles's view to establishing an organization of mathematical knowledge.

Other correlated features emerge here. The failure to perceive connections between propositions is correlated with the fact that the "true meaning of these numerous propositions" is missed and reading them is laborious. We can grasp here a contrast between such a situation and the ease yielded by an understanding of the connection between propositions stressed above.[9] Moreover, Chasles establishes a link between "understanding" propositions and relating them with each other in a way that connects them all to a more general fact. Identifying the "meaning" amounts to finding the most general theorem from which they derive. He concludes:

> This analysis of Pappus' 43 lemmas seems to me to allow us to grasp the *general idea* and to make their *reading easier*. One sees there that *many propositions* among them express the *same theorem*: it is because the statements of these propositions are applied to *specific*

---

[8] Similar reorganizations of ancient Greek source material are frequent in the *Aperçu historique*.

[9] Chasles returns to this contrast when he presents Desargues's contribution and, especially, the introduction of the relation of involution between six points arbitrarily placed on a conic. On the one hand, he stresses that ancient geometers did not know such general propositions. On the other hand, he explains why this accounts for the length and intricacy of their proofs and treatises (Chasles, 1837: 79–80).

figures and the few differences between them come from the *difference of position* of the points considered in these propositions (Chasles, 1837: 41, Section 35, my emphasis).

Besides emphasizing the two points made earlier, Chasles adds an important detail: from the new perspective, statements that in Pappus's book look different are seen to "express the same theorem." This is an essential reason for Chasles's interest in generality in geometry. He puts forward two reasons to explain *how* the relation between the various propositions was obscured.

On the one hand, propositions concerned figures that were not understood as being particular cases with respect to a single more general figure. In other words, the general diagram for which the proposition held true was not brought out. Here, Chasles shows that the various propositions concern four or five specific points, instead of the configuration of six points involved in the most general relation of involution. Chasles formulates the same diagnosis in other circumstances.

On the other hand,—and this leads us to another fundamental reason for which Chasles diagnosed a lack of generality in ancient geometry—there were as many propositions as there were possible relative positions of the various parts of a figure. Chasles contrasts this practice with the modern one, which introduced means to group into a single proposition all the statements about figures that differed from one another merely with respect to the positions of their components:

> It is one of the great advantages of modern Geometry over the ancient one that it (the former) can, through considering positive and negative quantities, *comprehend,* by means of the *same statement, all the various cases* that the same theorem can present when one varies the relative positions of the different parts of a figure.[10]

In his *Report on the progress in geometry,* written upon the request of the Ministry of Education and published in 1870, Chasles stressed this point again, associating it with Carnot's contribution in his *Géométrie de position* (Carnot, 1803). In addition, and this introduces a new element into the discussion, Chasles emphasized at that time the link between, on the one hand, the introduction in geometry of the tool of negative and positive quantities and, on the other hand, a change in the practice of proof which is essential to achieve generality in geometry. He writes:

> The *Geometry of position* contained a felicitous conception of the nature of positive and negative quantities, which allowed one to *generalize each question,* in the sense that *a single proof could suffice, whichever the relative positions* of the different parts of a figure may have been, whereas, until then, each question demanded as many proofs as there were varieties of position in the points and the lines of a figure. This conception appears to us to be the main idea that characterizes Carnot's book.[11]

---

[10]   Cf. Chasles (1837: 41, Section 35). My emphasis. The third edition deletes the key word "all" (Chasles, 1889: 41–42).

[11]   Chasles (1870: 4, my emphasis). In between, Chasles had introduced into geometry sign conventions for directed segments or angles and, although he credited Carnot with this idea, he had modified the implementation of this tool. Compare Chasles (1852: III–XI). Nabonnand (2011: 37–43) analyzes Chasles's introduction of signs and their relation to his practice of generality.

Clearly, the issue of achieving a greater generality in geometry also requires developing a reflection on how proof should be conducted for establishing a general statement in a general way. This was one of the issues at stake in the early nineteenth century, one to which Chasles would return regularly.

One aspect of this question evoked here is the global device that Carnot introduced to reach this aim: a conception of positive and negative quantities. This device was one of the contributions on which Poncelet meditated before he introduced his "principle of continuity," one of the main tools crafted in the early nineteenth century to introduce some generality into geometry and, more specifically, into geometrical proofs. Poncelet's main goal with the "principle of continuity" was to change the practice and conception of proof along the same lines and for the same purpose.[12] It is interesting that in Chasles's note XXIV on the history of the "principle of continuity," he insists on the tension between rigor and generality:

> One cannot conceal to oneself that one owes the huge progress that the Moderns have achieved in Geometry to this relaxation of the (demand for) rigor of the Ancients. The Ancients, who were more keen to convince than they were to highlight, have hidden all the threads that would have led one to find the path towards their methods of discovery and invention and that could have guided those who continued their works. This has been the cause of the timorous and muddled course of Geometry, of the *incoherence of its methods* in questions that had the *same nature*, or, to speak more precisely, of the *absence of secure methods of its own* like those of modern Geometry *for entire classes of questions* that *possessed* a certain generality.[13]

In Chasles's discussion of the question of proof in ancient geometry, the lack of generality related to proving is approached from the viewpoint of the constraints imposed by rigor, which demand distinguishing between various cases. Chasles further stresses other dimensions of the question of the generality of a proof in the following conclusion he draws with respect to ancient geometry:

> Each *method* did *not* in effect possess *anything general* and was *limited to the particular question* that had given rise to it; *each* known *curve*—and their number was quite limited—had been *studied individually* (that is, in isolation from the others, note by KC), and *by means* that were *completely specific* to it, without any of its properties, or the procedures that had led to it, being possibly used to discover the properties of another curve.[14]

The lack of generality of proofs in ancient geometry is here analyzed from two different angles. The means used to deal with a given mathematical object brought into play particular features of that object. Facts that might have been general were approached by procedures specific to the objects under study. Moreover, as a result, the methods devised could not be used to study other similar objects, preventing the establishment

---

[12] On this question, compare Chemla (1998). We analyze below how Chasles reformulated the principle of continuity and renamed it as "principle of contingent relationships."

[13] Chasles (1837: 359). Poncelet had emphasized the same point before Chasles (Chemla, 1998: 173).

[14] Chasles (1837: 51, Section 1). The third edition has a typo here and indicates Section 9. Emphasis is mine.

of connections between the objects. The organization of mathematical knowledge that inspires Chasles's comments connects propositions in a way different from what is the main concern in connecting them within an axiomatic-deductive structure.[15]

Chasles goes on with the example of the problem of determining tangents to curves: in ancient geometry, for each curve, the problem was solved by a method that could not inspire the solution for another curve. These methods, he stresses, were "essentially different." Clearly, this comment is inspired by a contrast he has in mind with the general methods that were devised in the seventeenth century to solve this general problem. He comments on the methods in great detail, and precisely from this standpoint, in the following chapters.[16] One remark allows us to grasp better what, in Chasles's view, characterizes a genuinely general method. Indeed, when speaking of "general method" in relation to ancient geometry, the so-called "method of exhaustion," which Euclid and Archimedes regularly used in issues related to measuring surfaces or volumes, immediately comes to mind.[17] We know that the general name for what we might be tempted to perceive here as a general method was introduced only in the seventeenth century, by Grégoire de Saint-Vincent. In ancient texts, the "method" has the status of a pattern of proof recurring in various contexts. Chasles has an interesting comment on this, which casts light on a dimension of his approach to the problem of generality. He writes:

> The *method of exhaustion*, which rested on a main idea (KC: literally, "mother idea") that was completely general, did not deprive Geometry of its character of narrow-mindedness and specialization, since this conception, lacking general means of application, was becoming, in each particular case, a wholly new question, that did not find resources (KC: that is, resources to become efficient) but in the individual properties of the figure to which it was applied.[18]

In other words, despite the fact that a single pattern of reasoning was used, the ways in which it was *applied* were not general. The attention placed on the modalities of *application* of general methods, and especially their *uniformity*—another type of generality—inspires, as we shall see, many other comments in the *Aperçu historique*. By contrast with the specific uses of the "method of exhaustion" in antiquity, Chasles stresses the shaping of *uniform* means of applying a single and general method as characteristic of a turn that occurred in the seventeenth century.[19]

---

[15] We see here how in his view general methods connect the objects to which they are applied and the propositions yielded. For Chasles, the roots of mathematical knowledge are provided by "general propositions" from which the others are derived as mere "transformations" or "easy consequences." In contrast, an axiomatic-deductive organization places emphasis on another kind of starting point and stresses the rigor of the derivations rather than their generality. We return to these issues below.

[16] See Evelyne Barbin's chapter in this book (i.e., Chapter 15).

[17] Youschkevitch (1964).

[18] Chasles (1837: 51–2).

[19] In the specific case of assessing areas and volumes, Chasles attributes to Cavalieri, with his geometry of indivisibles, the creation of means allowing a uniform "application, or, rather, a transformation of the method of exhaustion" (Chasles, 1837: 57, Section 5).

Chasles's Chapter II deals precisely with the shaping of geometrical means that introduced some generality into geometry mainly in the seventeenth and eighteenth centuries. Chasles opens this chapter by stressing the contrast between geometry as practiced in that period and geometry as practiced until the fifteenth century, which he treated in his first chapter. In his view, the latter was as much "specific" as the former was "general," as much "concrete" as the former was "abstract." He announces:

> The characteristics of generality and abstraction that Geometry acquired from this period on became more and more prominent in the subsequent periods. At the present day, these (features) make a huge difference between modern Geometry and that of the Ancients.[20]

Let us hence turn to Chasles's account of the various ways in which geometry achieved greater generality from the sixteenth century onward, relative to the various directions identified as problematic in the previous time period.

2.2.2   The contrast with geometry in the seventeenth and eighteenth centuries

What Chasles calls the second and the third time periods, and treats in Chapters II and III, respectively, are in fact overlapping timespans. Chapter II deals more specifically with the introduction of new geometrical approaches and extends partly into the eighteenth century, whereas Chapter III begins with the radical change brought about by the publication of Descartes's *Géométrie*, as an Appendix to the *Discours de la méthode* (1637), and the introduction of analytical means into geometry. Let us examine, from Chasles's perspective, the related changes with respect to the question of generality, beginning with the impact of new geometrical approaches like those developed by Desargues and Pascal, among others.

We have already alluded to the emphasis placed by Chasles on the emergence, during these centuries and within this framework, of general methods, like the method of indivisibles or the methods of tangents, which allowed practitioners to deal with classes of problems *uniformly*. On which other aspects does Chasles dwell?

Chasles underlines two key changes with respect to generality and, perhaps even more importantly, highlights a *correlation* between them: they bear on, respectively, the kind of geometrical objects dealt with and the connections that can be established between the propositions related to them. In the *Aperçu historique*, these changes and the correlation between them are exemplified by the approach to conic sections presented in Desargues's *Brouillon projet d'une atteinte aux événements des rencontres du cône avec un plan (Rough draft for an essay on the results of taking plane sections of a cone,* hereafter abbreviated to *Brouillon projet)* (Desargues, 1639). In Chasles's terms, "Desargues's method allowed him to bring

---

[20] Chasles (1837: 52). Note the use of the term "specific" in opposition to "general," and that of the term "concrete," referring precisely to the specific, in opposition to the word "abstract." This conception of abstraction is worth another publication. To state it briefly, Chasles considers the objects of ancient geometry as "concrete" by comparison to the geometrical objects introduced in the seventeenth century. See below for Desargues's conception of conics or Descartes's approach to curves.

into the theory of conics, as he did in various other writings, new views of *generality* which widened the *conceptions* and *metaphysics* of Geometry."[21]

Chasles refers here to a shift in the conception and the treatment of the object "conics." In ancient geometry, in a tradition based on the approach to conic sections in Apollonius' *Conica*, the various conics were produced as curves cut in a cone by a plane placed in a specific and fixed position with respect to that cone—that is, perpendicular to a determined triangle attached to the cone (Chasles, 1837: 17–18). Depending on the nature of the cone, the curves obtained were ellipses, parabolas, or hyperbolas. According to Chasles's analysis, Desargues's key modification was to introduce the idea that the plane could be placed arbitrarily with respect to the cone. As a result, the various conics could be obtained as sections of one and the same cone, thereby being the perspective image of each other. The fact that the position of the plane became general is important, but it would not matter, were it not for the essential following remark, which was Desargues' correlated step: "conic sections, being formed by the different ways in which one cuts a cone which has a circle as its base, *should participate in the properties of this figure*" (Chasles, 1837: 75, Section 20, my emphasis).

Several consequences resulted from this method.

First, for each property of the circle, one could look for an analogous property for any other conic section. The properties of the simplest curve among them could serve as a guide for inquiring into the properties of the others.

Second, as a result, each property of any of them could be reflected in a related property of any of the others. The properties of the various conic sections were thereby brought into connection with each other, and were no longer isolated facts, in contrast to how, according to Chasles's diagnosis, propositions had been treated in ancient geometry. This strategy did not apply only to the ellipse, the parabola and the hyperbola. In the same vein, as Chasles stresses in the two following sections, "Desargues applied to systems of straight lines properties of curved lines" (Chasles, 1837: 76) and his ideas led him to "apply to conic sections various properties known for the system of two straight lines" (Chasles, 1837: 77). If such a circulation of properties between these geometrical situations was possible, this was linked to an essential fact on which Chasles dwelled. Indeed, the conic sections, being "formed by the various ways in which one cuts a cone having a circle as its base," included for Desargues not only the circle, but also a system of two lines.

This remark leads us to the third point emphasized by Chasles. In correlation with such an exploration of the connections between their *properties*, Desargues brought about a change in the nature of the *objects*:

> He thus considered, as *varieties of the same curve*, the various sections of the cone (the circle, the ellipse, the parabola, the hyperbola, and the system of two straight lines), which, *until then*, had always been treated separately and *by means specific* to each of these sections.

---

[21] Chasles (1837: 75, Section 21), my emphasis. Even though the sentence is not entirely clear, it indicates that Chasles perceives the techniques for achieving generality have their history. At the time when Chasles writes, Desargues's *Brouillon projet* was lost. We return to the method by means of which Chasles restored these ideas below. On Desargues's geometry, see Taton (1951a) and Field and Gray (1987).

> Descartes tells us Desargues also considered a system of several straight lines parallel to each other as *a variety* of a *system of lines* converging towards the same point (...).[22]

These are the changes in which Chasles saw "new views of generality, which widened the conceptions and metaphysics of geometry." Interestingly enough, he relates this issue to that of the *treatment* of the objects. The contrast between Desargues's approach and that of ancient Greek geometers in this respect brings us back to the question of the generality of proof, evoked in the previous section. Here, this issue is connected to that of the generality of the objects.

Clearly, for Chasles, studying objects in relation to one another was not merely a method for discovering new properties for each of them. Neither was it simply a way of eliciting general objects. Most importantly, such a technique would lead to shaping means for approaching objects, solving problems and proving properties in a *uniform* and hence *general* way. The "methods of indivisibles" or the "methods of the tangents," shaped in precisely the same time period to address different questions, corresponded to a concern of this kind. Designing them required that the problems for which they provided solutions could be formulated in a general way.[23] This, in turn, also required that curves be approached as general objects. These considerations explain, I think, Chasles's comment on how these methods compare with Descartes's analytical method: they "also bore, in their metaphysical principles, the seal of this generality (...)" (Chasles, 1837: 94). General objects and general methods appear to be, in his eyes, two sides of the same coin, emerging conjointly in the seventeenth century.

In the specific case under consideration, Chasles views the key operation allowing Desargues to carry out this mutation in the objects of geometry—, that is, bringing the properties of conics in relation to each other through linking all the conics cut in a cone to its circular base—as a particular case of a more general method: finding out methods of transforming figures into one another and thereby transferring the properties obtained about one to the others. In his Chapter II, devoted to the second time period in his periodization of the history of geometry, Chasles identifies this idea in various domains of sixteenth and seventeenth century geometry.[24] However, it must be stressed that these methods are connected, under the single heading of "transformations," only by Chasles's retrospective view. At that time, instead of being various applications of a general method, they appear to have occurred as distinct and unrelated ideas.

In Chasles's historical account, this was not the only way of connecting propositions in a geometrical fashion that appeared in the seventeenth century. What he could gather

---

[22] Chasles (1837: 75–6). Emphasis is mine.

[23] About Descartes's, Fermat's and Roberval's contributions to the problem of determining tangents, Chasles writes: "These three famous men share the glory of having solved, each in a different way, a problem which *no geometer* had yet *dared* tackle in its *generality*, that is, the problem of tangents to curves (...)" (Chasles, 1837: 57). This statement is to be compared to Chasles's comment on the determination of tangents by authors of antiquity, quoted in the previous section.

[24] He interprets in this way Vieta's introduction, in spherical trigonometry, of the triangle reciprocal of a given triangle, or the geometrical transformations of conics into one another, carried out for instance by Grégoire de Saint-Vincent (see, respectively, Sections 3 and 33 in Chasles, 1837: 54, 91–2).

from Pascal's complete treatise on conics led Chasles to identify another modality of carrying out the same operation at the time. The treatise was already lost when the *Aperçu historique* was written. Yet, Chasles could rely on a seven-page long excerpt from it, entitled *Essai pour les coniques*, which Pascal published in 1640 and which had been rediscovered and published by Bossut in 1779. Moreover, Chasles relied on references to the complete treatise found in the correspondence between scholars as well as on notes taken by Leibniz on it.

Evidence showed that Pascal had adopted Desargues's approach to the conics, evoked above (Chasles, 1837: 74), and used perspective methods. In addition, and this is the important point now, the *Essai pour les coniques* started with a "lemma," called the "mystic hexagram," a proposition "from which everything else had to be deduced" (Chasles, 1837: 72, Section 17). Chasles describes the various theorems that Pascal had derived from this fundamental proposition and had selected for inclusion into the summary that the *Essai pour les coniques* offered. As a geometer who had an intimate understanding of propositions on conics and their connection with each other, Chasles concluded:

> We *understand* perfectly well, from what we know of the *fruitfulness* of the theorems we just quoted, that, as Pascal announced it, he made them the *basis of Complete Conical Elements* and that, by deducing them from his mystic hexagram, he *derived from this single principle 400 corollaries*, as Father Mersenne said (...).
>
> We notice that these various main theorems each *expressed a certain property of six points placed on a conic*, which *explains how* Pascal could deduce them from his mystic hexagram, which was itself a *general property* of these six points. But each of these theorems had taken a *different form*, which made it suitable for particular uses which encompassed a huge number of properties of conics.
>
> It is *this art*, which is *infinitely useful* and consists of *deducing from a single principle* a large number of truths, of which the *writings of the Ancients* offer *no example* and in which our methods are more advantageous than theirs.[25]

This is a key statement, on which we shall dwell. Chasles emphasizes several points in what, he gathers, was Pascal's full treatment of the conics. These points will turn out to be essential features of Chasles's own practice of geometry, as we shall show in Section 2.3. Moreover, they represent key aspects of Chasles's approach to generality. Let us exhibit them.

First, Chasles derives his assumption regarding the organization of the treatise from Mersenne's allusion to it: a single "principle" would have been placed at the basis of the whole domain, and its "fruitfulness" would be clear from the wealth of propositions deriving from it. Such "principles" evoke those typical of analytical treatments of mathematical domains in the eighteenth century: statements from which, by the mere tool of algebraic transformations, all the propositions that can be formulated on a given topic derive.[26] These "principles" must be distinguished from other kinds of principles, like

---

[25] Chasles (1837: 73, Section 18). Emphasis is mine. The third edition has minor differences.
[26] On the case of analytical geometry, see Taton (1964). Chemla and Pahaut (1988) deals with the example of analytical treatments of spherical trigonometry. Further examples are evoked in Chemla (2003). Boyer

the "principle of continuity," evoked above. The latter kind, governing the nature of the conclusion that a proof allows to be stated, became central in the practice of geometry in the early nineteenth century (see Section 2.2.3). Yet, despite key differences, these principles all relate to questions of generality.

If we were to remain at this superficial level of description, the opposition drawn by Chasles between Pascal's treatment of conics and ancient geometry could not be grasped. We would not understand why such an organization of knowledge is different from the axiomatico-deductive organization presented in, for instance, Euclid's *Elements*. The other main point to be observed, Chasles adds, is the way the derivation was (allegedly) made.

In Chasles's understanding, the mystic hexagram is a "general property" of six points on a conic. He further perceives that each of the "main theorems" deriving from it expressed a property of six points. Note, *en passant*, that this relationship between the mystic hexagram and the main theorems appears to be of the same kind as that between the propositions Chasles had used to discuss the connections between Pappus's lemmas, as was outlined above. However, although in these two contexts Chasles reads propositions with a similar tool of analysis, consisting of "general relations" holding between kinds of arbitrary points, Pascal's own formulation of the mystic hexagram was different. For the sake of historical analysis, Chasles brings into play his personal mathematical expertise in guessing what the lost treatise looked like. Connections of this kind between propositions were a key tool in his hands. Quite revealing is the fact that, precisely in relation to this type of derivation, Chasles refers the reader here to his "Note XV," which was part of a treatise he was preparing on "rational geometry" and in which he used repeatedly this tool.[27] We shall explain this point in greater detail in Section 2.3.

On the other hand, although each of Pascal's "main theorems" were derived from the fundamental "general property," Chasles emphasizes, the fact that they had a "different form" was *precisely* what lent them their specific "fruitfulness" (each having "particular uses" and "encompassing a huge number of properties of conics"). This emphasis on the correlation between the "form" of a theorem and its "fruitfulness" also proved to be a key concern in Chasles's own research on geometry.

In addition, according to Mersenne's testimony, the deduction of properties of conics from the "main theorems" was performed in such a way that the propositions appeared as "400 corollaries" of the mystic hexagram. In other words, an innumerable number of "particular properties" were deduced from the "general property" with almost no proof.[28] The fruitfulness of the "main theorems" is correlated with the ease of the deduction they

---

(1956) seems to have overlooked this specific feature of research on analytic geometry in the second half of the eighteenth century.

[27] Chasles describes the treatise he first intended to compose and explains why he gave up writing it (Chasles, 1837: 253–4). In fact, in 1852, he started the publication of the treatise he had in mind (Chasles, 1852). In the introduction, he recounts the genesis of the book.

[28] Poncelet (1822: xli) uses the same word of "corollaries" to designate the properties of conics that Pascal derived from the "mystic hexagram." His assessment relies on his own perception of the part played by the "mystic hexagram" in the theory of conics. Poncelet (1822: 110) credits Brianchon with understanding the nature of Pascal's contributions to the study of conics.

allow. All these features characterize Chasles's approach to the "generality" of the mystic hexagram.

To conclude, for two essential ideas, Chasles imagined that Pascal, whose treatise was lost, had used means he himself considered as fundamental. This explains why he considered this writing to exemplify a geometrical practice different from what was known of ancient geometry. According to Chasles's intuition, Pascal's lost treatise shed light on a link that proved essential for Chasles: the link between the fruitfulness of some propositions and their ability to connect a large number of other propositions.

In the following pages (Chasles, 1837: 74–80), Chasles brings to bear historical evidence to suggest that all these ideas and practices (considering conics in the most general way, connecting propositions by means of transforming figures or by deriving them as particular cases from the most general proposition) came from Desargues's *Brouillon projet*, which by Chasles's time was lost. In it, Chasles gathered, these means and ideas were combined, and Pascal drew inspiration from this writing.[29] In Desargues's case, Chasles asserts that the single fundamental proposition on which the treatment of conics relied expressed a relation of involution between six points arbitrarily chosen on a conic. This reconstruction allows Chasles to place strong emphasis on the relationship between the various new means of achieving generality he had so far highlighted:

> However, in addition to its extreme fruitfulness, the theorem in question presents another feature that it is no less important to bring to light for a *philosophical examination of the course and spirit of methods concerning conics*. This (feature) is that this theorem, by its very nature, allowed Desargues to consider, on a cone with a circular base, sections that were *entirely arbitrary* (...) (Chasles, 1837: 79, Section 23, my emphasis).

In other words, Chasles suggests there is an essential link between the nature of the theorem Desargues placed at the basis of his treatment of conics and the general and uniform approach Desargues promoted for all conics. When we turn to Chasles's own treatment of conics, we shall understand better how he conceives of the relation between these various aspects of the mathematician's activity.

We can conclude from our discussion that Chasles consistently insists on the fact of connecting propositions. He does so negatively, in his account of Greek geometry, and now positively, in his discussion of (supposedly) seventeenth-century achievements.

Despite these essential changes in the methods and objects of geometry in the seventeenth century, these means remained less powerful than those provided by the other approach that took shape at roughly the same time, or even slightly earlier. In Chapter III,

---

[29] Note that the historical evidence Chasles adduces indicates that Desargues's contemporaries in the seventeenth century perceived the generality of the objects and propositions partly in the same way as Chasles did. Chasles (1837: 88) credits Poncelet with having been the first geometer to understand Desargues's contribution. However, interestingly enough, Poncelet's account of Desargues's and Pascal's work on conics does not insist on the point Chasles stresses here and elsewhere: the organization of knowledge about conics in the form of a single theorem from which every other property can be easily derived. This aspect of the value of generality in geometry appears to be specific to Chasles. Section 2.3 argues that this feature permeates his historical as well as his mathematical approach. Since the publication of the *Aperçu historique*, a copy of the *Brouillon projet* resurfaced thanks to the efforts of René Taton (see Taton, 1951a).

Chasles turns to a discussion of the characteristics and results of the latter, emphasizing right at the outset how Descartes' *Géométrie* (1637), from which it stemmed, "gave to geometry the character of abstraction and universality that distinguished it in an essential way from ancient geometry" (Chasles, 1837: 94).

As far as "universality" is concerned, Chasles stresses the superiority of the Cartesian "conception" with respect to earlier methods like those of Cavalieri, Fermat, and others. The latter were general, he grants, in their "metaphysical principles," in the sense analyzed above. However, only the Cartesian approach "provided the means of applying these methods in a *uniform* and *general* way."[30] Chasles thus emphasizes again, albeit in a new way, a dimension in which generality can be sought for in geometry: the "uniformity" with which one *applies* a method. We shall meet again with this concern in Section 2.3. In that respect, the Cartesian "doctrine" brought about a change of the older general methods.

As for the "abstraction" Chasles attaches to the Cartesian method, the value stands in contrast to the "concrete" approaches that were the hallmark of Greek geometry, in the sense emphasized above (footnote 20). In the way Chasles describes the turn Descartes brought about in geometry in this respect, we recognize he lays stress on exactly the same features as those he emphasized as essential about Desargues a few pages before. He writes:

> Descartes's Geometry (...) established, by means of *a single formula, general properties* of *entire* families of curves, so that one could not *discover* in this way *any property* of a curve that would not immediately make us know *similar, or analogous, properties in infinitely many* other lines. Until then, one had studied *only particular properties* of some curves, taken one by one, and always with *different means* that established *no connection between different curves*" (Chasles, 1837: 95, First section, my emphasis).

By contrast to ancient geometry, instead of working with "concrete" curves—we would rather say "particular"—, Chasles stressed, Descartes was dealing with "entire families of curves." Moreover, by the very fact that he proceeded with "formulas," Descartes could *establish general properties* for these families. Abstraction, in his sense, thus meshes with generality. Here too, Chasles further emphasizes the *connection* that Descartes's approach brought to light between properties of a vast number of geometrical objects. More importantly, however, the contrast Chasles draws in the following lines indicates the key point in his view here: *proofs* for these properties could be carried out *in the same way*. This is where the superiority of Descartes's method lay and where it allowed powerful developments in the general methods brought forward previously. Chasles's analysis thus highlights how, even though by different means, Desargues's and Descartes's approaches brought about, in some respects, *comparable* kinds of *generality*. Moreover, he indicates where they diverge in their treatment of generality.

In Chasles's eyes, the invention of the calculus, by Newton and Leibniz, gave geometers so powerful and easy a means in mathematics that it led most practitioners to neglect previous approaches, except Descartes's geometry, which had provided the true foundation

---

[30] Chasles (1837: 94, 1889: 95), first section. Emphasis is mine.

of the new calculus (Chasles, 1837: 142). Analytical treatments of geometrical problems developed, and they brought to geometry a form of generality that reached a level hitherto unheard of. As a result, they remained the mainstream of geometrical research until the end of the eighteenth century and overshadowed forays made by people like Desargues and Pascal.

Yet, some geometers kept developing purely geometrical approaches, and Chasles analyzes their contributions. He chooses this point of his exposition, in particular in relation to Tschirnhausen, to present his views on the advantages and shortcomings of the analytical approach as well as to outline his own conception of the practice of mathematics (Chasles, 1837: 114–15). He will then show how the shortcomings of analytical methods, the conception of which relates closely to Chasles's ideal for mathematics, are overcome by methods specific to the new geometry, which he presents in the following chapter. Further, Chasles expounds *how* the new geometry is also able to provide the same advantages as analytical methods, as far as generality is concerned. We shall return in the conclusion to the overall vision Chasles develops for mathematics. Let us now examine his views on the advantages of analysis in terms of generality—a feature that, in the eyes of the practitioners of projective geometry, gave it superiority over geometrical methods. We shall also outline how, in Chasles's view, these advantages were implemented in geometry, thereby allowing practitioners to practice geometry in a purely geometrical way.

2.2.3   A diagnosis regarding the power of analysis and the new geometrical methods

According to Chasles's analysis, a "new era" started in pure geometry after a century of "rest" in the eighteenth century. Its opening was marked by the development of Monge's descriptive geometry, published in book form as the *Géométrie descriptive*.[31] In his view, this "doctrine" played a role in, and exerted an influence on, geometry comparable to that of Descartes's *Géométrie* nearly two centuries earlier (Chasles, 1837: 189). As before, we shall not assess the historical claim. Nor will we attempt to determine whether Chasles retrospectively attributes to Monge ideas that emerged only later or whether he rightly grasps in Monge's book the origins of these ideas. Rather, we shall go on following Chasles' analysis of the various forms taken by generality in geometry and try from this viewpoint to interpret this claim. Clearly the parallel Chasles draws between the *Géométrie descriptive* and Descartes's geometry derives from his perception of the advantages brought by analysis to geometry and also of his interpretation of how Monge's contribution introduced similar benefits in geometry per se. In that respect, we shall concentrate on two fundamental contributions which Chasles reads in the *Géométrie descriptive* and which embody a turn in the way of working in pure geometry: first, the use of a "method of transmutation of figures" (Chasles, 1837: 194–5) and, second, the introduction of what Chasles calls the "principle of contingent relationships" (Chasles, 1837: 204). Chasles subscribes to the view, also held by Poncelet, that these two contributions represent the

---

[31] A critical edition of the book can be found in Belhoste et al. (1992).

main new sources of generality introduced in pure geometry at the end of the eighteenth century and the beginning of the nineteenth century. In the *Aperçu historique*, he will assess each of them by means of a comparison and a contrast with analytical methods.[32]

Before he begins highlighting these contributions, Chasles sketches the contents of the *Géométrie descriptive*, stressing two main ideas in it. On the one hand, he writes, Monge gave a method to "*represent*, in a plane area, all the bodies with a given form and thereby *transform*, in *plane constructions, graphical operations* that would be *impossible* to carry out *in space*." On the other hand, Monge "deduced, from this representation of the bodies, their mathematical relationships, resulting from their forms and respective positions" (Chasles, 1837: 189, Section 1, my emphasis). Chasles grasps in these ways of proceeding in geometry several essential consequences.

First, it opened the way to a new form of reasoning. Having a fixed correspondence between figures actually drawn in the plane and bodies in space allowed the geometer to draw inferences about the latter by means of a reasoning on the former. Space geometry could thereby develop. Further, more generally, the "correlation" established between plane and space geometry led to the possibility of proving propositions in one by means of a piece of reasoning in the other. Introducing perspective in addition to the projections used by descriptive geometry—to transform the plane figure associated to a spatial configuration into another plane figure—constituted a systematic means of transferring a property established in the former to the latter.

Chasles introduced the expression "*method of transmutation of figures*" to designate what, in his view, constituted the essence of such operations. By means of this general expression, he stressed where according to his own diagnosis the import of these new proofs beyond the new theorems proved lay (Chasles, 1837: 195). The comparison with the use of algebra in geometry proved to be the most telling way of formulating this import, thereby pointing to one of the advantages of algebra with respect to generality as well as to the way Monge opened the path toward implementing this benefit in the new geometry. Chasles wrote:

> In addition, if one thinks about the procedures of algebra and seeks for the *cause* of the *immense advantages* that it brings into Geometry, does not one perceive that it owes part of these advantages to the *ease* of the *transformations* that one applies to the expressions that are first introduced in it?, transformations whose secret and mechanism constitute the analyst's true knowledge and which are the constant object of his research. Was it not natural to seek to *likewise introduce* within pure Geometry *analogous transformations* bearing *directly on the figures* proposed and on their *properties?*" (Chasles, 1837: 196, Section 8, my emphasis).

---

[32] In his preface to the *Traité des propriétés projectives des figures*, Poncelet had developed a similar diagnosis regarding the key changes in pure geometry with respect to generality, in the few decades before the publication of his own book (compare Poncelet, 1822: XXI–XXXV). However, Poncelet's way of recounting the historical evolution differs from Chasles's. Moreover, the comparison Poncelet develops between geometrical and analytical approaches presents nuances worth examining. Lastly, it is interesting that, although Chasles agrees with the main points in Poncelet's analysis, he renames and reformulates each of the two principles that Poncelet identified as essential. Explaining these points, and comparing the two authors in this respect, exceeds the scope of this chapter.

Chasles thus depicts the "method of transmutation of figures" as a tool in geometry that parallels the use of "transformations of expressions" in algebra and that could bring generality to that field by way of geometrical means. It is interesting for our purpose to interpret his comparative statement with respect to the two terms of the comparison.

First, as regards algebra, the way in which Chasles draws the comparison indicates a feature that is specific to its methods and responsible, in his view, for the power it brings to geometry: the systematic and convenient use, in proofs, of *transformations* applied to expressions, that is, to formulas. In addition to the generality proper to the formulas, which we stressed above, the techniques for transforming them yield a powerful, uniform, and general means of proving, which, further, establishes links between the propositions entering in the reasoning. One recognizes here several aspects of the connections between propositions, which, we emphasized above, Chasles assessed positively. Here they mesh with each other, Chasles placing emphasis on the fact that these transformations provide a specific way of proving. This feature of the method thereby distinguishes the knowledge gained in this way from that obtained in the context of ancient geometry, whose propositions, Chasles regularly deplored, remained isolated from each other.

As for the second term of the comparison, that is, the "method of transmutation of figures" in modern geometry, Chasles invites us to consider it as a geometrical counterpart to the analytical procedures just outlined. It transforms figures into each another *directly*, that is, *not* by means of algebraic tools. In correlation with this, it transforms properties of the original figure into properties of the resulting figure. If we limit ourselves to such a coarse description of the method, we may wonder why Chasles did not consider it to have been introduced, in an embryonic form, in the seventeenth century, through the treatments of the conics developed by Desargues and Pascal. In fact, the key point here appears to be, in Chasles's view, Monge's introduction of the idea of "*transmutation* of three dimensional figures into plane figures and conversely" (Chasles, 1837: 195). This new way of reasoning reveals how particular the transformations used in the seventeenth century were: by means of a particular construction, they linked together figures sharing the "same genre" (Chasles, 1837: 128, 195, 261), and they connected their related properties. In contrast, Monge's novel approach paved the way to a much more general conception of geometrical transformations. They could link together figures having different "genres" and could transform properties of some of them into wholly different properties of the others—Chasles speaks of an "immense generalization" of the perspective used by Desargues and Pascal (Chasles, 1837: 212). These transmutations could be employed within proofs to transform a situation into a different one and to deduce a property about the former from a related but inherently different property about the latter which would be much easier to prove. In this way, geometrical transformations could play a key part in proofs, just as transformations of expressions had been essential for analytical approaches (Chasles, 1837: 268). In terms of generality, they would bring to geometry the benefits that analytical methods derived from their use of transformations.

One thus understands why Chasles sees in the ideas presented in the *Géométrie descriptive* resources that could exert on geometry an influence similar to that of Descartes's geometry. Further, Chasles relates all kinds of transformations introduced in geometry since the publication of the book to the same general concern. In conclusion, he invites geometers more generally to develop such methods systematically (Chasles, 1837: 196). It is interesting for our purpose to notice that Chasles considers these methods from the most general viewpoint, that is, as belonging all to a single class of tools. This fact can be perceived from the status Chasles gives to the two topics to which he devotes the *Mémoire* introduced by the *Aperçu historique*: the duality and the homography of figures. In his view, they are "general principles of extension," from which geometers can derive diverse methods that are in fact "methods of transformation of that kind" (Chasles, 1837: 196). Duality and homography are even "general doctrines of *deformation* and *transformation* of figures" (Chasles, 1837: 254, Chasles's emphasis).[33] Chasles claims to have determined the single theorem from which any method of transformation between figures of the "same genre" could be derived (Chasles, 1837: 224, 228). In section 3, we return to this claim. Let us simply stress here that Chasles manifests an interest for the foundation and the organization of such methods.

Symptomatically, the diagnosis that developing these tools is of prime importance derives less from remarks regarding their efficiency than from some of Chasles's reflections on analytical methods. He identifies the *cause* of the power of algebra in the *ease* with which its "expressions," that is, the formulas, are transformed—in Section 2.3.1 we return to the emphasis Chasles places on determining the "cause" or the "source" of mathematical properties. Moreover, he takes the "true knowledge" in algebra to lie in the way transformations are carried out. Far from being "given" *a priori* by the very nature of the topic, as one might think, Chasles stresses that the transformations of expressions are *as such* a key topic of research for "analysts." The parallel leads him to plead for developing methods for the "transmutation of figures" in geometry.

In fact, from Chasles's perspective, these operations represented only a first way of systematically introducing into geometry transformations comparable to those characteristic of analytical methods. In Section 2.3, we shall meet with other types of transformations that play a key part in Chasles's practice of generality in geometry.

The second decisive contribution regarding generality that, in Chasles's eyes, the *Géométrie descriptive* made to geometrical methods relates to a principle that, in his *Traité des propriétés projectives des figures* of 1822, Poncelet explicitly brought to the foreground under the name of "principle of continuity." In fact, Chasles reads the principle back

---

[33] Chasles (1837: 268) explicitly compares all these means of transformation to the "formulas and general transformations of algebra." He concretely describes how they allow the practitioner to derive from any known proposition an almost infinite number of others, contrasting such a practice to that of ancient geometry, precisely because links are thereby established between propositions. As a result, "truths" are no longer isolated facts. Within this context, Poncelet's use of projections in his *Traité des propriétés projectives des figures* (1822) appears as one among several other means inspired by a similar concern. An important difference nevertheless must be noted. In the preface to his treatise, Poncelet explains his motivation for focusing on projections: a reflection on the *nature* of propositions led him to identify, and concentrate on, "projective properties," that is, properties that remain valid under projection. Because of this feature, these properties could be established by means of a proof carried out on a particular figure, the conclusion then being stated for the general figure. Chemla (1998: 181) discusses a context in which Poncelet shaped this approach. In the *Aperçu historique*, Chasles dissociates the two issues. As we saw, he deals with "transformations" as a topic in itself and with full generality. He raises the question of the methods for generalizing properties separately and also in general (Chasles, 1837: 262–4). I plan in another publication to compare Chasles's and Poncelet's conceptions of geometrical transformations.

into the *Géométrie descriptive*, published some decades earlier, by bringing to light how it implicitly underlaid proofs that Monge had carried out (Chasles, 1837: 198). Chasles also offers a new conceptualization of the principle, thereby possibly playing down once more Poncelet's contribution (Chasles, 1837: 197–207, 357–9).[34] In Chasles's view, the principle constituted a new method of proving in geometry—a general method—which was the counterpart, for geometrical methods, of some aspects of the application of analysis to geometry. So, again, Chasles discusses how geometrical reasoning should proceed, by reference to the advantages brought to proof by the use of analytical methods in geometry. However, in this case, the parallel with analysis is more fundamental than in the previous case, where it merely supported a plea in favor of developing research in a given direction. Now, Chasles derives the very validity of this new principle from the fact that it merely translates within geometry modes of reasoning that are common in the application of algebra to geometry.

Let us sketch the essence of the principle in Chasles's analysis, before we examine the latter question. Its formulation requires a concept and several key oppositions.

The concept is that of a (geometrical) "figure," in relation to the "circumstances of construction" according to which it is considered. The examples Chasles examines illustrate what he has in mind. A figure can be the combination of a straight line and a surface of the second degree, or it can be two circles. In other words, a figure combines various elements (points, lines, or surfaces). The introduction of the "circumstances of construction" aims at distinguishing, for the *same* figure, between the various "dispositions" that these elements can have with respect to each other. For example, the line can meet with the surface or not.

Among the circumstances of construction, Chasles opposes "particular" and "general" ones. The opposition is mainly explained by discussing what he means by "particular circumstances of construction": such circumstances occur when "points, lines, or surfaces happen to coincide with each other." In the examples mentioned, this is the case where the line is tangent to the surface or the circles are tangent to each other. With respect to the principle examined, Chasles's attention focuses mainly on the "general circumstances of construction of a figure." The key remark is that a figure can present several "cases" in its "most general construction": the line can either meet with the surface or not. Moreover, if we exclude the case of tangency, the cases where the line meets with the surface, or where it does not, have "the same generality," this feature being guaranteed by the fact that the line is drawn arbitrarily with respect to the surface. In Chasles's words for it, "the general conditions of construction of the figure remained the same."

The next opposition Chasles introduces allows him to characterize the differences between two such distinct cases of the same figure. Some parts in the distinct cases

---

[34] In his discussion on the validity of the principle, Chasles alludes to the criticism that Cauchy addressed to Poncelet's "principle of continuity." Cauchy formulated the criticism in his report on Poncelet's Memoir presented to the Académie des sciences, treating it as a mere induction (Chasles, 1837: 199). Poncelet (1822: vii–xvi) reproduces Cauchy's report, which was read on 5 June 1820. This attack probably explains why Chasles makes the relation between the principle and the analytical methods clear. It might also explain his attempt to reformulate the principle. This question exceeds the scope of this chapter. Here, I focus on Chasles's version of the principle.

are "*integrant and permanent parts* of the figure," that is, "they depend on its general construction" and, also, "they are always real." Other parts are "*secondary*, that is, *contingent* and *accidental*," in that they are "indifferently real or imaginary, without changing the general conditions of construction of the figure" (Chasles, 1837: 200, Chasles's emphasis).[35]

On the basis of these distinctions, Chasles further opposes "accidental and contingent *properties*" (my emphasis) of integrant parts of a figure to those that are "intrinsic and permanent." The latter allow in all the cases to construct the integrant parts of the figure, whereas the former can play this role only in some cases (Chasles, 1837: 205–6). The example given is essential in that it illustrates how the same geometrical object—the radical axis of two circles, whose existence is a permanent property of the figure of two circles—can be approached in both ways. Defining it as the common chord of the circles approaches it by means of a contingent property, whereas one can also define it by the permanent property of being the line of equal power.

The principle Chasles grounds on such an analysis of a figure concerns the proof of general theorems that bear only on permanent parts, and not on contingent parts, of the figure. The "principle of contingent relationships" states that such theorems can be proved with respect to *one* general circumstance of construction of a figure and, as a result, be claimed valid for *any* other general circumstance. One can hence choose a general disposition of the figure for which the proof is made easier. In particular, in the general disposition chosen, some contingent elements can be palpable and hence used in the proof, even though in other general circumstances of construction of the figure, they would no longer appear. The "principle of contingent relationships" claims that this will not affect the validity of the conclusion.

Note that generality plays a key part in the principle: the starting point of the proof *must* be a figure in a *general* circumstance of construction. This is an essential condition to be able to state the conclusion with full generality. The subtle analysis of the principle in terms of allowing a transfer of a property from the general to the general, without further proof, evokes Poncelet's analysis of the "principle of continuity" in his 1822

---

[35] Chasles (1837: 200.) Earlier, in relation to the example of the line and the surface of the second degree, considered by Monge, Chasles had been more specific, in relating these qualifications to the actual figure: "...A figure can present in its most general construction two cases; in the first one, some parts (points, planes, lines, or surfaces), on which the general construction of the figure does not necessarily depend, but that are *contingent* or accidental consequences of it, are real and palpable; in the second case, these same parts no longer appear; they have become imaginary; and yet the general conditions of construction of the figure remained the same" (Chasles, 1837: 198). The emphasis is Chasles's. The concept of figure Chasles describes recalls the one introduced in Carnot (1801). Carnot had analyzed such a figure as a varying system of quantities and he considered the correlations between a primitive system and any other one. The continuous deformation that allowed Carnot to transform the former into the latter played a key role in his theory. Poncelet adopted this feature of the *dispositif* in his formulation of his "principle of continuity." By focusing on various kinds of parts and properties, instead, Chasles wants to do away with considerations of continuity, that is, of infinity. In his interpretation of the consequences of his analysis of a figure for granting a geometrical meaning to the term "imaginary," Chasles refers explicitly to Carnot's contribution, denying, however, that Carnot considered imaginary correlations between figures. Compare the note XXVI "Sur les imaginaires en géométrie" (Chasles, 1837: 368–70). This is in fact incorrect (see the section "Des corrélations complexes et des imaginaires," in Carnot, 1801: 177–88).

treatise.[36] However, Chasles's interpretation of the *nature* of the opposition between the various states of the figure differs.[37] The *justification* of the validity of this mode of proving is also partly different. Its justification is of interest to us here, since it shows how the introduction of the principle into pure geometry derives from a diagnosis of the power of analytical methods with respect to generality.

Such a principle, Chasles admits, cannot be justified *a priori* with the means available to the geometers of his time in pure geometry. He concedes that the proof of the validity of this mode of proceeding in geometry could *only* be given, "for each case taken separately" and "*a posteriori*," by "a reasoning based on the general procedures of analysis" (Chasles, 1837: 200). The key point here is the fact of noticing that the different dispositions of a figure, depending on different general circumstances of construction, are not distinguished by analytical methods. This was one of the properties of Descartes's method Chasles praised—a property he characterized as deriving from a specific use of formulas. This is how by a single piece of reasoning, such methods reach conclusions that are applied to all general dispositions. In other words, this specificity of algebra is precisely what grants to its reasoning its generality, allowing one to avoid different proofs for different cases as ancient geometers did. The "principle of contingent relationships" is nothing else but an importation of this type of generality in geometry. The trust geometers placed in the way of proving was in fact, Chasles underlines, grounded on "their habits of using the analytical methods."

However, Chasles does not satisfy himself with this justification. Instead, for any truth, for the proof of which contingent features of a figure in a general state of construction were used, he invites practitioners of geometry to go beyond "Monge's somewhat superficial method" (Chasles, 1837: 205) and search for a proof that would only use the permanent properties of integrant parts of the figure. The latter proofs would thus hold for any general construction of the figure (Chasles, 1837: 204–6). In his 1852 treatise, Chasles systematically provides such proofs (Chasles, 1852: XV–XVI).

This remark concludes our survey of the diagnosis Chasles reformulates, after Poncelet, regarding the power of analytical methods in geometry and the reasons why the new ideas introduced a similar generality into pure geometry. We have so far examined Chasles's treatment of the *history* of geometry from the perspective of the value of generality. We shall now examine some ways in which Chasles's own practice of geometry reveals other elements of his view on the question of generality. Our analysis will highlight how his negative comments on the geometry of the Ancients and his praise for what he gathers

---

[36] As Poncelet emphasized, when proving projective properties, one can conclude the *general* statement from a proof based on a *particular* case. However, the "principle of continuity," like its reformulation as the "principle of contingent relationships," is a way to prove the general statement by relying on a figure in "general circumstances of construction." On this opposition, compare (Chemla, 1998: 179–81). See also Chasles (1837: 203).

[37] Poncelet had introduced "ideal elements" to refer to elements that were still extant on a figure, although the relationship between them and other parts of the figure had changed. Deleting any reference to ideal elements, Chasles distinguishes between permanent/contingents parts and properties. When Kummer introduces ideal elements in number theory, his use of the terms "contingent" and "permanent" discloses the influence of Chasles's interpretation of the "principle of contingent relationships," whereas the use of the term "ideal" clearly retains features of Poncelet's conceptual framework. See the chapter by J. Boniface (Chapter 18) in this volume.

from Desargues's and Pascal's approaches are both inspired by his own stand on the issue of generality.

2.3   Chasles's methods related to generality in geometry

2.3.1   Looking for the source of methods and theories

The full title of Chasles's *Aperçu historique* manifests a difference between the two parts composing it. The first half of the title, that is, *Aperçu historique sur l'origine et le développement des méthodes en géométrie, particulièrement de celles qui se rapportent à la géométrie moderne*, relates to the historical section of the book and refers to *methods* in geometry. By contrast, the second half, that is, *suivi d'un mémoire de géométrie sur deux principes généraux de la science : la dualité et l'homographie*, corresponds to the *Mémoire* and stresses that it bears on "two *general principles*," that of duality and that of homography. We saw that Chasles dates the introduction of "methods" in geometry from the seventeenth century and examines how such methods were further developed in modern geometry. In contrast, by his account, the principles appear to characterize mainly modern geometry.

Chasles is not the first geometer to bring forward principles in geometry. As was already indicated, in his *Traité des propriétés projectives des figures*, Poncelet had, for instance, introduced the "principle of continuity." Chasles, however, does more than just introduce principles. He provides personal reflections on their nature, their properties, and their justification. These principles are intimately related to his approach to the issue of generality in geometry. We shall hence sketch here some aspects of his conception of principles that relate more specifically to that issue.[38]

As we saw above, when discussing various methods of transformation of some figures into others, Chasles does not consider these methods independently from each other, but seeks to understand their relationships and to organize them as a set. His analysis of methods of transformation between figures of the "same genre" leads him to state that they "all derive from a single *fundamental principle*, with respect to which they are only particular applications" (Chasles, 1837: 219, 223–4, my emphasis). Chasles refers here precisely to the "principle of *homographic deformation*, or simply *principle of homography*" (Chasles, 1837: 261, Chasles's emphasis), which is one of the two main topics of his *Mémoire*. Here the principle brings together a variety of methods, introduced by various authors, and shows how they become particular cases of a general mode of operating. The same remark holds true for the principle of duality: Chasles reads different methods of transformation of one figure into another one, this time of a different "genre," as particular applications of a "principle that constitutes a complete doctrine of *transformation* of figures" (Chasles, 1837: 228, Chasles's emphasis). In addition to the relations thereby established between the

---

[38] We withhold for another publication a more detailed analysis of the various types of principles that were introduced in mathematics at the time.

figures, Chasles insists on the transformations of properties of the former into properties of the latter.[39] Such principles differ from a principle like the "principle of continuity" or, in Chasles's terms for it, the "principle of contingent relationships" in an essential way. The point here is not to allow practitioners to extend a property proved for one figure to other figures. Rather, these principles are general in the sense that they easily yield many other methods for carrying out this operation that were once thought unrelated to each other. We shall consider below examples of the generation of a set of methods.

Chasles's interest in bringing out such principles is an essential feature of his approach to generality—perhaps even a specific feature of his, by contrast to other geometers of his time. This interest extends to the organization of properties and appears to relate more generally to his conception of the ideal approach to geometry. Chasles devotes some pages to the description of his methodology, which he concludes with the following declaration:

> … We believe we can say that, in each theory, there must always exist—and we must iden-
> tify it—some principal truth from which all the others can be easily deduced, as simple
> transformations or natural corollaries; and that if this condition is achieved, it alone will
> be the hallmark of true perfection in science (Chasles, 1837: 115, my emphasis).

The faith Chasles expresses here strikes an echo with the shortcomings he recurrently exposed about Greek writings on geometry. In the passages quoted above, we saw Chasles emphasizing exactly the same virtues, yet deploring their absence in Pappus' or Apollonius' works. Instead of keeping propositions as independent statements, he suggests, the practitioner should look for a general truth from which all other truths derive, in a way that connects them with each other. Chasles's approach to history thus appears to be tightly correlated with his own values.

Here, as above, Chasles underlines the benefits in terms of proof. However, in this case, the expectations differ strikingly from the benefits drawn from the "principle of contingent relationships": in his view, bringing to light such a key statement reduces proof to almost nothing. The type of generality that emerges from Chasles's methodological creed differs from those he had attached to Monge's *Géométrie descriptive*. Interestingly enough, it brings us back to the structural properties Chasles attributed to the writings of Pascal and Desargues, on the basis of the information available to him. We remember he emphasized how these geometers had assigned, to a fundamental position in their treatment of conics, a general property, from which all the other propositions would be derived as "corollaries." In fact, we shall now see that the qualities Chasles prized in their approach to geometry mainly reflect the values he himself prized in his practice of geometry.

Chasles comes back to the same thesis in the concluding pages of the *Aperçu historique*. There, a quotation he gives from a letter Quetelet wrote to him reveals that he shares with this correspondent a common ideal in this respect.

> It is unfortunate, M. Quetelet wrote to me, that most mathematicians today have such
> an unfavorable judgment about pure Geometry … It has always seemed to me that what

---

[39] For the list of methods falling under that principle, compare Chasles (1837: 224–35, 237).

deters them most is the lack of *generality of the methods* they think they perceive there. However, is it really the fault of Geometry or of those who cultivated this domain? I am very inclined to believe that there exist *some higher truths* that must be so to say the *source of all the others*, more or less *like the principle of virtual velocities* is for mechanics (Chasles, 1837: 267, footnote 1, my emphasis).[40]

The words Quetelet uses, and which evidence an ideal comparable to Chasles's, make the relation between looking for the "source of all" propositions and the issue of generality explicit. Moreover, they reveal a striking comparison Quetelet makes: he draws a parallel between such a practice in geometry and a feature typical of analytical practices in the eighteenth century. As was mentioned above, practitioners of mathematics like Euler or Lagrange strove to offer analytical treatments of whole domains of mathematics, in the following sense: they attempted to identify the least amount of propositions possible, also called "principles," from which all the other propositions of the domains in question could be derived by the simple virtue of analytical transformations (Chemla and Pahaut, 1988: 151–2.) In this respect, Quetelet evokes the principle of virtual velocities, from which Lagrange derived the whole mechanics in his *Méchanique analytique* (Paris, 1788). The general propositions which should be identified as the "source" of all other truths in pure geometry appear to share key properties with such analytical principles: the other truths derive from both types of statements, by "transformations." However, a difference between the two practices is that, in the case of geometrical procedure, deduction is "easy," as Chasles emphasized, the other truths appearing as "simple transformations or natural corollaries." In the main text, Chasles develops Quetelet's comparison, continuing the list of domains that in the previous century were shown to derive from basic principles of that kind. In Chasles's view, the "general laws" (Chasles, 1837: 267), in which he is interested and from which he strives to derive geometry in the sense sketched above, thus constitute, within geometry, an ideal inherited from analytical treatments developed in the eighteenth century.

Chasles's sharing of this ideal can be correlated with the fact that he brought to the fore, and focused in a specific way on, the notion of cross-ratio and its conservation by all the transformations deriving from his two general principles.[41] He writes:

We shall present them (that is, the principle of homography and the principle of duality, note KC) ... with a greater *generality* than any of these methods (that is: specific methods for transforming a figure into another one either of the same kind or of a different kind, like in the transformation by reciprocal polars, note KC). The scope we shall give them will find its principal usefulness in a most simple *principle* of relations between magnitudes, which will make them *applicable* to *numerous* new questions.

   This principle rests on a *single relation*, to which it will always suffice to *reduce* all the others. This relation is the one we have called anharmonic ratio between four points or a pencil of four lines. This is the *single type* of *all* relations that can be *transformed using the*

---

[40] In fact, a few pages above, Chasles had endorsed exactly the same views and given them as what motivated his own research in geometry (Chasles, 1837: 253–4).

[41] Chasles is aware that some of his predecessors, like Poncelet, Brianchon, and even before, knew this property. His claim of originality with respect to cross-ratio lies in the position he grants to the concept and its conservation in geometry (Chasles, 1837: 34–5).

two principles that we shall prove. And the *law of correspondence* between a figure and its transformed one consists in the equality of corresponding anharmonic ratios.

The *simplicity* of this law, and that of the anharmonic ratio, makes this *form* of *relation* most adequate to play such an important part in the science of extension (Chasles, 1837: 255, my emphasis).

From Chasles's perspective, the importance of the cross-ratio—in his terms "anharmonic ratio (rapport anharmonique)"—rests on the fact that it is the single relation underlying all relations that continue to hold after a figure is transformed according to the two principles. More precisely, any equality that can be transferred in this way can be "reduced" to cross-ratios. Chasles adds the equality of the corresponding cross-ratios grounds the relationship between two such figures. Bringing this common origin to light in all such relations, that is, linking them to a general property from which they all derive, plays, in Chasles's view, a key role in the practice of geometry which he champions. This is carried out by the operation of "reduction," central in Chasles's definition in the following lines of the art of the geometer, in comparison with, as well as in contrast to, that of the analyst:

> When the relations to be considered appear at first sight not to fall under this formula, the *art of the geometer* will consist in *reducing* them to it, using different preparatory operations, in some respects *analogous to the changes of variables* and the *transformations* of *analysis* (Chasles, 1837: 255).

We have emphasized above how Chasles compared the transformations of figures into one another with the transformations of expressions characterizing the work of the analyst. The two principles of homography and duality encompass all the geometrical transformations of a comparable kind. Here, however, Chasles compares other operations carried out by the geometer to those proper to the analyst. These operations relate to generality in that they aim at bringing to light the "source" of a relation and thereby linking it to other similar relations. They hence play a part in the organization of properties.

In order to understand more concretely what Chasles means by these operations, *how* he puts them into play and, more generally what his conception of the "source" actually amounts to, we shall now follow part of his own treatment of the cross-ratio, in which many features of his practice of generality are exemplified.

2.3.2 The form of a definition and its generality

Chasles devotes his Note IX (Chasles, 1837: 302–8) to the definition and the main properties of the cross-ratio of four aligned points, or four converging lines. For four aligned points, $a, b, c,$ and d, he defines the function in question—which we shall denote as (a, b, c, d)—to be given by the expression:

$$\frac{ac}{ad} : \frac{bc}{bd},$$

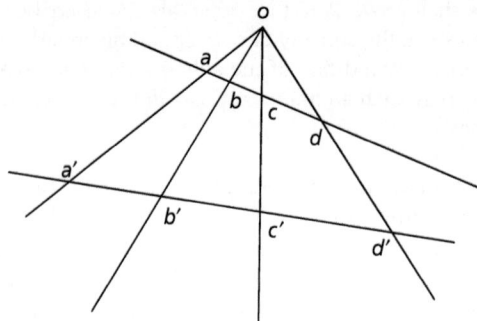

Figure 2.1 *The invariance of the cross-ratio by projection*

or any of the three similar expressions that one can form with these four points. As he stresses, the key property of the cross-ratio was long known, since it can be identified in Pappus' *Collection*. Yet, the formulation for which Chasles opts and which differs from Pappus' makes explicit a key reason for the central importance of the notion in the kind of geometry he seeks to develop. Chasles states:

> When four lines stem from the same point, any transversal meets them in four points whose anharmonic ratio has constantly the same value, whatever the transversal may be (Chasles, 1837: 302).

If one considers two transversals *ad* and *a′d′* cutting the four lines stemming from the point O (see Fig. 2.1),[42] the proposition yields the following relation:[43]

$$\frac{ac}{ad} : \frac{bc}{bd} = \frac{a'c'}{a'd'} : \frac{b'c'}{b'd'}$$

Chasles alludes to a proof of this relation, which in fact is to be found for the first time in Carnot's *Essai sur la théorie des transversales* (1806):[44]

> If, from a point taken arbitrarily, one draws four lines that end on four aligned points, the anharmonic function of these four points will have precisely as its value what this function becomes, when one substitutes, for the four segments occurring in it, the sines of the angles that the four lines comprising these segments make with each other (Chasles, 1837: 302).

---

[42] For modern geometry, Chasles calls for a style without figures (Chasles, 1837: 190, 207–8), the development of which is required by the change of methods. We cannot comment on this most interesting feature here, and will leave the topic for a future publication. However, to make things easier for our reader, we shall not conform to Chasles's style. Chasles himself included figures in his later treatises.

[43] Chasles shows that any such equality entails the equality between the other expressions of the cross-ratio (Chasles, 1837: 303).

[44] See Carnot (1806: 77), to which Chasles (1852: xxi–xxii) explicitly refers.

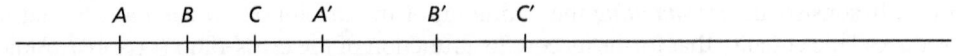

Figure 2.2 *Involution of six points*

This property leads Chasles to the introduction of the second notion he needs in Note IX:

> This function between the sines of the angles between the four lines stemming from the same point will be called *anharmonic function* of the four lines (Chasles, 1837: 302, his emphasis).

Chasles emphasizes right away that this approach to the cross-ratio shows why it is invariant under projection.

With this notion, we can illustrate how Chasles carries out the reduction of a relation to an equality between cross-ratios and the consequences resulting from this operation. Our example will be the definition of the "involution of six points," to which Chasles devotes his entire Note X (Chasles, 1837: 308–27). In a first part of the note, which presents the known properties of the involution, he provides the usual definition for it. It reads (see Fig. 2.2):

> When six aligned points corresponding to each other two by two, such as A and A', B and B', C and C', make segments with each other, in such a way that we have the relation

(A)
$$\frac{CA.CA'}{CB.CB'} = \frac{C'A.C'A'}{C'B.C'B'},$$

> one says that the six points are in *involution*, and the points corresponding to each other are said to be *conjugate* (Chasles, 1837: 309, Chasles's emphasis).

Chasles adds that any permutation on the names A, B, C gives rise to equalities that also hold. He further states that one can *deduce* from these equations other "expressions" of the involution, such as the following relations:

(B)
$$AB'.BC'.CA' = AC'.CB'.BA'$$
$$or \quad AB'.BC.C'A' = AC.C'B'.BA',$$
$$or \quad AB.B'C'.CA' = AC'.CB'.B'A',$$
$$or \quad AB.B'C.C'A' = AC.C'B.B'A'. \text{ (Chasles, 1837: 309)}$$

Chasles insists: "One could set out to prove, by means of computations, that any of the equations of (B) follow from the equations of (A), and conversely." Moreover, one could also prove in the same way the equivalence between any of the forms within (A) or within (B). Chasles discards this approach as well as the geometric interpretation which he mentions Poncelet and Brianchon used to establish these equivalences. Chasles opts for another approach to the question, which yields an "even simpler and more direct" proof. It consists in *reformulating* the definition of the involution, in such a way that it "reduces" the equality that forms its core to a function of the cross-ratio (or anharmonic ratio, in Chasles's terms). Indeed, the second part of the note introduces the following new definition:

> Six points, pairwise conjugate[45] with each other, are in involution, when four of them have their anharmonic ratio equal to that of their conjugates.
>
> Thus the six points A, B, C, A', B', C', three of which, A', B', C', are respectively conjugate with the first three, are in involution if the anharmonic ratio of four points A, B, C, and C' is equal to the anharmonic ratio of their conjugates, A', B', C', and C (Chasles, 1837: 318).

Writing down the three equalities that follow from this definition, Chasles obtains:

$$\frac{CA.CA'}{CB.CB'} = \frac{C'A.C'A'}{C'B.C'B'}$$
$$CA.A'B'.BC' = C'A'.AB.B'C$$
$$CB.B'A'.AC' = C'B'.BA.A'C$$

In other words, with this change of definition for the involution, which shows how it depends on the cross-ratio, we now understand that the equations (A) and (B), which formerly appeared to be *different* properties, in fact all state the *same* relation: an equality between cross-ratios. This type of phenomenon is at the core of Chasles's reflection on generality in geometry. It illustrates benefits that can be derived from developing the "art of the geometer" along the lines quoted above. The comments with which Chasles introduces the new definition explain what is at stake for him with this phenomenon:

> ... The involution of six points enjoys several other properties, and can be *expressed* in *various forms*, different from the equations (A) and (B), which will possibly be useful in various geometrical researches.
>
> The *most important property* of this relation of involution, the one that appears to us to be the *source* of all the others, rests on the notion of anharmonic ratio. This crucial property allows us even to give a new definition of the involution of six points, *a definition that encompasses*, at the same time, the *two kinds of equations* (A) and (B), and which *leads naturally* to different other expressions of the involution of six points (Chasles, 1837: 317, my emphasis).

Chasles's statement, which bears here on the ideal form for a definition, echoes his general methodological declaration mentioned in the previous section. We illustrate below in greater detail how it is more generally typical of his research. In this case, the cross-ratio highlights the property of the involution from which other properties either appear to be the same or to follow naturally. We saw that for Chasles, the ease brought by the approach is proof that one has identified the "source." On this basis in the following

---

[45] Here "conjugate points" simply refer to the constitution of pairs of points. The question is whether these pairs of conjugate points are in relation of involution or not (Note KC).

pages, Chasles sets out to derive new "expressions," or "forms," of the involution, using the same techniques as those he uses in other contexts. We illustrate this below. The set of new expressions sheds further light on how Chasles correlates finding the source of properties and linking properties to each other. Despite the close relationship between these expressions deriving from the fact that they all stem immediately from the source, Chasles insists that each of them brings its own benefits to geometrical research.

We note that the source was disclosed by a mere change in the "form" of the expression. The relationship established between the form of a statement and its fruitfulness appears in this case with respect to a definition. We shall now analyze a case where the same phenomenon presents itself with respect to a theorem.

2.3.3   The form of a theorem and its generality

In the Note XV, entitled "On the anharmonic property of points on a conic line—proof of the *most general* properties of these lines" (Chasles, 1837: 334, my emphasis), Chasles revisits some properties of conics from the perspective of the notion of cross-ratio. The note, which presents the key new ideas he developed about conics, will provide us with source material to analyze features of his specific approach to generality. Chasles's starting point is Desargues's theorem, which bears on the figure of a quadrilateral inscribed in a conic, across which a transversal arbitrarily placed runs (see Fig. 2.3).[46] In Chapter II of the main text, Chasles had quoted the theorem as follows:

> This relation consists in that: "The product of the segments on the transversal contained between a point on the conic line and two opposite sides of the quadrilateral is to the product of segments contained between the same point of the conic line and the two other opposite sides of the quadrilateral, in a ratio equal to that of the products made in a similar way with the second point of the conic line situated on the transversal" (Chasles, 1837: 77).

If we call A, A' (resp. B, B') the intersections of the transversal with two opposite sides of the quadrilateral, and C, C' its intersections with the conic, Desargues's theorem can be represented by the following formula:

$$\frac{CA.CA'}{CB.CB'} = \frac{C'A.C'A'}{C'B.C'B'}.$$

We have seen that in his Note X, Chasles developed a new approach to this relation called by Desargues the "involution of six points."[47] From his new viewpoint, the relation

---

[46] On Desargues' own statement of the theorem, see Field et al. (1987: 54–5, 106–7). See also Taton (1951a: 143–7).

[47] When we outlined features of Chasles's reading of Greek geometrical texts of antiquity above, we indicated how he used the involution to show that a great deal of theorems, stated by Pappus as different propositions, were in fact merely "particular cases of this general relation." For this, Chasles needed to reformulate the propositions to bring to light the hidden relation between the facts stated. Chasles uses another operation to highlight the link between propositions and the involution: reading the figure on which they bear in a different

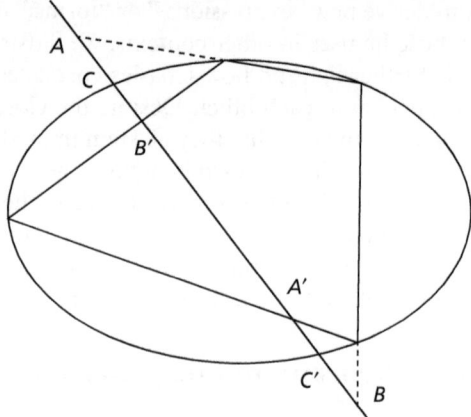

Figure 2.3 *Desargues' theorem*

amounts to stating that the cross-ratio of the four points A, B, C, C' is equal to that of the four conjugate points: A', B', C', C. At the beginning of Note XV (Chasles, 1837: 334–5), Chasles further reformulates the statement, using the fact that the cross-ratio of four aligned points is equal to the cross-ratio of four lines joining a point outside the line to these four points. From two opposite vertices of the quadrilateral, P and R, Chasles suggests drawing lines toward the points where the conic intersects the transversal (see Fig. 2.4). Desargues's theorem thus becomes an equality between the cross-ratios of two pencils of four lines:

$$(PA, PB, PC, PC') = (RA', RB', RC', RC).$$

Relying on the invariance of the cross-ratio by projection, Chasles proves the theorem easily, by considering that the conic is a circle (Chasles, 1837: 335). Moreover, its converse follows immediately:

way. For instance, through a reading of the figure for Proposition 130 of Book VII in Pappus's *Mathematical collection* in terms of a quadrilateral and its two diagonals cut by an arbitrary transversal, Chasles identifies it as an expression of the involution between six points (Chasles, 1837: 78). Similarly, a new reading of the figures for Propositions 127 and 128, where a line is read as a transversal having a specific position with respect to the quadrilateral, is needed to disclose that the propositions are particular cases of Proposition 130 (Chasles, 1837: 35). Yet, in his proofs, Pappus makes use of specific features of the figures, thereby concealing the reason why the propositions hold true and thus the connection Chasles identifies. For Chasles, Desargues derived the general statement about the involution through his attempt to "apply to conic sections various properties known for two lines" (Chasles, 1837: 77.) In Chasles's interpretation of Pappus's Proposition 130, the two diagonals of the quadrilateral can be read as a conic going through the four vertices. Seen from this viewpoint, Desargues's theorem is a generalization made possible by Desargues's general conception of conics alluded to above. This is one instance of a connection brought about by general conceptions.

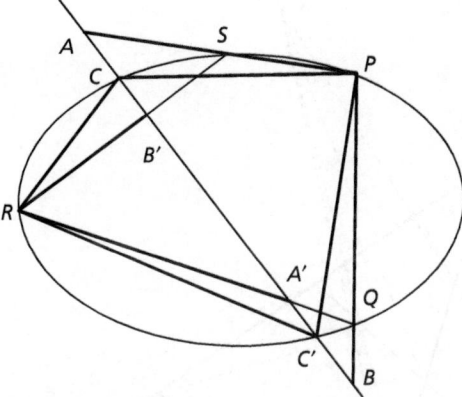

Figure 2.4 *Reformulation of Desargues' theorem*

When one has two pencils of four lines, corresponding one to one to each other, if the anharmonic ratio of the first four lines is equal to the anharmonic ratio of the other four, the lines of a pencil will meet, respectively, the corresponding lines in four points, which will be on a conic line going through the two points that are the center of the two pencils (Chasles, 1837: 335).

We may be tempted to consider that we simply have obtained an equivalent formulation of Desargues's theorem. Chasles's diagnosis is clearly different, and it reveals key features of the approach to generality in geometry to which he adheres:

> This theorem, as can be seen through the proof we just gave for it, is essentially only an *expression different* from that of Desargues; however, its corollaries, extremely numerous, *encompass part* of the properties of conic lines, to which Desargues' and Pascal's theorems did *not seem to possibly extend*. And indeed, *in addition* to the *benefits specific* to its *different form*, it has *something more general* than any of these two theorems; and these two theorems are deduced from it, no longer as *transformation*, but as *simple corollaries*. This is what we shall show in a moment, through indicating the nature of the applications to which this theorem lends itself (Chasles, 1837: 335, my emphasis).

In other words, equivalent formulations of the same fact—in Chasles's words, different "expressions" of the same theorem—do not have the same generality. This statement contradicts common beliefs about generality and is thus worth pondering.[48] In the case of Desargues's theorem, the key feature of the reformulation is that it highlights *how* the property relates to the cross ratio.[49] In Chasles's understanding, the important feature of the new "form" is that it allows practitioners to derive from it, as "corollaries," a wider range of propositions than its earlier form could apparently reach. As a result, a

---

[48] Gowers (2008), and the ensuing discussion on the same web page, deals with a similar issue.
[49] Chasles (1837: 81) evidences this is what is at stake for Chasles.

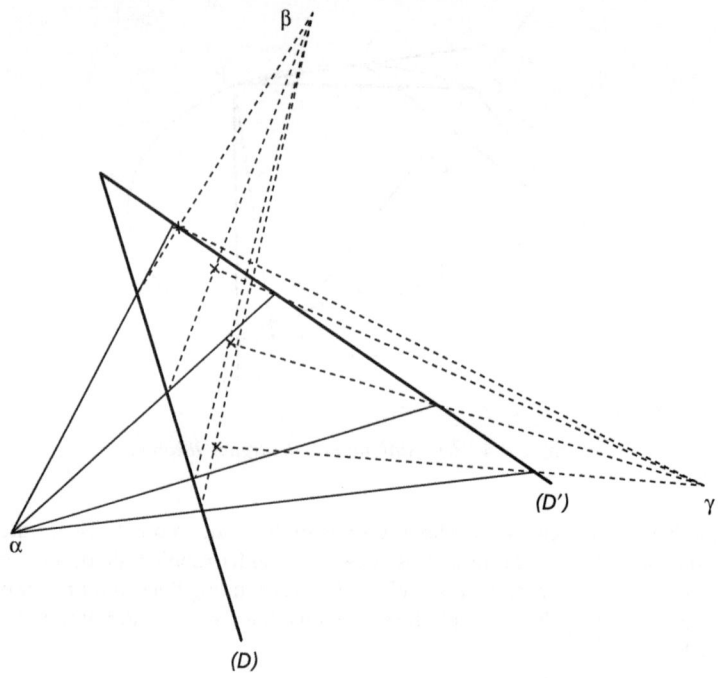

Figure 2.5 *Corollaries to the new formulation of Desargues's theorem*

greater number of theorems will thereby appear to derive from the same "source" and will thereby be connected. The ease brought into proof plays a key part in the *dispositif.* Observing how Chasles actually derives various propositions in this way will thus allow us to interpret this declaration.[50]

Chasles first derives from the theorem, by a mere change of perspective, a general proposition that gives rise to many different methods for generating conics. The derivation relies on a new way of reading and using Fig. 2.4. If we fix three lines in the first pencil, say, PS, PC, and PC', as well as the corresponding three lines in the second pencil (RS, RC, RC'),[51] any position of the line PQ determines a value for the cross-ratio of the first pencil, and hence a position for the fourth line in the second pencil. As a result, the latter two lines will intersect on the conic determined by the points P, S, C, R, C' (Chasles 1837: 336).

On this basis, Chasles suggests various devices that yield pencils of four lines with equal cross-ratios and thereby generate conics.

The first device uses a fixed angle ((D) and (D'), in bold lines on Fig. 2.5) and three fixed points (α, β, γ). Four lines drawn from α cut respectively (D) and (D') in two sets

---

[50] We simply give a glimpse of Chasles's style in Note XV, inviting the reader to consult the Note itself for a better appreciation of the elegance and fruitfulness of the approach.

[51] For the valid permutations of the terms of the cross-ratios, compare Chasles (1852: 30–3).

of four points having the same cross-ratio. Joining β (resp. γ) to the four points in line (*D*) (resp. (*D'*)) hence yields two sets of corresponding lines having the same cross-ratio. As a result, the four intersections of the corresponding lines, together with the points β, γ, belong to a conic. Another reading of this device yields yet another way of generating a conic, which, Chasles states, is "precisely Pascal's mystic hexagram, presented in another form" (Chasles, 1837: 336). In this way, Pascal's theorem is connected to the fundamental theorem placed at the basis of the treatment of conics.

The second device Chasles describes is represented in Fig. 2.6.

By construction, the two pencils drawn from α and β have the same cross-ratio. If they are placed differently with respect to one another, their corresponding lines intersect in points that belong, together with α and β, to a single conic. Chasles gives an example of this:

> Suppose that the two primitive pencils have kept their respective centers, through the motion; that is to say, that they turned around their centers; then the theorem that we have just stated *expresses precisely Newton's theorem* on the organic description of conic lines (Chasles, 1837: 336, my emphasis).

Again, thus, one derives *as a corollary* a statement that might have previously been thought as different from the others. In fact, Newton's theorem now appears connected straightforwardly to Pascal's and Desargues' theorems, through the immediate link each has with the theorem placed at the source of Chasles's treatment of conics. Chasles goes one step further:

> If two angles of arbitrary, but constant, magnitude, turn around their vertices, in such a way that the intersection point of two of their sides draws a conic line which goes through their vertices, the two other sides will intersect on a second conic line, which will also go through the two vertices (Chasles, 1837: 337).

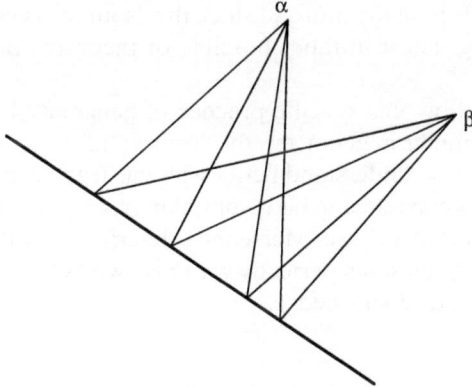

Figure 2.6 *Other corollaries to the new formulation of Desargues's theorem*

In this form, Chasles perceives the expression allows the geometer to connect different methods together and to yield infinitely more such methods:

> This theorem, which is a *generalization* of Newton's, is in fact only a *particular* way, among an *infinite number of similar other ways*, to draw conics through the intersection of two straight lines moving around two fixed points, or the intersection of two sides of two angles moving around their vertices (...). Therefore, Newton's theorem, which has had some fame and which seemed essential in the theory of conic lines, appears to be nothing but a very particular case of a general mode of drawing these curves.

This sequence of derivations further leads Chasles to formulate two general methodological comments, which we examine in turn. First:

> ... This circumstance appears to us fully appropriate to show two things: first, it is always *useful to return to the origin* of geometrical truths, to discover from this higher point the *different forms* that these truths are likely to take and that can *extend their applications*; for Newton's theorem, which some very distinguished geometers were not adverse to prove as one of the most beautiful of the theory of the conic lines, has nevertheless *not* yielded *important consequences*, since its *form lent itself only to few* corollaries. By contrast, the general theorem from which we deduce it lends itself to a great number of diverse deductions (Chasles, 1837: 337, my emphasis).

The previous examples help us substantiate and understand better the recurring declarations Chasles makes on the different forms a theorem can take. Even though all the properties stated above appear as simple transformations of one another, their different forms do not have the same fecundity. Newton's theorem would not allow geometers to derive as many properties as its reformulation—which derives from the fundamental property chosen—does, and certainly not as corollaries. Chasles insists: looking for the "source" is precisely looking for the viewpoint from which to uncover the different "forms" a given theorem can take. Since each form easily yields different applications, this research extends the range of truths covered by a theorem. Moreover, the expansion is carried out in a way that puts all derived propositions in relation to each other. It is all the more so since the "source" is common to all truths: it thereby connects the greatest number possible of theorems previously unrelated to each other.[52]

We now see clearly how this specific practice of generality compares to eighteenth century analytical treatments in geometry, the "source" of the former echoing the "principles" of the latter. We also understand better why this feature of Chasles's approach to generality in geometry is correlated to his assumptions regarding Desargues's and Pascal's geometrical practice, outlined above. Mersenne's description of the structure of Pascal's treatise, which Chasles quotes, fits perfectly with his own belief with respect to the ideal organization of mathematical knowledge.

---

[52] Chasles attributes all these properties explicitly to the fundamental theorem, compare Chasles (1837: 81, Section 26).

The second methodological comment Chasles offers bears on propositions that could be candidates as a "source." Note XV leads him to stress that they have other epistemological qualities:

> We, then, see here a proof of the following truth, that the *most general propositions* and the *most fruitful* ones are *at the same time* the *simplest* and the *easiest* to prove; since none of the proofs given for Newton's theorem can match the brevity of the one we gave for the general theorem in question (the theorem about the cross-ratios of the two pencils of four lines related to the quadrilateral inscribed in a conic line, note by K. Chemla); this proof has even the advantage that it requires no preliminary knowledge of any property of conic lines (Chasles, 1837: 337, my emphasis).

Chasles provides still other criteria to identify these highly important properties: they usually hold for the plane as well as for the sphere (Chasles, 1837: 240) or the three-dimensional space (Chasles, 1837: 45).

Another facet of Chasles's approach to generality appears essential: the search for the different "forms" of any given theorem. In this respect, Chasles manifestly devised *techniques* that could widen the applications of a theorem or connect it to a more general theorem. Observing his practice seems useful in interpreting how Chasles concretely understands the "art of the geometer." As I have argued, this art lies in shaping modes of transformation different from those that, according to Chasles, Monge developed in geometry. We have seen how Chasles attributed to the *Géométrie descriptive* the introduction of new ideas about the type of transformation of figures that would transfer properties from one figure to another. Chasles, by contrast, focused on the change of form of expression that could lead to connect theorems formerly thought as being unrelated.

These other transformations relate to two distinct tasks we have already evoked. First, as is the case for Pascal's "mystic hexagram," Chasles changes the form of its expression to bring to light its link to the fundamental theorem. Second, as was illustrated in the case of Newton's way of generating conics, Chasles derived from the fundamental theorem another form for Newton's method, which allowed him to generalize the modes of generation significantly. Perhaps Chasles also did perceive these transformations as similar to those that, in the analytical art, derive, from the principles lying at the basis of its organization and by the mere strength of the transformation of formulas, all the statements in a given domain.

Chasles further introduces in geometry techniques to pursue the same goals. Let us focus on one of them, which clearly has its origin in an observation of the analytical art: the introduction of geometrical indeterminates. The examples of Chasles's use of this technique that we can find in Note XV illustrate how it helps widen further the scope of the fundamental theorem. Continuing the sequence of transformations applied to the latter, Chasles writes (see Fig. 2.7):

> One can give another statement for the general theorem that is the topic of this note: (...)
> The first four lines will meet an arbitrary transversal in four points, and the other four lines will meet a second transversal in four points corresponding one by one to the first four

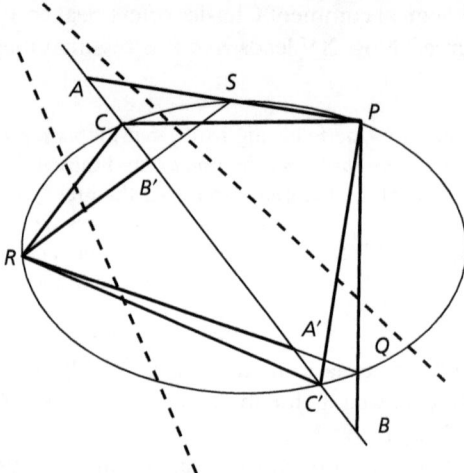

Figure 2.7 *Generalizing Desargues's theorem by the introduction of indeterminate transversals*

points; and the anharmonic ratio of the first four points will be equal to the anharmonic ratio between the other four (Chasles, 1837: 338).

The points where the lines *PS, PC, PC'*, and *PQ* meet the original transversal have the same cross-ratio as the intersections of these lines with any other transversal. This remark also holds true for the intersections of the lines *RC, RS, RQ*, and *RC'*. The original transversal can hence be duplicated, as shown with the two bold dashed lines in Fig. 2.7. Reading the same theorem with respect to the intersections on these two transversals yields the statement just quoted.

Chasles comments on it:

> This statement has the *greatest generality* possible, because of the *indeterminacy* of position of the two transversals (Chasles, 1837: 338, my emphasis).

In what follows, Chasles will suggest placing the two transversals in specific positions, which strongly evokes the formula of the analyst, in which one gives a variable a specific value. As a result, the theorem above will yield various known properties, the connection of which once more is thereby brought to light.

Chasles suggests the two following instantiations:

> (...) Suppose the first transversal is one of the four lines stemming from the second vertex of the hexagon (that is, *CSPQC'R*, addition by K. Chemla), and that the second transversal is one of the lines stemming from the first vertex; then the theorem obtained is precisely the first of the theorems Pascal stated in his *Essai pour les coniques*, as one that could be deduced from his hexagram.

(...) Now, suppose that the two transversals merge with one of the sides of the hexagon, the resulting theorem will be the very theorem by Desargues about the involution of six points (Chasles, 1837: 338).

We hence clearly see how the introduction of the indeterminate geometrical elements enables the geometer to connect properties.[53] In conclusion of this whole sequence of variations on the form, Chasles concludes:

> Thus, it is proved that the mystic hexagram, another theorem by Pascal also on the hexagon, that of Newton on the organic description of conic lines, that of Desargues on the involution of six points, and that of the Ancients *ad quatuor lineas*, are all corollaries of our theorem. We thereby understand the *great number of particular truths* to which this theorem (the one with which Note XV began, addition by K. Chemla) can *extend*, thereby showing *relationships* that had so far remained unnoticed and an origin both shared and satisfactory.
>
> We can thus consider this theorem as being in a way a *center* from which most properties of conic lines, including the most general, derive. It would be appropriate, due to this very significant fruitfulness and the extreme ease with which it can be proved, that this theorem served as the foundation of a geometric theory of conic lines (Chasles, 1837: 338–9).

The theorem "lying at the center" embodies a kind of generality that characterizes Chasles's approach to geometry in the first half of the nineteenth century by contrast to other French geometers, like Monge or Poncelet. This conviction regarding the organization of a domain of mathematics, which Chasles shares with Quetelet and which accounts for his practice of geometry, can easily be shown to be correlated with Chasles's personal mathematical contribution: bringing to the fore the notion of cross-ratio as that which lay at the source of transferable properties.

2.4 Conclusion

Chasles's historical, mathematical, and even philosophical reflection about generality in geometry appeared multi-faceted. It echoed several aspects of the generality actually achieved in the application of analysis to geometry and thus seems to have been derived from a reflection on the sources of generality in the latter approach.

To begin with, analytical approaches are characterized by their use of formulas and the transformations of formulas into one another. The formulas do not distinguish between various configurations of the same geometrical system. They also do not separate related

---

[53] It would be worth analyzing systematically the similar techniques that Chasles brought into play in his writings, in order to complete the description of his reflection on the transformations that can be used in geometry. Such transformations evoke the kind of operations by means of which the geometer emulates the analyst and also brings a kind of generality to the practice of geometry.

but different objects such as conics. As a result, they connect propositions that could be otherwise perceived as different. Moreover, they allow the geometer to deal with them conjointly in his derivations, thereby subjecting them to a uniform treatment. Finally, formulas establish connections between propositions through the transformations to which they lend themselves as formulas. In Chasles's view, Desargues had brought views to geometry that introduced a similar uniformity. When taking as object any section of a cone, he had modified the objects geometers traditionally took into consideration, in a way that allowed him to treat them conjointly. Further, in transferring appropriate properties of a given conic to other conics, he had introduced connections between properties that were once perceived as unrelated. Such connections were, till then, believed to be proper to algebra. Chasles identified in Monge's *Géométrie descriptive* key ideas that inspired generalizations of these moves within geometry. First, Monge had carried out reasoning simultaneously on different geometrical configurations, further widening the objects taken into consideration in geometry. In the *Aperçu historique*, Chasles formulates a "principle of contingent relationships" to synthetize subsequent geometers' reflections on this idea that he grasps in Monge's reasoning. Second, Monge had widened the conception of the type of transformations one could apply to a geometrical configuration to transfer its properties to a transformed configuration. Chasles compares this practice to the analyst's use of transformations, calling geometers to work systematically on transformations. Further, he develops a "doctrine" that highlights the connection between all these methods, in a way that characterizes specific features of his own approach to generality.

Chasles also focuses on another type of transformation in geometry, which a reflection on analytical treatments seems to inspire: the change of form of a theorem. The transformations of formulas into one another constitute indeed the heart of analytic procedures and account for its simplicity and efficacy in Chasles's view. However, these transformations cannot bring "understanding," since they do not shed light on the intermediary truths and thus do not show to the mind the connection between the result and the origin of the reasoning (Chasles, 1837: 114). As a result, they do not highlight "the true reason of things."

The theory Chasles develops about the change of forms of theorems has the goal of introducing transformations into geometry. However, in contrast to the analytical ones, these transformations are easy and simple, to the point that they can be explained to anyone on the street.[54] They are immediate and connect truths to one another in a way that enlightens the mind, by showing the "source" of each truth. This attempt makes sense in relation to a belief to which Chasles subscribes: in any theory, there is a principal truth, which lies at its source and from which all the other truths follow "as simple transformations or natural corollaries." The achievement of such simplicity yields the "criterion" that "genuine perfection" has been achieved in the theory (Chasles, 1837: 115). In this context, the change of form of a theorem is a tool by means of which the geometer can identify the truth which is the source of the others. This truth is characterized, among other criteria, by how simple it is and how easy it is to prove.[55] It is further characterized

---

[54] Chasles attributes this ideal to Gergonne, referring to a memoire by Quetelet (Quetelet, 1827: 88, note 1.)
[55] Chasles (1837: 115–116, Note 2) attempts to account for this feature.

by how easy it is to derive from it all other truths. It is thus most fruitful and general, in the sense that the other truths appear to be merely variants of it. The change of form of the source is also a tool to derive from it different expressions for it that enable extending the scope of truths that can be deduced from it. It is finally a tool used to establish connections between a given truth and the source, and in this lies part of the "art of the geometer." Accordingly, although in this context like in analytical treatments all truths are connected through transformations, in contrast to analytical approaches the connections are established in a way that the mind can understand.

This belief bears on the theory as a whole, and on the organization of knowledge that can be achieved for it. Transformations of the statement of theorems play a key part in shaping such a theory, in a way that is reminiscent of the organization of knowledge brought about by analytical treatments of a domain. As we have shown, actors make the parallel, and compare the "source" to the principles of analytical theories. Here too this practice of generality in geometry thus appears to be inspired by other features of analytic approaches. In both cases, the organization achieved is completely different from what an axiomatico-deductive structure would yield. A first difference is manifest: neither the "principle" nor the "source" of a theory is taken to be indemonstrable. In Chasles's case, the ease of its proof is an important criterion in its identification. Moreover, the reduction of proof to a minimum is a goal guiding the reshaping of the theory to reach its ideal form, whereas other goals characterize the project of organizing a theory in the context of axiomatico-deductive practices.

The parallel between Chasles's ideal for a theory, which captures the essence of his practice of generality, and the analytical treatments should not conceal key differences between the two. The propositions that are placed at the root of the theories are, Chasles insists, of a different nature (Chasles, 1837: 119, Note 1). Probably even more important, however, are two related properties of Chasles's take on generality: the geometrical theory shows the deep reasons for the truths of theorems and it establishes connections between the truths. These two properties derive mainly from the dissolution of proof that has been achieved in the theory. As a consequence, every statement is derived in such an easy way that understanding of the whole theory can be shared with everybody. After all, these differences account for why, despite the advantages that analysis brought to geometry, Chasles devoted much effort to develop a geometrical approach to geometry.

In that sense, Chasles's *Aperçu historique* appears to provide a synthesis, within geometry, of all the means of generality introduced by traditions that since the seventeenth century had remained in this respect unrelated to each other.

REFERENCES

Belhoste, B. (1998) 'De l'École polytechnique à Saratoff, les premiers travaux géométriques de Poncelet', *Bulletin de la SABIX* **19**: 9–29.

Belhoste, B., and Taton, R. (1992) 'Leçons de Monge', in *L'Ecole Normale de l'An III. Leçons de mathématiques. Laplace-Lagrange-Monge*, ed. J. Dhombres. Paris: Dunod, 266–459.

Boyer, C. B. (1956) *History of analytic geometry*. New York: Scripta Mathematica.

Carnot, L. (1801) *De la corrélation des figures de géométrie.* Paris: Duprat.

Carnot, L. (1803) *Géométrie de position.* Paris: Duprat.

Carnot, L. (1806) *Mémoire sur la relation qui existe entre les distances respectives de cinq points quelconques pris dans l'espace suivi d'un essai sur la théorie des transversales. Un appendice leur est adjoint : Digression sur la nature des quantités dites négatives.* Paris: Courcier.

Chasles, M. (1837) *Aperçu historique sur l'origine et le développement des méthodes en géométrie, particulièrement de celles qui se rapportent à la géométrie moderne, suivi d'un mémoire de géométrie sur deux principes généraux de la science : la dualité et l'homographie.* Bruxelles: M. Hayez.

Chasles, M. (1852) *Traité de géométrie supérieure.* Paris: Bachelier, Imprimeur-Libraire de l'Ecole Polytechnique, du Bureau des Longitudes.

Chasles, M. (1870) *Rapport sur les progrès de la Géométrie, publication faite sous les auspices du Ministère de l'Instruction Publique* Paris: Imprimerie Nationale.

Chasles, M. (1889) *Aperçu historique sur l'origine et le développement des méthodes en géométrie, particulièrement de celles qui se rapportent à la géométrie moderne, suivi d'un mémoire de géométrie sur deux principes généraux de la science : la dualité et l'homographie. Troisième édition, conforme à la première.* Paris: Gauthier-Villars et Fils, imprimeurs-libraires de l'Ecole Polytechnique, du Bureau des Longitudes.

Chemla, K. (1990) 'Remarques sur les recherches géométriques de Lazare Carnot', in *Lazare Carnot ou le savant-citoyen,* ed. J. P. Charnay. Paris: Presses de l'Université de Paris Sorbonne, 525–41.

Chemla, K. (1998) 'Lazare Carnot et la généralité en géométrie. Variations sur le théorème dit de Menelaus', *Revue d'histoire des mathématiques* 4: 163–90.

Chemla, K. (2003) 'Euler's work in spherical trigonometry: contributions and applications', in *Euler. Opera Omnia. Commentationes physicae ad theoriam caloris, electricitatis et magnetismi pertinentes. Appendicem addidit Karine Chemla,* eds. P. Radelet-de Grave and D. Speiser. Basel: Birkhäuser Verlag, CXXV–CLXXXVII.

Chemla, K., and Pahaut, S. (1988) 'Préhistoires de la dualité: explorations algébriques en trigono-métrie sphérique (1753–1825)', in *Sciences à l'époque de la Révolution Française,* ed. R. Rashed. Paris: Lib. Sci. Tech. Albert Blanchard, 151–201.

Desargues, G. (1639) *Brouillon projet d'une atteinte aux événements des rencontres du cône avec un plan.* Paris.

Field, J. V., and Gray, J. J. (1987) *The geometrical work of Girard Desargues.* New York: Springer.

Gowers, T. (2008) 'How can one equivalent statement be stronger than another one', http://gow-ers.wordpress.com/2008/12/28/how-can-one-equivalent-statement-be-stronger-than-another/ (accessed on 20 October 2010).

Koppelman, E. (1971) 'Chasles, Michel', in *Dictionary of scientific biography,* ed. C. C. Gillispie. New York: Charles Scribner's Sons, 212–15.

Kötter, E. (1892) *Die Entwickelung der synthetischen Geometrie von Monge bis auf Staudt (1847).* Jahresbericht der Deutschen Mathematiker-Vereinigung Vol. 5 (2). Leipzig, Stuttgart, Wiesbaden: B. G. Teubner.

Nabonnand, P. (2011) 'L'argument de la généralité chez Carnot, Lazare et Chasles', in *Justifier en mathématiques,* eds. D. Flament and P. Nabonnand. Paris: Editions de la Maison des sciences de l'homme, 17–47.

Poncelet, J.-V. (1822) *Traité des propriétés projectives des figures; ouvrage utile à ceux qui s'occupent des applications de la géométrie descriptive et d'opérations géométriques sur le terrain.* Paris: Bachelier, libraire, quai des Augustins.

Quetelet, A. (1827) 'Résumé d'une nouvelle théorie des caustiques suivi de différentes applications à la théorie des projections stéréographiques, présenté à l'Académie Royale, dans la séance du 5 novembre 1825', *Nouveaux mémoires de l'Académie royale des sciences et belles-lettres de Bruxelles* 4: 79–109.

Raina, D. (1999) 'Nationalism, institutional science, and politics of representation: Ancient Indian astronomy in the landscape of French enlightenment historiography', Ph.D. Thesis, Faculty of Arts, Göteborg University, Sweden.

Taton, R. (1951a) *L'œuvre mathématique de Desargues*. Paris: Presses Universitaires de France.

Taton, R. (1951b) *L'oeuvre scientifique de Monge*. Paris: Presses Universitaires de France.

Taton, R. (1964) 'L'Ecole polytechnique et le renouveau de la géométrie analytique', in *L'aventure de l'esprit, mélanges offerts à Alexandre Koyré. Vol. 1*. Paris: Hermann, 552–64.

Youschkevitch, A. P. (1964) 'Remarques sur la méthode antique d'exhaustion', in *L'aventure de l'esprit, mélanges offerts à Alexandre Koyré. Vol 1*. Paris: Hermann, 635–53.

3

Generality in Leibniz's mathematics

EBERHARD KNOBLOCH

3.1   Introduction: "Everything is ruled by reason"

On 2 February 1702, Leibniz wrote to Varignon: "This is the case because everything is ruled by reason, and because otherwise there would be no science or rule. That would not be in accordance with the nature of the highest principle."[1] In other words: "The great harmonious order of the universe that ideally exists in God is embedded in the creation as reflection of the supremely rational nature of the creator" (Pasini, 2001: 959). Mathematics can reflect that order and that harmony.

In principle, there is a philosophical-theological basis of Leibnizian mathematics. From the beginning everything that exists is to be found in an orderly relation. The general and inviolable laws of the world are an ontological *a priori* (Holz, 1983: 55). The universal harmony of the world consists in the largest possible variety being given the largest possible order so that the largest possible perfection is involved.

Thus, to find the general laws, the general theorems, the rules, and the methods means to discover the universal harmony in mathematics and elsewhere.

This was the origin of Leibniz's wish for such methods, theorems, rules, laws, objects, notions, notations, disciplines, and problems.

Even his language reflects the theological context—he called the theorems "divine":

> Sed omnium difficillima artis analyticae pars est, inventio theorematum ... cum theoremata quae scilicet res maxime dissitas inter se harmonia quadam ligant, praeclara sunt compendia rationis humanae.... V. g. nisi extarent divina illa theoremata de centro gravitatis, tot praeclaras quadraturas, fortasse non invenissemus. (But the most difficult part of all of the analytical art is the invention of theorems ... because the theorems that link together the most dispersed things by a certain harmony are excellent summaries

---

[1] "C'est par ce que tout se gouverne par raison, et qu'autrement il n'y aurait point de science ny regle, ce qui ne seroit point conforme avec la nature du souverain principe." (GM IV, 94; Knobloch, 2006: 382).

The Oxford Handbook of Generality in Mathematics and the Sciences. First Edition. Karine Chemla, Renaud Chorlay and David Rabouin. © Oxford University Press 2016. Publishing in 2016 by Oxford University Press.

of human understanding.... If, for example, there were not those divine theorems on the center of gravity, maybe we would not have found so many excellent quadratures).[2]

After having examined the relationship between the value of generality and the harmonies that are at the center of Leibniz's concern (Section 3.2), I shall turn to the relationship of generality to several other epistemic values: beauty (Section 3.3); conciseness/simplicity, a topic that I shall approach on the basis of the example of Leibniz's work on the divisors of a product (Section 3.4). The latter epistemic value will lead me to raise the question of how the interest in generality relates to another of Leibniz's major concern: notations. I shall take in this case the examples of determinants and sums of powers (Section 3.5). In order to explain the relationship of generality to utility and fecundity I shall refer to his so-called transmutation theorem (Section 3.6). Eventually I shall demonstrate how generality is connected with laws of formation (Section 3.7).

3.2 Generality and harmony

Those among the theorems that are general represent the order and reveal the order, the harmony. In short, every harmony implies a general theorem, implies generality.

Such a belief is manifest in Leibniz's remarks written in January 1675, when he discusses analytical developments of an algebraic fraction whose numerator and denominator are polynomials. He replaces y by $y + \beta$—he calls this operation "explicare," "to unfold"—and compares the two formulas term by term:

$$A = \frac{by^z + ca^1 y^{z-1} + da^2 y^{z-2} + ca^3 y^{z-3} + \cdots}{gy^z + ha^1 y^{z-1} + ka^2 y^{z-2} + la^3 y^{z-3} + \cdots}$$

a and its powers are factors of homogeneity that serve to always obtain terms of the same dimension z.

Leibniz replaces y by $y + \beta$ and calls the result P. In order to compare the coefficients of the corresponding powers of y he denotes the numerator by S, the denominator by I. PSy^z denotes the coefficient of y^z in the numerator of P, ASz refers to the coefficient of y^z in the numerator of A, etc.

Thus he gets:

$$PSy^z = b(ASz),$$

$$PSy^{z-1} = b(ASz)z\beta + ca(ASz - 1), \text{ etc.}$$

because $b(y+\beta)^z = b(y^z + zy^{z-1}\beta + \dfrac{z(z-1)}{2} y^{z-2}\beta^2 + \cdots + \beta^z)$.

---

[2] A VII, 1, 708. Summer of 1673.

Eventually he sums up:

> Atque ita habemus formulam quae una est ex utilissimis totius analyseos, continuata enim exhibet generaliter explicationem formulae cujuscunque per binomium. Unde facile habetur et explicatio trinomii, scilicet explicando rursus ipsum β. per binomium; et per consequens habetur explicatio formulae cujuscunque per polynomium quodcunque. Progressiones hic occurrunt et harmoniae quocunque te vertas, et sufficiet inspexisse Tabulam, ad eas advertandas. Quot autem harmoniae, tot deteguntur theoremata generalia omnibus formulis communia, quae manifestum est, ex ipsa combinationum natura suam originem habere. (And thus we have a formula that is one of the most useful of all of analysis. Because if it is continued, it generally represents the explanation of an arbitrary formula by means of a binomial. For that reason one easily gets also the explanation of a trinomial, namely by explaining again β by a binomial. And as a consequence one gets the explanation of an arbitrary formula by an arbitrary polynomial. Here progressions and harmonies occur wherever one turns and it will suffice to have a look on the table to notice them. In fact, one discovers as many general theorems that are common to all formulas as there are harmonies. Obviously, such theorems originate from the very nature of combinations.)[3]

The order, the general law, must be discovered. Leibniz calls them *arcana*, mysteries, thus reminding us again of the religious context in which for him this research develops. For example, in 1678, he says about the general algorithmic solution of linear equations:

> Ecce verum analyseos Arcanum (Behold, the true mystery of analysis).[4]

In January 1676, he elaborates a figure in order to find the law of distribution of prime numbers (see Fig. 3.1).

The multiples of 2, 3, 4, 5, etc. are marked one after the other on the horizontal parallels. The diagonals connect

2, 3, 4, 5, etc., that is, the natural numbers beginning with 2,

4, 6, 8, 10, etc., that is, the even numbers beginning with 4,

6, 9, 12, 15, etc., that is, the multiples of 3 beginning with 6, etc.

He comments:

> Figura notabilis, in qua primorum et multiplorum arcana latent (A remarkable figure wherein the mysteries of prime and multiple numbers are hidden).[5]

To discover these mysteries, this harmony, one needs a "key." This is what is evidenced on the basis of Leibniz's own comments, when he is looking for the representation of

[3] LKK, p. 24.
[4] LDK, p. 12.
[5] A VII, 1, 580; Knobloch (2004: 59).

1. 2. 3. 4. 5. 6. 7. 8. 9. 10. 11. 12. 13. 14. 15. 16. 17. 18. 19. 20. 21. 22. 23. 24. 25. 26. 27. 28. 29. 30.

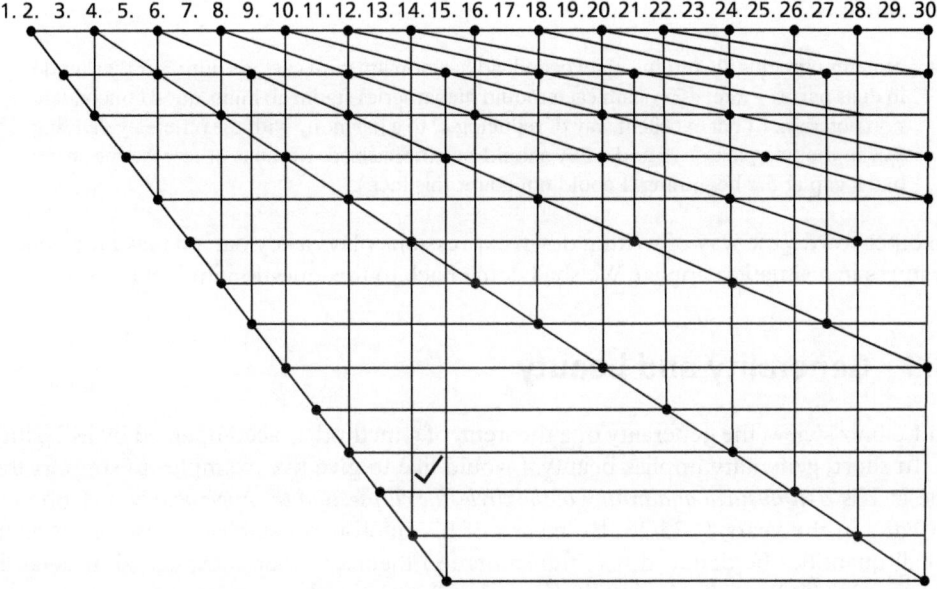

Figure 3.1 *The law of distribution of prime numbers*
(Credits: A VII, 1, 580)

symmetric functions by elementary symmetric functions $x, y, z, \delta, \varepsilon$, etc. Let $a, b, c, d, e,$ etc. be an arbitrary number of variables such that

$$x = a + b + c + d + e + \cdots$$
$$y = ab + ac + ad + bc + bd + \cdots$$
$$z = abc + abd + abe + bcd + \cdots \text{ etc}$$

are the elementary symmetric functions. The powers or products a^3, a^3b, a^3bc, etc. are meant to denote symmetric functions. He writes in September 1680:[6]

> Perficiendus est hic calculus, mira enim compendia, et omnino totius Algebrae clavem continet. (This calculus has to be perfected because it contains wonderful abridgements and, in general, the key of all of algebra.)

By contrast to the previous example, here it is no longer a theorem, but a "law of formation" of such calculations that reflects their order or their harmony. Such a law must comprehend all cases. It is reflected in the following representations that Leibniz deduces:[7]

$$a^3 = xx^2 - 2xy - yx + 3z$$
$$a^3b = yx^2 - 2yy - zx + 4\delta$$
$$a^3bc = zx^2 - 2zy - \delta x + 5\varepsilon$$

[6] LKK, p. 191.
[7] LKK, p. 190.

and remarks:

> Notetur egregius ille modus a^3 revocandi ad harmoniam cum caeteris dum $3yx$ divellendo in duas partes $-yx$ et $-2xy$ nam caeteroquin hiabat series statim ab initio, quod concoquere non poteram. (That excellent way of reducing a^3 to a harmony with the others by dividing $3yx$ in the two parts $-yx$ and $-2xy$ should be emphasized. Because otherwise the series had a gap at the beginning. I could not adapt this fact.)

In other words, the way of writing down expressions plays a key part to make the general features in a situation appear. We shall come back to this question in Section 3.5.

3.3   Generality and beauty

In Leibniz's view, the generality of a theorem, of a method, is accompanied by its beauty.

In short, generality implies beauty. I would like to give five examples to support this thesis. His *Arithmetical quadrature of the circle, the ellipse, and the hyperbola* was elaborated in Paris in the years 1675/76. By means of his rigorously established use of infinitely small quantities he deduced his "transmutation theorem" that is explained in detail in Section 3.6 ("Generality and utility/fecundity") and based on the construction of a new, integrable curve with regard to a given smooth curve.

Leibniz comments upon this method:

> Inde porro investigans Methodum reperi generalem admodum et pulchram ac diu quaesitam, cujus ope datae cuilibet curvae analyticae, exhiberi potest curva analytica rationalis aequipollens. (By continuing my researches I have found a completely general and beautiful method sought for since a long time. It enables one to exhibit an analytical, rational, equivalent curve to an arbitrary, given analytical curve.)[8]

Neither here nor elsewhere can the implication "generality implies beauty" be reversed.

In the same way, when Leibniz speaks about the elimination of a common unknown of two algebraic equations, he states:

> Sed possunt condi tabulae generales pulcherrimae. (But the most beautiful, general tables can be established.)[9]

By tables Leibniz means the general solution of elimination problems regarding two algebraic equations with several variables so that a special case can be reduced to the general case without that the special case has to be solved for itself.

The generality reveals the beauty of these tables, of general formulas. Likewise, when he is looking for the number of divisors of a product of prime numbers, he calls the theorem:

> theorema pulchrum, breve, generale (a beautiful, concise, general theorem).[10]

---

[8] Leibniz 2004: 138f.
[9] LDK, p. 160.
[10] LKK, p. 269.

Note here the link that Leibniz establishes to the value of conciseness, we shall come back to it in the next section. Again, when he deduces Cramer's rule for solving systems of linear equations, he concludes:

> Habemus ergo theorema pulcherrimum cuius vis se extendit in infinitum. (Thus we have a most beautiful theorem the force of which extends to infinity.)[11]

Here, the generality of the rule is captured by the extension it has. Lastly, about the elementary symmetric functions he remarks:

> Usus formarum praeter pulchritudinem et generalitatem contemplationis in eo consistit, ut ope earum inveniamus Radicem generalem aequationis affectae cujuscunque gradus. (If we leave aside the beauty and generality of the consideration, the utility of symmetric functions consists in the fact that we find with their aid the general root of a non-pure equation of an arbitrary degree.)[12]

Here too, beauty and generality are associated with a third value: utility, to which we shall return in Section 3.6.

3.4   Generality and conciseness/simplicity

In Leibniz's eyes, generality implies the two related values of conciseness and simplicity. If one really and completely penetrates the structure of the facts of a case, its description has to become concise, simple. The special details have disappeared in the statements of the facts, leading to a general theorem or method.

As a consequence, the generality of a theorem, of a method, is accompanied by its conciseness, by its simplicity. That is why Leibniz called theorems abridgements of the human understanding.

In short, generality implies conciseness and simplicity.

Such a link is precisely what Leibniz underlines in his comments about proposition 18 of his *Arithmetical quadrature of the circle, the ellipse, and the hyperbola,* in which $ax^n = y^m$ or $x^n y^m = a$ form simple analytical figures. Leibniz demonstrates that the ratio of the area between two ordinates, the arc of the curve and the axis to the area between the corresponding abscissas, the same arc of the curve and the conjugated axis is equal to m:n. He says:

> Hanc propositionem, novam ni fallor, credidi memorabilem, tum ob simplicitatem expressionis, quia facile retineri potest, tum ob usus generalitatem. (This proposition, new if I am not mistaken, seemed to me remarkable because of the simplicity of the expression, because it can be easily retained, and because of the generality of its application as well.)[13]

---

[11] LDK, p. 10.
[12] LKK, p. 55.
[13] Leibniz 2004: 157.

Likewise he greatly praises the power series of the cosine function, saying:

> Itaque hanc unicam seriem utique simplicissimam et retentu facillimam in animo haberi
> sufficit: $c = 1 - \dfrac{a^2}{2} + \dfrac{a^4}{24}$ etc. (Therefore it suffices to have in mind this single series that
> is definitely the simplest and the most easily retained: $c = 1 - \dfrac{a^2}{2} + \dfrac{a^4}{24}$ etc.)[14]

In both cases, he thus insists on the ease for memorization that these theorems represent: in the former case, the theorem is easy to remember, because of its simplicity, in the latter case, one can limit oneself to memorizing only the theorem, because everything else needed can be deduced from it. Somewhat later, Leibniz adds:

> Haec series ergo ... universalissima et omnium quas norim ad usum publicum aptissima
> est. (So this series ... is the most universal and, to the best of my knowledge, the most
> appropriate of all for a public application).

Leibniz mentions mechanics, engineering, geodesics, nautics, and astronomical calculations among the possible applications. Its special usefulness is based on the quick convergence of the series: the first three terms already lead to a high reliability of the value of c.

Yet the value of conciseness is sometimes in tension with that of generality, as it appears in the following example. In May 1678 he looks for an algorithmic solution of an algebraic equation of arbitrary degree. He uses symmetric functions and adds the revealing comment:

> Hac autem via simul et compendiosiorem et non minus generalem, reperiri posse non
> puto. (But I do not believe that a method can be found that is at the same time shorter
> and no less general.)[15]

As Leibniz emphasizes, the link between simplicity and generality depends on the way of writing. He creates a notation for the resultant of a system of linear equations in the following way (see Section 3.5, "Generality and notation"):

$$0 \cdot 1 \cdot 2 \cdot 3$$

denotes the product $10 \cdot 21 \cdot 32 \cdot 43$ that is the product of the elements of the main diagonal of the coefficient matrix

10	11	12	13
20	21	22	23
30	31	32	33
40	41	42	43

---

[14] Leibniz 2004: 337.
[15] LKK, p. 98.

$\overline{0 \cdot 1 \cdot 2 \cdot 3}$ denotes the 24 permutations having the sign + or − according to a special sign rule discovered by Leibniz in January 1684. In other words, $\overline{0 \cdot 1 \cdot 2 \cdot 3}$ represents the determinant of the coefficient matrix. He thus concludes:

> Atque ita habemus ex his modum simplicissimum et generalissimum, scribendi formulam hujusmodi quancumque. (And in this way we obtain the simplest and most general method of writing such an arbitrary formula.)[16]

We shall come back to the importance of notation to achieve conciseness in the next section. This is a key factor for generality as well, since for Leibniz, there is an essential link, between a genuine conciseness and universality, where conciseness enhances, instead of limiting generality. This is what he recognizes is at stake, when in 1686, he characterizes his differential calculus in the following way:

> Vix quicquam utilius, brevius, universalius fingi potest Calculo meo differentiali seu Analysi indivisibilium atque infinitorum. (Hardly anything can be imagined that is more useful, more concise, more universal than my differential calculus or analysis of indivisibles and of the infinite.)[17]

The number of divisors of a product is a good example to illustrate Leibniz's search for a simpler and more general method. Shortly after 1676, Leibniz studies Frans van Schooten's additive and recursive method of finding the number of combinations—in modern terms—the number of subsets of a finite set. Let be a^3b^2 a product whose divisors are looked for. Van Schooten elaborates the following table:

$$a$$
$$a \cdot aa$$
$$a \cdot aaa$$
$$b \cdot ab \cdot aab \cdot aaab$$
$$b \cdot bb \cdot abb \cdot aabb \cdot aaabb$$

Thus step by step the aliquot divisors can be found and counted: there are eleven. As van Schooten, Leibniz interprets the combinations of the elements as divisors of a product of numbers. After recognizing the multiplicative structure of van Schooten's recursive law of formation, he deduces the result that the number of divisors of the number $a^5b^4c^3d^2$ is equal to $6 \cdot 5 \cdot 4 \cdot 3 = 360$.

He correctly remarks that van Schooten's results regarding the numerical examples considered coincide with his own results and adds:

> Sed non cognorat theorema tam pulchrum, breve et generale ususque est via illa prolixiore additionis. (But he had not known such a beautiful, concise, and general theorem and used that more cumbersome method of addition.)[18]

---

[16] Knobloch (1972: 176).
[17] GM V, 230; Leibniz (1989: 136f.).
[18] LKK, p. 269.

3.5   Generality and notation

Leibniz always emphasized the importance of an appropriate notation for the art of invention. The characteristic art, which lay at the center of his preoccupations, had to occupy itself with such notations. In 1678 he wrote to Tschirnhaus:

> In signis spectanda est commoditas ad inveniendum, quae maxima est, quoties rei naturam intimam paucis exprimunt et velut pingunt, ita enim mirifice imminuitur cogitandi labor. Talia vero sunt signa a me in calculo aequationum tetragonisticarum adhibita, quibus problemata saepe difficillima paucis lineis solvo. (As far as signs are concerned one has to take care that they facilitate discovery. This facilitation is the greatest whenever they express, and so to speak paint, by little expenditure the intimate nature of a thing. Because the labor of thinking is wonderfully diminished. Now such signs have been used by me in the tetragonistic equations by means of which I often solve the most difficult problems in a few lines.)[19]

Leibniz thus makes here explicit the relationship between conciseness and ease of discovery. Leibniz refers to his differential calculus that is indeed the best known example. He explicitly said:

> Idem autem est ac si dixisses opus esse characteribus nihil aliud enim est calculus quam operatio per characteres, quae non solum in quantitatibus, sed et in omni alia ratiocinatione locum habet. (But this is the same as if you would have said that characters are needed. Because the calculus is nothing else but the operation by characters that takes place not only in quantities but also in every other reasonable argument.)[20]

To understand better how Leibniz worked on such kinds of notations and oriented himself in the tension between generality and conciseness, I shall evoke other examples where one can recognize how he changed the notation in order to get the most concise notation ("paucis") that nevertheless expresses the intimate nature of things.

3.5.1   The notation for determinants

We can identify four steps in Leibniz's production of his notation (Knobloch, 2001):

a) Leibniz replaces the letters with double indices by the indices themselves:

$$a_{10}a_{21} - a_{11}a_{20} = 0 \text{ leads to } 10 \cdot 21 - 11 \cdot 20 = 0$$

b) The left indices do not change their order. Thus they play a minor role, they are written somewhat smaller than the right indices:

$$_1 0 \cdot 21 - _1 1 \cdot 20 = 0$$

---

[19] A III, 2, 444f.
[20] A III, 2, 452.

c) The left indices are superfluous and can be left out: $0 \cdot 1 - 1 \cdot 0 = 0$

d) Even this notation can be simplified:

$$\overline{0 \cdot 1} = 0$$

The notation $\overline{0 \cdot 1 \cdot 2 \cdot 3 \, \dots \, n}$ denotes a determinant. We have only to know the meaning of this symbol in order to write down the algebraic expression explicitly. Every step of the procedure of abbreviation can be reversed. The conciseness of the notation reflects the conciseness of the solution and consequently of its generality. For example (see Section 3, 4, "Generality and conciseness/simplicity"),

$$
\begin{aligned}
\overline{0 \cdot 1 \cdot 2 \cdot 3} = {} & + 0 \cdot 1 \cdot 2 \cdot 3 - 0 \cdot 1 \cdot 3 \cdot 2 + 0 \cdot 2 \cdot 3 \cdot 1 - 1 \cdot 2 \cdot 3 \cdot 0 \\
& - 1 \cdot 0 \cdot 2 \cdot 3 + 1 \cdot 0 \cdot 3 \cdot 2 - 2 \cdot 0 \cdot 3 \cdot 1 + 2 \cdot 1 \cdot 3 \cdot 0 \\
& - 0 \cdot 2 \cdot 1 \cdot 3 + 0 \cdot 3 \cdot 1 \cdot 2 - 0 \cdot 3 \cdot 2 \cdot 1 + 1 \cdot 3 \cdot 2 \cdot 0 \\
& + 2 \cdot 0 \cdot 1 \cdot 3 - 3 \cdot 0 \cdot 1 \cdot 2 + 3 \cdot 0 \cdot 2 \cdot 1 - 3 \cdot 1 \cdot 2 \cdot 0 \\
& + 1 \cdot 2 \cdot 0 \cdot 3 - 1 \cdot 3 \cdot 0 \cdot 2 + 2 \cdot 3 \cdot 0 \cdot 1 - 2 \cdot 3 \cdot 1 \cdot 0 \\
& - 2 \cdot 1 \cdot 0 \cdot 3 - 3 \cdot 1 \cdot 0 \cdot 2 + 3 \cdot 2 \cdot 0 \cdot 1 + 3 \cdot 2 \cdot 1 \cdot 0
\end{aligned}
$$

Terms which emerge from an odd number of transpositions from one another have different signs, and in the case of an even number, they have the same sign.

3.5.2   The products of power sums[21]

Let us assume that we have to multiply three power sums

$$a^m + b^m + c^m + \cdots = \overset{..}{a}{}^m,$$

$$a^n + b^n + c^n + \cdots = \overset{..}{a}{}^n,$$

$$a^p + b^p + c^p + \cdots = \overset{..}{a}{}^p.$$

Again, Leibniz uses three notations step by step.

$$
\begin{aligned}
\overset{..}{a}{}^m \, \overset{..}{a}{}^n \, \overset{..}{a}{}^p = {} & \overset{..}{a}{}^{m+n+p} + \overset{..}{a}{}^{m+n} \, \overset{..}{b}{}^p + \overset{..}{a}{}^m \, \overset{..}{b}{}^n \, \overset{..}{c}{}^p \\
& + \overset{..}{a}{}^{m+p} \, \overset{..}{b}{}^n \\
& + \overset{..}{a}{}^{n+p} \, \overset{..}{b}{}^m
\end{aligned}
$$

a) First the sum $\overset{..}{a}{}^m = a^m + b^m + c^m + d^m + \cdots$ is composed of a finite number of elements (of a sum)

---

[21] LKK, pp. 236–42.

b) Then Leibniz leaves out the bases and writes only the exponents:

$$mnp + mn \mid p$$
$$mp \mid n$$
$$np \mid m + m \mid n \mid p$$

The vertical lines separate the different bases from each other.

c) Finally he describes these products by the partitions of the natural number that represents the number of power sums involved. In the case of 3, this yields

$$3 + ③2 \mid 1 + 1 \mid 1 \mid 1$$

The encircled number 3 is a true number and indicates the different possibilities of putting three objects (in our case the exponents m, n, p) into two classes in such a way that every class contains at least one object.

This is the most concise representation of our problem which at the same time reveals the hidden relations between an algebraic problem and a problem of additive number theory.

The not encircled numbers indicate the numbers of objects contained in a class. The vertical lines separate the classes of one partition from each other. In other words the original multiplication problem of power sums has been reduced to a partition problem of additive number theory. The general partition problem can be explained as follows. One has to find the number of partitions of n objects in a set X into a certain number of classes of cardinal 1, 2, ..., k.

Let $k = 1r_1 + 2r_2 + \cdots + kr_k$ be the type of the partition. k occurs on the right side of the equation if and only if $r_1 = r_2 = \cdots = r_{k-1} = 0$ and $r_k = 1$.

$$N = \frac{k!}{(1!)^{r_1} (2!)^{r_2} \dots (k!)^{r_k} \, r_1! r_2! \dots r_k!}$$

will be the number of partitions of that type.

3.6   Generality and utility/fecundity

In Leibniz's view, the more general a theorem or a method, the more useful it is. The great value of generality is based on its utility, its fecundity. We have to remember that Leibniz always emphasized that theory has to be combined with practice, that is, theory with its applications. This explains why generality plays a key part for developing the theory.

In short, generality implies utility and fecundity.

In 1673 Leibniz discovered the so-called transmutation theorem which we shall evolve in greater detail below. It can be interpreted in terms of integration by parts and enabled Leibniz to integrate large classes of curves. Consequently he said about it:

> Arbitror unam esse ex generalissimis, atque utilissimis, quae extant in Geometria, isque adeo enim universalis est, ut omnibus curvis, etiam casu aut pro arbitrio sine certa lege ductis, conveniat.... Sed et inter foecundissima Geometriae theoremata haberi potest. (I believe that it is one of the most general and most useful (propositions) which exist in geometry. Because it is universal to such a degree that it is appropriate for all curves, even those that are drawn by chance or arbitrarily without a certain law.... But it can also be numbered among the most fruitful theorems of geometry.)[22]

The statement underlines the fact that the method deals with all curves in the same way. It is the same value that Leibniz prizes, when he emphasizes further the fruitfulness of his method of using infinitely small quantities:

> Liberrimo mentis discursu possumus non minus audacter ac tuto curvas quam rectas tractare. Cujus specimen totus hic libellus erit si quis methodi fructum quaerit. (By the freest possible intellectual discourse we can no less boldly and securely handle curves than straight lines. This whole booklet will be an example thereof if one is looking for the fruit of the method.)[23]

In fact Leibniz always wanted to improve the art of invention by his methods. This is what is at stake when he later replaces the previous geometrical method by his differential calculus; we will come back to this issue. In 1691, he said:

> Sed animadverti, fontes non satis adhuc patuisse et restare interius aliquid, quo pars illa Geometriae sublimior tandem aliquando ad analysin revocari posse, cujus antea incapax habebatur. Ejus elementa aliquot abhinc annis publicavi, consulens potius utilitati publicae, quam gloriae meae, cui fortasse magis velificari potuissem methodo suppressa. Sed mihi jucundius est, ex sparsis a me seminibus natos in aliorum quoque hortis fructus videre ... methodos potius, quam specialia licet vulgo plausibiliora aestimavi. (But I noticed that the sources were not yet sufficiently clear and that something was lacking by which that higher part of geometry could sometimes finally be reduced to analysis, for which it was unsuitable in former times. Some years ago I published its elements caring more about public utility than about my glory that I presumably would have fostered more by suppressing the method. But for me it is more agreeable to see how the fruits grew up also in the garden of others after I had spread the seeds ... I rated methods higher than special problems though they usually are more pleasing.)[24]

In 1694, he characterizes the new general method introduced referring to God:

> Enfin notre méthode étant proprement cette partie de la Mathématique générale, qui traite de l'infini, c'est ce qui fait qu'on en a fort besoin, en appliquant les Mathématiques

---

[22] Leibniz (2004: 71).
[23] Leibniz (2004: 185).
[24] GM V, 257f.

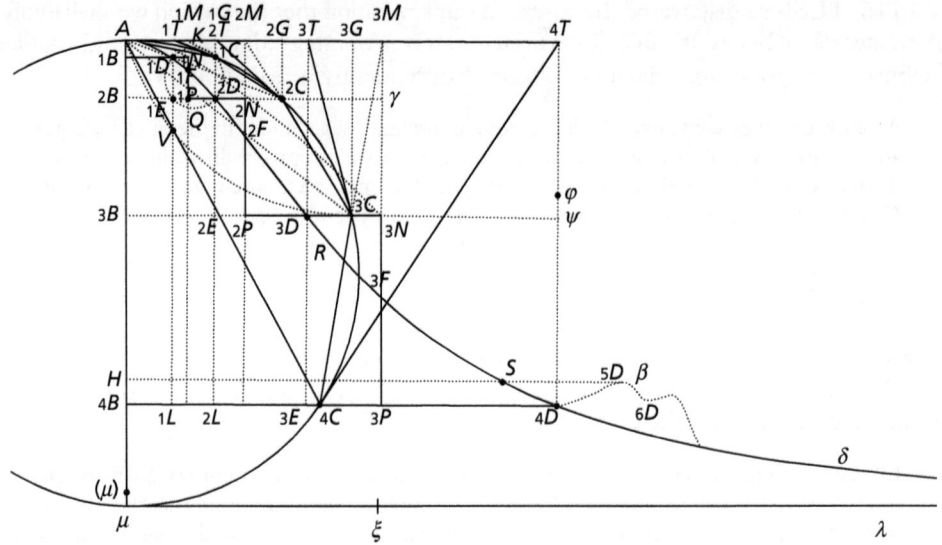

Figure 3.2 *The transmutation theorem*
(Credits: A VII, 6, 528)

à la Physique, parce que le caractère de l'Auteur infini entre ordinairement dans les opérations de la nature. (Finally our method is that part of general mathematics which occupies itself with the infinite. For that reason one needs it very much if one applies mathematics to physics because the character of the infinite Author usually enters the operations of nature.)[25]

There we have again the theological basis of Leibnizian mathematics, which accounts for another of its features.

The transmutation theorem may serve as a model that illustrates how the highly esteemed generality of a theorem can lose this high esteem on a new level of abstraction.

While Leibniz praises this theorem in the context of infinitesimal geometry—we came across this point earlier—he changed this assessment after the invention of his differential calculus. In other words, generality depends on the method applied: the geometrical method involved a certain level of generality while the calculus represented a new level of generality. But we must note that while every calculus, especially every algorithm, is a method, it is not true that, as the transmutation theorem illustrates, every method must be a calculus.

The transmutation theorem can be formulated as follows. Leibniz considers a curve through the points A, $1C$, $2C$, $3C$, etc. (see Fig. 3.2). He constructs a new curve A, $1D$, $2D$, $3D$, etc. in such a way that the ordinate BD (which is a general way of referring to $_iB_iD$,

---

[25] GM V, 308.

the same notation accounts for what follows) is equal to the segment AT on the y-axis: T is the point of intersection of the tangent in C with the y-axis AT. The theorem reads:

> The area Q of the mixtilinear quadrilateral $1D1B3B3D2D1D$ is twice as large as the area R of the sector $1CA3C2C1C$.

The theorem is a generalization of Cavalieri's method: a restriction, a limit has been eliminated. Cavalieri decomposed figures only into parallelograms. Leibniz replaces the parallelograms by triangles. As Leibniz explains, rectangles or parallelograms are only a special case of triangles: one can consider the parallels as convergent straight lines whose point of intersection, or common center, is infinitely far away, much like the second focal point of a parabola:

> Hinc jam ope theorematis hujus nostri feliciter evenit ut harum quoque novarum ordinatarum, nempe convergentium usus esse possit ad quadraturas, utque figurae non tantum per ordinatas parallelas in parallelogramma ... sed et per ordinatas convergentes in triangula ... resolvantur. (Thus thanks to our theorem it happily comes about that these new, namely convergent, ordinates are useful for quadratures, so that figures can be resolved not only by parallel ordinates into parallelograms ... but also by convergent ordinates into triangles.)[26]

The principle of such a generalization can be described as follows: The earlier objects used in the method are interpreted as special cases of new objects. The same principle is to be found in Leibniz's demonstration of his rigorous foundation of integration theory.

Instead of inscriptions and circumscriptions, in other words instead of extreme values of the integration interval, he takes an arbitrary intermediate value of such an interval. The extreme values that bound a certain interval are but intermediate values in that new interpretation.[27]

Such a method reminds us of the Leibnizian method of universality: this method shows how to find by a single operation general analytical formulae and geometrical constructions for different subjects and cases while every of them would need a special analysis or synthesis without that method.

> On peut juger par là que l'usage de la Méthode de l'universalité s'étend aussi loin que l'algèbre et l'analyse ... Car il arrive tous les jours qu'un même problème est de plusieurs cas, dont la multitude embarrasse beaucoup, et nous oblige à des changements inutiles et à des répétitions ennuyeuses dont cette méthode nous garantira à l'avenir. (Thus one can conclude that the use of the method of universality extends so far as algebra and analysis ... Because it happens every day that the same problem includes several cases the multitude of which is very embarrassing and forces us to make many changes and annoying repetitions. This method will protect us from them in the future.)[28]

---

[26] Leibniz (2004: 72.f).
[27] Leibniz (2004: 48–51).
[28] C, p. 123.

In 1690, about 15 years later, Leibniz explained this transmutation theorem to Bodenhausen, first with the aid of Archimedean methods, then with the aid of the differential calculus. In the end, he concludes:

> Fateor autem me Theorematis hujusmodi opus non habere, nam quicquid ex illis duci potest, jam in calculo meo comprehenditur; libenter tamen iis utor, quia calculum imaginationi quodammodo conciliant. (But I confess that I do not need such theorems (that is, the transmutation theorem, note of the translator) because whatever can be deduced from them is already contained in my calculus; yet, I use them willingly, because they connect in a certain way the calculus to the imagination.)[29]

And somewhat later, again addressing the issue of understanding, he writes:

> Sed haec omnia, ut verum fatear, non sunt nisi ad populum phalerae pro illis, qui analysin nostram non intelligunt; nam quadraturae talium figurarum ex nostro calculo inmediate deducuntur. (Truth to tell, all that is only a superficial popular ornament for those who do not understand our analysis. Because the quadratures of such figures are immediately deduced from our calculus.)[30]

In other words, the calculus makes possible the double aim of Leibnizian mathematics:

1. to liberate the arguments from an inspection of figures, that is, from recourse to the imagination (liberation);
2. to relieve the memory by a mechanization of the calculus, that is, by using an algorithm (mechanization).

3.7   Generality and the laws of formation

The regularity in the structure of an expression, the order of a series, the generality of a solution are mirrored by the general laws of formation that reveal their patterns and that are—according to Leibniz's conception—combinatorial laws: these laws render the tables, formulas, explicit calculation superfluous. We need not write down the terms of $0 \cdot 1 \cdot 2 \cdot 3$ if we know the structure of this expression (a determinant). We need not write down the representation of an arbitrary symmetric function (in order to elaborate a table of such representations), if we know the law of formation regarding such a representation. We need not write down the expansion of $f(x) = \sin x$ into an infinite series if we know its law of formation. The existence of such laws manifests the generality, as we have seen above. In short: the laws imply generality:

> Ars autem quaerendi progressiones, et condendi Tabulas formularum, est pure combinatoria .... Fateor interim nusquam pulchriora quam in algebra artis combinatoriae sive

---

[29] A III, 4, 637.
[30] A III, 4, 639.

characteristicae generalis specimina edita esse. (But the art of looking for progressions [note of the translator: that is, laws of formation] and of establishing tables of formulas is purely combinatorial…. In the meantime I confess that nowhere have more beautiful examples of the combinatorial or general characteristic art been published than in algebra.)[31]

Elaborating a general formula for the quotient (algebraic division) of two polynomials he concludes:

Et ces Loix étant trouvées, on n'auroit pas même besoin de la Formule, si ce n'est pour les mieux faire entendre (And if one has found these laws, one would not even need the formula unless it is to make them better understandable), he says in 1712.[32]

The expansion of the trigonometric functions like $f(x) = \sin x$ renders trigonometric tables superfluous. For that reason he calls his treatise *Arithmetical quadrature of the circle, of the ellipse, and of the hyperbola. A corollary thereof is the trigonometry without tables* (Leibniz, 2004). For the same reason the combinatorial art relieves the memory in another way than the calculus, while it is at the same time an essential instrument of the art of invention.

Let us consider two examples of the combinatorial art.

3.7.1 The general canon of division

Let us assume that we are looking for the quotient

$$20x^7 + 21ax^6 + 22a^2x^5 + \cdots + 26a^6x + 27a^7 : 10x^3 + 11ax^2 + 12a^2x + 13a^3$$

The quotient reads as follows:

$$
\frac{20x^4}{10}
+\frac{\begin{array}{l}+10\cdot21ax^3\\-11\cdot20\cdots\end{array}}{10\cdot10}
+\frac{\begin{array}{l}+10\cdot10\cdot22a^2x^2\\-10\cdot11\cdot21\cdots\\+11\cdot11\cdot20\cdots\\-10\cdot12\cdot20\cdots\end{array}}{10\cdot10\cdot10}
+\frac{\begin{array}{l}+10\cdot10\cdot10\cdot23a^3x\\-10\cdot10\cdot11\cdot22\cdots\\+10\cdot11\cdot11\cdot21\cdots\\-10\cdot10\cdot12\cdot21\cdots\\-11\cdot11\cdot11\cdot20\cdots\\+2\cdot10\cdot11\cdot12\cdot20\cdots\\-10\cdot10\cdot13\cdot20\cdots\end{array}}{10\cdot10\cdot10\cdot10}
+\frac{\begin{array}{l}+10\cdot10\cdot10\cdot10\cdot24a^4\\-10\cdot10\cdot10\cdot11\cdot23\cdots\\+10\cdot10\cdot11\cdot11\cdot22\cdots\\-10\cdot10\cdot10\cdot12\cdot22\cdots\\-10\cdot11\cdot11\cdot11\cdot21\cdots\\+2\cdot10\cdot10\cdot11\cdot12\cdot21\cdots\\-10\cdot10\cdot10\cdot13\cdot21\cdots\\+11\cdot11\cdot11\cdot11\cdot20\cdots\\-3\cdot10\cdot11\cdot11\cdot12\cdot20\cdots\\+2\cdot10\cdot10\cdot11\cdot13\cdot20\cdots\\+10\cdot10\cdot12\cdot12\cdot20\cdots\end{array}}{10\cdot10\cdot10\cdot10\cdot10}
$$

[31] A III, 2, 425.
[32] LDK, p. 325.

Leibniz enumerates four such combinatorial laws.[33]

a) The formal law of homogeneity
The members of the coefficient of a term like a^2x^2, for example, $10 \cdot 10 \cdot 22$, $10 \cdot 11 \cdot 23$, etc. are always the product of three quantities. They are divided by 10^3 so that the result is always a quantity of four dimensions.

b) The virtual law of homogeneity
The sum of the right figures of the factors of such a coefficient must always be the same in every number of the coefficient: $0 + 0 + 2 = 0 + 1 + 1 = 2$.

c) The veritable number (that is prefixed to every number)
This number is the number of permutations with repetitions of the quantities 11, 12, 13, etc. For example, one gets: $3 \cdot 10 \cdot 11 \cdot 11 \cdot 12$. (The factors 10, 20, 21, 22, etc. are left aside in this respect.)

d) The sign of such a combination

If the number of the dimension of a combination of quantities such as 11, 12, 13, etc. is even, then the sign is +, and otherwise −.

3.7.2 The determinants

In 1684, Leibniz developed his determinant theory. Theory means that he demonstrates/ deduces a set of general theorems, as he says in the title of his crucial treatise:

> In hac scheda theoremata generalia exhibentur pro aequationibus simplicibus. (In this sketch general theorems for linear equations are presented.)

And he remarks:

> Porro modus tollendi literas incognitas generalibus quibusdam Theorematibus ex arte combinatoria enuntiatis comprehendi potest. Itaque illis accurate constitutis, facilius ac distincta ratione expressis, maximam calculi algebraici partem abscindemus, ita ut desiderati valores sine calculo simplici quadam theorematum generalium ad exempla specialia applicatione statim scribi possint. (Furthermore, the method of eliminating unknown letters can be comprehended by some general theorems enunciated according to the combinatorial art. This is why, when those have been accurately established and expressed more easily and in a distinct manner, we will eliminate the greatest part of algebraic calculation, so that the values looked for can be at once written down without any calculation by a certain simple application of general theorems to special examples.)[34]

At the beginning of the twentieth century, determinant theory was indeed still a part of combinatorics. His studies of symmetric functions, of the polynomial theorem, of power sums, and related topics that all were related to the algorithmic solution of algebraic

---

[33] LDK, p. 325f.
[34] Knobloch (1972: 168).

equations led Leibniz to the conviction that the combinatorial art included algebra as a part of it.

As we have already pointed out, Leibniz invents a notation for a determinant by taking the product of the terms of the main diagonal:

$$\begin{vmatrix} 10 \ 11 \\ 20 \ 21 \end{vmatrix} = \overline{0 \cdot 1}$$

or more generally

$$1 \cdot 2 \cdot 3 \cdots n = \begin{vmatrix} a_{11} & \cdot & \cdot & a_{1n} \\ & \cdot & & \\ & & \cdot & \\ a_{n1} & \cdot & \cdot & a_{nn} \end{vmatrix}$$

He has full knowledge of the structure of such a determinant:

a. It consists of $n!$ terms that are generated by the permutations of the right indices.
b. The signs obey a combinatorial law.

3.8 Epilogue: the order of the disciplines in view of their generality

From the previous analysis, Leibniz deduces the following consequence. The combinatorial art comprehends algebra, or algebra is subordinated to the combinatorial art. Algebra needs the combinatorial art in order to be perfected. The more general discipline comprehends the less general discipline.

In May 1678, when he wrote to Tschirnhaus emphasizing the importance of symmetric functions, he made clear how disciplines were in his view organized:

Quae alios maximos habet usus, continet enim Algebrae totius arcana; Combinatoriae vero applicationem egregiam. Nam ego Combinatoriae subordinatam puto Algebram quia combinatoriam non habeo pro arte inquirendi numeros possibiles variationum; sed pro arte formarum seu pro scientia generali de Simili et Dissimili. Cujus regulas Algebra ad magnitudinem in universum, Geometria ad figuras applicat. (It [note of the translator: the table of symmetric functions] has other most important applications because it contains the secrets of the whole algebra, in fact an excellent application of the combinatorial art. For I believe that algebra is subordinated to the combinatorial art because I do not consider the combinatorial art as the art of asking for the possible numbers of variations but as the art of forms, or as the general science of the similar and the dissimilar. Algebra applies its rules to magnitude in general, geometry to figures.)[35]

[35] A III, 2, 425.

Hence he wrote to Thevenot on 24 April 1689 on his algebraic studies on the solution of equations of the fifth degree:

> Pour aller au cinquième il m'a fallu trouver la méthode générale pour toutes les autres je l'avois il y a longtemps, lors même que j'ay été en France ... et à fin de l'abréger je meditois de dresser certaines tables d'une progression réglée.... Après cela on pourra dire que ce qui est proprement l'Algèbre, vu de la résolution des équations par leurs racines, est achevé. (In order to go to the fifth degree I had to find the general method for all the others that I disposed of a long time ago when I was in France ... and in order to abridge it I thought of elaborating certain tables of an ordered progression ....After that one will be able to say that all that properly regards algebra as far as the solution of equations by means of their roots is concerned has been finished.)[36]

As we know today, Leibniz was too optimistic.

....

REFERENCES

The following abbreviations are used in footnotes:

A = Leibniz, G.W. (since 1923) *Sämtliche Schriften und Briefe*, ed. by the Prussian, now by the Berlin-Brandenburg and Göttingen Academy of Sciences. Darmstadt: Otto Reichl Verlag-Berlin: de Gruyter. (The series, the volume, and the page are cited in this order).

C = Leibniz, G.W. (1903) *Opuscules et fragments inédits, Extraits des manuscrits ...* par Louis Couturat. Paris: Alcan (Reprint Hildesheim: Olms 1961).

GM = Gerhardt, C.I. (ed.) (1849–1863) *Leibnizens mathematische Schriften*, 7 vols. Berlin: Halle (Reprint Hildesheim: Olms 1962).

LDK = Knobloch, E. (ed.) (1980) *Der Beginn der Determinantentheorie, Leibnizens nachgelassene Studien zum Determinantenkalkül*, Textband. Hildesheim: Gerstenberg.

LKK = Knobloch, E. (ed.) (1976) *Die mathematischen Studien von G.W. Leibniz zur Kombinatorik*, Textband. Wiesbaden: Steiner.

Holz, H.H. (1983) *Gottfried Wilhelm Leibniz, Eine Monographie*. Leipzig: Rororo.

Knobloch, E. (1972) 'Die entscheidende Abhandlung von Leibniz zur Theorie linearer Gleichungssysteme', *Studia Leibnitiana* 4: 163–180.

Knobloch, E. (2001) 'Déterminants et élimination chez Leibniz', *Revue d'histoire des Sciences* 54: 143–164.

Knobloch, E. (2004) 'Leibniz and the use of manuscripts: Text as a process', in *History of science, history of text*, ed. K. Chemla. Dordrecht: Springer: 51–79.

Knobloch, E. (2006) 'La généralité dans les mathématiques leibniziennes', in *VIII. Internationaler Leibniz-Kongress. Einheit in der Vielheit. Vorträge* 1. Teil, ed. H. Breger, J. Herbst, S. Erdner. Hannover: 382–389 (abridged French version of this article).

Leibniz, G.W. (1989) *La naissance du calcul différentiel, 26 articles des Acta Eruditorum*, Introduction, traduction et notes par Marc Parmentier. Paris: Vrin.

[36] A I, 5, 681.

Leibniz, G.W. (2004) *Quadrature arithmétique du cercle, de l'ellipse et de l'hyperbole et la trigonométrie sans tables trigonométriques qui en est le corollaire,* Introduction, traduction et notes de Marc Parmentier, Texte latin édité par Eberhard Knobloch. Paris: Vrin.

Pasini, E. (2001) 'La philosophie des mathématiques chez Leibniz. Lignes d'investigations', in *Akten des VII Internationalen Leibniz-Kongresses,* Berlin, 10–14 September 2001, Vol. II, Hannover: 954–63.

Section I.2

Actors' reflections on generality in science

4

The problem of a "general" theory in mathematics: Aristotle and Euclid

DAVID RABOUIN

4.1. Introduction

Science is the knowledge of the general and there is no science in the perception of the particular (αἰσθάνεσθαι μὲν γὰρ ἀνάγκη καθ᾽ ἕκαστον, ἡ δ᾽ ἐπιστήμη τὸ τὸ καθόλου γνωρίζειν ἐστίν).[1] This famous Aristotelian motto is often considered as an essential ingredient in any definition of what "science" is: there is no science without generality. But what does generality mean? A simple (but not exclusive) criterion, proposed by Aristotle, is that any truly scientific demonstration should hold "for all" (κατὰ παντὸς) the entities it concerns, exhibiting an attribute belonging "in itself" (καθ᾽ αὑτὸ) or "essentially" to these entities (*Analytica Posteriora* I, 4). To demonstrate that the sum of the interior angles of an isosceles triangle is equal to two right angles is correct. However, the demonstration would not satisfy the requisite for a truly scientific knowledge, since the property holds "for all" triangles and belongs to the triangle "in itself" (*An. Post.* I, 4, 73 b 38–9). As is obvious from this kind of example, these criteria suppose a clear delineation of the "domains of objects" under study—what Aristotle often calls a "genus" delineated in some "species" and which he considers to be the foundation of any scientific knowledge (see Section 4.3.1).

At first glance, this general condition fits very well into the general framework of Euclidean mathematics. Indeed, in the first nine books of the *Elements*, we deal only with two distinct domains of objects: geometrical entities, sometimes called "magnitudes," and "numbers." The delineation is so neat that Euclid never transfers a demonstration from one of these fields to the other. But at the beginning of Book X, he suddenly permits a

---

[1] *An. Post.* 87 b 37. Unless otherwise stated, all translations are mine.

The Oxford Handbook of Generality in Mathematics and the Sciences. First Edition. Karine Chemla, Renaud Chorlay and David Rabouin. © Oxford University Press 2016. Publishing in 2016 by Oxford University Press.

comparison between magnitudes and numbers—more precisely, he sets a certain proportion in which a ratio between magnitudes and a ratio between numbers are compared. How can this be, since ratios between magnitudes and ratios between numbers have been *defined* separately (and differently) in the preceding Books? Is there a "more general" point of view from which magnitudes and numbers might be treated without distinction? And if so, did Euclid violate the Aristotelian prescription by treating general features (typically the operative core of "proportions") in specific cases in Book V (for magnitudes) and in Book VII (for numbers), where a more general treatment was possible? This will be the specific figure of what I will call the "problem of generality" in Euclid's *Elements*.

In the first part of this chapter, I will scrutinize the general context of this problem and evoke some of the modern strategies proposed to solve it. These strategies, as I will explain, rely on historical reconstructions and often make use of evidence taken from Aristotle. But they usually have a purely utilitarian relationship to these sources. On the contrary, I will propose, in the second part of the chapter, to study these Aristotelian passages in their proper context. Indeed the problem of generality makes several appearances in Aristotle's texts and these occurrences open up very interesting perspectives on the problem.

4.2 Euclid

4.2.1   The domains of objects in Euclidean mathematics

At first glance, the domains of objects in Euclid's *Elements* are quite clearly delineated. Books I–IV deal with lines, triangles, rectangles, circles, polygons, etc., that is, what we would like to call "geometrical" objects. Books VII–IX deal with "numbers." Book V occupies a particular situation in this setting, since it proposes a unified theory of ratios existing between any kind of geometrical objects (through the basic relation "having the same ratio," usually rendered as "proportion")—a theory which is then used to demonstrate geometrical results (Book VI). On this occasion, a new term is introduced to designate geometrical objects (susceptible to enter in a ratio): "magnitude." This term is not defined, but conditions are specified for something to be of such a "magnitude."[2]

Contrary to "magnitude," "number" is defined: it is a multiplicity composed of units (τὸ ἐκ μονάδων συγκείμενον πλῆθος VII.Def2). Hence number is not simply a mere multiplicity (πλῆθος), but rather a multiplicity composed *of units* (μονάδων). This distinction is of tremendous importance when it comes to comparing magnitudes. For example, when Euclid says that a magnitude *AB* is "twice" another magnitude *DC*, he does not use a "number," but a "multiple." The language of "multiplicity," contrary to that of "number,"

---

[2] See V.Def4: "Magnitudes are said to have a ratio to one another which are capable, when multiplied, of exceeding one another" (The English translation of Euclid that I use is Heath, 1908). There is a first problem of generality in the introduction of this concept, since it is not clear that it can really serve as a general designation for all the geometrical objects of Books I–IV. One classic counter-example is given by the notion of "angle" (horn angle, problem of division of an angle in an arbitrary part, etc.). I won't engage with this discussion in this paper.

is treated as primitive in Books V and VII.[3] Another important aspect of the Euclidean definition of number is that a ratio between numbers is not a new kind of number and cannot be treated as such.[4]

There is absolutely no communication between these two blocks. Not only does Euclid not use "numbers" when dealing with "magnitudes," he never transfers a demonstration from one field to the other. This is true to the point that he sometimes gives two different demonstrations for what appear to us as similar, if not identical, propositions. We will see one famous example of this later on (the fact that proportions can be "alternated"). The only "general" point of view (i.e., holding for magnitudes and numbers) is given at the very beginning of the treatise by what are called "common notions" (κοινὰ ἐννοία). But if we except these common notions, which are deliberately set up before we enter our blocks, there seem to be (at least) two completely separate fields of objects.

One could propose a simple explanation for this fact by saying that in Euclid the separation between what we call geometry and arithmetic, ultimately grounded on distinct "domains of objects," "magnitude," and "number," could not be overcome. But this interpretation is challenged at the beginning of the third block (Book X). In this book a comparison between numbers and magnitudes is permitted. Moreover, it plays a crucial role in the characterization of a very important notion, that of "commensurability":

X.5: Commensurable magnitudes have to one another the ratio which a number has to a number.

X.6: If two magnitudes have to one another the ratio which a number has to a number, the magnitudes will be commensurable.

These propositions make it clear that a "general" point of view, allowing for the comparison between the different domains of objects, is possible. We could even consider this point of view as "universal" since it encompasses *any kind* of mathematical object appearing in the treatise (we will see later that there is evidence in Aristotle of this possibility). One immediate difficulty, however, is that the relation "to have the same ratio" has been *defined* separately for magnitudes and numbers. Let's recall these definitions to make the problem clear:

V.Def5: Magnitudes are said to be in the same ratio, the first to the second and the third to the fourth, when, if any equimultiples whatever be taken of the first and third, and any equimultiples whatever of the second and fourth, the former equimultiples alike exceed, are alike equal to, or alike fall short of, the latter equimultiples respectively taken in corresponding order.

VII.Def20: Numbers are proportional when the first is the same multiple, or the same part or the same parts, of the second that the third is of the fourth.

---

[3] On the way in which one can use "multiple" without using "numbers," see Vitrac (1990–2001, Vol. 2: 15–17). The basic term of comparison will be the relation of "equimultiple," that is to say the relation "being the same multiple as." Note that Heath's translation of Euclid in English does not respect this vocabulary and regularly introduces numbers to express "multiples."

[4] As most recent commentators have emphasized, this goes against the modern tendency to consider relations *as* objects. See Mueller (1981), Caveing (1998), and Vitrac (1990–2001, Vol. 2: 127–34, "Sur les présupposés du livre V").

The definition in Book V is a very specific and complex treatment of "proportion" relying on the stability of comparison between magnitudes through the "equimultiple" relation. The definition of proportion for numbers is much more straightforward and is directly derived from the basic relationship of measure. Considering these definitions, the possibility of a common point of view seems at first inconsistent.[5] There is no way to *compare* a ratio between magnitudes and a ratio between numbers, since to "have the same ratio" would not have the same meaning on both sides of the comparison. But still *there is* such a comparison in Book X. As will become clearer in what follows, this problem was of tremendous importance in the history of the modern commentaries to the treatise.[6]

4.2.2   Proportion and domains of objects

The independence of numbers and magnitudes is not solely grounded on the fact that Euclid treats them separately. He also seems to sometimes give separate demonstrations for the same property. This is precisely the case for the properties of proportions. One famous example, to which I will return later, is the case of permutability or, as the Greeks called it, "alternation" (ἐναλλάξ):[7]

> V.16: If four magnitudes be proportional, they will also be proportional alternately.
>
> VII.13: If four numbers be proportional, they will also be proportional alternately.

The propositions are here *strictly identical* in their formulation. This is a typical case where we moderns have the unfortunate tendency to ask ourselves: why doesn't Euclid set this property as being a property of proportion *in general*? An immediate (and accurate) answer would be that he had no notion of "proportion in general." But the real question is *why*? Is it because he does not "see" this level of generality, or because this level is not possible in his system?

To answer these questions, we have to take a closer look at the demonstrations. Demonstration of V.16 is a direct application of the tool introduced in V.Def5. But instead of taking equimultiples of the first and the third (resp. the second and the fourth) magnitudes, Euclid takes equimultiples of the first and the second (resp. the third and the fourth). These equimultiples remain of course in the same proportion as the given magnitudes. This means that "if the first be greater than the third, the second will also be greater than the fourth; if equal, equal; and if less, less." But this is precisely the condition imposed on equimultiples (in V.Def5) for the ratio of the first to the third to be the same than the ratio of the second to the fourth. Hence, the conclusion.[8]

---

[5] For the literature on this problem, see Vitrac (1990–2001, Vol. 3: 106, n. 57 and 58).

[6] In this paper, I will focus on the modern commentaries, but the problem of the expression of a "general theory" in Euclid's *Elements* comes from a very ancient tradition. On that point, I refer the reader to Rabouin (2005) and Rabouin (2009).

[7] In symbolic notations, permutability is expressed by the equivalence: $(A : B :: C : D) \Leftrightarrow (A : C :: B : D)$, where ":" denotes a *ratio* between magnitudes and "::" the relation "having the same ratio."

[8] Here is the demonstration in Heath's translation: "Let A, B, C, D be four proportional magnitudes, so that, as A is to B, so is C to D; I say that they will also be so alternately, that is, as A is to C, so is B to D."

The demonstration of VII.13 is completely different. It is, in fact, almost immediate since Euclid established in proposition VII.10 that the relation "being the same part or parts" can be alternated. He can then directly apply VII.Def20 to conclude: "since, as A is to B, so is C to D, therefore, whatever part or parts A is of B, the same part or the same parts is C of D also [VII.Def20]. Therefore, alternately, whatever part or parts A is of C, the same part or the same parts is B of D also [VII.10]. Therefore, as A is to C, so is B to D. Q. E. D."

Even if the propositions seem identical in their form, the demonstrations are very different. The fact that number and magnitude could enter in a same proportion like in X.5 and X.6 is then more mysterious than ever.

A natural hypothesis to clear up this mystery is to appeal to an historical claim. It is largely plausible that Euclid was, in the *Elements*, collecting results that he did not discover all by himself. It is, therefore, possible that he put side by side (in Books V and VII) different stages of theories of proportion, a general and a particular one. The more "general" theory could then allow for a proportion holding for numbers and magnitudes like in X.5 and X.6, even if it was not explicitly stated for both types of objects.[9]

This hypothesis is quite easy to test. We need simply insert the objects of one theory into the other in order to see if it works. Let us first try inserting magnitudes instead of numbers in the theory of proportion of Book VII. This will, of course, mean to restrict ourselves to a particular case in which the notion of proportion given in VII.Def20 holds. We will thus work with magnitudes which are "the same part or the same parts" of each other, that is to say which are "commensurable" with one another. Since this is the framework of X.5 and X.6, we can accept this restriction. We can then read proposition VII.13 as saying: "If four *magnitudes* be proportional, they will also be proportional alternately" ("magnitudes" meaning here "commensurable magnitudes"). The demonstration will still be valid, as would be VII.Def20, since they only make use of the basic relationships: "being a part of" or "being some parts of," which holds between magnitudes when they are commensurable. This conclusion is sometimes considered as supporting the existence of a kind of "general" theory of proportion (say, before one saw the importance of incommensurable magnitudes or even before one actually "discovered" them: I'll return to the context of this hypothesis later).

For of A, B let equimultiples E, F be taken, and of C, D other, chance, equimultiples G, H. Then, since E is the same multiple of A that F is of B, and parts have the same ratio as the same multiples of them, [V.15] therefore, as A is to B, so is E to F. But as A is to B, so is C to D; therefore also, as C is to D, so is E to F. [V.11] Again, since G, H are equimultiples of C, D, therefore, as C is to D, so is G to H. [V.15]. But, as C is to D, so is E to F; therefore also, as E is to F, so is G to H [V.11]. But, if four magnitudes be proportional, and the first be greater than the third, the second will also be greater than the fourth; if equal, equal; and if less, less. [V.14]. Therefore, if E is in excess of G, F is also in excess of H, if equal, equal, and if less, less. Now E, F are equimultiples of A, B, and G, H other, chance, equimultiples of C, D; therefore, as A is to C, so is B to D. [V.Def5] Therefore etc. Q. E. D.

[9] Another natural hypothesis, which I will treat separately in the next section, is that there was a third, more general, theory of proportion allowing for X.5 and X.6 to be formulated.

But the next occurrence of proportion in VII.19 suffices to ruin this hypothesis:

> VII.19: If four numbers be proportional, the number produced from the first and fourth will be equal to the number produced from the second and third; and, if the number produced from the first and fourth be equal to that produced from the second and third, the four numbers will be proportional.

In this case, there is no way one can replace numbers by magnitudes, even if we restrict ourselves to magnitudes commensurable with each other. This is grounded on a very important feature of "magnitudes": there is nothing like an internal multiplicative operation acting on them.[10] In Greek classical mathematics, the product of a line by a line was seen as being an area (the Euclidean term, introduced in Def 1 of Book II, being not "product," but "rectangle contained by two lines"). Hence this construction obeys constraints of homogeneity and dimensionality. For example, if we want to compare two ratios between surfaces in a proportion, there is no way to give sense to the "product" of the middle terms as in VII.19. We see clearly in these examples that the concept of "magnitude" involves specific features ("essential property," as Aristotle would say) quite different from that of "number."

Let us now try it the other way round and insert "numbers" into the theory of proportion of Book V. Here, too, we will not have difficulties with the definition, but with the demonstrations. One famous example is given by prop. V.5: "If a magnitude be the same multiple of a magnitude that a part subtracted is of a part subtracted, the remainder will also be the same multiple of the remainder that the whole is of the whole." Euclid calls *AB* and *CD* the given magnitudes, *AE* and *CF* the parts subtracted, and *EB* and *DF* the remainders. To achieve the demonstration, he introduces an auxiliary magnitude *GC* satisfying the following relation: *EB* is the same multiple of *GC* as *AE* is of *CF*. This relies on the possibility of taking an arbitrary part of any given magnitude—a postulate for magnitude which is implicit in Euclid's work and which is of course not true of "number."[11]

We could pursue this inquiry more carefully, but these two counter examples suffice to cast serious doubts on the simplistic view according to which "magnitudes" and "numbers" are in a relation of inclusion (in one way or the other). Our first idea of a separation is then stronger than before. It is not an external remark on the organization of the treatise: when we enter the particularity of the demonstrations, we see that they invoke "specific features" of their objects.

4.2.3   The reconstructive hypothesis

There is another way of approaching the problem. Instead of trying a direct comparison between Books V and VII, one can start directly from Book X and consider it as evidence of a *third* (hidden) theory of proportion. This strategy was, and still is, a very common

---

[10] There is, as we saw, an *external* operation given by the "multiple" relation (and the equivalence: "being the same multiple as").

[11] See Mueller (1981: 122).

point of view shared by authors engaged in a "reconstructive approach." Its origin can be traced back to the beginning of the last century when Zeuthen first published a reconstruction in Danish in 1917, which was offered again independently by Oskar Becker and published in his *Eudoxos Studien* of 1933.[12] The reconstruction is quite complex since it is grounded on internal and external evidence and can vary from one author to another, but we will focus first on their common basis.

As we saw earlier, one important step in Book X is to set a proportion between a certain class of (ratios between) magnitudes and (ratios between) numbers [X.5 and X.6]. Magnitudes in this kind of ratio will be called "commensurable." When a ratio between magnitudes is not equivalent to a ratio between numbers, the magnitudes will be called "incommensurable" [X.7 and X.8].[13] To enter this construction, we obviously need first to be able to recognize when some magnitudes are "commensurable" or "incommensurable." This is precisely the purpose of the beginning of Book X. A nominal definition is given in Def. 1: "Those magnitudes are said to be commensurable which are measured by the same measure, and those incommensurable which cannot have any common measure." But Euclid then proposes a *criterion* to distinguish the two cases: "If, when the less of two unequal magnitudes is continually subtracted in turn (ἀνθυφαιρουμένου) from the greater, that which is left never measures the one before it, the magnitudes will be incommensurable" (X.2). In the case of commensurable magnitudes, the same procedure would reach an end and could serve as a way of determining the greatest common measure of two or three magnitudes (Prop. X.3 and X.4). These propositions come just before the one considered before (5 and 6 about commensurability, 7 and 8 establishing conversely that incommensurable magnitudes do not have the ratio of a number to a number).

A striking fact is that the procedure used in these propositions is *exactly the same* as the one appearing for numbers at the beginning of Book VII and known today as the "Euclidean algorithm." In Greek, the verb used to designate the reciprocal subtraction is *anthyphairen* and this procedure is often called *anthyphairesis* by historians (a word not used by Euclid). It appears in the very first proposition of Book VII: "Two unequal numbers being set out, and the less being continually subtracted in turn (ἀνθυφαιρουμέ νου) from the greater, if the number which is left never measures the one before it until an unit is left, the original numbers will be prime to one another."[14] Not only are the

---

[12] Becker (1933) claims not to have been aware of Zeuthen's interpretation. The Zeuthen–Becker hypothesis has been taken over and modified by a great number of commentators such as Van der Waerden (1954) and Knorr (1975). Fowler (1999) has based a complete reconstruction of what he calls "The Mathematics of Plato's Academy" on it. Indeed, as Mendell (2007) puts it, "all modern reconstructions start with Becker" (p. 5, n.3). A detailed criticism of this hypothesis can be found in Vitrac (2002).

[13] It is very important to always keep in mind that "commensurability" is, according to Euclid, a relation and not a monadic property. It is also important not to confuse "incommensurable" (ἀσύμμετρα) and "irrational" (ἄλογοι) magnitudes.

[14] In modern notations, for any two numbers *A*, *B* (smaller than *A*), we subtract *B* from *A* a certain number of times (a certain "multiple") *Q*, until a remainder smaller than *B* is found. Hence $A = Q.B + R$. If *R* is not equal to nothing (*B* measures *A*), we repeat the same procedure for *B* and *R* so that $B = Q'R + R'$, with *R'* smaller than *R*. And so on and so forth until a null remainder is obtained. The proposition states that if the penultimate step gives a remainder equal to 1, the numbers are prime to one another.

formulations of the propositions similar, as for alternation of proportion, but in this case *the demonstrations follow exactly the same structure.*[15] This similarity led naturally to the hypothesis, that there could have been a stage at which mathematicians had a general procedure at hand allowing them to treat numbers and magnitudes *in a unified way.*

This hypothesis could certainly serve as a solution to our first difficulty, that is: how Euclid can *compare* magnitudes and numbers in Book X. But *anthyphairesis* does not intervene in the context of a definition or a comparison of *ratios.* It serves the formulation of a criterion allowing to distinguish incommensurable and commensurable magnitudes (depending on whether the algorithm reaches an end or not). So the reconstructive approaches need further evidence to support their view. These sources can vary from one author to another. With regards to external evidence, however, they all appeal to a very famous passage from Aristotle's *Topica*, which relates *anthyphairesis* and proportion:

> Many theses are not easy to argue about or tackle because the definition has not been correctly rendered .... It appears also in mathematics that the difficulty in constructing a figure is sometimes due to a defect in definition; e.g. in proving that the line which cuts the plane (τὸ ἐπίπεδον) parallel to one side, divides similarly both the line which it cuts and the area. If the definition be given, the fact asserted becomes immediately clear: for the areas have the same fraction subtracted from them (ἀνταναίρεσιν) as have the sides and this is the definition of 'the same ratio' (ἔστι δ' ὁρισμὸς τοῦ αὐτοῦ λόγου οὗτος)."
> (*Topica* VIII 158 b 24-35; transl. Aristotle 1984: 125)[16]

This piece of evidence, although very striking, is not as strong as it might seem. In fact, it gives rise to some difficulties, which I shall list briefly:

- First, the text does not present the definition as referring to ratios *in general.* It just presents an alternative definition, which could be useful in a particular situation. Bernard Vitrac proposes an interesting parallel.[17] In the *De Anima*, Aristotle explains that a good definition is supposed to express not only the fact, but the cause of the fact (*De An.* II, 2, 413 a 13–20). As an example, he proposes the definition of the squaring (τετραγωνισμός) of a plane figure. "Squaring" could be defined as a procedure in which a square is found to be equal to a rectangle or another figure, but this, according to Aristotle, is a conclusion. To present it as the finding of a pro-portional mean would better express the cause. Vitrac notices that the mathematical definition was certainly not the second: one can demonstrate that a squaring is equivalent to the finding of a proportional mean, but to define it this way would be very impractical in the context of Greek geometry.[18] Considering this parallel, one

---

[15] See Appendix 1 and Vitrac (1990–2001, Vol. 3: 103–4) for a general description of the parallelism between the beginning of Book VII and that of Book X.

[16] The fact that Aristotle is not using the term "anthyphairesis" (but "antanairesis") is a difficulty here, but one which can be solved by recourse to a passage in Alexander of Aphrodisias (1891/1999) where he explains that what Aristotle calls *antanairesis* is what the Ancients used to call *anthyphairesis* (*In Ar. Top. Com.* 545, 12–17).

[17] Vitrac (1990–2001, Vol. 2: 518–19).

[18] See also Mendell (2007: 9).

should be cautious before concluding that Aristotle is referring here to something like "the" mathematical definition of a proportion.[19]

- More importantly, it is very difficult to accept, especially if our only source is Aristotle, that the definition of proportion "in general" was based on the anthyphairetic procedure. The obvious reason is that it would lead, in the case of incommensurable magnitudes, to a comparison of *infinite sequences* (since a characterization of two incommensurable magnitudes, according to X.2, is precisely that their *anthyphairesis* does not reach an end). The most extreme "reconstructors" have no problem with this consequence.[20] But many of them, since they want their solution to fit with Aristotle's refusal of any use of an actual infinite in mathematics, are embarrassed by this objection. They thus try to find a way to escape the use of infinite sequences.[21]

- Notwithstanding the risk of anachronism, a last difficulty here is that the commentators referring to an "anthyphairetic" solution have to express the ratio between magnitudes through a sequence of *numbers* (that is: the sequence of "partial quotients" in what would later be called a "continued fraction"). Indeed, in this interpretation, a ratio is *identified* with a certain sequence of numbers. But this is absolutely not the way Euclid proceeds. For reasons already explained, Euclid cannot take the "quotient" between two magnitudes to be a number (although it could be a multiple, like the "double," the "triple," and so on).[22] He then has to tediously introduce a *new magnitude* for each step of the procedure of the "reciprocal subtraction" (see Appendix 1 for an example). Hence the problem is not only to find evidence for the use of *anthyphairesis* to define ratios, but for the particular way in which modern commentators render them as sequences *of numbers*.

4.2.4 General mathematics?

Suppose we pass over these difficulties, we will still have another problem to solve: why wouldn't Euclid have made any reference to this alternative theory of ratios in the *Elements*? Why, more generally, do we not have any direct traces of it (or of the criticisms it could have received) in Greek mathematics? Here another crucial passage of Aristotle comes into play, where the link with the problem of generality is made explicit:

Alternation (ἐναλλάξ) used to be demonstrated separately of numbers, lines, solids, and times, though it could have been proved of them all by a single demonstration. Because

---

[19] One needs other arguments, allowing for a restricted characterization to be considered as a definition. See Mendell (2007).

[20] See Gardies (1988).

[21] This means that they will try to find a way of treating what would now be called an "infinite quotient" by finite means. Usually, they will reason on the rank of the "partial quotient," using results discovered many centuries later by Euler and Lagrange. See Fowler (1999), especially chapter 9 about the history of "continued fractions."

[22] Since a magnitude taken n-times is not considered as multiplied by a "number," it is clear that the division of two magnitudes is not a "number" either.

there was no single name to denote that in which numbers, lengths, times, and solids are identical, and because they differed specifically from one another, this property was proved of each of them separately. Now, however, the proof is universal (καθόλου), for they do not possess this attribute as lines or as numbers, but as manifesting this character which they are postulated as possessing universally. (*An. Post.* I, 5, 74 a 17–25)

This text was Becker's starting point. One of his greatest achievements was to succeed in showing that if one tries to demonstrate the properties of proportion by anthyphairetic means, following the order of *Elements* Book V, the first difficulties will occur precisely with proposition V.16 (i.e., alternation of proportion) where one would have to give a "case by case" demonstration. He could then conclude that the theory of Book V had two tremendous advantages over the anthyphairetic one: it gave a way of treating magnitude "in general," without entering the distinction "commensurable/incommensurable (but with a price to pay: a very complex definition of proportion by "equimultiples," generally attributed to Eudoxus); and it permitted a *unique* demonstration for properties like alternation, in accordance with Aristotle's declaration in *An. Post.* I, 5. This second advantage would explain, in Becker's eyes, the eclipse of the "anthyphairetic" theory of proportion.

This type of demonstration being called "universal" by Aristotle gives us a natural paradigm for the discipline he refers to allusively elsewhere as being a "universal" or "common" mathematics:

> For one might raise the question whether first philosophy is universal, or deals with one genus or nature; for not even the mathematical sciences are all alike in this respect, geometry and astronomy deal with a certain particular nature, while universal mathematics applies alike to all (ἡ δὲ καθόλου πασῶν κοινή). (*Ar. Met.* E1, 1026 a 20 sq.; transl. D. Ross modified)[23]

To this testimony, one could add that of Proclus (1992) who states in his commentary on Euclid that Eudoxus was the first to "increase the number of the universal theorems, as they are called (τῶν καθόλου καλουμένων θεωρημάτων)" (*In Eucl.* 67.4). A scholium to Book V also presents its aim as that of giving a theory common to all of mathematics, and in particular to geometry and arithmetic.[24]

This interpretation supported a very nice (but potentially mythical) historical reconstruction: first, there would have been a theory of proportion grounded on numbers and relying on the fact that magnitudes could be treated "like numbers" (magnitudes being then restricted to "commensurable magnitudes"), a theory sometimes attributed to the "Pythagoreans;" then the "anthyphairetic" approach allowing a first general

---

[23] The link between *Metaphysics* E 1 and *An. Post.* I, 5 is explicitly made by authors like Ross in his translation of Aristotle's *Metaphysics* (Oxford, 1924). See also Heath (1949), who relates this passage, without further explanation, to the "eudoxean" theory of proportions. See also Cleary (1995: 289–92).

[24] *Scholia in Euclidis elementa* 5.1.1–4: Σκοπός τῶ πέμπτῳ βιβλίῳ περὶ ἀναλογιῶν διλαβῖν κοινὸν γὰρ τοῦτο τὸ βιβλίον γεωμετρίας τε καὶ ἀριθμητικῆς καὶ μουσικῆς καὶ πάσικης ἁπλῶς τῆς μαηματικῆς ἐπιστήμης.

treatment including incommensurable ratios (sometimes attributed to Theaetetus), but leading to case by case demonstrations; finally the truly "universal" theory proposed in Book V (sometimes attributed to Eudoxus) and related to the possibility of a "universal mathematics." My aim is certainly not to enter into the details of the complex arguments supporting this picture. What is of interest for us here is that the main thread of this reconstructive framework is clearly given by a search for a *general* theory, presumably forming the core of classical Greek Mathematics. Hence the question of "generality" is not a local problem emerging from the reading of Euclid's *Elements*. It is the very ground on which a widespread picture of the development of ancient Greek mathematics is built and around which the majority of the debates were centered until very recently.

One immediate difficulty with this reconstruction is that even if we give full credit to Aristotle's testimony, we have to keep in mind that according to him the καθόλου demonstration was supposed to hold, for "*numbers*, lines, solids, and times." We saw previously that *this is not the case in Book V* (which applies to what Aristotle would designate as the proper subject of geometry alone: "magnitude"). Becker escaped the difficulty by saying, without much justification, that "magnitude" of Book V "inherited" the kind of unity occurring first in the comparison of anthyphairetic developments (hence its crucial role in the reconstruction). "Magnitude" was thus, according to him, a first figuration of our "real numbers" and "anthyphairetic developments" something like developments of our real numbers.[25] This part of his argumentation is particularly weak: even if one can conceive "magnitude" as an abstract entity (and not only as a way of giving a collective name to geometrical entities), one can hardly go to the point of considering it as a "number."[26]

Another weakness of this type of reconstruction is the use of Aristotle as a mere witness—a feature which has less called for attention than the technical difficulties attached to it. It led to a paradoxical situation in which the Stagirite was supposed to hold views completely opposed to some of his most famous philosophical theses. The obvious case is given by the argument in favor of anthyphairetic developments as holding "in general," that is to say for incommensurable magnitudes too.[27] But even the idea of a univocal general theory in mathematics seems *directly opposed* to Aristotle's conceptions. That is why I would like to come back now to Aristotle theses and challenge the idea of what a "general" or "universal" point of view in mathematics may have been according to him. As I will try to show, this will give us a very nice context to understand many of the difficulties raised by the reading of the Euclidean text.

[25] This is made explicit in Becker (1954: chap. II.C).

[26] One could try to stay closer to the structure of the *Elements*, as proposed by Gardies (1988). Gardies refuses the idea that we can find a "universal mathematics" in the *Elements*, for the reasons already mentioned. His idea is that the anthyphairetic theory *was* the universal theory (which implies showing that alternation can be demonstrated without case by case demonstration—a task that Gardies fails to fulfill entirely) and that it was due to Eudoxus (misunderstood by Euclid).

[27] The fact that we can reduce the problem to a consideration on the finite rank of the development is not an issue here, since it would still have been necessary to define comparison of ratios in general through these developments, be they finite *or not*.

4.3 Aristotle

4.3.1 Aristotle's conception of scientific knowledge

Aristotle's epistemology is in large part directed against what he saw as an illusory position of philosophy as a "science of everything" (ἐπιστήμη ἡ ἐκείνων κυρία πάντων, see *An. Post.* I, 9, 76 a 17–20)—a program which is not difficult to recognize as being close to Plato's "Dialectic."[28] His basic argument is that there is nothing like a universal and univocal sense of "Being." According to Aristotle, a science is always based on a specific "subject matter" or "genus."[29] "Being" is not such a "genus," but a way to refer to the diversity of domains of objects. The "science of being" is then a study of the different ways of expressing "what is."

The important thing to retain is that principles *proper to the "subject" considered* will then occur in any truly "scientific demonstration." According to Aristotle, the paradigm of this situation is precisely to be found in mathematics, where arithmetic deals with numbers and geometry with magnitudes:

> Of the things they use in the demonstrative sciences some are proper (ἴδια) to each science and others common (κοινά)—but common by analogy, since things are useful in so far as they bear on the genus under the science. Proper: e.g. that a line is such and such, and straight so and so; common: e.g., that if equals are taken from equals, the remainders are equal. But each of these is sufficient in so far as it bears on the genus; for it will produce the same result even if it is not assumed as holding of everything but only for the case of magnitudes—or, for the arithmetician, for numbers.
>
> Proper too are the things which are assumed to be, about which the science considers what belongs to them in themselves—as, e.g., arithmetic is about units, and geometry is about points and lines. For they assume these to be and to be this. As to what are attributes of these in themselves, they assume what each signifies—e.g., arithmetic assumes what odd or even or quadrangle or cube signifies, and geometry what irrational or inflection or verging signifies and they prove that they are, through the common items and from what has been demonstrated. And astronomy proceeds in the same way.
>
> For every demonstrative science has to do with three things: what it posits to be (these form the genus of what it considers the attributes that belong to it in itself); and what are called the common axioms, the primitives from which it demonstrates, and thirdly the attributes, of which it assumes what each signifies. (*An. Post.* I, 10, 76 a 37–b 3; transl. Aristotle, 1984: 14)

This argument is particularly important for us, because it leads directly to the fact that one cannot transfer a demonstration from one science to the other. This gives us a very nice context for understanding the situation we began with in Euclid's *Elements*:

---

[28] Compare with the characterization of dialectic proposed in *Resp.* VII, 533 b 3/534 b-e.

[29] The beginning of *Analytica Posteriora* is explicitly dedicated to a determination of what ἐπιστήμη is and, more precisely, ἀποδεικτικὴ ἐπιστήμη. I translate the first term by "science" and the second by "demonstrative science," even if it is obviously in a different meaning than the modern one. I translate by "subject matter" or "subject" the Greek term ὑποκείμενον and by "genus" the term γένος.

One cannot, therefore, prove anything by crossing from another genus—e.g. something geometrical by arithmetic. For there are three things in demonstrations: one, what is being demonstrated, the conclusion (this is what belongs to some genus in itself); one, the axioms (axioms are the things on which the demonstration depends); third, the underlying genus of which the demonstration makes clear the attributes and what is accidental to it in itself. Now the things on which the demonstration depends may be the same; but of things whose genus is different—as arithmetic and geometry, one cannot apply arithmetical demonstrations to the accidentals of magnitudes, unless magnitudes are numbers. (How this is possible in some cases will be said later).[30]

Arithmetical demonstrations always include the genus about which the demonstration is, and so also do the others. (*An. Post.* I, 7, 75 a 38–b 6; transl. Aristotle 1984: 12)

As is clear in the two passages, "common" principles certainly exist in science. But these common principles go along with proper principles characterizing the "subject matter" (ὑποκείμενον) of each scientific domain. Of course, we can also study these principles for themselves and this will allow for a kind of "universal" knowledge. But to look at things at this level means *giving up* a certain kind of scientific knowledge. This is the basis of Aristotle's position on what a "universal science" is (be it Metaphysics, Logic, or Dialectics):

All the sciences associate with one another in respect of the common items (I call common (κοινὰ) those which they use as demonstrating from them—not those about which they prove nor what they prove); and dialectic associates with them all, and so would any science[31] that attempted to prove universally the common items—e.g. that everything is affirmed or denied, or that equals from equals leave equals, or any things of the sort. But dialectic is not in this way concerned with any determined set of things, nor with any one genus. (*An. Post.* I, 11, 77 a 25–35; transl. Aristotle 1984: 16)

4.3.2 The problem of the *katholou*

The passages quoted above give an overall clear image of the way generality is supposed to hold in mathematics. This is in concordance with a broader view of what makes a scientific statement general and the limits within which this generality is supposed to hold. The *Posteriora Analytica* explain at the very beginning that any "scientific knowledge" should rely on properties attributed to a subject "for every case, in itself and in general" (κατὰ παντὸς καὶ καθ᾽ αὐτὸ καὶ καθόλου [73 a 26–7]). The first criterion expresses the fact that a scientific conclusion should hold "for all" the objects of the domain it refers to (and not, for example, to a particular species subsumed under a more general genus). But this is not enough since the correlation between this "subject" and the property involved could

---

[30] I will come back later to the mysterious exception "unless magnitudes are numbers." The case mentioned of admissible transference is when one genus is included in another, giving rise to what Aristotle designates as a "subordinated" science—such as music with respect to arithmetic or optics with respect to geometry (*An. Post.* I, 9).

[31] For parallel statements see *Metaphysics* Γ 3, 1005 a 19–30 about "Philosophy" and 1005 b 3–7 about "Analytics."

happen by accident. Hence the two other criteria: the property demonstrated "for every case" should also belong "essentially" or "in itself" (καθ' αὐτο) to the subject considered. Moreover, this essential relationship should occur through a kind of reciprocity between the property and the subject considered (ᾗ‘ αὐτό says Aristotle: in itself and "as such"). When the property is attributed "for all" the objects (κατὰ παντὸς), "in itself" (καθ' αὐτο) and "as such" (ᾗ' αὐτό), it can truly be called "universal" (καθόλού).

At first glance, this description nicely expresses the kind of "generality" needed in scientific knowledge. In the next chapter of *Posteriora Analytica*, however, it appears that the hierarchy between these criteria was not as clear as it first seemed. Aristotle explains in effect that there are many ways to miss universality in the demonstrations. However, these mistakes do not correspond exactly to the violation of the criteria proposed in the previous chapter: the first mistake, he says, occurs when one cannot grasp something apart from the particulars considered; the second, when there is something to grasp, but "which has no name" (ἀνώνυμον); the third, when one takes as a whole what is only a part (of a larger whole).[32]

There have been numerous debates around these passages since Antiquity. Not everyone agrees, for example, on the way in which the examples given by Aristotle can be related to the different kinds of errors he described. Fortunately for us, the link between the second kind of mistakes and the second example given (alternation of proportion) is the least controversial, since the idea of a "nameless" subject is taken up again in the example. As we have seen, the fact that proportion can be alternated, Aristotle explains, was first demonstrated separately for numbers, lines, solids, and times, because there was *no name* to designate what they have in common, although a single and "universal" demonstration was possible. Here, the difficulty lies in the fact that the property in question and its demonstration are supposed to hold for different domains of objects or "genus." As we saw earlier, the characterization of "scientific demonstration" in terms of principles "proper" to a subject does not seem to allow for a demonstration holding for different genus (except if a domain of objects is included or "subordinated" to another—which is not the case, according to Aristotle, for numbers and magnitudes).

I will not enter into the numerous debates around these passages, but take them as such to indicate first that generality functions here at the same time as a requisite for demonstration *and* as a problem. It is not enough to reach conclusions holding "essentially" "for all" the elements of a domain, since there is a more profound difficulty appearing in the very delineation of these domains. The example of the "universal" demonstration for alternated proportion expresses this clearly: it seems to directly contradict the clear cut delineation of mathematical kinds of objects which Aristotle himself has set up in the previous development of the *Posterior Analytics* (I, 4). Another interesting aspect of these series of examples is that they rely on what is presented as existing mathematical demonstrations. This allows us to overcome the simplistic view according to which Aristotle would be fighting with paradoxes created by his own epistemological prescriptions. Quite to the contrary, Aristotle seems here to encounter the fact that the prescriptive view proposed at first fails to give an

---

[32] *An. Post.* I, 5, 74 a 2–23.

accurate view of the way generality *actually works* in some scientific demonstrations. This
is where the comparison with Euclid will reveal itself particularly interesting.

4.3.3   Universal propositions

As we have seen, Aristotle mentions very clearly in *An. Post.* I, 5 the existence of universal
demonstrations in mathematics.[33] The demonstration of alternation, he says, holds for
numbers, lengths, times, and solids. How can this be possible? How can a demonstration
hold for different kinds of subjects if principles proper to the subject or genus are, as he
explicitly stated, *always* required in any scientific demonstration? Things get even worse
when one reads in the *Metaphysics* (E1) that one can oppose a kind of mathematics dealing
with a particular genus and a "universal" one, "common to all!" What could this mean,
considering that a *mathêma* like Arithmetics was said to be *always* attached to a specific
subject (ἡ δ᾽ ἀριθμητικὴ ἀπόδειξις ἀεὶ ἔχει τὸ γένος περὶ ὃ ἡ ἀπόδειξις, καὶ αἱ ἄλλαι ὁμοίως)?

There are many hypotheses proposed to solve this riddle. My own attempt will rely on
a fact which has not received much attention: as emphasized above, *the difficulties emerging
in the Aristotelian text are strictly parallel to the one emerging in the Euclidean text*. On the
one hand, we had separate domains of objects (corresponding to the typical example of
Aristotelian genera in *Analytica Posteriora*: magnitude and number) and demonstrations
implying specific features ("essential properties") of these objects; as we have seen, there
was absolutely no communication (no demonstration "crossing" the borders) between
these domains (except for "common notions," paralleling the Aristotelian κοινά); but, on
the other hand, there was some sort of "general" point of view, provided from within the
theory and allowing us to compare in one and the same relation the separated domains
of objects in *Elements* X.5 and X.6. This might convince us that the usual strategy, which
consists in trying to solve what looks *to us* like difficulties, could be called into question.
The parallel between the two situations could indicate that there is not so much a problem
to solve here, as a more complex and subtle conception of generality to understand. This
is the path I will try to outline in what follows.

4.3.4   Aristotle on generality in mathematics

To set this hypothesis on more solid grounds, I propose not to consider isolated passages,
as the "reconstructive approach" mentioned in the first section did, but rather the entire
context of the treatment of "generality" in mathematics by Aristotle. Indeed, there are other
important occurrences of this theme than the one considered by the "reconstructors" in the
constitution of their corpus and these occurrences allow for a reading quite different from the
one they emphasize. A first example is given by a passage from book M of the *Metaphysics*:

> The universal [propositions and demonstrations][34] of mathematics deal not with some-
> thing which exists separately, apart from magnitudes and from numbers; but they deal

---

[33] This example is mentioned again in *An. Post.* I, 24 and II, 17.

[34] The Greek states here: τὰ καθόλου ἐν τοῖς μαθήμασιν, but the second part of the passage makes it clear
that the comparison is on the level of "propositions and demonstrations" (λόγους καὶ ἀποδείξεις).

with magnitudes and numbers, not however *qua* such as to have magnitude or to be divisible. It is clearly possible that in the same way there should also be both propositions and demonstrations about sensible magnitudes, not however qua sensible but qua having such and such characteristics. (*Met.* M 3, 1077 b 17-23, transl. Ross modified)

What is Aristotle explaining here? That a universal point of view in mathematics is possible—and not only on the level of principles—but that it does not subsist *outside* of the given domains of objects, numbers and magnitude. This argument is directed against Plato's conception. For the Platonists—at least in the way Aristotle pictures them—the universals were "Ideas" subsisting outside of the particular substances.[35] In other words, the fact that there are universal propositions and demonstrations does not allow to posit in the same breath the existence of a separate, all-encompassing universal domain of objects. But it is also very interesting with respect to the expression of generality in mathematics to which Aristotle had access. As in the case of *An. Post.* I, 5, Aristotle is not saying here that we could *interpret* mathematics as being like this or that. He relies on what is presented as an existing mathematical situation, which can even be used as an objection to his opponent. The fact is, according to Aristotle, that generality is not expressed "separately" in mathematics. And this fact is apparently so obvious to everyone that it could serve to support his view on the role of "abstraction" in sciences against that of his master Plato.

Another important Aristotelian mention of the καθόλου in mathematics is linked to the debate about the "elements." According to Aristotle, some philosophers (presumably Speusippus) defended the view that the "elements" in mathematics, the "One" and the "Point," could hold as "general principles" (the Greek word is καθόλου).[36] The counter-argument, developed in the *Metaphysics* relies on the fact that this type of generality is ultimately grounded on equivocity. There is no possibility of a universal theory in mathematics derived directly from these general principles, because "unity" does not function in a univocal way in the various domains of objects:

> The one is indivisible because the first of each class of things is indivisible. But it is not in the same way that every 'one' is indivisible, e.g., a foot and a unit; the latter is absolutely indivisible, while the former must be placed among things which are undivided with respect to our perception, as has been said already—only to our perception, for doubtless everything, which is continuous, is divisible.
> The measure is always homogeneous with the thing measured; the measure of magnitudes is a magnitude, and in particular that of length is a length, that of breadth a breadth, that of sound a sound, that of weight a weight, that of units a unit. (*Met.* I, 1. 1053 a 20-7)

The situation described in these passages conforms once again to what we have found in Euclid. On the one hand, we clearly had a separate treatment of numbers and magnitudes, based on specific features of these domains of objects; on the other hand, there was a parallel given by the anthyphairetic procedure. This treatment had no self-subsisting existence

---

[35] The passage was announced in M 2, 1077 a 10–14 in which *katholou* occurs to support a *reductio* ad absurdum against the separation of "ideas."
[36] *Metaphysics* Δ 3, 1014 b 6–9.

outside of the domains of objects. Each domain obeyed specific constraints as regards the status of measure and unity of measure. There was no way of bypassing these constraints. This will be a typical case of what Aristotle would describe as a unity "by analogy."

But at this point, our main problem remains acute: if we do not accept the "anthyphairetic" solution as forming an autonomous theory, subsisting outside of the treatment of the given domains, we still need a "general" conception of ratio allowing for propositions, such as X.5 and X.6, in which numbers and magnitude enter in one and the same proportion.

4.3.5 (Dis)solving the mystery

I shall begin with a side remark: it is highly noticeable that Aristotle systematically considers "commensurable" as an attribute *essential* to number and not, as one would expect, to magnitude ("incommensurable" being symmetrically presented as an essential attribute of magnitude).[37] From this point of view, if one had a way of distinguishing magnitudes behaving in a relation of "commensurability," part of the mystery could be easily solved: these magnitudes would behave "like" numbers. This could be a way of understanding the curious exception to the "incommunicability" mentioned in *An. Post.* I, 7: "in the case of two different genera such as arithmetic and geometry you cannot apply arithmetical demonstration to the properties of magnitudes *unless the magnitudes in question are numbers*" (εἰ μὴ τὰ μεγέθη ἀριθμοί εἰσι [75 b 5]). This would not imply any contradiction: Aristotle's epistemology does not allow us to transfer a demonstration about number *qua number* to magnitude *qua magnitude*, but it does not prevent us from comparing a relation between magnitudes to a relation between numbers.[38] It seems that this is precisely what we find in Euclid.

To support this view, one might first note that the demonstrations of X.5 and X.6 rely on arguments based on the whole/parts relation and not on the treatment of equimultiples. But one should also recall that, whereas modern commentators were puzzled by X.5 and X.6, we do not possess even one Greek testimony concerning difficulties in the reading of these propositions (Vitrac, 2002)—as if it were obvious that "commensurable" magnitudes could enter in a proportion with numbers. This is no mystery to us since we already know that *anthyphairesis* allows precisely such a comparison. Moreover, we saw that a famous passage from Aristotle's *Topics* mentions this procedure as a way to define ratio.

As we said earlier, there is no reason to move too quickly from this result to the conclusion that it serves as a definition of proportion *in general*. This is where we have to be cautious. In Euclid's approach, *anthyphairesis* is not used to express ratios, but to discriminate two classes of magnitudes, one which we can treat like numbers and another which we cannot.[39] What this means is that we can establish a partial equivalence between the way we treat *ratios* between magnitudes and the way we treat *ratios* between numbers. This is exactly the result demonstrated in X.5 and X.6. It does not contradict Aristotle's

---

[37] See, for example, *Metaphysics* Δ 15, 1021 a 3–8; Γ 2, 1004 b 10–13.

[38] Compare with *Physics* 220 a 27–32, where a line is considered "qua multiplicity."

[39] Note here than when we say that we can treat magnitudes like numbers, it can't be in the "Pythagorean" sense whereby magnitudes *are* numbers (like everything else) since some magnitudes are not in this class.

theses since it allows for a comparison on the level of relationships, but not directly on the level of the domains of objects themselves.

We saw that the reconstructive approaches went much further and conjectured a "hidden" theory based on an extended use of *anthyphairesis*. Moreover, *anthyphairesis* was used in this hypothesis to justify the possibility of a treatment of incommensurable magnitudes in a general theory of proportion. As Knorr puts it: "We can conceive *of only one reason* for the Ancients' invention of the *anthyphairetic* definition of proportion: to extend the former numerical definition so that proportions of incommensurable magnitudes may be included. That is argued by Becker, who proceeds to reconstruct the material of Book V in accordance to the alternative definition. But the fact that the necessity of the alternative definition derived from the existence of incommensurable magnitudes makes inherently plausible that the context of its origin was the theory of incommensurability" (Knorr, 1975: 258, my emphasis).

As can be inferred from such a statement, a large use is often made in these discussions of Plato's famous (and isolated) testimony about a first treatment of incommensurability in the *Theaetetus* (147 d-148 e). The interpretation of this highly allusive passage is a matter of controversy amongst scholars and I do not want to go into its interpretation here. I shall limit myself once again to a very simple remark: the *Theaetetus* makes no mention of *anthyphairesis* at all. However, one can certainly assert uncontroversially that Theaetetus' strategy, as described by Plato, consisted of treating incommensurable magnitudes via the surface one can build on them. Fortunately, we do not need more to pursue our inquiry.

Indeed, this general strategy, other important differences notwithstanding, *is precisely the one used in Euclid*. As we saw earlier, the first distinction presented in Book X is between commensurable and incommensurable magnitudes. We can then consider lines commensurable "in square only," that is to say when the square built on them are measured by the same area. Now, take a given magnitude, represented by a line segment, build the square on it and consider all the rectangles commensurable to this square: this constitutes a class of comparable magnitudes which Euclid calls "rhete" ("expressible"). When a magnitude is not "rhete," it will be called "alogos."[40] Book X consists for a large part in understanding the behavior of these classes of objects.[41]

As should be obvious from this very brief description, we simply do not need the hypothesis of a general "anthyphairetic" treatment to consider that this strategy would lead to "case by case" demonstrations: the very principle of this theory is in effect to dispatch objects into classes which do not exhibit the same behavior under the basic operations. To demonstrate a general property of proportion between these "magnitudes," say "alternation," would amount to demonstrating that it holds for each class. The basic case will be given by proportion between numbers, our point of comparison, then ratios of magnitudes which we can

---

[40] It is important to keep in mind that these properties are not attributed to magnitudes per se, but to magnitudes compared to one another (in this case, considering an arbitrary given line segment taken as reference).

[41] If we take the basic operations (sum, difference, and product) on "rhete" lines, the result is "stable" when the lines are commensurable in length (i.e., it will always lead to "rhete" magnitudes), but not necessarily when they are "commensurable in square only." This gives rise to three kinds of "irrational" lines corresponding to the sum, the difference, and the proportional mean of two "rhete" lines commensurable in square only (what Euclid calls: *binomial, apotome,* and *mesos*). The aim of Book X is to give a general classification of irrational lines by reasoning about the way one can combine these various types of magnitudes. For a presentation, see Vitrac (1990–2001, Vol. 3: 51–63).

compare to a ratio of numbers ("commensurable in length"), then lines commensurable in square only[42] (the *a-logoi* being that which is out of this realm of magnitudes having "expressible" ratios and that we want to describe with them). In each of these cases, we will have to *specify* the proper way to treat the "commensurability" between the objects considered (between numbers, between magnitudes, between squares built on lines).

4.3.6   The universal theory

Now, it is clear that a theory of proportion like the one presented in Book V is more *general* in the sense that it allows for a unified treatment for commensurable *and* incommensurable ratios between magnitudes. But how can we say that the "universal" demonstrations will in this case hold for "*numbers*, lines, solids, and times"? As I explained earlier, to consider that Aristotle's description of "universal" demonstrations holds on the level of the domains of objects will contradict not only what we find in Euclid, but the very basis of Aristotle's epistemology. However, if we pursue the same strategy as for the "unity by analogy" given by *anthyphairesis*—which may have led, as we saw, to a first general theory embarrassed by "case by case" demonstrations—it works quite well. Indeed, commensurable ratios between magnitudes are the one equivalent to ratios between numbers, so in a sense the properties of proportions involving this type of ratios could be considered as holding for numbers. As we saw earlier, this kind of unity *will not have any subsistence outside of the domains of objects considered, as an autonomous theory* (in accordance with *Metaphysics* M2 and 3). It will be given from within the theory of geometric magnitudes as a universal point of view in the following sense: the properties holding for commensurable ratios can be transferred to ratios between numbers by the very definition of what commensurable magnitudes means. Contrary to what happens in the case of anthyphairetic procedure, the unity will not rely here on equivocity. This is the crucial point if we want to distinguish "unity by analogy" from "universality."

To support this last point, I would like to mention another text from Aristotle about the καθόλου, which has received less attention. At the end of *Analytica Posteriora*, the example of alternated proportion is taken up again in a passage about the question of the unity of causality in a scientific problem: "Why can we alternate proportion (ἐναλλὰξ ἀνάλογον)?," asks Aristotle, "The cause is different for lines and for numbers, but it is also the same: as lines, it is different, but as involving a determinate increase (αὔξησιν τοιανδί), it is the same" (*An. Post.* II, 17 99 a 2–10). Against interpretations defending *anthyphairetic* theory as being the "universal" one (J.-L. Gardies) or as being the basis for a first *objective* unity between numbers and magnitudes upon which Book V would rely (O. Becker and his followers), we can first note that *anthyphairesis* could hardly be called a "determinate increase" (it is rather a "decrease"). Symmetrically, we should notice that this description fits Book V.Def5 very well. What are we doing in this very intricate definition? Instead of trying to directly compare magnitudes (through a relation of "measure")—which might seem natural, but which will lead to tremendous problems when the said magnitudes are

---

[42] Surfaces are not mentioned in *An. Post.* I, 5 by Aristotle, but appear in *An. Post.* I, 24, 85 a 30–b 3. The "solid" case is not treated by Euclid, but is mentioned in *Theaetetus* (148 b 2-3).

not commensurable with one another—we treat them through a "determinate increase," using only the universal relations of multiplicity and order: we take the same (arbitrary) multiple of the first and the third; the same (arbitrary) multiple of the second and the fourth, and we ask that the order relations between magnitudes be preserved under this transformation.

Hence the theory of proportion in Book V is not only general in the sense that it holds for commensurable and incommensurable magnitudes (commensurable magnitudes being the one having the ratio of a number to a number). It is general in the sense that its basic definition gives us the *reason* or the cause of proportion between any quantities, be they numbers or magnitudes. This relation is not based on measure (like VII.Def20)—hence subject to equivocity whether we deal with commensurable or incommensurable magnitudes—but on the stability of order relations under the truly universal relationship of multiplicity. This is in accordance with the intriguing statement of *Metaphysics* M3, in which magnitudes and numbers were supposed to be considered "not however qua such as to have magnitude or to be divisible." This level does not operate on the ground of the properties (divisibility, that is, the consideration of "parts" and "measure"), which allow defining and distinguishing between the two genera of quantity (discrete and continuous). But it is not accidental either. It exhibits a universal structure of mathematical objects as stability of the comparison relation (equality/inequality) under (equi)multiplicity. This will also conform nicely to another very famous passage from Aristotle about the category of "quantity." At the end of the *Categories,* chapter VI, where he describes the different types (genus) of quantities (discrete and continuous), Aristotle offers in effect a *unified* description in terms of a *proprium* common to *all* quantities: "what is really proper to quantities is to be said equal or unequal." From the equality relation derives that of multiplicity and from the inequality that of order, so that V.Def5 uses characteristics at the same time proper and common to all quantities.

4.4 Conclusion

As a conclusion, I shall first emphasize the fragility of a certain image of rationality, sometimes seen as inherited from "The Greeks," and which supports a picture of the way "generality" is supposed to hold in science. A very quick glance at Aristotle's *Analytica Posteriora* should convince anyone that behind the clear criteria exposed in chapter 4, all kinds of difficulties are lurking. As is obvious in chapter 5, "generality," in the guise of the καθόλου criterion, is precisely one locus where these difficulties appear. Moreover, Aristotle is very clear about the fact that these difficulties emerge from some irreducible mathematical *facts*. As I have tried to show, this leads him to a very complex and subtle description of what "generality in mathematics" (καθόλου ἐν τοῖς μαθήμασιν) is. The striking point is that this description concords very well with what modern commentators have considered as important mysteries in the structure of Euclid's *Elements*.

My second point will be more reflexive. If I am right in my reconstruction of what "generality in mathematics" means for Aristotle, a large part of the "defaults" found in the *Elements* could be seen as linked to *our* expectations about the way "generality" is supposed to hold in science. This teaches us two lessons: first, we have criteria for "generality" which do not necessarily coincide with what was at stake in ancient Greek mathematics; second, the kind of generality which is to be found in these mathematics may be not less, but indeed more subtle than the one which we project on the texts today.

....

APPENDIX 1

Here are, in Heath's translation, the first part of the "parallel" demonstrations of X.3 and VII.2 (determination of a common measure):

X.3	VII.2
Let the two given commensurable magnitudes be *AB, CD*, of which *AB* is the less; thus	Let *AB, CD* be the two given numbers not prime to one another; thus
it is required to find the greatest common measure of *AB* and *CD*.	it is required to find the greatest common measure of *AB* and *CD*.
Now the magnitude *AB* either measures *CD* or it does not.	
If then it measures it—and it measures itself also—then *AB* is a common measure of *AB, CD*. And it is manifest that it is also the greatest, for a greater magnitude than the magnitude *AB* will not measure *AB*.	If now *CD* measures *AB*—and it also measures itself— then *CD* is a common measure of *CD, AB*. And it is manifest that it is also the greatest, for no greater number than *CD* will measure *CD*.
Next, let *AB* not measure *CD*. Then, if the less is continually subtracted in turn from the greater, that which is left over will sometime measure the one before it, because *AB, CD* are not incommensurable.	But, if *CD* does not measure *AB*, then, the less of the numbers *AB, CD* being continually subtracted from the greater, some number will be left which measures the one before it. For a unit will not be left, otherwise *AB, CD* will be prime to one another, which is contrary to the hypothesis. Therefore some number will be left which measures the one before it.
Let *AB*, measuring *ED*, leave *EC* less than itself, let *EC*, measuring *FB*, leave *AF* less than itself, and let *AF* measure *CE*.	Now let *CD*, measuring *BE*, leave *EA* less than itself, let *EA*, measuring *DF,* leave *FC* less than itself, and let *CF* measure *AE*.
Since, then, *AF* measures *CE*, while *CE* measures *FB*, therefore *AF* also measures *FB*. But it measures itself also, therefore AF will also measure the whole *AB*.	Since then, *CF* measures *AE*, and *AE* measures *DF,* therefore *CF* also measures *DF.* But it also measures itself, therefore it also measures the whole *CD*.
But *AB* measures *DE*, therefore *AF* will also measure *ED*. But it measures *CE* also, therefore it also measures the whole *CD*.	But *CD* measures *BE*, therefore *CF* also measures *BE*. And it also measures *EA*, therefore it will also measure the whole *BA*. But it also measures *CD*, therefore *CF* measures *AB, CD*.
Therefore *AF* is a common measure of *AB, CD*.	Therefore *CF* is a common measure of *AB* and *CD*.

REFERENCES

Alexander of Aphrodisias (1891/1999) *In Aristotelis Topicorum libri octo commentaria.* ed. M. Wallies, Berlin. Reprinted in 1999. *Commentaria Aristotelem Graeca,* vol. 2(2). Berlin: De Gruyter.

Aristotle (1984) *Complete works of Aristotle.* The Revised Oxford Translation edited by J. Barnes, 2 vols. Princeton, NJ: Princeton University Press.

Becker, O. (1933) 'Eudoxios-Studien. Eine voreudoxische Proportionenlehre und ihre Spuren bei Aristoteles und Euclid' *Quellen und Studien zur Geschichte der Mathematik, Astronomie und Physik* 2: 311–33.

Becker, O. (1954) *Grundlagen der Mathematick in geschichtlicher Entwicklung.* Freiburg/ München: Alber.

Caveing, M. (1998) *L'irrationnalité dans les mathématiques grecques jusqu'à Euclide.* Lille: Septentrion.

Cleary, J. J. (1995) *Aristotle and mathematics.* Leyde: E.J. Brill.

Euclid, *Euclidis Elementa,* ed. Heiberg. Leipzig: Teubner, 1977; French transl. (Vitrac, 1990–2001); English transl. (Heath, 1908)

Fowler, D. (1999) *The mathematics of Plato's Academy: A new reconstruction.* 2nd ed. Oxford: Clarendon Press.

Gardies, J.-L. (1988) *L'héritage épistémologique d'Eudoxe de Cnide.* Paris: Vrin.

Heath, T. L. (1908) *The thirteen books of Euclid's Elements translated from the text of Heiberg with introduction and commentary.* Cambridge: Cambridge University Press.

Heath, T. L. (1949) *Mathematics in Aristotle.* Oxford: Clarendon Press.

Knorr, W. (1975) *The evolution of the Euclidean Elements: A study of the theory of incommensurable magnitude and its significance for early Greek geometry.* Dordrecht: Reidel.

Mendell, H. (2007) 'Two traces of two-step Eudoxan proportion theory in Aristotle: A tale of definitions in Aristotle, with a moral' *Archive for History of the Exact Sciences* 61: 3–37.

Mueller, I. (1981) *Philosophy of mathematics and deductive structure in Euclid's Elements.* Cambridge, Mass.: The MIT Press.

Proclus (1992) *In Primum Euclidis Elementorum Librum Commentarii ex Recognitione Godofredi Friedlein.* Leipzig: Teubner; 2nd ed. Hildesheim: Olms-Verlag.

Rabouin, D. (2005) 'La 'mathématique universelle' entre mathématiques et philosophie d'Aristote à Proclus' *Archives de Philosophie* 68/2: 249–68.

Rabouin, D. (2009) *Mathesis universalis. L'idée de 'mathématique universelle' d'Aristote à Descartes.* Paris: P.U.F.

Van der Waerden, B. L. (1954) *Science awakening.* Groningen: P. Noordhoff.

Vitrac, B. (1990–2001) *Euclide. Les Eléments.* French translation with commentary by B. Vitrac. Paris: P.U.F (4 vols.: 1990; 1994; 1998; 2001).

Vitrac, B. (2002) 'Umar al-Hayyam et l'anthyphérèse' *Fahrang. Quarterly Journal of Humanities and Cultural Studies* 14/39–40: 137–92.

Zeuthen, H. G. (1917) 'Hvorledes Mathematiken i Tiden fra Platon til Euklid blev rationel' *DVSSkr* 1 (8. ser.): 1–184.

5

Generality, generalization, and induction in Poincaré's philosophy

IGOR LY

This chapter develops some remarks about, and suggestions for interpretation of, Poincaré's philosophical conceptions of generality in mathematics and physics. First, we will try to explain what constitutes, according to Poincaré, the specificity of the mathematical way of thinking, that is, of reasoning and constructing concepts. While doing so, we will notice that generality and generalization are necessarily implied in these issues. We will then show that generality in mathematics and physics is construed by Poincaré in a very specific way. Finally, we will illustrate these issues in greater detail by examining the way that Poincaré analyzes empirical induction in physics. To do so, we will focus on the main form that, according to Poincaré, induction takes in physics: a curve-fitting—or interpolation[1]—operation. One of the aims of this paper is to give some hints for interpreting the following claim of Poincaré, when he wonders why, "in physical science, generalization so readily takes the mathematical form."[2]

To carry out this program, four main topics will be considered. We will consider the relationship among generality, generalization, and the infinite, showing why, when one analyzes Poincaré's philosophy, these topics cannot be studied independently from each other. This point will lead us to examine the contrast between actual and potential infinity. We will also see that mathematical generality must be distinguished from what we will call "predicative generality." The latter concept can be coarsely defined as follows: according to "predicative generality," a generality judgement consists either in asserting a property common to individuals or in grouping individuals together in a class. According to Poincaré, mathematical and consequently, physical, generality[3] has a form which differs

---

[1] In accordance with Poincaré's vocabulary, we will use "interpolation" rather than "curve-fitting."

[2] Poincaré (1952: 158). The text in which this sentence occurs is quoted below. In all the quotations from Poincaré (1952) and Poincaré (1958), we have made some changes in the translation.

[3] The indispensable use of mathematics in physics is one of Poincaré's main philosophical claims. It follows from this claim that several specific features that Poincaré attributes to mathematics have direct consequences in physics. This is the case, as we shall see, for generality and generalization.

The Oxford Handbook of Generality in Mathematics and the Sciences. First Edition. Karine Chemla, Renaud Chorlay and David Rabouin. © Oxford University Press 2016. Publishing in 2016 by Oxford University Press.

from that. On this basis, the bulk of this chapter will be a comparison between Poincaré's concern regarding empirical induction and Goodman's "new riddle of induction."

In several texts, Poincaré claims that there are links among science, infinity, and generality. For instance, Poincaré writes about mathematical induction:

> In this domain of Arithmetic we may think ourselves very far from the infinitesimal analysis, but the idea of mathematical infinity is already playing a preponderating part, and without it there would be no science at all, because there would be nothing general (Poincaré, 1952: 11).

This remark reveals that Poincaré establishes two important relations: (1) a relation between science and generality; (2) a relation between infinity and generality. The first relation looks like an Aristotelian claim. However, it differs from Aristotle's idea in that Poincaré means that knowledge about particulars is not yet science, whereas Aristotle claims that there is no possible knowledge about particulars as such. A statement of the same relation can be found in Poincaré's writings about interpolation, which is the main example of physical induction chosen by Poincaré:

> I require to determine an experimental law; this law, when discovered, can be represented by a curve. I make a certain number of isolated observations, each of which may be represented by a point. When I have obtained these different points, I draw a curve between them as carefully as possible, giving my curve a regular form, avoiding sharp angles, accentuated inflections, and any sudden variation of the radius of curvature. This curve will represent to me the probable law ... Why, then, do I draw a curve without sinuosities? Because I consider, *a priori*, a law represented by a continuous function (or function the derivatives of which to a high order are small), as more probable than a law not satisfying those conditions. But without this conviction, the problem would have no meaning; interpolation would be impossible; no law could be deduced from a finite number of observations; science would cease to exist (Poincaré, 1952: 204–6).

One can notice that the issue here is generalization rather than generality, if one understands the former as an operation which leads to the latter. We will try to show in the next section that this distinction is not relevant in Poincaré's philosophy: according to Poincaré, generalization *is not a means* to reach generality, since generality is instead defined in terms of generalization. For that purpose, it is necessary to understand the link between infinity and generality in Poincaré's thought.

5.1   Infinity, generality, and generalization

As was mentioned above, Poincaré never speaks about generality independently from generalization: in his view, mathematical generality must be understood as meaning generalization. The idea is that generality does not refer to an assertion about a collection of objects which have already been given; the term generality refers to sequences of operations which can be understood as generalizations. Generality *is not the result* of a generalization, but rather the operation of generalization itself, conceived in a way that will be explained in this section.

Poincaré often refers to the opposition between the idea of a collection of objects already given and the idea of a sequence of operations, which can consist of the construction of such objects, when he distinguishes actual infinity and potential infinity. Several texts show that Poincaré considers the idea of actual infinity to be irrelevant in mathematics:

> There is no actual infinity, and when we speak about an infinite collection, we mean a collection to which one can always add new elements (similar to a subscription list which will never be closed waiting for new subscribers).[4]
>
> When I speak of all the integers, I mean all the integers that have been invented and all those which will be invented one day; when I speak of all the points of space, I mean all the points the coordinates of which can be expressed either by rational numbers, or by algebraic numbers, or by integrals, or in any other way that we will be able to invent. And it is this *"we will be able"* which is the infinity.[5]

The latter quotation shows that Poincaré links together the issue of infinity and the theme of generality: it is an issue related to generality which is introduced when one raises the problem of the meaning of the word "all" in mathematics. In a way, Poincaré's position here is close to constructivism or intuitionism; but one should say instead that Poincaré meets those positions while developing his own and original philosophical conception of mathematics. In this conception, not only do constructions take a central place, but so does the infinite, in a specific way that we now turn to examine.

For this purpose, it is convenient to start by reading the first chapter of *La science et l'hypothèse*. In it, Poincaré (1902) analyzes the principle of mathematical induction in order to show that mathematical reasoning differs from logical reasoning. Here again, it is the link between generalization and the infinite which is at the core of Poincaré's analysis. According to him, the principle of mathematical induction allows both generalization and the introduction of the infinite in mathematical reasoning. By contrast, logical operations (from which "analytical verifications" are constituted) cannot achieve these operations:

> If, instead of proving that our theorem is true for all numbers, we only wish to show that it is true for the number 6, for instance, it will be enough to establish the first five syllogisms in our cascade [i.e., the cascade of syllogisms that lead from the theorem stated for the number 1 and the statements of the theorem for the subsequent integer when one knows it for an integer to the theorem for 6, see below]. We will require 9 if we wish to prove it for the number 10; for a greater number we will require more still; but however great

---

[4] "Il n'y a pas d'infini actuel, et quand nous parlons d'une collection infinie, nous voulons dire une collection à laquelle on peut sans cesse ajouter de nouveaux éléments (semblable à une liste de souscription qui ne serait jamais close dans l'attente de nouveaux souscripteurs)" (Poincaré, 1913: 104).

[5] "Quand je parle de tous les nombres entiers, je veux dire tous les nombres entiers qu'on a inventés et tous ceux qu'on pourra inventer un jour; quand je parle de tous les points de l'espace, je veux dire tous les points dont les coordonnées sont exprimables par des nombres rationnels, ou par des nombres algébriques, ou par des intégrales, ou de toute autre manière que l'on pourra inventer. Et c'est ce *'l'on pourra'* qui est l'infini" (Poincaré, 1913: 131).

the number may be we will always reach it, and the analytical verification will always be possible. But, however far we went we should never reach the general theorem applicable to all numbers, which alone is the object of science. To reach it we should require an infinite number of syllogisms, and we should have to cross an abyss which the patience of the analyst, restricted to the resources of formal logic, will never succeed in crossing.

I asked at the outset why we cannot conceive of a mind powerful enough to see at a glance the whole body of mathematical truth. The answer is now easy. A chess-player can combine for four or five moves ahead; but, however extraordinary a player may be, he cannot prepare for more than a finite number of moves. If he applies his faculties to Arithmetic, he cannot conceive its general truths by direct intuition alone; to prove even the smallest theorem he must use reasoning by recurrence, for that is the only instrument which enables us to pass from the finite to the infinite. This instrument is always useful, for it enables us to leap over as many stages as we wish; it frees us from the necessity of long, tedious, and monotonous verifications which would rapidly become impracticable. However, when we take in hand the general theorem it becomes indispensable, for otherwise we should ever be approaching it, thanks to the analytical verification, without ever actually reaching it (Poincaré, 1952: 10–11).

The idea is that mathematical induction consists of an *indefinite* succession of syllogisms, each of them being a *logical* operation. The infinity thus introduced cannot be reduced to any logical procedure: that is why mathematical induction belongs to mathematics and not to logic. Poincaré introduces this idea in the following way:

The essential characteristic of reasoning by recurrence is that it contains, condensed, so to speak, in a single formula, an infinite number of syllogisms. We will see this more clearly if we enunciate the syllogisms one after another. They follow one another, if one may use the expression, in a cascade. The following are the hypothetical syllogisms: The theorem is true of the number 1. Now, if it is true of 1, it is true of 2; therefore it is true of 2. Now, if it is true of 2, it is true of 3; hence it is true of 3, and so on. We see that the conclusion of each syllogism serves as the minor of its successor. Further, the majors of all our syllogisms may be reduced to a single form. If the theorem is true of $n-1$, it is true of n.

We see, then, that in reasoning by recurrence we confine ourselves to the enunciation of the minor of the first syllogism, and the general formula which contains as particular cases all the majors. This unending series of syllogisms is thus reduced to a phrase of a few lines (Poincaré, 1952: 9–10).

It is worth noting that the method of achieving generality for the theorem thus obtained differs from the way in which one gives the generality of the theorem underlying the major premise of each syllogism, that is: "$\forall nP(n) \rightarrow P(n+1)$" As a matter of fact, the latter theorem can be proved by applying the logical generalization rule:

$$\frac{F(a)}{\forall xF(x)},$$

once $F(a)$ has been proved for any a. Now, it is precisely when such a method of reasoning for the theorem $P(n)$ is not available that a proof by mathematical induction is required.

The specificity of the latter is that it is a generalization which cannot be reduced to the use of a formula like the one above: in contrast to the generalization rule, a proof by mathematical induction cannot be reduced to a formula written in a finite number of symbols.[6] Poincaré stresses the non-logical nature of the mathematical induction principle by considering the way that we get *convinced* by it:

> Why then is this principle imposed upon us with such an irresistible weight of evidence? It is because it is only the affirmation of the power of the mind which knows it can conceive of the indefinite repetition of the same act, when the act is once possible. The mind has a direct intuition of this power, and experiment can only be for it an opportunity of using it, and thereby of becoming conscious of it (Poincaré, 1952: 13).

The point is that the intuition which makes the principle obvious differs from the intuitions that are brought into play in logical operations.[7] The latter allow us to admit and use logical rules such as *modus ponens* while the former makes us aware of our faculty to conceive the indefinite repetition of acts such as the analytical operations allowed by logical rules. In the case of mathematical induction, we conceive of the indefinite repetition of logical acts, each of them being a *modus ponens*. What is of importance for the question of generality is that the "power of the mind" described here is a capacity for conceiving a kind of generality that is specific, according to Poincaré, to mathematical thought. Two points must be noted. First, the "power of the mind" is put into play not only within mathematical induction but also for the construction of some fundamental mathematical concepts. Secondly, Poincaré's approach to mathematical activity allows us to understand why Poincaré develops considerations about generality, generalization, and the infinite conjointly, without studying them independently. Let us make these two points explicit.[8]

Poincaré refers to the text quoted above at several points in his philosophical writings, noticing that the same conceptual capacity—"the power of the mind"—occurs in the construction of some fundamental mathematical concepts. These concepts include the concept of natural number, the concept of infinite set, the concept of mathematical continuum, and the concept of continuous groups in geometry. Natural numbers are constructed by conceiving of the indefinite repetition of the act which consists in adding a unit to a finite collection of units. Such an act is linked to the operation of counting. An infinite set is constructed by conceiving of the indefinite addition of new elements to a finite set. The first-order continuum (which corresponds to the power of rational magnitudes[9]) is constructed by conceiving of the indefinite repetition of the act which consists in measuring an empirical given (which is composed of sensations) with devices of increasing precision. As a result, that repetition corresponds to the operation consisting of intercalating a sensible element between two others, which, at a lower degree of

---

[6] This point is developed in Heinzmann (1988).

[7] As a matter of fact, according to Poincaré, the acceptance and use of logical rules are due to a specific kind of intuition, as he claims in the first chapter of *La valeur de la science*.

[8] The following section does not go into all the details of the philosophical reconstructions of the natural numbers, the mathematical continuum, and geometric groups. It only sketches the idea which governs them.

[9] In modern mathematical terminology, we speak of the *density* of rational numbers in the set of real numbers, and not of a continuum. However, Poincaré refers to any set formed in the same way as the rational

measurement precision, would be indistinguishable by our senses. A continuous group in geometry is constructed, on the one hand, by conceiving of the indefinite composition of displacements and, on the other hand, by conceiving of the indefinite division of "physical" displacements in the same way in which a continuum is constructed. (These displacements are also given as sensations whose properties are, roughly speaking, those of mathematical groups.) The following texts show the implication of the power of the mind in this repeated use of the operations. The first of these texts concludes a discussion on the various kinds of intuition used in mathematics. Poincaré states:

> Finally, we have the intuition of pure number, whence arose the second of the axioms just enunciated, which is able to create real mathematical reasoning (Poincaré, 1958: 20).

Since the axiom mentioned here is the mathematical induction principle, this text shows that it is the same capacity of the mind which enters into the construction of the concept of natural number. The first chapter of *La science et l'hypothèse* shows that this capacity is the "power of the mind," understood as the capacity of conceiving the indefinite repetition of an act.

In the following text, from chapter II of *La science et l'hypothèse*, the description of the way natural numbers are conceived confirms that point. In addition, the text draws a parallel between the construction of the integers and the construction yielding the first-order continuum:

> But it is not only to escape this contradiction contained in the empirical data that the mind is led to create the concept of a continuum, formed of an indefinite number of terms.
>
> All happens as in the sequence of whole numbers. We have the faculty of conceiving that a unit may be added to a collection of units. Thanks to experience, we have occasion to exercise this faculty and we become conscious of it; but from this moment we feel that our power has no limit, and that we can count indefinitely, though we have never had to count more than a finite number of objects.
>
> Just so, as soon as we have been led to intercalate means between two consecutive terms of a series, we feel that this operation can be continued beyond all limit, and that there is, so to speak, no intrinsic reason for stopping (Poincaré, 1952: 24–5).

Once again, it is the conception of the indefinite repetition of an act which gives rise to a mathematical concept. This act consists of "intercalating means between two consecutive terms of a series"[10]—the terms mentioned are elements of a "physical continuum."[11] This analysis of the mathematical continuum is important with regards to the question of the necessary use of mathematics in physics: the reconstruction made by Poincaré can be understood as a way to understand how the mathematical concept of continuous

---

numbers as a *continuum* of the first order. He writes: "As an abbreviation, let me call a mathematical continuum of the first order every aggregate of terms formed according to the same law as the scale of commensurable numbers" (Poincaré 1952: 25).

[10] "Intercaler des moyens entre deux termes consécutifs d'une série."

[11] By "physical continuum," Poincaré does not mean the continuum as it would be conceived by physicists, but the sensible given. The latter is characterized by the fact that, at one level of measurement precision, there are always sensible elements which cannot be distinguished. However, they could be distinguished if we used

magnitude can be applied to sensible givens. Such a given cannot be, as such, directly considered as a magnitude (either intensive or extensive). The reason for this is that it is governed by the "formula of the physical continuum" (Poincaré, 1952: 22):

$$A = B; \quad B = C; \quad A < C,$$

in which the symbol "=" means the sensible indiscernibility between two sensible elements observed at a certain level of measurement precision, and "<" means a sensible difference of intensity. It is in order to solve the contradiction inherent in that formula and to be able to apply the concept of magnitude to the sensible given that, according to Poincaré, the mathematical concept of continuum is constructed as described above:

> Do we find [mathematical magnitude] in nature, or have we ourselves introduced it? And if the latter be the case, are we not running a risk of coming to incorrect conclusions all round? Comparing the rough data of our senses with that extremely complex and subtle conception which mathematicians call magnitude, we are compelled to recognize a divergence. The framework into which we wish to make everything fit is one of our own construction; but we did not construct it at random, we constructed it by measurement so to speak; and that is why we can fit the facts into it without altering their essential qualities (Poincaré, 1952: XXV).

As a matter of fact, Poincaré stresses the use of the mathematical continuum both in relation to measurement and for the elaboration of physical laws. Thus, the specific way by which the mathematical concept is constructed has important consequences in physics, notably concerning the generality of physical laws. We will come back to an aspect of that issue below.

As it is well known, the concept of group is at the core of Poincaré's analysis of geometry. In the following text, Poincaré refers to the same passage in chapter I of *La science et l'hypothèse* to talk about the "homogeneity law" which is related to the main property of groups, that is, the closure property. The text shows that Poincaré also understands this property in relation to the "power of the mind":

> We may also say that a movement which is once produced may be repeated a second and a third time, and so on, without any variation of its properties. In the first chapter, in which we discussed the nature of mathematical reasoning, we saw the importance that should be attached to the possibility of repeating the same operation indefinitely. The virtue of mathematical reasoning is due to this repetition; hence it is thanks to the law of homogeneity that mathematical reasoning applies to geometrical facts (Poincaré, 1952: 64).

Let us notice that the case of geometry suggests that the "power of the mind" is not only the capacity of conceiving the repetition of one act but must be extended to the capacity of conceiving the indefinite composition of *similar* acts. Another text describes

---

a more precise measurement device. The mathematical continuum is constructed in order to have a concept which allows us to deal with that property of the sensible given. The expression "physical continuum" is chosen by Poincaré because the phenomena studied in physics are given in sensible experiences.

the conception according to which studying a geometrical space consists in studying a continuous group. What is important for us in this is that it draws a comparison between that situation and the construction of the concept of mathematical continuum. Poincaré writes:

> The group of displacements as it is given to us directly through experience is something of a coarser nature. We may say that it is to continuous groups, strictly speaking, what the physical continuum is to the mathematical continuum. We first study its form in accordance with the physical continuum formula, but since there is something in this formula which is offensive to our reason, we reject it and replace it with the continuous group formula that potentially pre-exists in us, but that we initially know by its form.[12]

In the case of the mathematical continuum, the idea was to conceive the indefinite repetition of the act which corresponds to that of intercalating terms between two consecutive sensible terms of a physical continuum. In the case of *continuous* groups, similarly, the idea is to conceive the indefinite repetition of the act which corresponds to that of dividing a physical displacement into smaller displacements. "Dividing" here is understood according to the composition law which defines the group law: a displacement A is divided into two smaller displacements B and C when A is obtained by composing B and C.

To summarize, the "power of the mind" described in *La science et l'hypothèse*, chapter I, appears at the core of the construction of the main mathematical concepts studied by Poincaré—natural numbers, infinite sets, the continuum, space—as well as in the "raisonnement mathématique par excellence," namely mathematical induction.[13] Thus, according to Poincaré, a mathematical concept or reasoning is a twofold instance: on the one hand, it implies the intuition[14] of some acts; on the other hand it implies the conception of the indefinite repetition of those acts. This conception is made possible by a capacity of the mind, of which we also have an intuition.[15]

---

[12] "Le groupe des déplacements tel qu'il nous est donné directement par l'expérience est quelque chose d'une nature plus grossière; il est, pouvons-nous dire, aux groupes continus à proprement parler ce que le continu physique est au continu mathématique. Nous étudions d'abord sa forme conformément à la formule du continu physique et comme il y a quelque chose qui répugne à notre raison dans cette formule, nous la rejetons et nous y substituons celle du groupe continu qui en puissance préexiste en nous, mais que nous ne connaissons initialement que par sa forme" (Poincaré, 2002: 30).

[13] One can note that in those cases the "power of the mind" is applied to acts which are not *mathematical* ones: they are logical or sensible ("physical") acts. It is only when their indefinite repetition is put into play that we deal with mathematics. This remark allows us to understand in which sense Poincaré speaks about *foundations* in his philosophy of mathematics: looking for foundations means, according to Poincaré, understanding the acts which are at the origin of a mathematical concept. In other terms, it means determining acts, the conception of the indefinite repetition of which produces a mathematical concept. It is worth noting that this philosophical inquiry is not a mere historical or psychological one about scientific discovery. Rather, it deals with the true meaning of the concepts studied: the acts in question as well as the conception of their indefinite repetition must be grasped in order to understand the mathematical concept itself. In other words, for instance, one cannot truly understand the concept of the mathematical continuum without making use of sensible intuition.

[14] The intuitions concerned can be of different natures: they can be a logical intuition, a symbolic intuition, or a sensible intuition. We will not develop this point here.

[15] "The mind has a direct intuition of this power" (Poincaré, 1952: 13).

On this basis, let us now study the link between that special characterization of mathematics, on the one hand, and the notions of the infinite, of generality, and of generalization, on the other.

The way Poincaré describes the "power of the mind" allows us to understand why Poincaré claims that the mathematical infinite is a potential infinite and not an actual infinite. As a matter of fact, actual infinity deals with *things* already given. Such cannot be the case of *acts* that are indefinitely repeated or combined. The intuitions by which these sequences of acts are given do not allow us to consider them as an already given whole, because these intuitions would then be limited to the finite. About sets, for instance, Poincaré writes:

> No proposition concerning infinite collections can be made obvious by intuition.[16]

In other words, the acts by means of which mathematical concepts are constructed through applying the "power of the mind" are given by intuitions which deal only with the finite. That does not mean, for example, that we cannot deal with infinite sets. However, these sets are only *conceived* and cannot be *given* as a whole which would be grasped by *intuition*. That is why the way Poincaré understands the construction of mathematical concepts and reasoning—namely, in terms of acts indefinitely repeated or combined—implies that mathematical infinity is a potential infinity.

Now, we have seen that Poincaré connects considerations about generality and considerations about the infinite. From the fact that the mathematical infinite is a potential infinite, it follows that mathematical generality cannot be understood independently from the operation of generalization. As a matter of fact, Poincaré describes the conception of the indefinite repetition of an act as a kind of generalization, specific to mathematics: it follows that the generality involved in mathematical reasonings and concepts thus conceived cannot be understood independently from that special kind of generalization.

5.2   Generality in mathematical physics

This point has important consequences regarding generality in mathematical physics. We will now examine this point, by going back to the text in which Poincaré makes the statement about generalization in physics which we started with. Indeed, in this important text in which Poincaré stresses the role of mathematics in physics, Poincaré relates generalization in physics to the specificity of mathematical generalization which we have examined above and which Poincaré relates to "the power of the mind." The full text reads:

> *Origin of Mathematical Physics*—Let us go further and study more closely the conditions which have assisted the development of mathematical physics. We recognize at the outset

---

[16] "Aucune proposition concernant les collections infinies ne peut être évidente par intuition" (Poincaré, 1913: 138).

that the efforts of men of science have always tended to resolve the complex phenomenon given directly by experiment into a very large number of elementary phenomena …

The knowledge of the elementary fact enables us to state the problem in the form of an equation. It only remains to deduce from it by combination the observable and verifiable complex fact. That is what we call *integration*, and it is the province of the mathematician. It might be asked, why, in physical science, generalization so readily takes the mathematical form. The reason is now easy to see. It is not only because we have to express numerical laws; it is because the observable phenomenon is due to the superposition of a large number of elementary phenomena which are *all similar to each other*, and in this way differential equations are quite naturally introduced. It is not enough that each elementary phenomenon should obey simple laws; all those that we have to combine must obey the same law; then only is the intervention of mathematics of any use. Mathematics teaches us, in fact, to combine like with like. Its object is to divine the result of a combination without having to reconstruct that combination element by element. If we have to repeat the same operation several times, mathematics enables us to avoid this repetition by telling the result beforehand by a kind of induction. This I have explained before in the chapter on mathematical reasoning. But for that purpose all these operations must be similar; in the contrary case we must evidently make up our minds to work them out in full one after the other, and mathematics will be useless. It is, therefore, thanks to the approximate homogeneity of the matter studied by physicists, that mathematical physics came into existence. In the natural sciences the following conditions are no longer to be found: homogeneity, relative independence of remote parts, simplicity of the elementary fact; and that is why the student of natural science is compelled to have recourse to other modes of generalization (Poincaré, 1952: 153, 158–9).

This text can be interpreted by referring to the way Poincaré characterizes differential equations in other texts. For instance, he writes:

Newton has shown us that a law is only a necessary relation between the present state of the world and its immediately subsequent state. All the other laws since discovered are nothing else; they are in sum, differential equations (Poincaré, 1958: 87).

Poincaré describes here the infinitesimal transformations corresponding to what we now call the *flow* of a differential equation.[17] Thus, the elementary phenomena can be understood as conceived from observed phenomena, which are finite transformations, in the same way that continuous groups are conceived from sensible displacements in geometry. The elementary phenomena are mathematical concepts—infinitesimal transformations of the flow—conceived from observed phenomena in that way. The reference to the "power of the mind" in the text quoted above confirms that remark. It also allows us to understand why Poincaré speaks of *mathematical generalization*, since, as we have seen,

---

[17] The *flow* of a differential equation is the *group of transformations* defined as follows:

Suppose F: $R^n \rightarrow R^n$ is a C^1 vector field.
Then for each initial condition $X_0 \in R^n$, the ordinary differential equation $X' = F(X)$ has a unique solution, which we denote by $X(t)$.
Thus $X(0) = X_0$ and $X'(t) = F(X(t))$.
The flow $\varphi: R \times R^n \rightarrow R^n$ of $X' = F(X)$ is defined by $\varphi(t, X_0) = X(t)$.

the generalization or idealization allowed by that faculty is specific to mathematics. In the text quoted above, mathematical generalization is explicitly distinguished from those generalizations used by "naturalists": Poincaré does not develop this point, but one can try to characterize further the difference between these two kinds of generalization.

5.3   Mathematical generalization *versus* "predicative generality"

We have seen above that mathematical induction gives rise to generalizations which must be distinguished from the one expressed by the logical generalization rule:

$$\frac{F(a)}{\forall x\, F(x)}.$$

This allows us to propose an interpretation of the distinction suggested by Poincaré between generalizations used in mathematical physics and generalizations used by "naturalists." We have seen that in the same text Poincaré refers to the "power of the mind," namely the capacity to conceive the indefinite repetition of an act. Now, it is precisely the fact that repetition means the successive combination of the same act or of similar acts which prevents mathematical reasoning by induction from being reduced to such a logical rule. In the text that we are analyzing, that is precisely how Poincaré characterizes mathematics. Let me repeat his words:

> It is not enough that each elementary phenomenon should obey simple laws; all those that we have to combine must obey the same law; then only is the intervention of mathematics of any use. Mathematics teaches us, as a matter of fact, to combine like with like (Poincaré, 1952: 159).

It follows from this that the specificity of that kind of generalization rests on the idea of combining acts. This idea does not appear in the logical generalization rule. We can extend this remark to all the cases in which the "power of the mind" is involved and thus suggest that it characterizes the specificity of mathematical generalization.

The idea which arises then is that mathematical generalization differs from what we will call "predicative generality." The latter appears in the case of the logical generalization rule but it is a very general logical form. It can be described in two ways, namely in terms of properties or predicates, or in terms of classes. In this sense, a general proposition corresponds to the statement that *some property is common to several individual things or cases, or to the grouping of several individual things or cases into one class.* Generalization by abstraction, for example, belongs to the domain of predicative generality, for it consists in not taking into account differences among several cases or things in order to group them instead into a class or to subsume them under a generic concept which corresponds to a common property that they possess. One can suggest that Poincaré refers to generalizations of that kind when he speaks about naturalists, referring to zoological or botanical classifications. We can notice that predicative generality belongs to the logic of the subject/predicate[18] logical form or to the logic of classes. This does not mean that, for instance, zoological classifications should be reduced to logic, but that they are conceived using logical forms which, according to Poincaré, belong to logic and differ from mathematical thinking. Indeed, as we have suggested above, the generalization implied by the "power of the mind" is very different from that, notably due to the fact that operations of successive combinations of the acts implied play a central role. Predicative generality has no place for the idea of combining individuals subsumed under a genus or grouped into a class, if "combining individuals" had a meaning. Combining acts has nothing to do with grouping of things or cases in a class or asserting a common property.

As a matter of fact, Poincaré links up predicative generality with formal logic, including finite set theory. This fact appears clearly in the following description which he gives of formal logic:

> Formal logic is nothing but the study of properties common to all classifications. It teaches us that two soldiers who belong to the same regiment thereby also belong to the same brigade, and consequently to the same division. The whole theory of syllogism is reduced to that.[19]

The classifications described here are obviously linked with predicative generality. The analysis made above thus allows us to discover a link between Poincaré's opposition to logicism and his claim that mathematics is necessary for physics. The idea is that the nature of physical phenomena calls for kinds of generalizations which differ from the kind of generality conveyed by formal logic. A physical law is not a general proposition by which we assert that a common property belongs to individual cases or things; such a property being possibly a polyadic predicate relating individual cases or things. It is rather a mathematical equation by which phenomena are understood as a combination of transformations governing magnitudes.

5.4   Generality in experimental physics: Poincaré and the "new riddle of induction"

As we will see below, the idea that mathematical (and correlatively physical) generality differs from what we have called *predicative generality* can be confirmed by an examination of the way Poincaré analyzes empirical induction. In this respect, we will reach, in the present section, a conclusion similar to the one drawn above. We will not stress here the "power of the mind." However, it is clear that the mathematical concepts involved in Poincaré's

---

[18] These must be understood in a broad sense, including polyadic predicates.

[19] "La logique formelle n'est autre chose que l'étude des propriétés communes à toute classification; elle nous apprend que deux soldats qui font partie du même régiment appartiennent par cela même à la même brigade, et par conséquent à la même division, et c'est à cela que se réduit toute la théorie du syllogisme" (Poincaré, 1913: 102).

analysis of interpolation—continuity, differentiability, smoothness of a function—can be construed, in his philosophy, as depending upon that faculty.

As the quotation made at the beginning of this paper shows, Poincaré refers to interpolation when he speaks about empirical induction: by this operation a physical law appears as a generalization of empirical data, namely of a finite set of measurements. After mathematics and physical mathematics, we thus tackle experimental physics, and more precisely Poincaré's analyses of the way empirical data are generalized into a physical law. This kind of generalization belongs to the operation of induction, but we will show that, as understood by Poincaré, induction cannot relevantly be described as the extension of a property to a whole set of individuals from the observation of the fact that this property belongs to some of these individuals, which is the way to understand induction within the logic of predicative generality.

First, one must emphasize that Poincaré does not set mathematical physics and experimental physics in opposition: in both domains, according to Poincaré, mathematics plays a central role *and* a genuine physical law is characterized by the fact that it must be empirically confirmed in order to be considered true. Nevertheless, the way Poincaré analyzes the way a law is constructed within those domains differs, even if general conclusions drawn from these analyses tally with each other, notably regarding the specificity that the use of mathematics gives to generalizations. In mathematical physics, we have seen that Poincaré emphasizes the differential form of the laws; this is not exactly the case in his considerations about experimental physics, where laws, which are generalizations of empirical data, are merely described as mathematical functions or curves (which represent mathematical functions). The link between those two ways of describing physical laws can be seen when one notices that, for a given phenomenon, the empirically constructed function must be a solution of a differential equation whose study belongs to mathematical physics.

If, as we will present it, induction in experimental physics cannot be described in a relevant way as "ordinary induction," that is, extension of the domain of a predicate, it is because the former consists in *interpolation*, an operation which Poincaré refers to each time he mentions empirical induction. Let us start by making some comments about the texts in which Poincaré describes that operation. These comments will allow us to bring to light the specificity of generalization in experimental physics.

5.4.1 Interpolation

In accordance with the way Poincaré describes *objective reality* as constituted by mathematical relations expressed by laws, *empirical* construction of laws brings to light relations among empirical data by using mathematical concepts. After a geometrical frame is instituted—which defines the measurement operation—the construction of an empirical law consists in setting relations among measurement results, that is, among observed values of mathematical magnitudes which are associated with sensible data by means of the geometric frame. It follows from this that a law is a *mathematical function which expresses relations among magnitudes*. Now, this empirical construction of a law consists in a *generalization* and thus falls under *induction*, if induction is understood as the operation which consists in generalizing the results of an empirical observation. That is the reason Poincaré mentions interpolation in the main texts in which he considers induction, since interpolation consists in constructing a mathematical function (a curve) from a scatter plot obtained by measuring correlated values of different magnitudes. Let us read some texts in which interpolation appears as the main instance of induction in experimental physics. Here is the first one:

> However timid we may be, there must be interpolation. Experiment only gives us a certain number of isolated points. They must be connected by a continuous line, and this is a true generalization. But more is done. The curve thus drawn will pass between and near the points observed; it will not pass through the points themselves. Thus we are not restricted to generalizing our experiment, we correct it; and the physicist who would abstain from these corrections, and truly content himself with experiment pure and simple, would be compelled to enunciate very extraordinary laws indeed (Poincaré, 1952: 142–3).

In this text, the expression "true generalization" must be underlined, for it is the one Poincaré uses to designate physical laws. By this use, associated with interpolation, Poincaré stresses the empirical nature of physical laws as he often does. On the other hand, this expression suggests that other kinds of generalizations would not be *genuine* ones. According to one interpretation, these other kinds of generalizations refer to hypotheses which are not empirical laws (i.e., conventional principles, heuristic principles, or metaphysical assumptions). According to another interpretation, Poincaré refers to empirical generalizations that are not the ones specifically used in physics. In what follows, we develop this latter interpretation, which is by no means incompatible with the former.

We choose not to comment directly on those texts but to do so in the context of a comparison between Poincaré's analysis and Goodman's third chapter of *Facts, fictions, and forecast*, in which Goodman (1973) develops what he calls "the new riddle of induction," for this comparison allows a fruitful presentation of Poincaré's thought. Here, the reference to Goodman's book is only a starting point for analyzing Poincaré's texts, which are the main object of this paper: thus we will not develop a detailed analysis of Goodman's thought.

Let us start by making some remarks about the texts quoted in this section.

As mentioned above, Poincaré's considerations about interpolation are in line with a general reflection about induction in physics, that is, about the operation of generalizing empirical data which leads to a law. Thus those texts about interpolation have a large philosophical scope and importance in spite of their apparently narrow subject. This scope and importance are underlined by Poincaré in the last sentence of the following text, which we have already partly quoted above:

> Let us pass on to an example of a more scientific character. I require to determine an experimental law; this law, when discovered, can be represented by a curve. I make a certain number of isolated observations, each of which may be represented by a point. When I have obtained these different points, I draw a curve between them as carefully as possible, giving my curve a regular form, avoiding sharp angles, accentuated inflexions, and any sudden variation of the radius of curvature. This curve will represent to me the

probable law, and not only will it give me the values of the functions intermediary to those which have been observed, but it also gives me the observed values more accurately than direct observation does; that is why I make the curve pass near the points and not through the points themselves.

Here, then, is a problem in the probability of causes. The effects are the measurements I have recorded; they depend on the combination of two causes—the true law of the phenomenon and errors of observation. Knowing the effects, we have to find the probability that the phenomenon shall obey this law or that, and that the observations have been accompanied by this or that error. The most probable law, therefore, corresponds to the curve we have drawn, and the most probable error is represented by the distance of the corresponding point from that curve. But the problem would have no meaning if before the observations I did not have an *a priori* idea of the probability of this law or that, or of the chances of error to which I am exposed. If my instruments are good (and I knew whether this is so or not before beginning the observations), I shall not draw the curve far from the points which represent the rough measurements. If they are inferior, I may draw it a little farther from the points, so that I may get a less sinuous curve; much will be sacrificed to regularity.

Why, then, do I draw a curve without sinuosities? Because I consider *a priori* a law represented by a continuous function (or function the derivatives of which to a high order are small), as more probable than a law not satisfying those conditions. But without this conviction the problem would have no meaning; interpolation would be impossible; no law could be deduced from a finite number of observations; science would cease to exist (Poincaré, 1952: 204–6).

In other words, the existence of science itself depends upon the possibility of carrying out interpolation. Consequently, according to Poincaré, interpolation is not only an instance of induction but the main form of induction, at least in physics. Additionally, the affirmation quoted above also occurs in the following text:

I do not at all wish to investigate here the foundations of the principle of induction; I know very well that I shall not succeed; it is as difficult to justify this principle as to get on without it. I only wish to show how scientists apply it and are forced to apply it.

When the same antecedent recurs, the same consequent must likewise recur; such is the ordinary statement. But reduced to these terms, this principle would be of no use. For one to be able to say that the same antecedent was reproduced, it would be necessary for the circumstances all to be reproduced, since no one is absolutely indifferent, and for them to be exactly reproduced. And since this will never happen, the principle could have no application.

We should therefore rephrase the statement and say: if an antecedent A once produced a consequent B, an antecedent A' a little different from A, will produce a consequent B' a little different from B. But how can we recognize that A and A' are "a little different?" If any of the circumstances can be expressed by a number, and if in both cases this number has nearly the same value, the meaning of "a little different" is relatively clear; the principle then means that the consequent is a continuous function of the antecedent, and as a practical rule we are led to believe that interpolation is allowed. It is in fact what scientists do every day and no science would be possible without interpolation.

However we notice something. The law we are looking for can be represented by a curve. Experiment made us aware of some points on this curve. According to the principle we have just stated, we believe that these points can be connected by a continuous line. We draw this line with the naked eye. New experiments will provide new points on the curve. If those points are outside the line drawn in advance, we will have to modify the curve, but not abandon our principle. Any points, as numerous as they may be, can be joined by a continuous curve. Of course, if this curve is too irregular, we may be shocked (and even suspect experimental mistakes), but the principle won't be directly faulted.[20]

Another topic of those texts is the continuity and smoothness of functions or curves occurring in interpolation. These characteristics are described by Poincaré in terms of simplicity and by referring to the principle of sufficient reason. It is worth noting that continuity is not only mentioned by Poincaré when discussing interpolation, but also when discussing the principle of empirical induction directly. He reformulates the latter principle, in the text above, in terms of continuity while associating to it the topic of *approximation*,[21] which also occurs in those texts.

The text we have just quoted is an excerpt of the eleventh chapter of *La science et l'hypothèse*, which deals with probability. Poincaré's considerations regarding probability within his reflections about induction and interpolation differ considerably from conceptions like those of Carnap and Reichenbach. According to the conceptions of the latter, induction rests upon probability theory in the following way: *induction would consist in choosing the most probable hypothesis*. In the following paragraph, we will show that, according to Poincaré, induction and probability are related in a completely different way.

Finally, it is important to note that Poincaré claims that his goal *is not* to seek a *foundation* of the induction principle, that is, to try to prove that induction, under certain conditions, leads to generalizations whose truth is guaranteed *a priori*. His philosophical concern about induction is not, in that sense, a foundational concern. That is the reason that it is interesting to compare Poincaré's and Goodman's conceptions: as a matter of fact we claim that both deal with the same question.

5.4.2   "The new riddle of induction" in Poincaré's philosophy

We are going to try to show that "the new riddle of induction," to quote the title of the third chapter of Goodman's book *Facts, fictions, and forecast*, was not new in 1954, when it was first published, since it had been formulated by Poincaré half a century earlier.

---

[20] "... si un antécédent *A* a produit une fois un conséquent *B*, un antécédent *A'* peu différent de *A*, produira un conséquent *B'* peu différent de *B*. Mais comment reconnaîtrons-nous que les antécédents *A* et *A'* sont « peu différents »? Si quelqu'une des circonstances peut s'exprimer par un nombre, et que ce nombre ait dans les deux cas des valeurs très voisines, le sens du mot « peu différent » est relativement clair; le principe signifie alors que le conséquent est une fonction continue de l'antécédent. Et comme règle pratique, nous arrivons à cette conclusion que l'on a le droit d'interpoler. C'est en effet ce que les savants font tous les jours et sans l'interpolation toute science serait impossible" (Poincaré, 1905: 177).

[21] As mentioned above, approximation of measurements plays a central role in Poincaré's philosophical reconstruction of the mathematical concept of continuum.

Nevertheless, if the question raised by Poincaré and Goodman is the same, their answers differ considerably. Examining Poincaré's answer and the reasons that it differs from Goodman's answer enables us to present the main characteristics of Poincaré's conception of induction in experimental physics.

Goodman's formula for the "new riddle of induction" refers, not to the search for a foundation of induction (which is the "old" problem of induction), but to the search for a way of distinguishing a hypothesis which can legitimately be induced from a hypothesis which cannot, in the case where both hypotheses are confirmed by the same set of empirical data. The point is that this problem does not deal with the truth of the hypothesis which is legitimately induced. It is not the *truth* (the *success*) of an induction which is the concern, but its *legitimacy*. As a matter of fact, it may be legitimate to induce a hypothesis from a set of empirical data even if that hypothesis is wrong—which can be established by a subsequent empirical observation. In addition, given a set of empirical data, all the hypotheses confirmed by these data are not construed as legitimate generalizations and are not actually taken into account. Goodman's construction of the "grue" predicate aims to establish the existence of that problem and at proving that the criterion to distinguish a hypothesis which can be legitimately induced (a "lawlike" hypothesis) from one which cannot does not rest upon a formal or syntactic[22] difference between the hypotheses which are compared (or between the predicates which are projected[23]). In Goodman's work, the riddle is related to the topic of projection and can be expressed by means of different equivalent formulas:

- search for a difference between "lawlike statements" and "accidental statements"
- search for an answer to the question: "which hypotheses are confirmed by their positive instances?"
- search for a definition of the predicate "projectible" based upon the predicate "projected," which is equivalent to project the predicate: "projected."

Goodman's analysis of the predicate "grue"[24] leads him to a solution of the riddle that is a rule of induction which rests upon the concept of *entrenchment*: the idea is to say that a predicate or a hypothesis are more entrenched when they have *actually* been more induced. According to Goodman, the determination of a lawlike hypothesis is extrinsic: it can never rest upon intrinsic characteristics of the hypotheses. The fact that a hypothesis is more entrenched than another hypothesis is indeed independent from any of their intrinsic characteristics. Goodman's rule of induction is formulated as following:

> Among [supported, unviolated, and unexhausted] hypotheses, *H* will be said to override
> *H'* if the two conflict and if *H* is the better entrenched and conflicts with no still better
> entrenched hypothesis (Goodman, 1973: 101).

---

[22] One of Goodman's concerns is to criticize the idea of an inductive logic.

[23] We prefer to speak in terms of hypotheses rather than in terms of predicates, in order to keep a larger generality. Indeed, we will see that it is not relevant, in Poincaré's philosophy, to describe physical laws—which are the induced hypothesis—as consisting of associating a predicate to a subject.

[24] Goodman defines "grue" in the following way:

> "Suppose that all emeralds examined before a certain time *t* are green. At time *t*, then, our observations support the hypothesis that all emeralds are green; and this is in accord with our definition of

It is worth noting that Goodman underlines the fact that his solution consists in saying that what distinguishes a legitimate induction and an illegitimate induction *is only* the deeper entrenchment of the former *and nothing else.* That is the reason why authors who have tried to refute Goodman's argument have sought to show that *something else,* other than mere entrenchment, enables us to make that distinction. As we will see, precisely such a "something else" enables us to make the distinction in Poincaré's philosophy.

After Goodman's book was published, several authors—Hempel, Priest, and Hullett and Schwartz—noticed that the "new riddle of induction" can be raised in a natural way within the context of interpolation. Here is how Hullett and Schwartz present that point:

> If, as Goodman sees it, the "old riddle" of induction asked the question, "Why does a positive instance of a hypothesis give any grounds for predicting future instances?" and the "new riddle" asks instead, "What hypothesis are confirmed by their positive instances?" then much of the "new riddle" is not so new after all. Consider the interpolation situation: given a set of data points on a graph, which might be connected by countlessly many different curves, which is the best curve that might be drawn through these points, recognizing that each point in the curve selected (other than those points originally obtained as data points) constitutes a prediction about unexamined cases? (Hullett and Schwartz, 1967: 109).

Now, about the use of the principle of induction in physics, Poincaré raises the problem in very similar terms, as can be read in the following text:

> ... those who do not believe that natural laws must be simple, are still often obliged to act as if they did believe it. They cannot entirely dispense with this necessity without making all generalization, and therefore all science, impossible. It is clear that any fact can be generalized in an infinite number of ways, and it is a question of choice. The choice can only be guided by considerations of simplicity. Let us take the most ordinary case, that of interpolation. We draw a continuous line as regularly as possible between the points given by observation. Why do we avoid angular points and inflexions that are too sharp? Why do we not make our curve describe the most capricious zigzags? It is because we

confirmation. Our evidence statements assert that emerald *a* is green, that emerald *b* is green, and so on; and each confirms the general hypothesis that all emeralds are green. So far, so good.

"Now let me introduce another predicate less familiar than 'green.' It is the predicate 'grue' and it applies to all things examined before *t* just in case they are green but to other things just in case they are blue. Then at time *t* we have, for each evidence statement asserting that a given emerald is green, a parallel evidence statement asserting that that emerald is grue. And the statements that emerald *a* is grue, that emerald *b* is grue, and so on, will each confirm the general hypothesis that all emeralds are grue. Thus according to our definition, the prediction that all emeralds subsequently examined will be green and the prediction that all will be grue are alike confirmed by evidence statements describing the same observations. But if an emerald subsequently examined is grue, it is blue and hence not green" (Goodman, 1973: 73–4).

Obviously, only the first of these two incompatible predictions is a lawlike (legitimate) hypothesis. The new riddle of induction consists in asking how one can distinguish such a hypothesis from a hypothesis like the one which contains "grue," which is not legitimate and lawlike in spite of the fact that all the observations made in the past are positive instances of it.

know beforehand, or we think we know, that the law we have to express cannot be so complicated as all that (Poincaré, 1952: 145–6).

The question asked by Poincaré is the same as the one asked by Goodman. As a matter of fact, the *choice* he refers to in that excerpt is not the choice of the *right* hypothesis (i.e., the hypothesis which will be confirmed by any future measurement). The reason for this is that nothing guarantees that an induced hypothesis will never be refuted by a future experiment. In other words, the remark implies that there is no point searching for a foundation of induction. Consequently, the choice is the choice of the most *legitimate* hypothesis. Let us confirm that point by answering two objections to that interpretation of Poincaré's texts. These objections derive from Goodman's remarks about the search for a foundation of the principle of induction.

About such a search, Goodman writes: "Nor does it help matters much to say that we are merely trying to show that or why certain predictions are *probable*" (Goodman, 1973: 62). The texts quoted above, as well as Poincaré's general philosophical conceptions about probabilities, show that Poincaré's conceptions would not be similar in any way to what Goodman dismisses here. Indeed, Poincaré claims that all probability consideration implies that the probability of some candidates must be asserted "*a priori.*"[25] Now, that is precisely what is at stake when Poincaré refers to smooth[26] curves in interpolation. Far from seeking to show that the confirmation of the hypothesis of a smooth curve would be more probable, Poincaré claims that we assert *a priori* that such a curve is more probable when we make a interpolation. This point is expressed clearly in the following part of the statement already quoted above:

[25] Just before a text quoted above, Poincaré explains in terms of *conventions* the necessity of asserting a probability *a priori*:

First of all, I recall that at the outset of all problems of probability of effects that have occupied our attention up to now, we have had to use a convention which was more or less justified; and if in most cases the result was to a certain extent independent of this convention, it was only the condition of certain hypotheses which enabled us *a priori* to reject discontinuous functions, for example, or certain absurd conventions. We shall again find something analogous to this when we deal with the probability of causes. An effect may be produced by the cause *a* or by the cause *b*. The effect has just been observed. We ask the probability that it is due to the cause *a*. This is an *a posteriori* probability of cause. But I could not calculate it, if a convention more or less justified did not tell me in advance what is the *a priori* probability for the cause *a* to come into play—I mean the probability of this event to someone who had not observed the effect. To make my meaning clearer, I go back to the game of *écarté* mentioned before. My adversary deals for the first time and turns up a king. What is the probability that he is a sharper? The formulae ordinarily taught gives 8/9, a result which is obviously rather surprising. If we look at it closer, we see that the conclusion is arrived at as if, before sitting down at the table, I had considered that there was one chance in two that my adversary was not honest. An absurd hypothesis, because in that case I should certainly not have played with him; and this explains the absurdity of the conclusion. The function on the *a priori* probability was unjustified, and that is why the conclusion of the *a posteriori* probability led me into an inadmissible result. The importance of this preliminary convention is obvious. I shall even add that if none were made, the problem of the *a posteriori* probability would have no meaning. It must be always made either explicitly or tacitly (Poincaré, 1952: 203–4).

[26] In what follows, *smoothness* doesn't mean *infinite differentiability*, but refers to characterizations given by Poincaré in terms of reinforcement of the first derivative or slow change in the curvature of the curve.

The most probable law, therefore, corresponds to the curve we have drawn, and the most probable error is represented by the distance of the corresponding point from that curve. But the problem has no meaning if before the observations I did not have an *a priori* idea of the probability of this law or that, or of the chances of error to which I am exposed.

...

Why, then, do I draw a curve without sinuosities? Because I consider *a priori* a law represented by a continuous function (or function the derivatives of which to a high order are small), as more probable than a law not satisfying those conditions (Poincaré, 1952: 205–6).

In other words, it is not because the smoothest curve is the most probable that we choose it; it is the choice of the smoothest curve which institutes, *a priori*, its probability. Thus, Poincaré's position is in no way similar to the one which is discarded as useless by Goodman in the excerpt quoted above.

The second objection is directed toward attempts at solving the old problem of induction by making it rest upon a principle of uniformity of nature. Poincaré refers to such a principle in the following text:

Induction applied to the physical sciences is always uncertain, because it is based on the belief in a general order of the universe, an order which is external to us (Poincaré, 1952: 13).

It is important here to note that the principle of a general order of the universe is not a principle asserted by Poincaré to justify or to found induction, but a belief or a hypothesis which is stated to be coextensive to induction. In other words, Poincaré only says that we believe in a general order of the universe when we make inductions, without trying to justify such a belief and found induction upon it. Here again, Poincaré's position differs from that which Goodman's objection takes as its target.

In conclusion, the question raised by Poincaré regarding interpolation is thus similar to the "new riddle of induction" as it is set by Goodman: the point is to understand why we give priority to a certain kind of hypothesis when we make inductions—namely the smoothest curve which passes through or close to the points of the scatter plot—without asserting anything *a priori* about the truth of that hypothesis.

Let us examine the answer given by Poincaré to that question, which enables us to compare his position to Goodman's. Poincaré's answer can be read in the texts quoted above. For him, the criterion which characterizes hypotheses which are legitimately induced from a set of measurements is not entrenchment. Rather, it is related to the simplicity or smoothness of the function which is chosen. Once again, let us stress that by "smoothness" of a function or a curve, Poincaré means the property of having small high-order derivatives, or a slow change of the curvature. Without going into the details of the mathematical definition of smoothness, one can note that the properties mentioned by Poincaré mean that the function's behavior is not so different from a polynomial function's behavior.

To justify the criterion for selection used in interpolation, Poincaré refers to the *principle of sufficient reason* and to a belief in *simplicity*. Because the latter criterion raises problems of definition, we will try to understand how the former occurs in that context. The point is to answer the following question: *in the context of induction, do we have reasons to choose the smoothest function, regardless of the success of the induction (i.e., the fact that our hypothesis will be confirmed by future measurements)*? A form of the principle of sufficient reason occurs in the last part of the text quoted above. In it, Poincaré claims that it is meaningless to ask about the probability of an induced hypothesis unless we have already stated, *a priori,* that the curve which represents the real phenomenon is probably the smoothest one. Indeed this can be formulated as follows: without having stated that probability *a priori, we would have no reason* not to choose *any* curve which passes close to the given points, in which case interpolation would make no sense. In other words, interpolation would be meaningless if we had no reason to choose, *a priori,* between the possible hypotheses which fit the empirical data (the finite set of points).

A way to confirm that interpretation and to examine Poincaré's solution more closely is to start by paying attention to his rephrasing of the principle of induction in which a reference to interpolation occurs. Let us read again the passage quoted above:

> ... if an antecedent *A* once produced a consequent *B*, an antecedent *A'* a little different from *A*, will produce a consequent *B'* a little different from *B*. But how can we recognize that *A* and *A'* are "a little different?" If any of the circumstances can be expressed by a number, and if in both cases this number has nearly the same value, the meaning of "a little different" is relatively clear; the principle then means that the consequent is a continuous function of the antecedent, and as a practical rule we are led to believe that interpolation is allowed. It is in fact what scientists do every day and no science would be possible without interpolation.
>
> However we notice something. The law we are looking for can be represented by a curve. Experiment made us aware of some points on this curve. According to the principle we have just stated, we believe that these points can be connected by a continuous line. We draw this line with the naked eye. New experiments will provide new points on the curve. If those points are outside the line drawn in advance, we will have to modify the curve, but not abandon our principle. Any points, as numerous as they may be, can be joined by a continuous curve. Of course, if this curve is too irregular, we may be shocked (and even suspect experimental mistakes), but the principle won't be directly faulted.[27]

As we will see, the consideration of empirical data in terms of continuous magnitudes (which result from a measurement operation) plays a crucial role. It must also be noted that in using the expression "*continuous* function," Poincaré refers to *smooth* functions. As a matter of fact, however one defines that *A* and *A'* are "not so different," it is always possible to find a *continuous* (in the modern meaning of the word) function *f* which fits the scatter

---

[27] "... si un antécédent *A* a produit une fois un conséquent *B*, un antécédent *A'* peu différent de *A*, produira un conséquent *B'* peu différent de *B*. Mais comment reconnaîtrons-nous que les antécédents *A* et *A'* sont 'peu différents'? Si quelqu'une des circonstances peut s'exprimer par un nombre, et que ce nombre ait dans les deux cas des valeurs très voisines, le sens du mot 'peu différent' est relativement clair; le principe signifie alors que le conséquent est une fonction continue de l'antécédent. Et comme règle pratique, nous arrivons à cette conclusion que l'on a le droit d'interpoler. C'est en effet ce que les savants font tous les jours et sans l'interpolation toute science serait impossible" (Poincaré, 1905: 177).

plot and such that *f*(*A*) and *f*(*A'*) are very different. Now, since induction—interpolation—aims at allowing predictions, one must notice that a criterion that would demand only continuity (as distinguished from smoothness) would not be enough. Indeed, one can always find continuous functions passing by the same point P and taking very different values for another point located inside any neighborhood of P's abscissa. If we have no *reason* to choose one or another of these functions when fitting a curve, then induction loses all meaning since it does not enable predictions at all. Consequently, restricting the choice of the function induced to mere continuous functions is not enough. Otherwise, one could find among them two functions which fit the scatter plot and which are to each other what "grue" is to "green" in the case of Goodman's emeralds. This difficulty vanishes if one adopts the stronger criterion of smoothness as suggested by Poincaré: if, given a margin of error,[28] several different functions fitting the data are chosen among the smoothest ones, these functions give "not-so-different" predictions and, contrary to "grue" and "green," are thus not in conflict with each other. That is precisely the requirement for the operation of induction to enable predictions and to have a meaning. So, the smoothness criterion is such that, even if it allows several different hypotheses which fit the empirical data, these hypotheses lead to not-so-different predictions. In that way, it is not possible, within the limits imposed by that criterion, to construct hypotheses fitting the empirical data which lead to incompatible predictions, contrary to what "grue" and "green" do. By reason of the approximate nature of every measurement in experimental physics, two predictions that are not-so-different are not incompatible. Let us repeat again here that the point is not to know whether the prediction will be right, but only to have a criterion for choosing a hypothesis which enables predictions. To summarize, the choice of the smoothest curves as a legitimate (lawlike) hypothesis in the context of interpolation corresponds exactly to the principle of induction as it is rephrased by Poincaré.[29]

Thus, the criterion of choice proposed by Poincaré seems to be the only one which gives a meaning to interpolation itself, regarding the aim of this operation, namely *prediction*. The principle of sufficient reason occurs in the following way: if there is no reason to make a choice between two hypotheses which lead to very different predictions, there is no point making an induction. This is a way to understand Poincaré's claim according

---

[28] This margin depends upon the precision of the measurement device:

> If my instruments are good (and I knew whether this is so or not before beginning the observations), I shall not draw the curve far from the points which represent the rough measurements. If they are inferior, I may draw it a little farther from the points, so that I may get a less sinuous curve; much will be sacrificed to regularity (Poincaré, 1952: 205–6).

[29] In the text about the principle of induction, why does Poincaré mention only continuity and not smoothness though he obviously refers to the latter? It can be answered that he uses an old meaning of "continuity," which occurs in Euler's writings, for instance, according to which the continuity of a function means its differentiability and smoothness. As a matter of fact, in *Science et méthode*, Poincaré distinguishes two meanings of the continuity of a function—continuity "according to the *analytical* meaning of the word" and "continuity understood in a *practical* sense" (here, the remark is made about some law of probability):

> There is, then, no means of representing the law of probability of the effects by a continuous curve. I do not mean to say that the curve may not remain continuous in the analytical sense of the word. To

to which, without the "belief in continuity" (made precise in terms of smoothness of the functions induced), induction would not be possible, *would be meaningless*. In other words, the smoothness of the induced hypothesis and the possibility of prediction (regardless of the success of it) are one and the same thing.

The idea can be presented in another way. What we expect from a law is that a case observed in the future is not so different from the prediction that we can make using the law. Thus, *distance* between magnitudes measured plays a central role. Now, the smoothest curve (among the curves which fit the scatter plot) is the one which "changes the most slowly." If we take into account the approximate character of measurement, the faster a curve varies at a point, the less predictive the law (represented by the curve) is in the neighborhood of that point. Consequently, the smoothest function induced is the one which accomplishes the best predictability.[30]

The relevance of Poincaré's criterion lies in the fact that it rests directly upon one central aim of induction, namely prediction: this criterion enables us to choose, not necessarily the hypothesis which will lead to the best predictions, but the one which is the most likely to achieve that. Thus that criterion can be construed as expressing a condition of possibility of induction itself.

Thus it seems that the remarks above show that the smoothness criterion is an *intrinsic* criterion by which legitimate hypotheses can be distinguished from illegitimate ones. Once again, this criterion does not ensure at all that the hypothesis chosen according to it is *true* (in which case it would solve the old problem of induction). The setting of this criterion does not rest upon any successful induction: the issue here is not the success of induction but its meaning. This criterion rests upon neither a metaphysical principle nor future observations. Consequently it fulfills the conditions for solving Goodman's riddle. By examining what amounts to the "new riddle of induction" in the context of interpolation, Poincaré suggests an answer to the riddle that is opposed to Goodman's answer. For, according to Goodman, there are no intrinsic criteria which enable us to understand why some induced hypotheses are viewed as legitimate and why some other hypotheses are not. Still, according to Goodman, only an extrinsic characteristic of the hypotheses enables that, namely *entrenchment*. Indeed, as Poincaré claims it, the smoothness criterion enables us to choose not the "true" function, that is, the function which actually describes the phenomenon, but legitimate induced functions (lawlike hypotheses):

> According to the principle we have just stated, we believe that these points can be con-
> nected by a continuous line. We draw this line with the naked eye. New experiments will

infinitely small variations of the abscissa there will correspond infinitely small variations of the ordinate. But practically it would not be continuous, since to very small variations of the abscissa there would not correspond very small variations of the ordinate. It would become impossible to trace the curve with an ordinary pencil: that is what I mean (Poincaré, 1908: 73–4; 1914: 82–3).

This text shows that the continuity referred to in the rephrasing of the principle of induction is the continuity understood in a *practical* sense, which is described in other texts quoted above as smoothness.

[30] But not necessarily the *true* prediction! That is not the issue here: the problem is not to determine the induced hypothesis which will be confirmed by future measurements (the old problem of induction) but to determine the one which is the most legitimate (the new riddle of induction).

provide new points on the curve. If those points are outside the line drawn in advance, we will have to modify the curve, but not abandon our principle. Any points, as numerous as they may be, can be joined by a continuous curve. Of course, if this curve is too irregular, we may be shocked (and even suspect experimental mistakes), but the principle won't be directly faulted.[31]

Nevertheless, one objection to that solution of the "new riddle of induction" must be taken into account. In the papers mentioned above, the authors,[32] contrary to Poincaré, do not claim that criteria such as the smoothness criterion solve the riddle. They do not always refer to that criterion exactly but also to similar criteria expressed in terms of simplicity, against which they raise a single objection which can be summed up saying that a mere change of variables leads to another choice of the simplest curve for the function. In doing so, they try to show that one can "goodmanize" (we take this neologism from Hacking, 1993) simplicity (or smoothness) criteria, that is, that even within the scope of such a criterion, one can always construct hypotheses fitting empirical data which are incompatible with each other because they lead to very different predictions. Here is how Hullett and Schwartz explain that idea:

> Here, it may seem that the analogy between the "new riddle" and the curve-fitting situation breaks down. For it might be claimed that in the latter case what is projectible is the smoothest curve ... since it would seem possible to construct a mathematical definition of "smoothest curve," which would give us the correct results in our intuitively clear cases, it would seem that we have a purely syntactical way of distinguishing bizarre or unprojectible from projectible curves. However, the problem with this proposal is that in each case the curve that is the smoothest curve is smoothest only with respect to some particular plotting of data points, i.e., with respect to some representation of our evidence. But in the plotting of our data, there are no fixed points. How we plot our data to begin with depends on the sorts of properties or units of measurement we have available. By changing the units of measurement or the properties we plot along either or both our x and y axes, we can always construct alternative graphs whose smoothest curves gives us projections that are incompatible with those of our original graph. Thus, our definition of "smoothest curve" is not enough. We would need, in addition, some means for choosing from among competing graphs that differently represent our data (Hullett and Schwartz, 1967: 112).

Similarly, Priest writes:

> We can summarize the general situation thus: if f is the curve from the simplest family which fits the data S and if θ is virtually any transformation of the Cartesian plane into

---

[31] "En vertu du principe que nous venons d'énoncer nous croyons que ces points peuvent être reliés par un trait continu. Nous traçons ce trait à l'œil. De nouvelles expériences nous fourniront de nouveaux points de la courbe. Si ces points sont en dehors du trait tracé d'avance, nous aurons à modifier notre courbe, mais non pas à abandonner notre principe. Par des points quelconques, si nombreux qu'ils soient, on peut toujours faire passer une courbe continue. Sans doute, si cette courbe est trop capricieuse, nous serons choqués (et même nous soupçonnerons des erreurs d'expérience), mais le principe ne sera pas directement mis en défaut" (Poincaré, 1905: 177).

[32] None of them seem to have noticed that Poincaré sets the riddle in the case of interpolation exactly as they do.

itself, then the image of f under θ will not in general be the curve from the simplest family which fits the image of S under θ. Thus an appeal to simplicity will not help us with this problem of induction (Priest, 1976: 155).

In the same paper, Priest gives an example by using this simple mathematical development (Priest, 1976: 153–4):

Let a set of data be $S = \{(x_i, y_i), i = 1 - n, n \in N\}$.

The point is to predict the value y_0 of y which corresponds to the value x_0 of x, with $x_0 \in S$.

Let us call f_1 the simplest curve (for example the smoothest curve) which fits the data S.

Then our prediction is: $y_0 = f_1(x_0)$.

Let us call f_2 another curve which also fits the data S and such that $f_2 \neq f_1$ and $f_2(x) \neq 0$ for any x.

We have: $y_i = f_1(x_i)$,   for any i

and $y_i = f_2(x_i)$,   for any i

Now let us suppose that we do not deal with the variables x and y but with x and y', with $y' = y / f_2(x)$ (here is the change of variables)

Then for any i, $y_i' = y_i / f_2(x_i)$

So: $y_i' = f_1(x_i)/f_2(x_i)$

and $y_i' = f_2(x_i) / f_2(x_i) = 1$

Consequently, the following curves fit the data S:

$y' = f_1(x) / f_2(x)$

$y' = 1$

Since the second one is simpler (or smoother) than the first one, it will be chosen.

With that choice, the prediction is: $y_0' = 1$

It follows that: $y_0 / f_2(x_0) = 1$

Thus $y_0 / f_2(x_0)$ and not $y_0 / f_1(x_0)$ as was the case before the change of variables.

To sum up, depending on the variables which are correlated, the choice of the simplest curve leads to arbitrary different predictions.

In other words, with an appropriate change of variables, any curve or function fitting the data can be made the simplest one and thus fulfills the criterion by which we choose the most legitimate hypothesis induced.

Let us argue against that objection.

One must notice first that the objection introduces an asymmetry. For it is x and y that are measured, and not y'. To make the objection symmetric, let us consider that the change of variables is a change of measurement device; then the objection can be formulated as follows: if we measured y' instead of y, we would induce another hypothesis from the data, even though using the same criterion to choose the curve. As a matter of fact, we would choose f_2 and not f_1, which are incompatible since the predictions $f_2(x_0)$ and $f_1(x_0)$ are different.

Our claim is that this objection can be answered within Poincaré's philosophy. The example developed by Priest can be used to present the main idea governing that answer: since in both cases (first case: the data are measurements of x and y; second case: the data are measurements of x and y') we choose the simplest or the smoothest curve, we can say that we do not induce two competing hypotheses but one and the same hypothesis, namely the simplest or smoothest curve fitting the data. Therefore the objection is not valid because it rests upon a misleading link between the use of the criterion and the predictions made by hypotheses chosen in different measurement contexts. The idea consists in showing that the fact that f_1 and f_2 compete regarding their predictions does not imply that the criterion used to choose them does not solve the new riddle of induction. The fact that future measurements will show that one of the two hypotheses is right and the other wrong (since they lead to different predictions, such measurements are necessarily possible, provided that one of the two hypotheses is right) does not change anything about the fact that f_1 is *the most legitimate* hypothesis if the data are given by measurements of x and y, and that f_2 is *the most legitimate* hypothesis if the data are given by measurements of x and y'.

Let us develop that answer. The objection could be valid only if the choice of the measurement framework were indifferent or irrelevant, that is, only *if the measurements were merely representations of physical magnitudes which do not depend upon them.* That is precisely what Priest presupposes: "which description is best [i.e., the simplest] depends not on the situation but on how you describe it" (Priest, 1976; 152). The point is that, in Poincaré's philosophy, "the situation" can in no way be considered independent from the measurement framework, because one cannot define physical magnitudes independently from a measurement framework, which therefore cannot be reduced to a mere descriptive system. According to Poincaré, a set of measurements is not a representation or a description of an independent reality constituted of physical magnitudes which are themselves independent from the measurement framework—a reality which would be represented or described by physical laws. Such a view is presupposed by Priest. It is not Poincaré's view. According to Poincaré, objective reality—which we have a knowledge of thanks to physical laws—consists in mathematical relations between measured physical magnitudes, in such a way that this reality depends upon these magnitudes which themselves depend upon the choice of the measurement framework.[33] As a matter of fact, according to Poincaré, the choice of a framework for measuring lengths coincides with the choice of a geometry: it would have no meaning to speak about "lengths" independently of a geometry, and

---

[33] One must recall that this view is not a *nominalist* view, as Poincaré underlines it. The reason is that, even if we choose the measurement framework—and therefore the definition of the magnitudes — *the values of* the magnitudes which are measured do not rest upon our choice. They are imposed by sensible data which constitute experience.

therefore independently of a measurement framework. Moreover, it is only after such a geometric/measurement framework is chosen that physical magnitudes can be defined. Poincaré's geometric conventionalism means exactly this: there are no independent physical magnitudes which would be measured in different ways depending on the choice of geometry. It is the chosen geometry which institutes both the measurement framework and the definition of physical magnitudes. Now, because of the dependence between physical magnitudes and measurement/geometrical framework, there is no meaning in asserting that y and y' are two different representations or descriptions of one physical magnitude whose values would not depend upon that framework.

Therefore, the situation in experimental physics is the following one: the starting point is a set of measurements which presupposes the choice of a geometrical/measurement framework from which physical magnitudes are defined; from the set of measurements, the relations between physical magnitudes—the physical laws—are induced. Physical laws do not predict phenomena which could be arbitrarily measured, but the results of measurement operations whose nature and application are not independent from the way the laws are induced. Thus f_1 and f_2 compete only within a same framework and not outside measurement frameworks. And we are right to induce f_1 if x and y are measured, and we are also right to induce f_2 if x and y' are measured: in both cases, the same hypothesis is induced, namely the smoothest curve or function fitting the data on which we work. If a future measurement of y_0 confirms one of the hypotheses and refutes the other one, then it will only show that induction led to a *wrong* law in one of the two cases, but not that it was *illegitimate* to induce that law. Now, the new riddle of induction deals with the latter issue and not with the former. That is the reason that the objection raised by Priest and Hullett and Schwartz is not valid within Poincaré's philosophy.

From this point one can ask the following question: if we admit that Goodman's and Poincaré's arguments about the new riddle of induction are both valid, how can one explain that their solutions differ? We suggest that this difference rests upon the logical forms according to which Goodman and Poincaré analyze the induced hypothesis. Goodman's argumentation about "grue" takes place within a logical framework governed by the relation [individual subject/predicate]. It implies that the world is already divided into *fixed individual entities* (here: emeralds). Now, one concern of Poincaré's—notably within his critics of ontological commitments—is to challenge the relevance of that logical framework in philosophy of mathematics and physics; more precisely, Poincaré objects to the idea that mathematical objects and operations could be relevantly described according to the distinction between subject and predicate.[34] The whole discussion above consists precisely in underlining the idea that it is not relevant to construe a physical law as an independent reality (i.e., as a logical subject, namely as a set of related values independent from a measurement framework) whose properties, such as the form of the curves describing it, would depend upon a system of representation or description, which is presupposed by Priest. We have tried to show that such a view is incompatible with Poincaré's philosophy. Now, the argument of a change of variables, as well as Goodman's

---

[34] We have already suggested this point about the way Poincaré analyzes generality and generalization.

reflections about "grue," necessarily implies that one *logically distinguishes* given entities on one hand, and some of their properties on the other hand, respectively: "real" law and emerald on the one hand, and representative curves and colors on the other hand. Without that logical independence, there is no possibility of predicating different properties (which depend upon the framework of representation) to a same subject. That is precisely what Priest does: he considers a logical subject—the "physical magnitude"—and constructs two frameworks of representation (which can be understood as measurement frameworks) such that the properties that we predicate to that logical subject—the functions or the curves chosen by interpolation within each of these frameworks—are incompatible in spite of the fact that their choice rests on the same criterion. According to Poincaré's conception of physical laws, the situation is very different: sensible experience is structured by the choice of a geometrical space which institutes both a measurement framework and physical magnitudes. Objective reality consists in mathematical relations between the latter. These relations are curves or functions which are induced from actual measurements according to smoothness criterion. We have tried to show that only this criterion gives a meaning to induction (to interpolation) and that one cannot "goodmanize" that criterion for there is no relevant way to describe objective reality and physical laws according to a [subject/predicate] logical form, such as the one occurring in what we called 'the predicative generality' in the previous sections. Therefore, here again, empirical induction in experimental physics (interpolation) is a generalization which cannot relevantly, according to Poincaré's philosophy, be understood within the logical context of predicative generality. One can notice that Goodman, on the contrary, understands statements such as "All emeralds are green" in terms of predicative generality.

To summarize, according to Poincaré, induction in experimental physics does not consist in extending the domain of a predicate. In other words, a mathematical function should not be understood as a *predicate*. In that case, interpolation could be understood as a *projection* (in the sense Goodman uses this word) of a predicate. That is what authors quoted above do to apply Goodman's analysis to interpolation. But this is relevant only if one can logically distinguish a reality and the function expected to represent or describe it: the function is then construed as a predicate of that reality, which depends upon the representation framework. We tried to show that Poincaré's conception of objective reality and physical laws is completely different from that view and doesn't allow such a philosophical argumentation. Therefore we can draw a conclusion similar to the conclusion of the previous section: due to the use of mathematics in physics and similarly in the case of generalization in mathematical physics, empirical induction in physics differs from induction understood as the extension of the domain of a predicate, which occurs in other empirical sciences.

REFERENCES

Primary sources

Poincaré, H. (1902) *La science et l'hypothèse*. Paris: Flammarion, reed. Flammarion, 1968.
Poincaré, H. (1952) *Science and hypothesis*. New York: Dover Publications.

Poincaré, H. (1905) *La valeur de la science*. Paris: Flammarion, 1970.

Poincaré, H. (1958) *The value of science*. New York: Dover Publications.

Poincaré, H. (1908) 'Science et méthode', *Philosophia Scientiae 1998–99 Cahier Spécial 3*. Paris: Kimé, 1999.

Poincaré, H. (1914) *Science and method*. London, Edinburgh, Dublin and New York: Thomas Nelson and Sons.

Poincaré, H. (1913) *Dernières pensées*. Paris: Flammarion; rev. ed. Flammarion, 1930.

Poincaré, H. (2002) *L'opportunisme scientifique*, compiled by L. Rougier, ed. L. Rollet. Basel, Boston, Berlin: Birkhäuser.

Secondary literature

Goodman, N. (1973) *Facts, fiction, and forecast*. New York: The Bobbs-Merrill Company.

Hacking, I. (1993) *Le plus pur nominalisme*, traduit de l'anglais par R. Pouivet, Combas: Editions de l'éclat.

Heinzmann, G. (1988) 'Poincaré's philosophical pragmatism and the problem of complete induction', *Fundamenta Scientiae* **9**/1: 1–19.

Hullett, J., and Schwartz, R. (1967) 'Grue: some remarks' in *The philosophy of Nelson Goodman*, Tome II: *Nelson Goodman's new riddle of induction*, ed. C.Z. Elgin. New York & London: Garland Publishing, 1997: 109–21.

Priest, C.S. (1976) 'Gruesome simplicity' in *The philosophy of Nelson Goodman*, Tome II: *Nelson Goodman's new riddle of induction*, ed. C.Z. Elgin. New York & London: Garland Publishing, 1997: 152–7.

Part II

Statements and concepts: the formulation of the general

Section II.1

Developing a new kind of statement

6

Elaboration of a statement on the degree of generality of a property: Poincaré's work on the recurrence theorem

ANNE ROBADEY*

6.1 Introduction

If ABC is a right-angled triangle at A, then $BC^2 = AB^2 + AC^2$.

This statement of Pythagoras's theorem is formed of two parts: a hypothesis and the ensuing conclusion. Many theorems are constructed in such a manner. A set of hypotheses defines a set of situations for which a property is satisfied. However, these hypotheses may be stated in different forms, which are not always interchangeable. Therefore, one can of course state:

If $a \neq 0$, then the equation $ax + b = 0$ admits a unique solution.

But one might also readily say:

Equation $ax + b = 0$ admits a unique solution, unless $a = 0$.

* This chapter was prepared by Anne Robadey on the basis of her Ph.D. Dissertation (Robadey, 2006), before she entered the Cistercian Abbey Notre Dame d'Igny, in which she now lives. She could not revise the translation carried out by Théodora Seal and prepared for publication by Karine Chemla, Renaud Chorlay, and Jonathan Regier. We refer the reader to her dissertation for a French version and further developments of the ideas presented in this chapter (http://tel.archives-ouvertes.fr/tel-00011380/).

On the contrary, barring a very particular context, the following statement seems incongruous:

In the triangle ABC, $BC^2 = AB^2 + AC^2$, unless ABC is not right-angled at A.

Thus, there is a difference between the two conditions "if $a \neq 0$" and "if ABC is right-angled at A."

In view of these simple examples, the difference between the two kinds of hypotheses seems clear. The case where $a = 0$ constitutes a *particular case*, an *exception; in general*, equation $ax + b = 0$ admits a unique solution. By contrast, the fact that a triangle is not right-angled is not exceptional, on the contrary. Therefore it is essential to specify in the statement of Pythagoras's theorem that the triangle must be right-angled.

No doubt, this difference has been perceived by mathematicians for a long time. This can be seen from the fact that Pythagoras's theorem has always been presented as stating a property of right-angled triangles, whereas the limits of validity of statements such as "equation $ax + b = 0$ admits a unique solution" have not always been given such attention.

But it seems that it is at the end of the nineteenth century that a new type of statement—today widespread in many areas of mathematics—appears. It states that a property is true for "almost all" considered objects, in a precise mathematical sense. Several characteristics of such a proposition deserve further development. First, the property is not necessarily satisfied by all the objects, it may admit exceptions. Further, a mathematical tool is introduced in order to quantify these exceptions and characterize their rarity. Finally, no attempt is made to find a criterion allowing identification of the exceptions.

The history of the theorem commonly known today as the "recurrence theorem"[1] bespeaks Poincaré's conception of such a statement—probably one of the first. For certain systems of differential equations, this theorem ensures the existence of *recurring* solutions (or trajectories):[2] for any region of the studied domain, as small as it might be, there exist trajectories that return to it infinitely often. But Poincaré also studied solutions that do not have this property. Therefore, there are recurring trajectories and others that are not. Poincaré states that the first are more general. But in what sense?

Poincaré progressively clarifies his answer to this question through a series of texts.

The theorem appears in Poincaré's memoir *Sur le problème des trois corps et les équations de la Dynamique*, which competed for the prize offered in 1888 by King Oscar II of Sweden. The original manuscript has not been found;[3] however, there are two printed versions of the memoir. Following the discovery of an important error, the first, (Poincaré, 1889; hereafter: [Pa]), was not distributed. The corrected memoir, (Poincaré, 1890; [Pb]), was published in *Acta mathematica* the next year. It was in this amended version of the text that Poincaré added a corollary to the recurrence theorem. He created a statement that gives a precise mathematical meaning to the comparison between the

---

[1] Poincaré does not himself give it this name. This term is introduced by Birkhoff to designate a slightly stronger property than that studied by Poincaré (but very similar).

[2] Rather, Poincaré considers them as "stable" trajectories, or "stable in the sense of Poisson."

[3] See Barrow-Green (1997).

degrees of generality of the two types of trajectories. The non-recurring trajectories are "exceptional" and this term is defined using a concept built on probability theory. By this, Poincaré means that the probability is equal to zero for a trajectory randomly chosen not to be a recurring one.

The 1890 memoir is soon followed in 1891 by two summaries without proofs. In 1894, Poincaré again mentions the recurrence theorem, still without proof, in a very different context: the kinetic theory of gases. Finally the theorem, with a proof, is integrated into the third volume of Poincaré's master work on celestial mechanics *Méthodes nouvelles de la mécanique céleste*.[4] The following table gives an overview of this corpus.

Sur le problème des trois corps et les équations de la Dynamique			
	1888—	original version (not found)	manuscript
[Pa]	1889—	first printed version, with appended notes	not published
[Pb]	1890—	corrected version of [Pa], including appended notes	published
January	1891—	Two summaries of [Pb], without proof: (Poincaré, 1891b) *Bulletin astronomique* (Poincaré, 1891a) *Revue générale des sciences*	
	1894—	*Sur la théorie cinétique des gaz* (Poincaré, 1894) *Revue générale des sciences*	
	1899—	*Méthodes nouvelles de la mécanique céleste* (Poincaré, 1892–99)	(volume III)

Why does Poincaré feel the need in 1890 to thematize the exceptional character of non-recurring trajectories? What makes this thematization possible?

The comparison between the successive versions of the memoir gives access to the context in which Poincaré introduces this new statement. We shall study more particularly the relations between the addition of the corollary and the other alterations of the text following the discovery of the error. This will allow us to identify the motivations which render Poincaré's reflections on the generality of recurring trajectories necessary.

The existence of successive texts will also help us to better understand Poincaré's work. Indeed, the progressive improvements of the text shed light on the questions, which drive Poincaré's reflection and on which he concentrates his efforts. We shall see that this piece of work focuses, of course, on proving the result, but also on the formulation of the statement itself and the means set forth to formulate it. In particular, Poincaré adapts the calculus of probability so as to make it a tool corresponding to the encountered problem. To highlight this part of Poincaré's work, alongside the study of the group of texts on the recurrence theorem, we shall also consider Poincaré's course on probabilities, given in 1894 at the Sorbonne and published in 1896. The latter will

---

[4]   I shall abbreviate this title as *Méthodes nouvelles*.

be compared to Bertrand's work (1888); Bertrand's text is a source of inspiration for Poincaré, although he disagrees on several points, in particular, those related to his research on the recurrence theorem.

6.2 *"Exceptional"* trajectories: introduction of a new language and concept into the formulation of the recurrence theorem

The recurrence theorem concerns the trajectories of a system of differential equations of the form $\dfrac{dx_i}{dt} = X_i$, in dimension three (i = 1, 2, 3). In other words, let us suppose that the motion of a point P is governed by these equations and consider the different curves, which are described by this point according to its initial position. The theorem is identically given in both memoirs [Pa] and [Pb]:

Theorem 1. Let us suppose that the point P remains at finite distance and that the volume $\int dx_1 dx_2 dx_3$ is an integral invariant; if any region r_0 is considered, as small as this region might be, there will be trajectories which pass through it infinitely many times.[5]

It is then extended to the case of a point P moving in a space which is of any dimension, and where there exists an integral invariant (not necessarily the volume), the point P remaining at finite distance. In other words, as Poincaré announces himself in both memoirs, this theorem states that, given the above hypotheses, there are an infinity of trajectories, which are stable in a way named "stability in the sense of Poisson"[6] by Poincaré in [Pb].

From [Pa] onward, Poincaré claims a stronger result, which concerns the nature of this infinity: "I shall even add that the trajectories which have this property are more general than those that do not, precisely just as incommensurable numbers are more general than commensurable numbers." But this assertion is given without proof.

6.2.1 Explicit choice of a specific vocabulary

The idea of comparing—from the standpoint of their generality—recurring trajectories with those that are not can again be seen in the later texts. But the formulation of this comparison changes. By taking one by one the successive statements of the recurrence theorem, I shall point out an interesting process. Starting in 1890, not only does Poincaré develop

---

[5] [Pa, p. 31] and [Pb, p. 314] (*Œuvres*). The statement, identical, is entitled "Theorem 1" in both texts. Only the typography presents slight variations.

[6] In the third memoir *Sur les courbes* (Poincaré, 1885, p. 94), Poincaré already had defined a very similar notion of stability.

a stable terminology to state the contrast between the two types of trajectories; he also insistently emphasizes the technical meaning given to the terms he introduces. The making of statements expressing the results is also part of the mathematician's work. I shall first examine the *formulation* of the comparison between the recurring and non-recurring trajectories, before considering secondly the mathematical tools used by Poincaré. This analysis will then allow us to highlight the correlation between the development of results and that of statements.

In [Pb], Poincaré announces, right at the beginning of the sections that interest us, the reinforcement of the theorem: "I shall add that the first [the trajectories that are not stable in the sense of Poisson, AR] can be regarded as exceptional and I shall seek later on to clarify the precise meaning I give to this word" [Pb, p. 314].

The formulation has changed in comparison to [Pa]. The accent was placed on trajectories that are "more general" than others. The emphasis is now placed on the "exceptional" character of the latter. We shall soon see that this change of standpoint is related to what Poincaré's proof establishes in [Pb]. Further, right from this introductory announcement, Poincaré draws the reader's attention to the "word" he is using. The notion of precision that was already present in 1889 is found again "precisely just as." But, in that case, mathematical objects were being directly compared. In [Pb], it is a word that is going to receive a precise mathematical specification. Therefore, in 1890, Poincaré thematizes a mathematical *notion* by attaching to it a *vocabulary*.

Following the statement of Theorem 1 and its proof, which were kept unchanged in relation to [Pa], it is in the form of a corollary that Poincaré introduces the announced clarification [Pb, p. 316]:

Corollary 1. *It follows from what precedes that there exist an infinity of trajectories that pass through the region r_0 infinitely many times; but others may exist that only pass through this region a finite number of times. I now intend to explain why the latter trajectories can be regarded as exceptional.*

Poincaré immediately points out that "this expression does not have any precise meaning in itself," and gives a "definition" of it in probability terms. Moreover, Poincaré mentions that he takes a term from everyday language and gives it a technical meaning. In the rest of the memoir, he reserves the term "exceptional" for this precise meaning. Besides, on occasion he recalls how the term is defined: "if initial conditions are not exceptional *in the sense given to this word in the corollary of Theorem 1*" ([Pb, p. 323]; emphasis added).

If alerted by the author's insistence on this term, one reads the introduction of [Pb] again, it appears that Poincaré already uses the same word, although not yet insisting on the technical meaning he will give to it in the main part of the memoir. Indeed, he describes his work in the following way: "I studied more particularly a particular case of the three-body problem, that in which one of the masses is zero and the motion of the other two is circular; I recognized that in this case the three bodies pass infinitely many times as close as one wants to their initial position, unless the initial conditions of the motion are exceptional" [Pb, p. 264].

In the texts that interest us, the vocabulary of "exceptions" is simultaneously used in other senses. However, it can be shown that from 1890 Poincaré systematically uses this terminology when referring to the corollary of the recurrence theorem. Further, we shall see that, almost always, he carefully points out the technical meaning in which he uses it, referring to probability calculus.

Therefore, in the summary (Poincaré, 1891b) of the results published in the memoir [Pb] written for the *Bulletin astronomique*, one reads:

> Thus, from this point of view, one can say that there are an infinity of unstable particular[7] solutions and an infinity of stable particular solutions.
>   But there is more: one can say that the first are the exception and the latter are the rule, just as rational numbers are the exception and incommensurable numbers are the rule (Poincaré, 1891b, p. 490).

The sentence is built on the same pattern as the one in [Pa]: the distinction between the two types of trajectories is compared to that between commensurable and incommensurable numbers. But Poincaré now uses the term of "exception," according to the convention he defined in [Pb], rather than the expression "more general" found in [Pa]. He then clarifies the meaning taken by the term of "exception," which he has just used:

> I prove, indeed, that the probability that the initial circumstances of the motion are those corresponding to an unstable solution, I say, that this probability is equal to zero. This word has no meaning by itself: I give in my Memoir a precise definition that I do not think useful to reproduce here [...] (Poincaré, 1891b, p. 490).

Therefore, even without the intention of going into any details, Poincaré takes care to use the vocabulary introduced in [Pb]. Further, even though he does not go over it again in 1891, he insists on the fact that he gave a proof of this statement. Finally, let us note that the word of which Poincaré announces the definition is that of "probability" rather than "exception."

The same phenomenon can be observed in the article written for the *Revue générale des sciences* (Poincaré, 1891a). The formulation is similar, with the same insistence on the technical character of the terminology:

> I shall add that the first [the unstable solutions, AR] are exceptional (this allows to say that in general there is stability). Here is what I mean by this, because by itself this word has no meaning. What I want to say is that there is a probability equal to zero that the initial conditions of the motion are those corresponding to an unstable solution (Poincaré, 1891a, p. 68).

Here again the term "exceptional" is found with its definition in probability terms. The parenthesis inserted by Poincaré explains the main reason which makes him clarify

---

[7] That is, non-recurring (AR).

the meaning of the term and prove the resulting theorem, on the probability of unstable trajectories. The purpose is to go further than the result proved since [Pa], which only establishes the existence of an infinity of stable trajectories in any region, or, in modern terms, the density of stable trajectories. The characterization of unstable trajectories as "exceptional" allows us to establish mathematically the assertion of stability "in general." We will come back to this point in Section 6.4.2.

In 1894, Poincaré again refers to the recurrence theorem in another article for the same journal (Poincaré, 1894), where he reminds the reader of his results in celestial mechanics for his argumentation on the kinetic theory of gases.

Finally, in *Méthodes nouvelles*, Poincaré again employs the same terms. In this work, he states and proves the recurrence theorem by using a language with more imagery. He compares the trajectories of a system of differential equations to those of molecules, the motions of which are defined by the equations.[8] Here is how he announces the corollary of the recurrence theorem:

> We have seen in n°291 that there are molecules which pass through U_0 infinitely many times. On the other hand, in general, others only pass through U_0 a finite number of times. I propose to show that the latter must be regarded as exceptional or, to be more precise, that the probability that a molecule passes through U_0 only a finite number of times is infinitely small, *if one admits* that this molecule is inside U_0 at the origin of time (Poincaré, 1892–99, t. III, p. 151).

The term "exceptional" is used many times in the rest of the *Méthodes nouvelles*, always with the same technical meaning, and often accompanied by the words "in the sense that I gave to this word above."

Therefore, a clear contrast can be observed between [Pa], on the one hand, and [Pb] and all the later texts on the other.

In 1889, Poincaré merely makes a brief remark, without proof, on the greater generality of recurring trajectories. The reader must himself deduce what led Poincaré to such an affirmation. The very meaning of the remark is not obvious. The comparison with "incommensurable numbers" is supposed to shed light on the nature of the generality of recurring trajectories, but it can be understood in several manners. Is Poincaré considering a characterization of recurring trajectories by the irrationality of a parameter? Is he referring to the opposition, proved a few years earlier by Cantor, between the set of rational numbers, which is countable, and the set of irrational numbers, which is uncountable? Or is he already thinking of the proof he gives in [Pb]? The question remains open.

From 1890, Poincaré introduces a new vocabulary. He uses it systematically when talking about non-recurring trajectories, and he insists on the technical character of this terminology. Further, he almost always mentions the definition, in terms of zero

---

[8] This image was most certainly suggested by the comparison between the recurrence theorem and questions of kinetic theory of gases.

probability, that he gives to the word "exceptional," even if he does not go into the details of the probability definition.

6.2.2 The mathematical content of this terminology

Poincaré does not merely create a technical vocabulary to address the generality of certain trajectories; he explicitly associates with this terminology a precise mathematical definition.

This remark might seem at first sight tautological. However, it is not: in Poincaré's work, there are other expressions of a technical nature, which are not accompanied by a definition. This is the case for "the most general polynomials of their degree." This expression appears, several times, in italic characters in two successive memoirs, without any variation—not even small—in the words composing it, that is, Poincaré (1878) and Poincaré (1881). Thus, the way in which this expression is used in these texts leads to the same conclusion as the one for the term "exceptional": this is an expression that seems to come from everyday language, but its stable use and the emphasis placed on it by the author—via, in this case, the use of italic characters—reveal its technical nature.

But Poincaré, in these texts, does not define what he means by "the most general polynomials of their degree." To me, it seems that the difference between Poincaré's attitude toward "the most general polynomials of their degree" and that toward "exceptional trajectories" shows the novelty of his characterization of the latter. Poincaré could rely on his reader's mathematical education for the understanding of "the most general polynomials of their degree." However, he draws the reader's attention to the new definition he gives to the exceptional character of non-recurring trajectories.

Let us now consider the definition of the term "exceptional" such as it is given in [Pb]:

> Let us adopt the convention that the probability that the initial position of the moving point belongs to a certain region r_0 is to the probability that this initial position belongs to another region r_0' in the same ratio as the volume of r_0 to the volume of r_0'.
>
> The probabilities being defined in this way, I propose to establish that the probability that a trajectory coming from a point of r_0 does not pass through this region more than k times is equal to zero, however large k is and however small the region r_0 is. This is what I mean when I say that the trajectories that only pass through r_0 a finite number of times are exceptional [Pb, p. 316].

Poincaré thus chooses a probabilistic characterization to clarify what he means by "exceptional trajectories." This invites us to study the relation between the memoir on the three-body problem and Poincaré's work on probability. It is a fact that all of Poincaré's texts specifically on probability calculus come after [Pb]. Therefore, the memoir on the three-body problem is the first witness of Poincaré's reflections in this area. This circumstance however does not deprive us of means of investigation. Indeed, one can compare the manner in which probability calculus is employed in the work on the three-body problem with Poincaré's presentation of the subject in the following years. The continuity that we shall point out shows that the use of probabilities, in 1890 and 1891, is the opportunity for Poincaré to develop his own conception of this area of mathematics. Therefore, the issue at stake in this project is not only the emergence of a new type of statement. The project also sheds light on Poincaré's development of probability calculus. This is the reason I shall devote an entire section to this topic (see Section 6.6).

Our interest will concern, in particular, Poincaré's first work on probability calculus (Poincaré, 1896). It contains the lectures given during the second semester of 1893–94 at the Sorbonne. Poincaré then published articles with a more philosophical aim and, in 1912, an augmented new edition of *Calcul des probabilités*.[9]

I would like to now show that the way in which Poincaré defines probability in [Pb] follows the theoretical requirements set out in Poincaré (1896). Therefore, this already is a technical use of probabilities. The first words of the technical definition of [Pb], "*let us adopt the convention*[10] that the probability [...]" indeed find a clear echo in the first of the *Leçons sur le calcul des probabilités*: "The complete definition of the probability is therefore some kind of *petitio principii*: how can one recognize that all the cases are equally probable? Here, a mathematical definition is not possible; we must define *conventions*[11] for each application, specify that we consider such-and-such a case equally probable. These conventions are not entirely arbitrary, but they are concealed from the mind of the mathematician, who will not have to examine them once they are acknowledged" (Poincaré, 1896, p. 5). In 1890, Poincaré indeed gives a *convention*, which clarifies the cases assumed to be equally probable: the belonging of the initial position to two regions is assumed to have the same probability whenever the regions have the same volume.

In both summaries of the results of [Pb] published in 1891, Poincaré adds a remark concerning the validity of the corollary. For us, it is interesting, because he insists on the fact that the choice of how the probability is defined is free, to a large extent:

> One might object that there is an infinity of manners to define this probability; but this [the probability that the initial conditions of the motion correspond to a non-recurring solution is equal to zero, AR] remains true whatever the definition one takes, provided: let x and y be the coordinates of the third mass [this is in the context of the three-body problem, AR], x' and y' the components of its speed. I call $P\,dxdydx'dy'$ the probability that x lies between x_0 and $x_0 + dx$, y between y_0 and $y_0 + dy$, x' between x'_0 and $x'_0 + dx'$, y' between y'_0 and $y'_0 + dy'$. We can define the probability as we want and, consequently, arbitrarily give P as a function of x_0, y_0, x'_0, and y'_0. Well, the result I stated above remains true, whatever this function P, *as long as it is continuous* (Poincaré, 1891a, p. 537).

The formulation is briefer in Poincaré (1891b), but it concerns the same idea (p. 490). The summary for the *Bulletin astronomique* (Poincaré, 1891b) and the article in the *Revue générale des sciences*, which has been examined above, are texts that were mainly intended

---

[9] As both editions, dated 1896 and 1912, show few differences with regard to the main body of the text and as the second is much easier to consult, I shall quote the lectures with reference to the pagination of the second edition, that of 1912, except for the passages which were altered by Poincaré in between the two editions. The major changes consist in the addition of a long introductory chapter and a supplementary final chapter.

[10] Emphasis added.

[11] Here, the italic characters are due to Poincaré. Poincaré often used this term in his philosophical texts (see, for example, Poincaré, 1902). I shall only consider Poincaré's use of it in the field of probabilities.(AR).

for the popularization of the ideas in [Pb]. In fact, these texts are much shorter and it might seem *a priori* that precision was not the most sought-after value. And yet, it is in these texts that a supplementary detail appears in comparison with the 1890 analysis, namely that the result remains unchanged even if the definition of the probability is changed.

This point is extremely revealing. On the one hand, it can be concluded: Poincaré realized that the probability definition could be given a large scope precisely between the writing of [Pb] and the writing of the 1891 articles. On the other hand, this property is significant for Poincaré, since he mentions it in short summary articles.

This fact is confirmed in *Méthodes nouvelles*, where this addition is inserted from the moment Poincaré defines what he means by probability.

> [...] First, I must explain the meaning I give to the word *probabilities*. Let $\varphi(x,y,z)$ be any positive function of the three coordinates x, y, z; I shall say that the probability that at instant $t = 0$ a molecule is inside a certain volume is proportional to the integral
>
> $$\mathcal{J} = \int \varphi(x,y,z)dx\,dy\,dz$$
>
> extended to this volume. It is therefore equal to the integral $\mathcal{J}$ divided by the same integral extended to the entire vase V.
>
> The function φ can be arbitrarily chosen, and the probability is thus completely defined: as the trajectory of a molecule only depends on its initial position, the probability that a molecule behaves in such-and-such a manner is an entirely defined quantity from the moment the function φ is chosen (Poincaré, 1892–99, t. III, pp. 151–2).

He then proves the corollary of the recurrence theorem for the case where $\varphi = 1$, which corresponds to the convention adopted in [Pb] where the probability is proportional to the volume. Finally, he returns to the case of any function φ, and announces that "the same results remain when, instead of taking $\varphi = 1$, an entirely different choice is made for the function φ." He then proves this remark: this is its first proof, since in the 1891 articles he merely stated the results of [Pb]. Let us point out that it is in the proof that he assumes φ continuous.

Poincaré will mention in his 1894 lecture the existence of similar problems, in which the probability of an event does not depend on the adopted definition of the probability, as long as it is given by a continuous function φ. We shall come back to this point in more detail in Section 6.6; already this shows that, in Poincaré's eyes, the addition introduced in 1891 belongs to a technical use of probabilities.

6.2.3   The proofs of the recurrence theorem and the corollary

So far our interest was focused on the vocabulary introduced by Poincaré to formulate the corollary added in 1890, and then on the probabilistic concepts, which Poincaré uses to define this terminology. We now must examine the proof of the affirmation according to which the trajectories that are not stable in the sense of Poisson are exceptional, that is, of probability equal to zero. Thus, we shall consider precisely what the tools are that Poincaré uses to establish this property. The comparison of the proof with that of the recurrence theorem will allow a better understanding of the relations between the initial theorem and the corollary.

Let us first consider the proof of the recurrence theorem proper, as it is given in [Pa], and again, almost given word for word in [Pb].

Poincaré first notes the first hypothesis of the theorem, namely that there exists, in the phase space, a region R containing the trajectories and of finite volume V. He then considers "a very small region" r_0 in R, of volume v. Arbitrarily setting a time scale τ, he considers the images $r_1, r_2, ..., r_n$ of the region r_0 under the flow of the differential equations at times $\tau, 2\tau, ..., n\tau$. He names r_n the nth *consequent* of the region r_0. Conversely, r_0 is called the nth *antecedent* of r_n.

The second hypothesis stipulates that the volume is an integral invariant. It follows that a region and its consequents have the same volume.

From this, Poincaré deduces that any region r_0 has a part in common with one of its consequents $r_1, r_2, ..., r_n$, from the moment that $n > \dfrac{V}{v}$. Indeed, $r_0, r_1, ..., r_n$ are all of equal volume v, and in the interior of R of volume V. If they had no common point, their total volume would satisfy the two incompatible conditions of being greater than nv, but smaller than V, the total volume of the region R containing them. Therefore, at least two of these regions have a common part when n is large enough. If r_p and r_q are these two regions (with $q > p$), if follows that their $(q - p)$th antecedents, r_0 and r_{q-p}, also have a common part: r_0 indeed shares a common part with one of its consequents.

Poincaré then notes a generalization of this property that he had not used in 1889, but which will be of use in [Pb] for the proof of the corollary: for any k, there exists a common part to at least k regions $r_0, r_1, ..., r_n$, from the moment that n is large enough. The proof is similar to the previous one.

Thanks to the first property, Poincaré builds two sequences of nested regions $r_0, r'_0, r''_0, ..., r^n_0, ...$ and $s'_0, s''_0, ..., s^n_0, ...$ in the following way. Let r_{p_0} be the first consequent of r_0 which has a part in common with r_0. Poincaré writes r'_0 this common part, and s'_0 the p_0th antecedent of r'_0. Then s'_0 is included in r_0, similarly to r'_0, which is its p_0th consequent. r''_0 and s''_0 are then built in the same way from r'_0.

By induction hypothesis, let us suppose r^n_0 and s^n_0 built such that s^n_0 is included in r_0, just as all its consequents of orders $p_0, p_0 + p_1, ..., p_0 + p_1 + \cdots + p_{n-1}$.

Let $r^n_{p_n}$ be the first consequent of r^n_0 that has a common part with r^n_0. r^{n+1}_0 is defined as this common part, and s^{n+1}_0 as the antecedent of order $p_0 + p_1 + \cdots + p_n$ of r^{n+1}_0. Then s^{n+1}_0 is included in r_0 just as all consequents of order $p_0, p_0 + p_1, ..., p_0 + p_1 + \cdots + p_n$.

Poincaré then considers the non-empty intersection—Poincaré states it without justification—of all regions s^n_0, that he writes σ. By construction, σ and all its consequents of order $p_0, p_0 + p_1, ..., p_0 + p_1 + \cdots + p_n, ...$ are in r_0. Therefore, the trajectory coming from a point of σ goes through r_0 infinitely many times.

Let us now consider the proof of the corollary in [Pb]. Poincaré seeks to calculate the probability that a trajectory coming out of a point of r_0 does not go $k + 1$ times through the region r_0 between time zero and time $n\tau$. He uses the property proved above, which ensures that if $n > \dfrac{kV}{v}$, then we can find $k + 1$ regions $r_0, r_{\alpha_1}, ..., r_{\alpha_k}$ that share a common part s_{α_k}. Let s_0 be the antecedent of order α_k of s_{α_k}. One can see that the trajectory coming from a point of s_0 passes through the region r_0 at least $k + 1$ times between time zero and time $n\tau$. In other words, Poincaré shows here that if the volume v of a region r_0 satisfies the inequality $n > \dfrac{kV}{v}$, then there exist points in r_0 (that it suffices to take in the region s_0) such that the trajectory coming from it goes through r_0 at least $k + 1$ times between time zero and time $n\tau$.

Poincaré is now interested in the region σ_0 of r_0 defined by the following property: the trajectories coming from the points of σ_0 do not go through the region r_0 more than k times between time zero and time $n\tau$. Let us remark that this region can be defined from r_0 and its n first consequents by a finite number of unions and intersections. Moreover, Poincaré indicates how to do it: all the "analogous" regions to the region s_0 just defined are considered, that is, for each combination $\alpha = (\alpha_1, ..., \alpha_k)$ of k indices $\alpha_1 < \cdots < \alpha_k$ between 1 and n, one considers s_0^α, the α_kth antecedent of the intersection of regions $r_0, r_{\alpha_1}, ..., r_{\alpha_k}$ (some of the s_0^α can be empty, but Poincaré showed that there exist some that are non-empty). Then σ_0 is obtained by removing from r_0 the union of these s_0^α, which are in finite number. Thus, to consider the volume of σ_0 does not require having a measure of volume that is countably additive.

The volume of the region σ_0 is denoted by w. The previous reasoning, applied to σ_0 instead of r_0, shows that $w < \dfrac{kV}{n}$, since no trajectory coming from σ_0 passes through r_0, nor *a fortiori* σ_0, $k + 1$ times between time zero and time $n\tau$.

It follows that the probability that a trajectory coming from r_0 does not pass through this region more than k times between these two times—probability given, by definition, by the ratio $\dfrac{w}{v}$—is upper bounded by $\dfrac{kV}{nv}$. This bound can be made as small as one wishes by taking n large enough. If one no longer limits oneself to time $n\tau$, we see that the probability that a trajectory coming from r_0 does not pass through it more than k times is smaller than $\dfrac{kV}{nv}$, whatever n is. Therefore, it is equal to zero.

In other words, the key property, which allows one to prove the theorem as well as the corollary, is the following remark: when a region has a large enough volume, it can be deduced that it has a non-empty intersection with some of its consequents. The property is used in this form to prove the theorem, and in the contrapositive form to prove the corollary: an upper bound of the number of intersections of a region with its consequents is rewritten as an upper bound of its volume.

6.3 A piece of explanation for this novelty: the change of status of the recurrence theorem

6.3.1 How to interpret the changes between [Pa] and [Pb]?

The above analysis highlighted that a continuity exists in the formulation of the corollary of the recurrence theorem in [Pb] and in all the following texts. Indeed, the development of the concept of "probability equal to zero," which Poincaré uses to characterize "exceptional" trajectories, goes on after the publication of [Pb]. But, the 1891 and 1899 additions only deepen what was already present in [Pb], and they are expressed by means of the same technical vocabulary.

On the other hand, the formulation between [Pa] and [Pb] undergoes a mutation, and in [Pa] Poincaré does not explicitly explain what is covered by his remarks. Therefore, the question arises whether the development of a vocabulary associated with a precise mathematical definition, in [Pb], is only the clarification that came with the rewriting of the memoir—in order to detail what had only been outlined—or whether Poincaré really deepened his understanding of the phenomenon in reworking his memoir.

One might be tempted to tend toward the first interpretation and point out that, from the first version, Poincaré claimed the existence of a precise characterization, although without saying so explicitly: "I could even add that the trajectories that have this property are more general than those that do not, *just as much* as incommensurable numbers are more general than commensurable numbers."[12]

It is difficult to know what Poincaré was aiming at when he wrote "just as much." We saw that the proof of the corollary is built on the same kind of considerations as the proof of Theorem 1 itself, namely the study of a region and its consequents at times $\tau, 2\tau, ..., n\tau$. If the trajectories remain in a bounded region, some of these consequents will have a non-empty intersection; this is the central idea of the proof of Theorem 1, as well as that of the corollary. Moreover, we have seen that Poincaré shows from 1889 that not only every region has a non-empty intersection with one of its consequents, but that one can even find a common intersection to k consequents. Therefore, it is not by introducing a new technique in the proof that Poincaré gives a precise meaning to the idea that certain types of trajectories are more general than others. In [Pb], Poincaré merely develops techniques already present in [Pa]. The real novelty is the *formulation*, in terms

---

[12] Poincaré (1889, p. 31). Emphasis added.

of probability, of the generality of stable trajectories in the sense of Poisson and the care brought to giving a proof of it.

Further, Poincaré is known to have used, at times, quite an allusive style of writing.[13]

Another fact can be added to this point: Poincaré himself mentions, in the introduction to [Pb], the clarification of a number of things that he had not had the time to make sufficiently explicit in the first version of his text [Pb, p. 263].

These considerations all lead to the indication that the corollary, added in 1890, might be part of the results, given without proof in 1889, that Poincaré completes in 1890.

To me this answer seems insufficient. Indeed, this "result" was "stated" in a rather vague manner in the first version of the memoir, where it had more the form of a commentary in the course of the discussion than that of an important corollary. So it should be further understood why Poincaré did not consider it useful, in 1889, to formulate it more precisely, even without a proof. The thesis I propose to defend is that the introduction of the corollary to the recurrence theorem is the consequence of a crucial modification which led this theorem, in 1890, to play a new role in the memoir.

Following questions raised by Phragmén, who was preparing the publication, Poincaré became aware, at the end of year 1889, of an important error. At that time, [Pa] was already printed—with notes given in the appendix to ease the reading. But the scope of this error required a fairly important reshaping of the memoir[14]. Poincaré and Mittag-Leffler's correspondence shows that it is the latter who suggests to the author to integrate the notes, as he makes the necessary corrections, and to announce these two modifications in the introduction, by first pointing out the former (Poincaré and Mittag-Leffler, 1999, letter 92, 5/12/1889). It is quite obvious that Mittag-Leffler suggests this behavior to minimize the importance of the error. Indeed, he was in a delicate position: the memoir had won a prize without the inaccuracy of its contents being noticed. Mittag-Leffler imagined, in his first letter to Poincaré after the discovery of the error, the "scandal" which his "adversaries, gained by the success of the *Acta*," might cause "because of this case" [letter 91, 4/12/1889]. As a result, he did everything in his power to surround the modifications of the memoir with great discretion. This context invites to read carefully what Poincaré says, in consultation with Mittag-Leffler, about the differences between the original memoir and the published version [Pb].

However, considering the importance of the stake, namely the correction of the memoir, gives a criterion of analysis for the differences between [Pa] and [Pb]. The time spent by Poincaré for the rewriting was relatively short. The correspondence shows that he noticed his error during the first days of December, and he announced to Mittag-Leffler the following 5th of January that he was sending Phragmén, by the same mail service, the final writing of his memoir. Therefore, it is very probable, that from the moment he became aware of the error, Poincaré spent most of his energy on its correction, rather than on the clarification of secondary points. This gives us a reading key of the recurrence theorem

---

[13] See, for example, Barrow-Green (1994, p. 65).
[14] Barrow-Green (1997) gives a general presentation of the organization of the competition and the contents of the memoirs of Poincaré, including the differences between the two versions.

and its rewriting. If it were completely independent of the error, we would be led to think that the change of formulation is more probably a clarification made on the occasion of the rewriting, but the idea of which was already clear to Poincaré. Nevertheless, the clear difference that we have shown between the vague formulation of 1889 and the precise vocabulary, which appears in 1890 and is kept in the later texts, points toward another interpretation. The following study will confirm that the recurrence theorem plays an important role in [Pb], if not directly in the correction of the error, at least indirectly as a compensation for a result, which showed to be false.

The point here is not only to determine the date at which Poincaré might have had the idea to use probabilities to clarify the recurrence theorem. By studying the place of this theorem in the arrangement of the two memoirs [Pa] and [Pb], we will be able to trace very closely the context in which Poincaré develops the statement of the corollary and his motivations, when he develops the technical terminology described above. Therefore, this is a privileged access to understand how in history, a statement of the very particular type that is of interest to us here—that is, a statement which *quantifies* the generality of a property—appeared.

6.3.2 The first memoir: the issue of stability and the recurrence theorem

To grasp the impact of the error and understand what were the resulting modifications between [Pa] and [Pb], it will be useful to describe the structure of the initial memoir. Further, we shall see that the error particularly affects the proof of stability. This is why the place of the stability issue in [Pa] will carefully be examined, in order to consider, in a second step, the way it changes in [Pb].

6.3.2.1 Structure of [Pa]

Poincaré presents in the introduction of [Pa] the orientation of his work and the main tools he developed. He starts by placing things in a historical perspective, which allows him to reconsider the results obtained by his predecessors and to show how his work is linked to this history of celestial mechanics and the *n*-body problem.

He thus mentions the successive attempts to integrate the differential equations of celestial mechanics using trigonometric series and their limits.

First, Laplace and Poisson were able to show that the major axes of the planets orbits are only subject to "periodic variations"[15] when only the first and second powers of the masses are considered. In other words, the expansions obtained are sums of terms of the form $a \cos nt$, called "trigonometric," or "mixed" terms of the form $at^m \cos nt$. But if

---

[15] In the introduction of [Pa], Poincaré uses this term in a very wide sense. In [Pb] and in the *Méthodes nouvelles*, he makes his words more explicit and distinguishes the purely periodic terms from the "mixed" terms of the form $at^m \cos nt$. The latter appear in the expansions obtained by Poisson, who goes up to the second order of the masses, whereas Lagrange and Laplace stopped at the first.

one takes into account the third powers, "secular" terms appear, that is, terms of the form αt^m, in which time does not appear in any trigonometric function.

New expansions were then obtained, in particular by Gyldén and Lindstedt, that only contain trigonometric terms. However, the convergence of these series was not proved, and Poincaré announces a first result—negative—of his memoir, namely that these series diverge.

This historical overview enables Poincaré to show the limits of expansions in trigonometric series and the necessity of finding new methods to study the problems of celestial mechanics. He then introduces a second type of tool, imagined by Cauchy to study differential equations and developed by Weierstrass, then by Briot, Bouquet, Kowalevski, and Poincaré himself: the "calculus of limits" (*calcul des limites*). This theory allows one to prove the convergence of some series expansions of solutions. Here too, he points out its weaknesses: "these series, proceeding according to the ascending powers of some variables, can be used to prove the existence of the integral, or even to calculate its numerical value; but most of the time, they do not reveal its properties."

Poincaré then mentions a last method: the geometrical method that he developed in his memoirs *Sur les courbes définies par les équations différentielles*. He summarizes the approach, which will be adopted in this new contribution, in the following terms:

> In this work, I shall use the three methods which I have just mentioned to work toward my goal; from the ancient methods of celestial mechanics, I shall take the trigonometric form of the expansions; from "the calculus of limits" the proof of their convergence; finally it is the geometrical method of M. Poincaré[16] that I shall use to prove stability. [Pa, p. 7–8]

He then presents the study of the restricted three-body problem, to which he confined himself, and announces two results: the rigorous proof of stability in the restricted three-body problem and the "complete theory" of periodic solutions. He finally draws attention to a notion, which he introduced in order to apply his geometrical method to the equations of dynamics: "integral invariants," to which we will come back.

The memoir is then composed of two parts.

The first opens with a chapter where Poincaré gathers the notations and definitions that he will use afterward. Here, he clarifies the form of the equations which will be studied and the meaning he gives to *solutions, integrals, trajectories*. He defines "surface trajectories" which will play a major role in the proof of stability:

> Let us consider any skew curve. A trajectory passes through each point of this curve; the set of trajectories forms a surface that I shall name *trajectory surface* [Pa, p. 11].

The important property of these *trajectory surfaces* is that they cannot be cut by other trajectories. Therefore, they lie at the center of the analysis of the stability question, which was proposed immediately after their definition by Poincaré [Pa, pp. 11–12].

---

[16] The 1889 memoir retains the anonymous form of the original memoir that was submitted for the prize offered by the King of Sweden.

In this way, Poincaré introduces, from the beginning of his memoir, his main objective: to prove stability (meaning here: that trajectories are bounded). He announces at the same time the reasoning that will lead him to this result, namely the search for closed trajectory surfaces.

The two other chapters of the first part present the mathematical tools developed by Poincaré to study the restricted three-body problem. They are organized in two theories, which first appear as two autonomous developments, the "theory of integral invariants" and the "theory of periodic solutions." It will not be necessary here to go into the mathematical details of these two chapters; it will suffice to remember two points:

- It is in the chapter on integral invariants that Poincaré states and proves the recurrence theorem, in the section on the "use of integral invariants."

- The heart of the chapter on periodic solutions consists in using the calculation with limits to search for periodic solutions. Knowing a periodic solution to non-perturbed equations (for example, in the three-body problem, when the third mass is equal to zero), Poincaré studies the conditions for the existence of a periodic solution to the perturbed equations as well (when the third mass is no longer equal to zero, but is very small). Poincaré further studies the stability properties of periodic solutions.[17]

In the second part, Poincaré discusses the subject proper of his memoir: the study of the equations of dynamics and the *n*-body problem. He first considers the case of two degrees of freedom. The equations of Hamilton are then composed of four equations and four unknowns, but the energy invariance allows reduction of the system to a system of three equations with three unknowns. Poincaré thus starts by presenting several geometric representations, which allow reduction of the problem to that of the integration of a vector field in a portion of the three-dimensional space. In particular, he gives the example of the restricted three-body problem. Once this framework is set, the chapter on the theory of periodic solutions gives an infinity of unstable periodic solutions (in the sense given in note 17). Poincaré also showed that through each closed trajectory representing these periodic solutions, there pass two trajectory surfaces called *asymptotic* "on which are traced trajectories in infinite number, which go asymptotically closer to the closed trajectory curve" [Pa, p. 114].

Poincaré then studies in detail these asymptotic surfaces, establishes their equations, and shows—or rather, believes he shows—that they join exactly, therefore forming a

---

[17] The question of the stability of a periodic solution, which depends on the behavior of neighboring solutions, should not be confused with the stability of a trajectory of a system of differential equations, which concerns the behavior of this trajectory in the long term. These two questions are never mixed by Poincaré, despite the use of a single term. In our study, our interest on the question of stability mainly lies within the framework of the second context, that is, the analysis of the behavior in the long term. We shall see that Poincaré uses several different characterizations of stability and that the main emphasis changes between [Pa] and [Pb]. This will lead us to distinguish a first and a second meaning of the term stability, but it will always be the stability concerning the behavior in the long term.

closed trajectory surface.[18] In accordance with the characterization given at the very beginning of the memoir, this leads him to the conclusion of stability [Pa, p. 143].

The memoir ends with two short chapters. The first gives a summary in two parts of the results obtained for the problems with two degrees of freedom: positive results and negative results. This chapter, just as the introduction, is particularly interesting for us, since it enables us to see how Poincaré conceives the assembling of the different results of his memoir. The last chapter presents the difficulties encountered by Poincaré, when he tried to extend his results to a greater number of degrees of freedom.

In the summary of the positive results, Poincaré first reminds his readers of the entire reasoning that leads to the proof of stability: the solutions of the equations of dynamics can be represented in the case of two degrees of freedom—in particular for the restricted three-body problem—by curves, named trajectories, in space. Among these trajectories, one distinguishes periodic trajectories, in infinite number. In the neighborhood of the unstable ones, one can show the existence of asymptotic trajectories, which together form a trajectory-surface. This surface divides the space into three regions, which again contain closed unstable trajectories, thus asymptotic surfaces, and therefore can in turn be subdivided.

Poincaré explains—without having developed this point before, and therefore without giving a proof:[19]

> This subdivision can be continued to infinity and it can be pushed far enough for the volume of each partial region to be as small as we want.
>
> As no trajectory can go from one region to another, it must be concluded that stability is rigorously proved; that it is possible, given the initial position of the point representing the situation of the system, to find a region from which this point will not be able to go out and to assign to this region, if not *its precise boundaries*, at least boundaries as close as one wants to these precise boundaries [Pa, pp. 154–5].

He concludes by recapitulating the different types of trajectories: closed, asymptotic, and others. On this occasion, he refers to the recurrence theorem by stating the conjecture that there exist closed trajectories in any portion of space, or, in modern language, that the closed trajectories are dense, and thus allow to approach any trajectory.

6.3.2.2 *The issue of stability in [Pa]*

The comparison between the introductions of memoirs [Pa] and [Pb] is significant to understand the mutation of the place given to the issue of stability in these two memoirs. The term of stability does not figure in the introduction of the memoir [Pb], whereas it appears from the very beginning of the introduction of memoir [Pa].

Indeed, it is according to this line of study that Poincaré reads again the work of his predecessors in order to link his memoir with this tradition of celestial mechanics: "Among

---

[18] See Fig. 6.1, where one easily imagines how the surface might close up.
[19] Poincaré considers again this question, with others, in note B of [Pa]. In this note, he tries, on Mittag-Leffler's request, to reformulate the main results in the usual language of astronomy [Pa, pp. 179, 183].

the results they obtained, one of the most remarkable is the one concerning the stability of the solar system."

As we have seen, Poincaré announces that "the three methods will contribute to [his] goal," of which origin he reminds the reader in the introduction: he will use trigonometric expansions, will show their convergence using the calculus of limits and will finally prove stability using his geometric method. The proof of stability, in the context of the restricted three-body problem, appears therefore to be the ultimate goal of the memoir.

This emphasis is reinforced in the following part of the introduction. Only one full sentence is underlined by the use of italic characters, and it concerns it: "*In this particular case,*[20] *I rigorously proved stability.*" Further, two other expressions are also given in italic characters. These are "*periodic solutions,*" for which Poincaré announces a complete theory, and "*integral invariants,*" a "new notion" introduced "to be able to apply to the equations of dynamics the geometric method," which was developed in the memoirs on curves defined by differential equations (Poincaré 1881, 1882, 1885, 1886). The theory of periodic solutions as well as that of integral invariants are treated in the first part, to be then used in the second part, where Poincaré "considers the problem itself" [Pa, p. 8]. So, these two theories, although they have their distinctive coherence and significance, are placed, in the 1889 memoir, as preliminaries vis-à-vis the heart of the study containing in particular the proof of stability.

As for the first result announced in the introduction, the divergence of the known series, it is also directed toward the proof of stability, since it shows the inadequacy of the methods available until then. This is a negative result that gives a motivation to introduce new tools, more than a result in itself. It appears in the body of the text as a consequence of the absence of an analytical and uniform integral other than the energy integral, and it is this last result which is highlighted. Therefore, the divergence of the known series is mentioned in the introduction incidentally, and not as an essential result.

Of all the results announced in the introduction, it is indeed the rigorous proof of stability that emerges as the central result of the memoir.

The recurrence theorem is also associated with the issue of stability. More precisely, to introduce this result, Poincaré defines a second meaning for the term stability:

> We have defined stability above by saying that the moving point P must stay at finite distance; sometimes it is understood in a different sense. For there to be stability, the point P must come back after a long enough period of time, if not to its initial position, at least to a neighboring position as close as we want to this initial position [Pa, p. 31].

But this meaning of the term stability is only used in the section of the memoir where it is introduced. On the contrary, in all the passages in which Poincaré insists on this essential result, namely his proof of stability, it concerns indeed the first sense, introduced from the beginning of the memoir: the sense referring to the confinement of the trajectories inside closed trajectory surfaces.[21] When he is questioned about his characterization of

---

[20] Poincaré has just presented the restricted three-body problem, to which he limits his study (AR).

[21] The exchange of letters with Mittag-Leffler supports this conclusion. See Barrow-Green (1997, pp. 88, 136) and Poincaré and Mittag-Leffler (1999, letter 70 15/11/1888, note 7, letter 74, 21/12/1888, letter 75, 25/12/1888, in particular note 8, and letter 76, 15/01/1889).

stability, Poincaré only refers to the first sense, the one he defines at the beginning of the memoir in the following terms: "There will be *stability*, […] if the trajectory of the point *P* remains entirely in a limited region of space." Stability in the second sense, to which Poincaré will give in 1890 the name of Poisson stability, does not come up at any moment in this discussion.

6.3.2.3   *The recurrence theorem in [Pa]*

If we now consider the second sense of stability, the one of the recurrence theorem, we see that it is presented as a secondary sense, relative to the first.

The recurrence theorem is presented, in [Pa], as a supplementary property of certain trajectories, a consequence of the first form of stability [Pa, p. 31]. The principal property of stability is indeed the first; it applies to all the trajectories of the three-body problem. As it will soon be seen, the situation is different in [Pb] and in the following texts.

To complete our analysis of the place of the recurrence theorem in the arrangement of [Pa], let us examine the way this theorem is exploited. It is first applied, immediately after having been stated, to Hill's problem, a model for the study of the motion of the Moon. We shall come back to this point.

As noted in our presentation of the outline of [Pa], the recurrence theorem will only be cited again at the end of the memoir, in the summary of the positive results. On this occasion, the term "stability" does not appear:

> We can state (theorem I, chapter II, part i[22]) that if one considers any portion of space, however small it might be, there will always be trajectories that pass through it infinitely many times.
>
> It is even very likely that any portion of space is crossed by an infinity of closed trajectories. If it is so, and if one considers any trajectory of the third category, we can always find a trajectory belonging to one of the first two categories[23] which for a period of time as long as we want, moves away from it as little as we want, allowing us to find the equation with the desired approximation [Pa, p. 155].

Thus, we see that, if Poincaré expresses his hope to improve the result given by the recurrence theorem, it is not in the sense of the corollary introduced in [Pb]. The latter asserts that recurring trajectories are not only *dense*, but also *more general* than the others. The result conjectured in [Pa], would say that not only *recurring trajectories* are dense, but *periodic trajectories* are too.

This conjecture was already formulated earlier, at the end of the study of systems with two degrees of freedom, made in [Pa] just before the summary of the results:

> It can be demonstrated that in the neighborhood of a closed trajectory representing a periodic solution, either stable or unstable, there passes an infinity of other closed trajectories. This is not sufficient, strictly speaking, to conclude that any region of space, however small it might be, is crossed by an infinity of closed trajectories,[24] but this is sufficient to give this hypothesis a high degree of likelihood [Pa, p. 153].

---

[22] This is the recurrence theorem (AR).

[23] Poincaré has just recapitulated the types of trajectories that he has pointed out. The first two categories are respectively the closed and the asymptotic trajectories; the third category gathers all the other trajectories (AR).

[24] Here, Poincaré inserts a note on which we will come back later: "The recent work of Mr. Cantor has indeed taught us (to use the language of this learned geometer) that a set can be perfect, without being continuous."

Therefore, the recurrence theorem appears, in the results summary, as one of the supporting elements of this conjecture on the density of periodic trajectories.

To summarize, in [Pa], the principal result is stability in the first sense regarding the restricted three-body problem. Stability in the sense of the recurrence theorem is only a consequence of it, and therefore has little to do with the issue of stability. This theorem is rather part of the attempt to characterize the different kinds of trajectories, which inspires the conjecture on the density of periodic solutions.

6.3.3 The impact of the error

Thanks to Phragmén's questions, Poincaré realized there was an error in his proof of the closure of asymptotic surfaces. These surfaces are composed of sheets which lean on a periodic trajectory, as can be seen in Fig. 6.1. Poincaré believed that, by using a simple—but wrong—geometric argument, he was showing that these sides join again to form a toric surface, which leaned on the considered periodic trajectory (this surface can easily be imagined by extending the pieces of surface represented in Fig. 6.1 all along the periodic trajectory). But at the end of 1889, Poincaré discovers that the two pieces do not exactly join. If the intersection of the pieces of asymptotic surfaces are represented with the section plane represented in Fig. 6.1, we do not get a closed curve, but two branches of curves, which intersect almost as tangents, as can be seen on the left of the Fig. 6.2.

This error does not directly concern the recurrence theorem. Indeed, on the one hand, the validity of the recurrence theorem is not affected by the discovery of this error. On the other hand, the ways Poincaré uses this theorem, in [Pa] and [Pb], do not require the

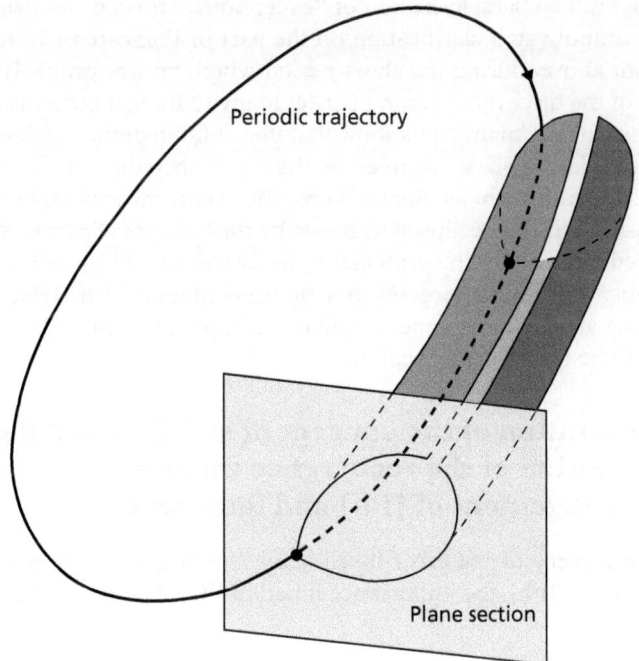

Periodic trajectory

Plane section

Figure 6.1 *Asymptotic surface.*

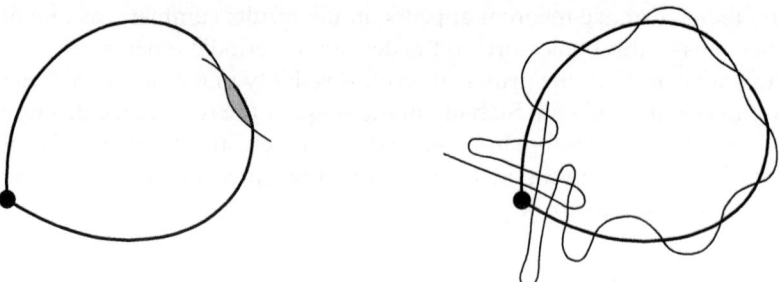

Figure 6.2 *Section of an asymptotic surface.*

corollary added in [Pb]. In fact—we have seen it for [Pa]—Poincaré does not use much the recurrence theorem. He first gives, in [Pb] just as in [Pa], a direct application: the recurrence theorem allows one to specify a result due to Hill on the theory of the Moon. In [Pb], Poincaré extends this point to a generalization of Hill's problem proposed by Bohlin.

The difference which might seem the most significant is that Poincaré uses the recurrence theorem for the study of asymptotic surfaces in [Pb]. Having shown that the two branches of the curve presented on the left side of Fig. 6.2 intersect at least twice, Poincaré applies the recurrence theorem to establish that the two curve branches intersect infinitely often. And as each of the two branches, taken individually, cannot have a double point, the result is an extremely complex pattern that is represented today by a figure of the type of Fig. 6.2, to the right. But Poincaré uses the initial theorem and has no need of the corollary at this point. Therefore, the introduction of the corollary in [Pb] is not made necessary by this correction.

So, at first sight, the characterization of "exceptional" trajectories using probabilities seems to be an unmotivated clarification on the part of Poincaré in 1890. And, this has been pointed out above, during the short period where he was probably occupied with the corrections of the discovered error, in order to give Mittag-Leffler, as fast as possible, a correct memoir. In fact, many signs show that the addition of the corollary accompanies a profound change in the place occupied by the recurrence theorem in the arrangement of the memoir. Even though, as already seen, the recurrence theorem had a relatively subordinate position in [Pa], eclipsed as it was by the "rigorous" proof of stability in the first sense, it becomes in 1890 the principal result of stability. The interest of the corollary becomes obvious: indeed, it expresses in a rigorous manner that stable trajectories (in the second sense) are the most general, when the initial theorem only stated that stable trajectories could be found in any region.

6.3.4   The mutation of the concept of stability and the new signification of the recurrence theorem in the arrangement of [Pb] and later texts

Following the discovery of the error invalidating the "rigorous proof of stability," the stability issue loses, in [Pb], the importance it had in [Pa]. The introduction of [Pb] gives a new position to several results that were, in [Pa], subordinate to the proof of stability in various ways. So, from now on, the recurrence theorem is stated in the introduction, with its corollary: "I recognized that in this case [the restricted three-body problem, AR] the three bodies will pass infinitely often as close as one wants to their initial position, unless the initial conditions of the motion are exceptional" [Pb, p. 264]. While in [Pa] this theorem was a result derived from stability in the first sense, Poincaré recognizes in [Pb] that it has its own significance justifying its mention in the introduction.

The presentation of the recurrence theorem in the body of the memoir is modified in a similar way. In [Pa], it appeared to be a consequence of stability [Pa, p. 31]. On the contrary, in [Pb], it is stated directly as a property satisfied by the restricted three-body problem [Pb, p. 313].

Even more, Poincaré now gives to the recurrence theorem the position that he gave, in [Pa], to the stability proof by confinement of trajectories.

He performs this substitution only a few days after the discovery of the error, as can be seen from his letter to Mittag-Leffler on 9 December 1889: "There is stability in the sense that unless the initial conditions of the motion are exceptional, the planet will pass infinitely often as close as one wants to its primitive situation" (Poincaré and Mittag-Leffler, 1999, letter 93). But it is only gradually that this change affects the organization of Poincaré's memoir.

In [Pb], the replacement of the first meaning of the term stability by the second is not too obvious. But, the recurrence theorem takes almost stealthily the place previously occupied by the proof of stability—in the first sense—in the arrangement of the memoir.

The status of the recurrence theorem in [Pb] vis-à-vis the stability question is rather hesitant. The reason for this is probably that [Pb] is the amendment of an erroneous text. Poincaré rewrote the incorrect passages and, as we started to show, reinforced the exposition of the recurrence theorem that allowed him to replace, as best he could, what he had to abandon. But he leaves without modification a great number of pages of his memoir. Therefore, the two concepts used to characterize stability, the first sense and the stability "in the sense of Poisson" lie side by side in [Pb].

However, from the standpoint of the arrangement of the memoir, the recurrence theorem takes the place of stability in the first sense. The latter is always defined first, but is no longer proved. One can even show a tendency to truly present it as a stability result, and at the same time to replace the first stability concept, which is now a fossil definition, by stability "in the sense of Poisson."

This change is only confirmed in the following texts, in particular in 1891. Indeed, in the two summaries of the memoir on the three-body problem for the *Bulletin astronomique* and the *Revue générale des sciences*, Poincaré presents the recurrence theorem no longer in the section on integral invariants—as it is the case in [Pa] and [Pb]—, but under the heading "stability."

Further, in the first summary (Poincaré, 1891b), Poincaré draws again attention on the progress that this result represents compared to Hill and Bohlin's work, in terms which confirm my hypothesis:

> In this case [the restricted three-body problem, AR], Messrs Hill and Bohlin have proved
> that the radius vector of the small planet always remains inferior to a finite limit. However,

> this is not sufficient for stability; the small mass must also pass infinitely often as close as one wants to its initial position (Poincaré, 1891a, p. 537).

This passage is particularly interesting because it shows how stability in the sense of Poisson became for Poincaré, at the moment he writes these lines, *stability*. Indeed, Poincaré does not present here Poisson stability, that is, the fact of passing "infinitely often as close as one wants to [the] initial position," as a definition among others of stability. It becomes a necessary condition to be able to talk legitimately about stability. The exact same idea can be found, almost in the same terms, in Poincaré (1891b, pp. 489–90).

Therefore, the notion of stability is, in 1891, identified with the property of recurrence.

6.4   In search of a suitable definition of stability

6.4.1   Two underlying concepts that are present throughout the studied period: confinement and recurrence

We have showed a major evolution between 1889 and 1891. In [Pa], the confinement inside the trajectory surfaces seemed to be the key property for the definition of stability. In 1891, recurrence is presented as *the* condition for stability. However, continuing the reading of Poincaré's texts on celestial mechanics, one finds that the definition adopted in 1891 does not definitely overshadow all other conceptions. The confinement condition of trajectories—this time, not linked to trajectory surfaces—appears again in the article *sur la stabilité du système solaire* (Poincaré, 1898, p. 540). We shall see that it is also found in the *Méthodes nouvelles*.

The reflections of Lakatos (1976, 1984) and the vocabulary he proposes seem likely to help us better describe the process under consideration. Indeed, we see how the definition of stability changes from one text to another, in a way that is very much linked to the proof that is given. The comparison between [Pa] on the one hand and [Pb] and the 1891 summaries on the other reveals a phenomenon similar to that expressed by Lakatos: "three proofs, three theorems, a single ancestor"[25] (Lakatos, 1984, p. 83). We have here two proofs, two theorems and a single ancestor. As in Lakatos' example, each couple proof–theorem corresponds to a peculiar definition of stability. Stability by confinement (or stability in the first sense) gave rise to the theorem proved in [Pa], which "rigorously" establishes stability in the restricted three-body problem. Stability in the second sense is studied by the recurrence theorem. But the two concepts are linked to the historical exploration of the stability of the solar system by expansion in perturbation series of the solutions of the equations of motion.

---

[25] Lakatos introduces this expression to replace "the usual expression '*different proofs of the same theorem*'," which "is misleading, for it hides the vital role of proofs in the formation of theorems." The three proofs he refers to in this passage are those of Euler's formula for polyhedra, respectively due to Cauchy, Gergonne, and Legendre.

However, the phenomenon under study is more complex in certain respects than that of Lakatos.

First, as we have just shown, the two meanings of the term stability are not used exclusively. More than two distinct definitions of stability, among which we choose depending on the proof we want to use or the study context, these two meanings appear as two components of the "familiar"[26] concept of stability.

Further, each of these two aspects of stability behaves like a concept such as those studied by Lakatos, liable of "extension" and successive definitions that do not exactly overlap. Regarding the second concept, that of recurrence, this phenomenon is of particular interest to us,[27] because these extensions are linked to the treatment of exceptional trajectories. We shall come back to this point in detail in Section 6.4.2.

Poincaré already associated these two concepts, confinement and recurrence, to study stability in his memoirs *Sur les courbes définies par les équations différentielles* (1881, 1882, 1885, 1886). In particular, the third of these memoirs opens with a chapter on "stability and instability." To my knowledge, it contains Poincaré's first reflection on the definition of stability. This text deserves a thorough study that cannot be done here. A few remarks will suffice for our purpose.

In the chapter mentioned, Poincaré starts by giving five examples, from which he will define what is meant by stability. In the first example, all the trajectories are closed, and Poincaré describes this case as "complete stability."[28] Further, he then points out that the property of recurrence, finally adopted to define stability, characterizes "a certain periodicity of a particular nature" (Poincaré, 1885, p. 92).

The last four examples are particularly interesting because they precisely illustrate the four possible combinations between the two characteristics according to which Poincaré examines the trajectories of the considered systems: the recurrence property and the question whether the trajectories remain—or not—in a bounded region of the plane. Therefore, while choosing the recurrence concept to define stability, Poincaré systematically examines the behavior of trajectories vis-à-vis the other concept. He also considers that "stability defined in such a way," that is, by the recurrence property "has only theoretical importance."[29]

In the 1891 summary and in the article on the stability of the solar system, two sentences seem to echo. First Poincaré states in 1891 that the confinement of trajectories

---

[26] Lakatos suggests this term to designate the concept as long as it has not been defined mathematically (Lakatos, 1984, p. 21). Before this gives rise to any objections, we suppose the ability of distinguishing what is stable from what is not.

[27] Concerning the first concept, a distinction should be made between the confinement within asymptotic surfaces of 1889 and the requirement according to which the system does not move away indefinitely. Indeed, the peculiarity of asymptotic surfaces was that they existed with arbitrarily small diameters. This is what made their superiority compared to Hill and Bohlin's result.

[28] To study in detail the formation of these two concepts, one should link this model, where all the trajectories are closed, with the tradition of trigonometric expansions of the solutions of the equations of the solar system.

[29] Tatiana Roque (2001, chap. 7) studies these texts, and the way the stability question is treated, under a slightly different angle. She examines the relation between mathematics and physics, theoretical development, and inspiration from celestial mechanics.

within a bounded region "is not sufficient" without the recurrence property. Conversely, in 1898, he believes that recurrence "is not sufficient" if one does not know how to show that the trajectories remain at finite distance. So, even when recurrence overrides in 1891, from the standpoint of the proof, the characterization in terms of confinement, the two concepts remain inseparable in Poincaré's reflection on stability. This close union between the two characterizations of stability is found more distinctly still in 1899 in the *Méthodes nouvelles*.

The coexistence of the two concepts of recurrence and confinement seems to express a tension between what the proof allows to prove ("*proof-generated concept*") and what is expected when talking about stability, in other words the "familiar concept" of stability. It seems that this tension is translated here into a tension between mathematics and physics, between theory and practice. From a mathematical and theoretical point of view, the most interesting concept of stability is the one that can be proved. But from a physical and practical point of view, a technical property more or less *ad hoc* remains unsatisfactory vis-à-vis what is expected of the use of the term stability.

6.4.2   The exceptional character of non-recurring trajectories

These two concepts of stability are not settled during this period. In particular, as noted above, the corollary of the recurrence theorem is progressively reinforced. Let us now examine in more detail the evolution of the concept of Poisson stability in these texts. I shall also consider the manner in which the mathematical definition of the term "exceptional" is used. The definition itself changes between 1890 and 1899. These successive amendments lead to making the most of the proof: its result, the concept of Poisson stability, is really a "*proof generated concept*" in the sense of Lakatos.

The first evolution of the use of the term "stability," which can be seen between [Pa] and the later texts, is related to the phenomenon that has been shown in Section 6.3.4. In 1889, the aim of the recurrence theorem was, above all, to show the existence, in any region of space, of trajectories of a particular type, the recurring trajectories, called "stable in the second sense." The question of the stability of the restricted three-body problem was treated elsewhere. Therefore, stability in the second sense appeared merely as a property of certain *trajectories*. From 1890, and even as early as Poincaré's letter to Mittag-Leffler on 9 December 1889, stability—in the sense of Poisson—is also predicated, by extension, of the *problem* defined by the studied differential equations, or the body (the Moon for example), of which they model the motion.

The novelty here is that the term "stability" then does not only cover recurring trajectories, but also the exceptional trajectories—that are not stable in the sense of Poisson—that the studied problem may present.

The already mentioned passage of the letter to Mittag-Leffler illustrates well this extension of the meaning of the term "stability" when it is not referring to a particular trajectory: "Stability remains, in the sense that unless the initial conditions of motion are exceptional, the planet will pass infinitely often as close as one wants to its initial position." Poincaré does not say here that the trajectory of the planet is stable *except for* certain exceptional initial conditions, but that the motion is stable, *in the sense that* the trajectories which are not Poisson stable are exceptional. These two uses of stability in the sense of the recurrence property are not presented by Poincaré as two different meanings of the term stability, in the same way he distinguishes, between 1889 and 1890, stability in the first sense from stability in the second sense, still named in 1890, stability in the sense of Poisson. So, the concept of recurrence is not mathematically immutable; it extends as the argumentation develops. From the moment Poincaré defined what is meant by exceptional trajectories, he allows himself to talk about stability in a more general sense. It is an *extension* rather than a transformation of the sense of Poisson stability. Indeed, Poincaré continues to talk about solutions that are stable—or not—in the sense of Poisson, at the same time that he says the problem as a whole is stable.

The shift in meaning that we have just mentioned is not particular to the mentioned letter: it is found in the entire corpus [Pb, pp. 313, 320].

In the *Méthodes nouvelles*, Poincaré defines Poisson stability by the condition "that the system passes infinitely often as close as one wants to its initial position" (see citation in Poincaré, 1892–99, t. III, p. 141). He then announces that this condition is fulfilled in the case of the restricted three-body problem. Of course, it must here also be understood that the trajectories that do not have this property are exceptional.

Further on, Poincaré also explicitly links the exceptional character of non-recurring trajectories with the legitimacy of affirming stability by including them in the statement:

> To summarize, the molecules that only pass through U_0 a finite number of times are exceptional in the same way that commensurable numbers are only an exception in the series of numbers, whereas incommensurable numbers are the rule.
>
> So, if Poisson thought he could answer positively the question of stability as he had posed it, although he had excluded the cases where the ratio of the mean motions is commensurable, similarly we shall have the right to consider that stability, as we define it, is proved, even though we are forced to exclude the exceptional molecules mentioned above (Poincaré, 1892–99, t. III, p. 154).

This allows us to clarify the brief commentary that is found in Poincaré (1891a): "I shall add that the first [the unstable particular solutions, AR] are exceptional (allowing us to say that there is stability in general)." Grammatically, the remark in parentheses can be analyzed in several ways, depending on whether "in general" is related to the noun "stability," or to one of the verbs "allow," "say," or "there is." These different possibilities seem to me to have two meanings. Either one understands, in general, that is, most often but not always, there is stability. Or, one understands that there is "stability in general": stability can be predicated, in the sense of a general property, as opposed to a universal property. Consideration of the other texts of the corpus shows that one can settle for the second interpretation. The exceptional character of unstable trajectories allows us to affirm stability, which would not have been justified if only the recurrence theorem had been established. Indeed, it only states the existence of stable trajectories in any region of the plane, but does not give any information on the quantitative importance of these trajectories compared to those that are unstable.

The mathematical quantification of the exceptional is thus presented by Poincaré as a central element of his argumentation. It allows him to limit the importance of certain marginal trajectories and to extend the property satisfied by the other trajectories to the problem considered globally. The two levels on which stability can be predicated, however, are never confused, and it is only when Poincaré talks about the stability of a problem that he excludes exceptional trajectories. Yet, he most often mentions—this can be seen in the above passages—in what sense he asserts stability: to say that the problem of the Moon, the restricted three-body problem, etc. is stable, means that the trajectories have the recurrence property unless the initial conditions are exceptional.

6.4.3   Three correlated changes

Our analysis allows us now to come back to the question posed at the beginning of Section 6.3.1: How can we interpret the changes made to the recurrence theorem and its formulation between [Pa] and [Pb]?

In 1890, Poincaré introduces a new formalism, which comes from probability calculus; this allows him to extend the recurrence theorem with a corollary. In the following texts, the corollary is gradually reinforced, in a process that can be described, according to Lakatos, as the search for the domain of the proof. Therefore, the introduction of the language of probabilities is part of a larger movement, in which Poincaré increases the scope of the recurrence theorem.

This evolution occurs precisely when Poincaré gives to the recurrence theorem the place of the proof of stability, which proved to be erroneous. Thus, there is a *correlation* between the change of status of the recurrence theorem and its mathematical strengthening by the corollary.

Finally, a third aspect of the texts of our corpus changes during the same period: the manner Poincaré links his work with that of the astronomers preceding him, in particular Lagrange and Poisson.[30] Poincaré draws a parallel between his theorem and the proof by Poisson of the stability of the solar system precisely on the two evolving points: the exclusion of certain trajectories in the statement of a stability result, which is the subject of the corollary added in 1890, and the fact of considering the recurrence property as a form of stability.

These three changes are clearly closely related. The aim of the legitimation—using Poisson as a reference—is both the form of the stability studied by the recurrence theorem and the aspect of the quantification given by the corollary. Moreover, the strengthening of the theorem is necessary to be able to present it as a stability result. In its original form, it appears as a technical existence condition of solutions that have a particular property. But, for it to be valid as a stability result, one must be able to say what type of trajectory

---

[30] This point is developed in Robadey (2005).

is the most widespread. Therefore, the information given by the corollary is particularly crucial. It is only this quantitative precision on the importance of recurring trajectories which gives the theorem its strength in the tradition of the stability results established by Lagrange, Laplace, and Poisson.

Now, Mittag-Leffler's report, which announced the outcome of the competition in honor of the King of Sweden, mentioned explicitly stability as heading the list of the questions treated by the prize-winning memoir:

> The most important and difficult questions, such as the stability of the world system, the analytical expression of the planets coordinates by series of sines and cosines of time multiples, and the very remarkable study of asymptotic motion, the discovery of forms of motion where the distances of the bodies remaining between fixed limits, their coordinates cannot however be expressed by trigonometric series, other subjects that we do not mention, are treated by methods that open, it is fair to say, a new era in celestial mechanics (*Rapport de la Commission du Prix offert par S.M. le roi Oscar* ii *de Suède,* 1889, p. 287).

This context must have encouraged Poincaré to keep a stability result in his memoir, even if it was to be weaker than that originally aimed at.

Therefore, it is possible to say, without committing oneself much, that it is precisely with the aim of replacing the erroneous proof of stability that Poincaré reshapes his recurrence theorem between [Pa] and [Pb], by giving it a greater mathematical scope and by presenting the obtained result in the tradition of the founding research on stability in celestial mechanics, even if he employs entirely new techniques.

In other words, it is indeed a new need that leads to a real change of viewpoint in the manner Poincaré considers stability "in the second sense." It is in this context that Poincaré resorts to probability calculus to strengthen the result given by the recurrence theorem.

6.5 Where does the idea of using probabilities come from?

In the study above, we were able to identify a precise moment, when Poincaré introduces a new type of statement following the discovery of his error. To do this, the tool he resorts to is probability calculus. At first sight, it seems to be a very distant area of mathematics. What made Poincaré adopt such a tool to study trajectories? How does probability theory answer his needs?

Asking such a question might lead to an anachronistic interpretation. Indeed, one might be tempted to consider the problem studied by Poincaré from a contemporary standpoint and seek to show that he did not have the choice: he was forced to introduce a new way of apprehension to evaluate the importance of recurring trajectories. Or even that he necessarily had to use probability calculus or a similar tool. But what does one mean by "probability calculus or a similar tool?" The modern reader of course thinks of measure theory. But, this is a retrospective way of thinking. More broadly, the state of development of theories which we today have to describe accurately what Poincaré did and possibly judge what was best absolutely does not correspond to what Poincaré actually had at his disposal. The result is that it is extremely difficult to pose the question of Poincaré's choice.

However, one can identify uses of the notion of probability in contexts close to Poincaré's work on the three-body problem: these facts might have suggested to Poincaré the use of probability calculus. This is what I shall strive to identify.

Further, I shall try to show elements of continuity between the proof of the recurrence theorem and that of the corollary. Therefore, we shall see that Poincaré exploited the work which had led him to the proof of the theorem and developed it to establish the corollary. It appears that, probability calculus, such as Poincaré formulates it in order to introduce it here, is an appropriate tool for the strengthening of the recurrence theorem. Our author does not just stick an exterior tool onto a given problem; he adapts them both to fit each other.

Finally, I shall show some specificities of Poincaré's work using probability calculus in the framework of the recurrence theorem. They highlight the novelty of this approach compared to other ways of studying sets of trajectories, or compared to other uses of probabilities.

6.5.1 General solution and particular solutions

Let us first start by examining the first work of Poincaré in the field of celestial mechanics.

This research is interesting for us in two respects. First, in these texts, Poincaré mentions the *probability* of these periodic trajectories—however, in a very different sense than the one he gives to this word in 1890, as will be seen. More broadly, Poincaré uses here other tools to characterize sets of trajectories.

The texts I propose to examine are a note published in the CRAS (Poincaré, 1883) and an article that was published a year later in the *Bulletin astronomique* (Poincaré, 1884), which develops the content of the former. Poincaré considers "some particular solutions to the three-body problem," of which he shows the existence and studies the properties.

The article *Sur certaines solutions particulières du problème des trois corps* (Poincaré, 1884) starts by opposing the "general solution to the three-body problem," which "is still to be found," and "some particular solutions" that must be brought to the fore. While the convergence of the trigonometric expansions proposed for the general solution of the three-body problem is not proved,[31] Poincaré points out a major interest of some particular periodic solutions that he is going to study: they can be represented by convergent series. The opposition made here by Poincaré between "general" and "particular" solutions is classical in the study of differential equations and Poincaré reminds the reader of it at the beginning of both versions of his memoir on the three-body problem, [Pa] and [Pb]. If one is interested in the equations

---

[31] In his 1890 memoir, Poincaré also proves, among other things, the non-convergence of some of these series.

$$\frac{dx_i}{dt} = X_i \quad (i = 1,...,n),$$

the formulas

$$x_i = \varphi_i(t,C_1,...C_n) \quad (i = 1,...,n)$$

"represent the *general solution* if the constants C_i remain arbitrary." They "represent a *particular solution* if the constants C are given numerical values" [Pb, p. 266]. In other words, the "general solution" is given by a single formula that contains all the particular solutions; the latter are obtained by fixing the values of the arbitrary constants of the general solution.

Poincaré continues his 1884 article by the presentation of his method showing the existence of particular periodic solutions. He applies it successively three times to highlight three "kinds" of periodic solutions. Concerning the first kind, he specifies the quantity of periodic solutions shown in this way: "therefore there are periodic solutions of the first kind; there are even a quadruple infinity of them." In fact, Poincaré shows that the periodic solutions of this kind are given by formulas, in which remain four arbitrary constants. These four parameters give rise to the announced "quadruple infinity" of solutions.

These quantitative considerations on the periodic solutions, of which Poincaré proves the existence, were already present in the 1883 note:

> In the particular solution considered, if it is imposed on the three bodies to move in a plane, there still remain four arbitrary parameters; if they move in space, there remain eight parameters. Thus, in both cases, four conditions must be imposed on the initial elements of motion for it to have the periodicity that we have just mentioned (Poincaré, 1883, p. 252).

In 1884, Poincaré proves the existence of three "kinds" of periodic solutions. The first two correspond to motion in a plane, the third to motion in space. Poincaré only gives the enumeration of the arbitrary constants for the solutions of the first kind. The conclusion of the study of the solutions of the third kind simply states: "this proves the existence of an infinity of periodic solutions of the third kind." The reader must himself notice that there remain eight arbitrary parameters, which can be set at will to obtain these periodic solutions, as Poincaré had announced in 1883.

Similarly, Poincaré shows how to calculate, by successive approximations, the coefficients of series representing the solutions of the first kind. Specifically, he calculates the first approximation, by this indicating the procedure to follow. But, he only presents this calculation in the case of the solutions of the first kind.

It is therefore clear that Poincaré chooses, in 1884, only to detail the study of the solutions of the first kind, for the enumeration of the arbitrary constants as well as for the calculation of solutions by approximations. Concerning the other kinds, he only gives the parts of the analysis which present the acknowledged differences in relation to the first study, and leaves to the reader the task of completing the work, in order to convince himself of the results announced in the 1883 note.

However, even if the enumeration of the arbitrary constants is given only in the first example of the 1884 article, one can see that the considerations on the number of constants to set in order to perfectly determine a solution of a given type form the last third of the short note of 1883. This shows Poincaré attaches importance to this quantitative evaluation of the sets of solutions that he highlights.

Clearly, the mode of characterization, which Poincaré uses here in order to estimate the importance of a set of solutions, is linked to the opposition between general solution and particular solutions. The sets of solutions presented in these pieces of work appear to be a transition between general solution and particular solutions. They are indeed given by formulas, in which some constants remain arbitrary, whereas others must be imposed.

The passage of the 1883 note that I quoted calls for another remark. In this passage, Poincaré first considers the number of arbitrary constants which remain in the expression of the solution. In modern terms, this corresponds to the dimension of the considered set of solutions. Today, it is natural to understand such an indication as an element of evaluation of the generality degree of the suggested solutions. Such a relation between the number of arbitrary constants in a formula and the estimation of the generality of the solutions represented by this formula is not given explicitly in these texts of Poincaré. But the link is confirmed by a passage of the memoir *Sur le problème des trois corps*. In this text, Poincaré comments on a formula that he gives to represent certain solutions, in the following terms:

> This solution is not the most general, since it only contains three arbitrary constants, but it is the most general among those that can be written in the form (3).[32]

The form of the solutions which are studied here matters little to us. The important point is that Poincaré himself relates the number of arbitrary constants to the degree of generality of the solution.

After having paid attention to the number of conditions that have to be set to obtain periodic solutions, Poincaré remarks in 1884:

> *Application of periodic solutions*—It seems at first look that these periodic solutions are of no practical use, since they correspond to *particular* values of the initial elements, values of which the probability is equal to zero. But, if the initial elements are very close to those corresponding to a periodic solution, the real positions of the three masses can be linked to the positions that they would occupy in this periodic solution, and consequently we can use this solution as we would use an *intermediate orbit* (Poincaré, 1884, p. 260).

One sees that Poincaré insists on the term "particular," which echoes the opposition between general solution and particular solutions. We have just analyzed how Poincaré

---

[32] This passage is in [Pa, pp. 81–2], and is given again in [Pb, p. 360].

developed this opposition into a gradation of generality. He then notes that the probability of the initial conditions, which gives rise to these periodic solutions, is equal to zero. But, contrary to the use of probabilities that we studied in [Pb], he does not at all specify how this probability should be calculated. The fact that the probability is equal to zero seems here to stress the preceding indication on the particularity of these periodic solutions rather than adding new information. Thus it seems that the term of probability is used here in its everyday meaning and not in a technical sense.

Therefore, Poincaré employs, in his work on celestial mechanics, two very different means to consider sets of trajectories.

The attention paid to the dimension of classes of solutions appears already in the first pieces of work on celestial mechanics. This dimensional approach is still present in 1889 and 1890, as we have seen.

However, in 1890, Poincaré introduces a new means of comparing different kinds of trajectories. For this, he uses probability theory, employed in a technical way.

These two tools refer to a common approach, which consists in quantitatively evaluating a set of solutions. The use of the term "probability" in 1884, to me, seems rather to reflect the similarity between the two modes of apprehension than to be a reference to probability theory. The connection between these two ideas might have contributed in 1890 to give to Poincaré the idea of using probability calculus in a technical way. But both approaches are well distinguished in the subsequent work of Poincaré, which suggests he is fully aware of their differences. We shall soon come back to the specificity of the problem of recurring trajectories: this is another factor that might have led Poincaré to use a new tool.

6.5.2 Cantor's set theory

Poincaré borrows from Cantor's work a third type of consideration to study sets of trajectories. This mathematician forged specific concepts to study sets: countability, density, derived sets, closed sets, "perfect" sets, etc. Poincaré refers to this in the memoir *Sur le problème des trois corps*, as we have seen, to discuss his conjecture about the density of trajectories [Pa, p. 153; Pb, p. 454]. He shows that in the neighborhood of any closed trajectory, there passes an infinity of other trajectories. Poincaré remarks that this is not sufficient to conclude that closed trajectories are dense, and he justifies this remark using a result of Cantor: "a set can be perfect without being continuous."

A perfect set, in Cantor's theory, is a closed set, of which any element is the limit of a sequence of distinct elements of this set. Poincaré only showed that the set of closed trajectories satisfies the second of the conditions characterizing perfect sets. He therefore is not faced with a perfect set. Poincaré's commentary must probably be understood as referring to the properties of closure of the set of trajectories. The latter is closed by definition, and always satisfies the second condition, thus is a perfect set. Poincaré conjectures that all the trajectories are attained in this manner—in other words, the entire space of initial conditions. Here too, the point is to evaluate the importance of a set of trajectories. Since in the neighborhood of a periodic solution there exists an infinity of other periodic solutions, there are many closed trajectories and one might think they are dense. However, Cantor's result leads to moderate this first impression in the absence of any supplementary argument.

Poincaré's contacts with Cantor's work date back to the beginning of year 1883. At this time, he contributed to the translation into French of a memoir of this mathematician, requested by Mittag-Leffler for the *Acta Mathematica*[33] (Cantor, 1883). The correspondence between Mittag-Leffler, Hermite, and Poincaré shows that the Frenchmen involved in the translation are not convinced of the interest of this research in itself. But Poincaré does not hesitate, from 1883, to use notions defined by Cantor, when they allow him to clarify his understanding of phenomena that he is studying (Poincaré and Mittag-Leffler, 1999, letter 28, note 4). The memoir *Sur le problème des trois corps* is a new example of this.

This shows that Poincaré studies sets of trajectories in all their aspects, using all available means: dimensional approach, probabilities, set theory. More generally, we see that Poincaré is open to using various methods. We have already remarked that he willingly gives a geometric interpretation of the studied phenomena. The influence of the kinetic theory of gases on the rewriting of the proof of the recurrence theorem is another example of the links he readily makes between various fields of study. It seems to me that the introduction of a probabilistic tool also reflects the manner in which Poincaré works, prompt to make various areas of his mathematical activity interact.

6.5.3   Gyldén's use of probabilities

Poincaré already had used several tools to study the sets of trajectories: the dimensional approach and set theory. And although he had mentioned probabilities in 1884, he had not used probability calculus proper, neither in his previous work nor in [Pa]. A circumstance can however be noted that might have drawn his attention to this tool. The Swedish astronomer Gyldén had used it shortly before. He sought to evaluate the probability that the integers appearing in the continued fraction expansion of an irrational number taken at random are very large. This work is presented in two notes in the *Comptes rendus de l'Académie des Sciences* (Gyldén, 1888c, 1888d), which summarize two articles in Swedish of the *Comptes rendus de l'Académie des sciences de Stockholm*, (Gyldén, 1888a, 1888b).

Several factors converge to suggest that Poincaré had knowledge of Gyldén's work. For example, he mentioned this astronomer's work in the introduction of [Pa]. So he had knowledge of the work of Gyldén that was related to his own centers of interest. Further, right after the announcement of the award, before the memoir had been printed in its first version, Gyldén had claimed priority over certain of Poincaré's results, referring to his previous work (Gyldén, 1887). Mittag-Leffler had called on Poincaré to prepare his response, giving rise to a series of long letters at the beginning of 1889 (Poincaré and Mittag-Leffler, 1999, letters 79–87). In these letters, Poincaré comments on the 1887 memoir of Gyldén and discusses the convergence of the series that are defined in it.

Gyldén's motivation to study the decomposition of irrational numbers in continued fractions is related to these questions of celestial mechanics and convergence of series.

---

[33] See Dugac (1984) and Poincaré and Mittag-Leffler (1999, letter 28, note 3).

He develops it in the first memoir in Swedish. The underlying problem is that of the convergence of trigonometric series, which intervene most particularly in celestial mechanics in the calculation of the perturbations of the Keplerian motion of a planet due to the attraction of the other planets. Gyldén concludes: "the probability that the series representing the perturbations of the planets diverge is less than any given value" (Gyldén, 1888b, p. 83).[34] This aim is mentioned in the conclusion of the second note in the CRAS: "From these results follows a thesis of great importance to judge the convergence of certain trigonometric series used in the calculation of perturbations, namely: the probability of finding a value of *a* outside a given limit is in reciprocal ratio to the number signifying this limit"[35] (Gyldén, 1888d, p. 1781). This is the only mention of the orientation of this research toward celestial mechanics, somewhat elliptical, in the communications addressed to the Academy of Paris. But, Poincaré knew Gyldén and the aim of the latter was possibly clear to him, having read—or not—the Swedish memoirs.

6.5.4 An important characteristic of the recurrence theorem: a non-constructive proof

We have already noted the continuity between the proof of the recurrence theorem proper and the corollary that Poincaré adds in 1890. Poincaré uses the same elements of proof, that is, consideration of the sequence of consequent regions of a given region and their common parts, as well as evaluations of their volumes. In fact, in [Pb], he only refines his analysis of these regions. Probability calculus is therefore an appropriate means to take a greater advantage of the proof method already used for the theorem.

This continuity is deeper. Considering the technical details, another characteristic shared by the theorem and the corollary can be uncovered. It shows a radical difference between the question of recurring trajectories and the problem of periodic solutions that has just been considered (see Section 6.5.1). The recurrence theorem states the existence of trajectories which have a certain property, but absolutely does not allow us to show explicitly such a trajectory. This is a non-constructive result. This particularity is not without consequence on the means by which one can evaluate the qualitative importance of trajectories of a given type, as Poincaré does in the corollary. In the article on periodic solutions studied above (Poincaré, 1884), Poincaré made an analysis of the conditions under which a trajectory is periodic of the first [respectively second or third] kind. From this, he drew considerations on the respective importance of these types of solutions among each other and in relation to the set of solutions of the problem. Here, this is not the point: it is not possible to characterize by a condition the initial positions which give rise to recurring trajectories.

The non-constructive character of the proof of the recurrence theorem is very strong.

[34] Cited by von Plato (1994).

[35] The italic type is that of Gyldén. Here, the a_i are the integers that appear in the expansion of a given irrational number into continued fraction. Gyldén shows that the probability for an a_i to be greater than n is proportional to $\frac{1}{n}$ (AR).

In fact, it is not even possible to define precisely what it means for a given trajectory to be recurring—or not—in the sense of the theorem, such as it is stated in [Pa] and [Pb]. In 1885, Poincaré gave "a precise definition of stability" for a trajectory:

> We shall say that the trajectory of a moving point is stable, when describing a circle or a sphere of radius *r* around the starting point, the moving point, after having left this circle or this sphere, will enter it an infinity of times, and this, however small *r* is (Poincaré, 1885, p. 94).

But he does not prove the existence of such trajectories neither in [Pa], nor in [Pb]. In both these memoirs, stability in the second sense is defined only by a sentence of everyday language, without any mathematical notions:

> For there to be stability, the point *P* must come back after a long enough time if not to its initial position, at least to a position as close as one wants to this initial position [Pa, p. 31; Pb, p. 313].

In everyday language, the order of the components of the sentence is much freer than in the technical language. It is only in the statement of the theorem that the position of the quantifier "a position as close as one wants" is fixed in relation to the determination of a stable trajectory:

> If one considers any region r_0, however small this region might be, there will be trajectories that pass through it infinitely often.

The introduction of the notation r_0 is preceded by the expression "if one considers" which marks the definition of a mathematical object. Therefore, it is indeed the region considered that is chosen first. Then only a trajectory is fixed in order to examine its properties. In the first formulation, in everyday language, the point *P* appears without being defined, and the sentence does not impose the reading according to which stability would be defined for a given trajectory as the fact of coming back infinitely often in any region containing the starting point. It admits several interpretations, which is not the case for the theorem.

Thus, the definition of stability is not fixed technically except in the statement of the theorem, and it is inseparable from this statement. The theorem does not predict the existence of stable trajectories. But, for each region, the theorem predicts the existence of recurring trajectories with regard to this region. It is this that is interpreted as the existence of stable trajectories.

The same goes for the corollary. In [Pb] one can see that the definition of the term "exceptional" is neither isolated from that of stability nor from the statement of the corollary. First, this is a consequence of what we have just seen: there is no set of non-recurring trajectories, the probability of which could be calculated, but only recurring trajectories with regard to a region. Further, the probability calculated by Poincaré in 1890 is not that of the non-recurring trajectories with regard to a given region, but that of trajectories which do not come back in this region more than *k* times. The result is that the meaning of the term "exceptional" is only defined by the statement that Poincaré is going to establish. More precisely, Poincaré writes:

> The probabilities being defined in this way, I propose to establish that the probability that a trajectory coming from a point of r_0 does not cross this region more than k times is equal to zero, however large k might be and however small the region r_0 might be. This is what I mean when I say that the trajectories that only cross r_0 a finite number of times are exceptional [Pb, p. 316].

Thus, he claims not to give a mathematical sense to the term "exceptional" in itself, but to clarify what must be understood by the fact that non-recurring trajectories—with regard to a given region—are exceptional.

This can also be seen by a quite strange text organization. The section, which is marked as being the corollary, comes before the passage that I have just cited. It is a paragraph given in italic characters, preceded by the heading "corollary":

Corollary—*It follows from the preceding that there exists an infinity of trajectories crossing infinitely often the region r_0; but there can exist others that only cross this region a finite number of times. I propose now to explain why the latter trajectories can be regarded as exceptional* [Pb, p. 316].

At this point in the text, Poincaré has not yet defined the term "exceptional;" he has not even mentioned probability. He announces an explanation and a proof, at the same time as a definition of what he means by exceptional.

Therefore, a new characteristic common to the recurrence theorem and its corollary emerges: the close connection between these two statements and the definition of the terms by which they are translated. Stability—that is, Poisson stability—is not an intrinsic property of certain objects, just as the exceptional character of certain types of trajectories is not an intrinsic property of a given set. These terms allow Poincaré to reformulate complex technical properties.

6.6 Probabilities

6.6.1 Poincaré and probability calculus

Probability calculus is not an area in which Poincaré wrote much. In fact, the use of probabilities in [Pb] is the oldest evidence of his reflections on the subject. In 1890, Poincaré has been holding the Chair of "mathematical physics and probability theory" at the Sorbonne for three years, but it is only during the second semester of year 1893–4 that he chooses to teach probability theory. This course is written by Albert Quiquet, former student of the École Normale Supérieure (graduation year 1883), who had been working since 1886 as an actuary in an insurance company. It was published two years later (Poincaré, 1896). The book is composed of chapters of similar lengths, only identified by the lesson numbers. The content probably reproduces very faithfully Poincaré's teaching of year 1894. It is only in the second edition, published in 1912, that Poincaré reconsiders the structure and organizes the chapters thematically. The modifications of the body of the text are quite minor with regard to what interests us here. Between the two editions, Poincaré published a less specialized article (Poincaré, 1899) that will be integrated into *La science et l'hypothèse* (Poincaré, 1902), and an article (Poincaré, 1907) that forms the introduction, mentioned above, of the second edition of the course.

The period that particularly interests us is that of the rewriting of the memoir on the three-body problem and the period of the publication of the summaries for the *Bulletin astronomique* and the *Revue générale des sciences*, that is, the 1889–91 period. The work explicitly on probability calculus comes later. But the 1894 course presents a mature version of Poincaré's reflections of the preceding years. In fact, we have already noted above (see Section 6.2.2) striking correlations between the statement of the corollary of the recurrence theorem and the content of this course. Pursuing this analysis, we shall be able to suggest a chronology of Poincaré's work in the field of probabilities. More than a chronology, the study of the research on the three-body problem reveals some of Poincaré's motivations for the elaboration of his personal approach to probability calculus.

The course published in 1896 mentions very few sources. The most frequently cited author is clearly Joseph Bertrand, whose book on probability calculus had just come out (Bertrand, 1888). Besides, Poincaré knew certain points of Bertrand's work from the talks given by the latter at the meetings of the Academy while he was writing the text. He might also have heard of Bertrand's reflections on probabilities at the École Polytechnique.

Poincaré refers to Bertrand several times in his course. He recognizes his paternity of several examples,[36] questions,[37] results,[38] and remarks.[39] Concerning these constant references to Bertrand's book, one can note that when Poincaré considers some of the explanations of the paradox of Saint-Petersburg, he omits naming their authors, whereas Bertrand precisely traced the origins of the various interpretations. The circumstance seems to me to indicate that Poincaré uses Bertrand's book as his main source. He takes advantage of the latter's work of compilation.

The only mathematician that Poincaré refers to, not already cited by Bertrand, is Maurice d'Ocagne. Poincaré attributes to him the solution of a problem of geometric probabilities[40] (Poincaré, 1896, p. 113). But, this is an isolated contribution and bears no comparison with that of Bertrand.

Therefore, it seems that Poincaré thought about probability calculus essentially on the basis of Bertrand's book.

---

[36] For instance, the first examples of probability problems. The reference is explicit: "Let us cite two other examples of Bertrand." Even before this reference, Poincaré had chosen, just as Bertrand, the question of the probability of point *4* on the throw of a dice as the simplest example of probability theory. Similarly, one then finds several examples—not indicated—common to Bertrand and Poincaré: the pistol shot, Maxwell's error, and the polling problem (Poincaré, 1896, pp. 41–9; Bertrand, 1888, pp. 29–31, 18–20).

[37] Let us particularly note Bertrand's paradox, solved by Poincaré, to which we shall come back.

[38] Poincaré (1896, p. 73): "J. Bertrand calculated the *probable moment of his ruin*."

[39] Poincaré (1896, p. 173): "J. Bertrand presents the following objections: [...]"

[40] We shall see that it is in this area that Poincaré stands out the most in comparison to Bertrand's book.

This teaches us two things. First, Poincaré read Bertrand's text, at least before 1894, and probably earlier, maybe as early as 1889; it is not impossible that this reading—and the reflections it raised—is one of the sources of inspiration suggesting the formulation of the recurrence theorem in probability terms. More precisely, I propose to show that the work on the recurrence theorem between 1889 and 1891 is for Poincaré the occasion of shaping his own conception of probability calculus.

6.6.2 Bertrand and continuous probabilities

One of the points on which Bertrand and Poincaré's texts differ considerably concerns the treatment of the problem of continuous probabilities, that is, the problems for which the set of possible outcomes of a probability experiment is a continuum, and thus an infinite set. It is important for us to study the way Bertrand and Poincaré treat these problems, since the probability evaluated by Poincaré for the corollary of the recurrence theorem is of this type. Indeed, the set of possible initial conditions that determine a trajectory is a domain of the space of n dimensions.

From the very beginning of his book, Bertrand gives a word of warning against this type of problems, showing the contradictions that may arise:

> Another remark is still necessary: infinity is not a number; one must not, save an explana-
> tion, introduce it into the reasoning. The illusory accuracy of words might give rise to
> contradictions. To choose *at random* among an infinity of possible cases is not a sufficient
> indication (Bertrand, 1888, p. 4).

First, let us remark that Bertrand is not aiming here at all the questions involving continuous probabilities, but the problems implying a "random" choice among an infinite set of possible choices. Therefore, the chapters on the calculus of errors involve distributions on continuous probabilities, but do not relate to this initial warning.

Bertrand immediately gives a first example of the difficulties that arise in the treatment of these problems: "One asks for the probability that a number, integer or fractional, commensurable or incommensurable, chosen *at random* between 0 and 100, is greater than 50. The answer seems obvious [...]. The probability is ½" (Bertrand, 1888, p.4). Now a different result is found if one considers the square of this number and one evaluates the probability that this square is between 2,500 and 10,000: the natural answer is then ¾.

Bertrand then gives another problem: "One *randomly* traces a chord in a circle. What is the probability that it is smaller than the side of the inscribed equilateral triangle?" (Bertrand, 1888, p.4). He presents three arguments aiming at calculating this probability: each of them gives a different result. Bertrand concludes: "Among these three answers, which is the correct one? None of the three is false, none is exact, the question is ill-posed" (Bertrand, 1888, p. 5). It is this example that Poincaré chooses to open his chapter on continuous probability. But instead of rejecting the problem because it is badly formulated, he solves the paradox: each of these arguments, introducing different coordinates to locate a chord of the circle, leads to the definition of a different probability distribution on the set of chords.

Bertrand does not just give these two examples to warn against a difficulty. Throughout his book, he shows a clear attitude of rejection toward the problems that involve a random choice among an infinity of possible cases. Each time, he reaffirms that the question is ill-posed. We shall see several examples of this.

The interpretation of Bertrand's attitude concerning these paradoxes is a complex historical question, since his interest in probabilities mingles with a polemicist attitude that tends toward skepticism. Further, there is not serious study on the man and his philosophical positions, important for the conception of probabilities. Without claiming to complete this gap, I propose to highlight some aspects of Bertrand's thought on probability theory that emerge from the way he treats this problem.

The first thing to note is that Bertrand is not the first to point out a paradox of this type. In the second half of the eighteenth century, Lambert and Laplace had proposed two formulas, both of them plausible, although contradictory, to express what it means to choose at random a point on a sphere. The paradox is discussed by Cournot (1843), and also at the meetings of the Philomatic Society at the end of the years 1840. Bertrand is present (Bru et al., 1997, note 22, p. 161, note 58) and so from this period he is confronted with these difficulties of probability theory.

But it seems that his attitude toward these problems was at this time quite different from the one he adopts in 1888 in his book on probability theory. The accounts of the meetings of the Philomatic Society, although very brief, indeed lead to this interpretation. The only meeting where we find a sufficiently explicit reference to these paradoxes is that of 12 July 1851, on which it is stated:

> Mr. Bravais reports Mr. Cournot's opinion on the manner of calculating averages in the three-dimensional case. He also indicates the research that he has undertaken to determine the average form of the skull. Mr. Quetelet also studied these questions.
>
> Mr. Villarceau replies.
>
> Mr. Bertrand and Mr. Bienaymé add a few remarks. Mr. Bienaymé, to show how indeterminate the problem is, reminds the auditors of the previous discussions on average life[41].
>
> Mr. Bertrand thinks that it is possible to clearly define what is called average direction (*Comptes-rendus des séances de la Société philomatique de Paris*, Ms 2090, pp. 27–8).

Some explanations are required to situate these few lines. The problem reported by Bravais is slightly different from that of the indetermination of continuous probabilities, but it is linked to it. The origin of this discussion is a remark by Cournot in *Exposition* (Cournot, 1843; pp. 143–4 in the Vrin edition, Cournot, 1973–): can one define an average triangle? Cournot shows that this question is not obvious; in particular, if the three vertices are drawn at random and the averages of the three angles are made separately, the sum is not 180°, and yet this is always the sum of the angles. He uses this argument to criticize the notion of average man proposed by Quetelet in his important essay (Quetelet, 1835).

---

[41] In particular, Bienaymé is probably referring to the meeting of 29 March 1845 for which we read: "Mr. Bienaymé mentions research on the length of human life and on the manner the families disappear. This communication gives rise to a discussion that Mr. Bienaymé will summarize in a note" (Ms 2089, p. 71, verso). The archives of the Philomatic Society contain, in addition to manuscript records of the minutes of the meetings, about forty boxes (n°s 123–64). I have not found Bienaymé's note.

Following Quetelet's answer, Bravais (who has done his thesis on the localization errors of a point in space) continues this discussion at the Society. The account suggests that it tends toward the issue that interests us here, more specifically, because of the mention of the average direction. Indeed, this refers to Cournot's reflections on the various ways of defining the coordinates of an astronomical body, resulting in different behaviors of the average (Cournot, 1973–, pp. 150, 179).

The context having been clarified, let us remark that it is Bienaymé who insists on the indeterminate character of the question, whereas Bertrand considers that the problem can be sufficiently defined in order to remove the indetermination.

This reading is consistent with the use Bertrand makes of probability calculus in his course on integral calculus. Indeed, at this point, he is considering a continuous problem (Bertrand, 1870, pp. 483–91). More precisely, he wants to prove the following formula, attributed to Crofton:[42]

$$\int d\sigma(\omega - \sin \omega) = \frac{L^2}{2} - \pi\Omega,$$

"*dσ* designating any element of the plane exterior to a convex curve, of which the area is Ω and the length L, and ω the angle from which this curve is seen from a point situated on the element *dσ*." The proof uses the probability that a convex disc, randomly thrown on a plane cut by parallel lines, meets one of these lines. Bertrand then calculates the probability that two chords chosen at random on a convex curve meet inside the curve. Finally, he evaluates the probability that they meet outside in a given region. In other words, in 1870 Bertrand considers, apparently without it posing him any theoretical problem, the random choice of a chord of a convex surface. What is more, he gives his preference to this probability proof to establish the Crofton formula.

So, at this time, Bertrand does not show any reluctance to consider probability experiments in which the number of possible cases is infinite. Two problems addressed in this proof are considered again in 1888: the disc thrown on a series of parallel lines and the chord of a circle. The result is a particularly striking contrast between the 1870 course and the book on probability theory: while these two problems did not seem to pose any difficulty in 1870, they appear as sources of difficulties in the 1888 text. We have already seen this with the problem of the probability that a chord of a circle be smaller than the side of the inscribed triangle that Poincaré significantly named "Bertrand's paradox." Let us now turn to the problem of the random choice of the chord of a circle. It gives rise, indeed, to a very interesting passage in the 1888 text.

Bertrand is interested in this problem for the treatment of the classical problem of Buffon's needle: let equidistant parallel lines be traced on a plane; a needle is thrown at random on the plane and one seeks the probability that the needle meets one of the parallel lines. This classical example is given in the third chapter of the 1888 book.

---

[42] See Seneta et al. (2001) for a more detailed study of the history of this problem.

Bertrand solves it by considering the probability that a circle of radius R, thrown at random on a plane, meets one of the parallel lines. Therefore it is the same calculation than that opening the proof of Crofton's theorem in 1870. This is how Bertrand proceeded in 1870:[43]

> This said, the distance between the parallels being $2a$, if the disc is a circle of radius R, the probability that they meet is clearly R/a, because it is both necessary and sufficient for this that the center is at a distance less than R from one of the parallels between which it has fallen, and that, consequently, over a distance $2a$ of which all the points are equally possible, the center of the circle is in a portion where the total length is $2R$ (Bertrand, 1870, p. 484).

The problem of Buffon's needle is so classical and its solution so obvious, that Bertrand, apparently, is not able to resign himself to consider it as insufficiently determined—and yet this would be consistent with the initial warning against the choice among an infinite number of possible cases. He thus presents it, but it is the only example of continuous probabilities, and it appears among a collection of finite probability problems. Bertrand even tries to bring his resolution back to that of a choice among a finite number of possible cases (Bertrand, 1888, p. 55).

The result is correct, but the attempt at discretization is very unsatisfactory and attests to Bertrand's predicament. The problem is only apparently reduced to a finite number of cases. Bertrand was certainly quite aware of this discrepancy. This passage shows that he forbids himself to choose randomly among an infinite number of cases, however without rejecting, in fact, all the problems of this type.

One can suggest several reasons, which may be linked, to explain Bertrand's hesitation in this matter. I shall concentrate myself here on the question of the determination of the probability, even if there are probably other factors that could be taken into account.[44] This will allow me to show that this difficulty, on which Bertrand insists in his course, is resolved in Poincaré's course; Poincaré's resolution is linked to the way he introduces probability calculus in order to strengthen the recurrence theorem.

In order to better understand Bertrand's problem, we must closely study how he expresses the deficiencies of the random choice among an infinite set of cases. He points out, from the beginning of his indictment, that it is "not sufficient information" (Bertrand, 1888, p. 4).

Part of his analysis aims at denouncing a badly founded application of the principle of composite probabilities to prove—wrongly—that the probability distribution is necessarily Gaussian. It is the initial remark—according to which the data are insufficient—that interests us. Indeed, it agrees with the judgment on the first paradox: the information

---

[43] I have slightly modified the notations in order to harmonize them with those of 1888. In 1870, Bertrand wrote the radius of the circle r rather than R.

[44] B. Bru considers that Bertrand's position in his course on probabilities is revealing of his attitude: as he got older, he became very skeptical about the role of mathematics and mathematicians.

is insufficient. We shall see that Poincaré takes the same example and the comparison between the two passages will be significant regarding the different approaches of the two authors.

In all these passages, Bertrand always comes back to the way the problem is described, seeking there either the origin of the paradox or of the difficulties encountered: the question is ill-posed, not precise enough, the information is insufficient.

Bertrand seems to consider that the probability distribution must be determined by the description of the experiment that is being conducted. In other words, the probability is necessarily objective. It is the result of the conditions of the experiment that must be described very precisely.

This suggests an explanation to the radicalness with which Bertrand rejects, in 1888, the use of probabilities in the problems containing a choice among an infinite number of possibilities.

Let us first remark that Bertrand considers such problems mainly in two places in his book—almost always to criticize the solution that one has tried to give to them. This concerns the first chapter, on the enumeration of chances. Then, Bertrand comes back to the insufficiently determined character of some problems in his chapter on the probability of causes. In this area of probability calculus, one seeks to determine the probability of the different events that may have led to a same observed outcome. The first example given by Bertrand is the following:

> Peter bet he would obtain with three dice a value greater than 16; he won: this is the event. The value obtained can be 17 or 18: these are the possible causes of success (Bertrand, 1888, p. 143).

To solve this type of problem, one must know *a priori* the probability distribution of the experiment, of which the outcome is known. In the above problem, it is natural. Peter is playing with three dice. Of course, it is assumed that the dice are not loaded, so that for each dice, the probabilities of the six possible outcomes are, before Peter has thrown the dice, equal. But, the initial distribution is often not given in problems where it is nevertheless not obvious. This is the first example that Bertrand gives after having introduced the theoretical resolution of the problem of the probability of causes:

> An urn contains μ balls: some are white, others black, but one does not know in what proportion. k balls are drawn, each time replacing the drawn ball into the urn. Only white balls are drawn. What is the probability that the urn only contains white balls? (Bertrand, 1888, p. 146).

Bertrand starts by noting that "the question is ill-posed" and that "further, one ought to say what is the *a priori* probability" of each hypothesis about the composition of the urn. The experiment that determined the composition of the urn must be described. It is this precise description that allows us to know the probability *a priori*:

> If all the possible combinations having been prepared in urns of identical appearance, chance decided among them, the conditions are different than if one drew randomly balls

> from an urn of agreed composition, in order to constitute, with the drawn balls, the new urn that we are talking about (Bertrand, 1888, p. 146).

Bertrand carries out the calculation in different cases where the *a priori* probability is determined in this way; then his tone becomes more controversial:

> Without reason, one has put questions into the same category as the preceding problem that are in fact very different.
>
> When the observations contradict predictions, the probability of which seemed great, naturally, the influence of a disruptive cause is assumed and one is led to seek the probability of its existence.
>
> The question is insoluble. On the one hand, one does not have the necessary data. On the other hand, the dilemma, either there exists a cause or there does not exist a cause, does not have the clearness promised by the form of the statement (Bertrand, 1888, p. 156).

He gives several examples: the probability that a coin toss favors heads or tails according to the outcomes given by a large number of throws, reason for the gap between the numbers of girls and boys born in one region compared to another etc. In all these examples, one can see that the main problem is that of the *a priori* probability definition. The latter can no longer be defined by a sufficient description of experiment, because of the nature of the problems considered. One could say how the urn of unknown composition had been chosen among a set of urns, or filled with balls by drawing lots. But, to determine the *a priori* probability that a coin toss favors heads or tails, one would have to enumerate the factors that can create a disequilibrium of the coin, their contribution to this disequilibrium and their probability of occurrence.... Most of the problems that Bertrand mentions to fight the improper use of probability calculus are problems of probability of causes: the causes do not come from experiments of probability. The result is that one cannot precisely describe an experiment that would have given rise to them. Therefore, the data remain insufficient in Bertrand's eyes.

Thus, in Bertrand's book a link can be made between the rejection of the choice among an infinite number of possible cases, which is not a sufficient indication, and his vigorous assault against a great number of improper uses of probability calculus, with the aim of giving back to the latter its possibly lost legitimacy.

This link may be interpreted at least in two ways. An objective conception of probabilities may constitute the common origin of these two attitudes: the probability of the events must be determined by the conditions of the experiment. When this is not the case, the data are insufficient and the use of probability calculus is not founded. But, it might also be suggested: it is Bertrand's contentious intention to reject improper uses of probabilities that leads him to insist so firmly on the objectivity of probabilities. So, from the very beginning of his book, Bertrand adopts a radical attitude, to the point of putting himself in a predicament at the time of the treatment of the classical—and non-debated—problem of the needle of Buffon.

Whatever the profound reason of this offensive attitude may be, it led him to cast doubt on the legitimacy of the use of probability theory from the moment it did not concern game theory. This concerns justice,[45] but also physics and astronomy:

> Applied to dice, cards, to the game of red and black, to odd and even numbers, to heads and tails, the theory of chance is indisputable: nothing alters the rigor of the proofs; Algebra carries out more rapidly the counting which, with time and patience, could be carried on the fingers of one's hands [...]
> Physics, astronomy, social phenomena, seem to be, in many cases, governed by chance. Can one compare rainy and sunny weather, the appearance and absence of shooting stars, health and illness, life and death, crime and innocence with white and black balls drawn from a same urn? The same disorder appears in the details, does it conceal the same uniformity regarding the means? Will one find in the differences the known traits and the physiognomy of the effects of chance?
> [...] Reasoning cannot anticipate experiment; carefully discussed observations condemn at the same time the rebellious skeptics of any reconciliation and the exclusive minds that want everything to be subject to calculation (Bertrand, 1888, p. xxiii).

Earlier in the same preface (pp. xviii–xix) Bertrand had already mentioned the use of probabilities in celestial mechanics to point out its misuse. He had first explained the "too bold use" that "an astronomer whose name remained obscure without injustice, the Archbishop Mitchell" had made of an correct and ingenious idea: the latter had considered the distribution of the 3,000 known stars across the vault of the heavens. He had shown, among other things: one could bet 500,000 to 1 that a cause, apart from chance, had brought the six stars of the Pleiades closer. Bertrand does not definitely condemn this type of calculation, but he remains extremely careful and judges:

> In proposing the precise measurement of such vague assertions, Science can be jeopardized. [...] An assessment without figures is not binding, a number commits science, and this is with no right. The application of calculation to questions of this type is an illusion and an abuse (Bertrand, 1888, p. xviii).

The problem is developed in the body of the text (pp. 169–71), and its analysis ends with the same mistrust toward the application of probability to causes in celestial mechanics.

The only application of probabilities that Bertrand admits in astronomy concerns the theory of errors, uncertainty calculations, errors of observation. Celestial mechanics, from a theoretical point of view, is presented by Bernard as a field in which probability calculus has no reason of being used.

---

[45] The last chapter of his book is a "critical summary of the attempts made to apply probability theory to judicial decisions," where he shows the flaws of Condorcet's theory.

6.6.3   Use of probabilities in celestial mechanics: the probability distribution presented as a convention by Poincaré

In counterpoint to this rejection of "random" choices among an infinite number of possible cases, which concerns in particular continuous probabilities, in Bertrand's entire book, Poincaré devotes two chapters of his course to the latter.[46] In it, he exposes Bertrand's paradox, explains why one obtains different results for the same problem and shows how to avoid this difficulty.

The way Poincaré solves these paradoxes is rooted in a conception of probability calculus different from that of Bertrand. Indeed, Poincaré distinguishes two stages in the resolution of any probability problem. A first stage consisting in the determination of the cases considered equally probable, that is, setting the probability distribution; this is the choice of a "convention," and this stage is not in Poincaré's view, a mathematical stage. However, once this convention has been fixed, there remains only calculation. So, whereas Bertrand included, in probability calculus, the evaluation of the probability distribution determined by the conducted experiment, Poincaré places this stage of the problem resolution outside the scope of mathematics.

In his book, Poincaré deals very little with the part of the study that consists in fixing the probability distribution, if not precisely to explain the "paradoxes" given by Bertrand: they are only due to the fact that the various calculations are based on different probability distributions. In other words, it is not the calculation that causes the paradox, but the incompatibility of the conventions used in two different evaluations of the probability.

This evolution in the conception of probabilities coincides with a reassessment of the possible links between probability calculus and celestial mechanics in relation to Bertrand's position.

The latter challenged the legitimacy of applying probability calculus to physics or astronomy. Poincaré, on the contrary, gives in his course several examples of problems taken from celestial mechanics.

Twice, he takes the example of stars placed randomly on the celestial sphere—that Bertrand studied, we have seen, only to criticize the legitimacy of the calculation suggested by Mitchell. In the chapter where Poincaré presents several applications of continuous probabilities, he calculates the average distance between two stars assuming N stars randomly arranged on the celestial sphere. In the preceding examples, he had already considered figures randomly placed on a sphere and he had chosen the probability distribution to be proportional to the surface measurement of the sphere. He continues to work using this convention. For him, it is only a calculation problem, and the link with the effective arrangement of the astronomical bodies in the sky is not considered. He then looks at the way a great number of planets are spread across the zodiac after a relatively long lapse of time, whatever their initial positions were: we shall come back to this point.

---

[46] It is, in the second edition (1912), chapter VII, entitled "Continuous probabilities" and the chapter "Various applications;" these two chapters form, in the first edition, the second part of the eighth lesson, the ninth, and the tenth lessons.

He also proposes a problem of probability of causes regarding astronomical bodies: the divergence from Bertrand's attitude is particularly clear here. The problem is stated as follows: "Let N represent the *total number of small planets*; a certain number M of them are known. During one year, n planets are observed of which m are known. One asks for the probable value of N." Poincaré makes an additional hypothesis in order to be able to apply probability calculus: "we assume the probability known that, during the observation year, an existing planet was observed; let p be this probability: we admit that it is the same for known and unknown planets." He then considers the case where the probability p is assumed unknown and where we give ourselves a probability distribution satisfied by the value of p. Here too, it is quite clear that Poincaré is not concerned by the relation between the calculations he does and the true number of small planets. The calculations have no pretense of saying something about them; they are the result of the initial hypothesis.

Thus, Poincaré's approach allows him to apply probability calculus to objects that belong to celestial mechanics, without Bertrand's criticisms being relevant here. No more than in Bertrand's book, is Poincaré's physical world governed by probability laws. Poincaré introduces problems whose objects come from celestial mechanics. Whereas Bertrand denounced the improper use of probability calculus to avoid its devaluation, Poincaré presents this theory as a mathematical tool that may prove useful in various fields.

The 1894 course treats several problems coming from celestial mechanics, contrary to the position taken by Bertrand. One is tempted to ask whether this is not a clear sign of the use that Poincaré made of probabilities in his research in celestial mechanics, more particularly for the formulation of the corollary of the recurrence theorem.

One of the examples, given by Poincaré in the chapter concerning the applications of continuous probabilities, reinforces this hypothesis in a striking way. Poincaré shows that, for a differential Hamiltonian system, if a probability distribution on the initial conditions of the system is given and if this distribution is proportional to the volume, this distribution will be the same for the final values when one makes the system evolve during a given time. This is a variant of the invariance property of the volume for Hamiltonian systems. Poincaré had announced this last property in his work on celestial mechanics. Even more, this is precisely the property that allows us to apply the recurrence theorem. Therefore, we see that Poincaré relates probability calculus and celestial mechanics, and that he reinterprets the results obtained in the second field of research by using the terminology of probability calculus.

Further, we have observed that Poincaré already makes use, in 1890, of the conception of probabilities that allows him to solve Bertrand's paradoxes. Indeed, in [Pb], he carefully clarifies from the beginning the *convention* that he adopts to evaluate the probabilities involved. He uses the same terminology: "*let us adopt the convention* that the probability [...]" [Pb, p. 316] (italics added). Thus, his reflection on probabilities seems to show indeed, from 1890, the superiority that we have just mentioned in comparison with Bertrand's book.

In this, it differs from Gyldén. We have seen that the latter also uses probability calculus, with a view to celestial mechanics. But he does not set the probability distribution on the segment [0; 1] in which he chooses "*randomly*" a number μ. He first considers the continued fraction representing μ, then reasons on the numbers that appear during the calculation of this continued fraction, and supposes that a quantity formed as a function of these coefficients follows a uniform distribution. But, this distribution corresponds to a non-uniform distribution on the initial number. This is the problem that was highlighted by Bertrand's first example: the square of a number seems naturally to have a probability $\frac{3}{4}$ of being between 2,500 and 10,000, whereas a number only has a probability $\frac{1}{2}$ to be between 50 and 100. But, Gyldén does not show cautiousness in this regard.

On the contrary, when Poincaré uses probability theory in 1890, he first defines the adopted probability distribution. By this, he shows, vis-à-vis Bertrand's paradoxes, a rigor that announces that of his 1894 course and reveals itself in the same way. So if the 1890 work might have been inspired by Gyldén, it shows nevertheless Poincaré's personal reflection.

Moreover, I shall soon come back to the clarification added by Poincaré in 1891: the probability that a trajectory is non-recurring is equal to zero, *whatever the chosen probability distribution*. We shall see that Poincaré presents this addition as being the answer to an "objection." I shall show that one can recognize here an allusion to Bertrand's paradoxes, and the origin of an important theoretical development: the method of arbitrary functions.

The significance of probability calculus for physics is indicated even more clearly by Poincaré in the second edition of his course, in 1912. He then adds the introduction, which is based on an article published in 1907, and the last chapter. In the latter, Poincaré develops the analysis of the problems he presented in the introduction: the shuffling of cards, the distribution of the digits in a numerical table and the mixing of liquids. Therefore, Poincaré broadens the scope of his book at the time Borel—and some other young scientists—becomes the promoter of the use of probability theory in kinetic theories.

And yet, well before 1907, signs of Poincaré's orientation toward the use of probabilities in physics can be seen. I have shown the considerable difference, on this point, between Bertrand's book and the first edition of Poincaré's course on probabilities, in 1896, probably in accordance with the course given in 1894. The latter can further be placed in the continuity of the 1890 work on exceptional trajectories.

Poincaré's reflection continues in his first philosophical article on probabilities, published in 1899. In this text, Poincaré gives two examples of problems in which the initial state of the system is not known, and where one seeks nevertheless to say something about the present state: the problem of the small planets and that of the kinetic theory of gases. He concludes: "Only, probability theory allows one to predict the average phenomena that will be the result of the combination of these speeds." The irreplaceable role of probability calculus is thus already clearly stated in 1899.

On this point an evolution can be seen between the 1894 course and the 1899 article. In the first edition of his course, Poincaré mainly insisted on the *convention* that was a prelude to any probability calculation in these fields. He writes the following: "We know, for example, the distribution of the motion of molecules; if we knew exactly their initial position, we would be able to say where they will be at a given time; the probability that these molecules occupy such-and-such a final position depends therefore on the probability that we will give *by convention* to such-and-such an initial position. In each case, a particular hypothesis will be necessary."[47] One cannot fail to see the difference between this passage, in which Poincaré simply admits the usefulness of probability calculus, and the already mentioned conclusion of the 1899 article, in which probability calculus is presented as the "only" tool giving predictions.

Hence, the study of the use of probabilities to formulate the corollary of the recurrence theorem sheds new light on the first developments of Poincaré's reflection in this field. But, this is only a first work, focused on the early 1890s. It would be interesting to continue, by studying in particular Poincaré's writings on the kinetic theory of gases, starting from the mid-1890s. One could then better evaluate the respective importance of celestial mechanics, the kinetic theory of gases and Borel's reflections in the development of the conception of probability calculus and its role for Poincaré.

6.6.4   Poincaré's method of arbitrary functions

A second point can be mentioned: Poincaré's book constitutes, in relation to his work on the three-body problem, an important improvement in comparison with Bertrand's book. Indeed, Poincaré introduces a new method, the method of arbitrary functions, by which the arbitrariness of the choice by "convention" of the probability distribution is removed.[48]

This method was already used by von Kries in 1886 (Von Kries, 1886)[49] to show the equiprobability of red and black in a simplified roulette game. He asserts that for any regular enough probability distribution on all mechanical states of the system, the total probability of the color red and that of the color black will be approximately equal. However, he does not give the sufficient mathematical conditions to achieve this result. The same idea appears from 1891 in Poincaré's work on the three-body problem: whatever the—sufficiently regular—probability distribution adopted by convention on the space of initial conditions, the probability that a trajectory is nonrecurring is equal to zero. The arbitrariness that is present in the initial choice of the convention disappears during the calculation and the result no longer depends on it.

It is difficult to know with certainty if Poincaré knew of von Kries' work or not. The absence of his book in the Parisian libraries (and even the French libraries, except for Strasbourg) suggests that Poincaré apparently never had this book in his hands. Further, von Kries is a physiologist, and thus, belongs to a different milieu than that of Poincaré.

In fact, the clarification given by the 1891 summaries appears to be more an answer to Bertrand's paradoxes than a way of conceiving probability calculus in the manner of

---

[47] This passage is found, identical, in the first lesson of the 1896 edition and in the first chapter of the 1912 edition, p. 31.

[48] Von Plato (1983) proposes a more complete presentation of the history of the method of arbitrary functions, in particular the developments made by several mathematicians after Poincaré: Fréchet, Hostinsky, etc. The bibliography given by von Plato can be completed by Fréchet (1921, 1938). Bru (2003) also gives interesting historical information on the history of probabilities in the first half of the twentieth century, with a section on Poincaré's research.

[49] A discussion of von Kries' probability theory can be found in Kamlah (1983). More recently, the preprint of Shafer and Vovk (2004, pp. 13–14) briefly presents von Kries' book (1886).

von Kries' developments. The formulation of Poincaré (1891b) is too brief to be able to draw conclusions, but that of Poincaré (1891a) may give us an interesting clue. Indeed, Poincaré writes: "One will object that there is an infinity of ways of defining this probability; but this remains true whatever the adopted definition [...]." Therefore, Poincaré asserts he gives an answer to the objection saying there is no unique way to define probability in a continuous space. One recognizes here the difficulty raised by Bertrand concerning problems where one chooses "at random" among an infinite number of possibilities, in other words, problems in which one claims to calculate a probability when the set of possible causes is infinite. In 1890, was Poincaré aware of this weakness of his corollary expressed in probabilistic terms, without knowing in those days how to solve it? Or did he only realize it after the marathon of the correction of the memoir? Did he indirectly hear of von Kries' work, which might have inspired this addition? In any case, the clarification is presented as an answer to the problem that the paradoxes of Bertrand gave rise to. In other words, Poincaré's reflection on probability calculus—stimulated by the use of it induced by the discovery of the error in his 1889 memoir—is certainly based, as early as 1890–1, on a study of these paradoxes.

This reflection, which first appears in relation to the corollary of the recurrence theorem, leads Poincaré to point out, in his book on probability theory, a class of particular events. This concerns events, the probability of which does not depend on the chosen probability distribution. After having given many examples to show how the distribution transforms itself in the case of a change of variable, Poincaré concludes by reminding the reader of the importance of the initial choice of this distribution. He then insists on the existence of problems that do not depend on it. Therefore, in his book on probabilities, he develops the remark that he had made as early as 1891, according to which the exceptional character of non-recurring trajectories does not depend on the probability distribution.

> From the preceding, it follows that *great care must be given to the definition of the choice of the adopted probability distribution.*
>
> The probability that x is between x_0 and x_1 is expressed by an integral
>
> $$\int_{x_0}^{x_1} \varphi(x)dx;$$
>
> $\varphi(x)$ will be a function about which hypotheses will have to be made in order to know the probability distribution, but, in general, we will have to consider $\varphi(x)$ as being continuous.
>
> In general, the probability that x satisfies a given condition will depend on the choice of φ; however, this is not always the case, and *some problems are independent of the probability distribution.*
>
> *Example*—The probability that x is incommensurable will always be equal to 1, whatever the chosen continuous function φ, and the probability that x is commensurable will always be infinitely small.[50]

---

[50] Poincaré (1896, lesson n°X; pp. 147–8 in the 1912 edition). One can recognize in the last remark, although in a very embryonic form, a precursory sign of the sets of measure zero that will be studied by Borel and Lebesgue.

Following the example of rational and irrational numbers, Poincaré presents two other examples of such events. First the problem that had been studied by von Kries: "Let a wheel be divided into a great number of equal parts, alternately red and black; set it in a rapid rotating motion. When it stops, one of the divisions will be facing a fixed point of reference: what is the probability that this division is red or black?" (Poincaré, 1896, p. 148) Poincaré's analysis is mathematically more precise than that of van Kries. Poincaré shows that the probability approaches ½ for each color when the number of divisions increases indefinitely, under the hypothesis that the probability density has a bounded derivative. The second problem is inspired by celestial mechanics: "Let us consider a great number of *planets*, the orbits of which are approximately circular. [...] I say that after a very long time the planets will be equally distributed in all the signs of the zodiac. The probability that l [the longitude] is between given limits will therefore be independent of φ [the probability density]" (Poincaré, 1896, pp. 150–1). So, in Poincaré's book, the method of arbitrary functions is related to celestial mechanics and may help to justify the use of probabilities in it, contrary to the attitude adopted by Bertrand. The reasoning that I have just mentioned allows one to legitimize the adoption of the uniform distribution on the arrangement of these planets: whatever the initial distribution, after a sufficiently long time, the planets will be situated according to a uniform distribution.

6.7 Conclusion

In 1890, Poincaré introduces a statement of a new type, thanks to which he formulates mathematically the remark that he had previously made in vague terms: "the trajectories that have this property [of stability, AR] are more general than those that do not." The discovery of an error, which calls into question part of the results that seemed certain in 1889, motivates the search for mathematical foundations to this statement whose importance becomes crucial. Thus, this construction is not wanted for itself, but meets a need.

The mathematical problem at stake is quite singular. The recurring trajectories—and those that are not—cannot be designated, Poincaré only proves their existence. In order to show that the non-recurring trajectories are exceptional, he must use a suitable tool.

To answer this question, Poincaré turns toward probability calculus. Reading may have given him the idea of this resource, in particular the texts of Bertrand and Gyldén. But, this is not an instrument ready to meet Poincaré's needs. He adapts it to his subject and develops it according to his requirements. While Gyldén worked on a numerical problem by calculating averages, Poincaré bases himself on the volume estimates which are central to his proof of the recurrence theorem and reinterprets the volume as measuring the probability. He adopts the rigor requirement that made Bertrand highlight his paradoxes, but he works to solve them and to prove beyond doubt the relevance of his use of probabilities.

In this way, the correction work of his first memoir is accompanied by a reflection on the probability concepts, introduced on this occasion. Poincaré develops at this time a personal approach to probabilities that can be found in his lectures at the Sorbonne a few years later. In particular, the study of recurring trajectories is behind the method of arbitrary functions.

....

REFERENCES

Barrow-Green, June (1994). 'Oscar II's prize competition and the error in Poincaré's memoir on the three body problem'. *Archive for history of exact sciences* **48**, pp. 107–31.

Barrow-Green, June (1997). *Poincaré and the three body problem*. Providence, RI: AMS-LMS.

Bertrand, Joseph (1870). *Traité de calcul différentiel et intégral. Calcul intégral.* Paris: Gauthier-Villars.

Bertrand, Joseph (1888). *Calcul des probabilités.* Available online in Gallica. Paris: Gauthier-Villars.

Bru, Bernard (2003). 'Souvenirs de Bologne'. *Journal de la Société Française de Statistique* **144**, pp. 135–226.

Bru, Bernard, Marie-France Bru, and O. Bienaymé (1997). 'La statistique critiquée par le calcul des probabilités: deux manuscrits inédits d'Irénée Jules Bienaymé'. *Revue d'histoire des mathématiques* 3.2, pp. 137–239.

Cantor, Georg (1883). 'Fondements d'une théorie générale des ensembles'. *Acta Mathematica* 2. Translation from a German article carried out with Poincaré's help, pp. 381–408.

Comptes-rendus des séances de la Société philomatique de Paris. Manuscripts. Bibliothèque de la Sorbonne, Ms 2081–99. In particular: Ms 2089, years 1842–49 and Ms 2090, years 1850–57.

Cournot, Antoine-Augustin (1843). *Exposition de la théorie des chances et des probabilités. Œuvres,* (Cournot, 1973–, t. 1). Paris: Hachette.

Cournot, Antoine-Augustin (1973–). *Œuvres complètes.* Paris: Vrin.

Diacu, Florin and Philip Holmes (1996). *Celestial encounters: the origins of chaos and instability.* Princeton, NJ: Princeton University Press.

Dugac, Pierre (1984). 'Georg Cantor et Henri Poincaré'. *Bolletino di storia delle scienze matematiche* 4, pp. 65–96.

Fréchet, Maurice (1921). 'Remarques sur les probabilités continues'. *Bulletin des sciences mathématiques* 45, pp. 87–8.

Fréchet, Maurice (1938). *Méthode des fonctions arbitraires: théorie des évènements en chaîne dans le cas d'un nombre fini de cas possibles.* Paris: Gauthier-Villars.

Gilain, Christian (1977). *La théorie géométrique des équations différentielles de Poincaré et l'histoire de l'analyse.* Ph.D. Thesis, Université Paris I, France.

Gray, Jeremy J. (1992). 'Poincaré, topological dynamics, and the stability of the solar system'. In: *The investigation of difficult things. Essays on Newton and the history of the exact sciences.* ed A. E. Shapiro and P. M. Harman. Cambridge: Cambridge University Press, pp. 503–24.

Gyldén, Hugo (1887). 'Untersuchungen über die Convergenz der Reihen, welche zur Darstellung der Coordinaten der Planeten angewendet werden'. *Acta mathematica* 9, pp. 185–294.

Gyldén, Hugo (1888a). 'Om sannolikheten af att påträffa stora tal vid utvecklingen af irrationela decimal-bråk i kedjebråk'. *Öfversigt af Kongliga Vetenskaps-Akademiens Förhandlingar* 45, pp. 349–58.

Gyldén, Hugo (1888b). 'Om sannolikheten af inträdande divergens vid användande af de hittills brukliga methoderna att analytiskt framställa planetariska störingar'. *Öfversigt af Kongliga Vetenskaps-Akademiens Förhandlingar* 45, pp. 77–87.

Gyldén, Hugo (1888c). 'Quelques remarques relativement à la représentation des irrationnels au moyen des fractions continues'. *Comptes rendus de l'Académie des Sciences* 106, pp. 1584–7.

Gyldén, Hugo (1888d). 'Quelques remarques relatives à la représentation des irrationnels au moyen des fractions continues'. *Comptes rendus de l'Académie des Sciences* 106, pp. 1777–81.

Hald, Anders (1998). *A history of mathematical statistics from 1750 to 1930.* Wiley.

Kamlah, Andreas (1983). 'Probability as a quasi-theoretical concept: J. v. Kries' sophisticated account after a century; methodology, epistemology and philosophy of science'. *Erkenntnis* 19, pp. 239–51.

Lakatos, Imre (1976). *Proofs and refutations: the logic of mathematical discovery.* ed. John Worrall, et Elie Zahar. Cambridge: Cambridge University Press.

Lakatos, Imre (1984). *Preuves et réfutations. Essai sur la logique de la découverte mathématique.* Presentation by John Worrall and Elie Zahar. Translation by Nicolas Balacheff and Jean-Marie Laborde. Paris: Hermann.

Lambert, Jean-Henri (1801). *Lettres cosmologiques sur l'organisation de l'univers, écrites en 1761.* Translated from the German language by M. Darquier. Amsterdam: Gerard Hulst van Keulen. Facsim. published by A. Brieux in 1977.

Oxtoby, John C. (1971). *Measure and category: a survey of the analogies between topological and measure spaces.* New Yprk: Springer.

Poincaré, Henri (1878). 'Note sur les propriétés des fonctions définies par les équations différentielles'. *Journal de l'Ecole Polytechnique* 45, pp. 13–26. Also in *Œuvres* (Poincaré, 1916–54), t. 1, pp. xxxvi–xlviii.

Poincaré, Henri (1881). 'Sur les courbes définies par une équation différentielle *(première partie)*'. *Journal de Liouville (3rd series)* 7. Also in *Œuvres* (Poincaré, 1916–54), t. 1: pp. 3–44, pp. 375– 422.

Poincaré, Henri (1882). 'Sur les courbes définies par une équation différentielle *(deuxième partie)*'. *Journal de Liouville (3rd series)* 8, pp. 251–96. Also in *Œuvres* (Poincaré, 1916–54), t. 1: pp. 44–84.

Poincaré, Henri (1883). 'Sur certaines solutions particulières du problème des trois corps'. *Comptes rendus de l'Académie des Sciences* 97, pp. 251–2. Also in *Œuvres* (Poincaré, 1916–54), t. 7, pp. 251– 2.

Poincaré, Henri (1884). 'Sur certaines solutions particulières du problème des trois corps'. *Bulletin astronomique* 1, pp. 65–74. Also in *Œuvres* (Poincaré, 1916–54), t. 7: pp. 253–261.

Poincaré, Henri (1885). 'Sur les courbes définies par les équations différentielles *(troisième partie)*'. *Journal de Liouville (4th series)* 1, pp. 167–244. Also in *Œuvres* (Poincaré, 1916–54), t. 1: pp. 90–158.

Poincaré, Henri (1886). 'Sur les courbes définies par les équations différentielles *(quatrième partie)*'. *Journal de Liouville (4th series)* 2, pp. 151–217. Also in *Œuvres* (Poincaré, 1916–54), t. 1: pp. 167–222.

Poincaré, Henri (1889). 'Sur le problème des trois corps et les équations de la Dynamique. *Mémoire couronné du prix de S.M. le roi Oscar ii de Suède*'. This memoir, augmented by several lengthy notes, was planned for publication in *Acta mathematica*, but it was eventually replaced by Poincaré (1890). See Barrow-Green (1997) for greater details about the two versions.

Poincaré, Henri (1890). 'Sur le problème des trois corps et les équations de la Dynamique. *Mémoire couronné du prix de S. M. le roi Oscar ii de Suède*'. *Acta mathematica* 13. Also in *Œuvres* (Poincaré, 1916–54), t. 7, p. 262–479. In fact, an extended and corrected version of Poincaré (1889, pp. 1–270).

Poincaré, Henri (1891a). 'Le problème des trois corps'. *Revue générale des sciences* 2, pp. 1–5. Also in *Œuvres* (Poincaré, 1916–54), t. 8, pp. 529–37.

Poincaré, Henri (1891b). 'Sur le problème des trois corps'. *Bulletin astronomique* 8. Also in *Œuvres* (Poincaré, 1916–54), t. 7, pp. 480–90. Presentation, without demonstration, of the main results from Poincaré (1890).

Poincaré, Henri (1892–99). *Méthodes nouvelles de la mécanique céleste.* 3 volumes. Paris: Gauthier-Villars.

Poincaré, Henri (1894). 'Sur la théorie cinétique des gaz'. *Revue générale des sciences pures et appliquées* 5. Also in *Œuvres* (Poincaré, 1916–54), t. 10, pp. 246–263. Quoted from *Oeuvres*, pp. 513–21.

Poincaré, Henri (1896). *Calcul des probabilités, leçons professées pendant le deuxième semestre 1893–1894*. From notes taken by A. Quiquet. I quote the 1900 edition. A second edition, corrected and enlarged by Poincaré, was published in 1912. The main modifications were the addition of (Poincaré, 1907), as an introduction, and of a last chapter, which deals with problems tackled in this chapter.

Poincaré, Henri (1898). 'Sur la stabilité du système solaire'. *Revue scientifique* 9, pp. 609–13. *Œuvres* (Poincaré, 1916–54), t. 8, pp. 538–47.

Poincaré, Henri (1899). 'Réflexions sur le calcul des probabilités'. *Revue générale des sciences pures et appliquées* 10. This paper was included with slight modifications into Poincaré (1902, pp. 214–44, 262–9).

Poincaré, Henri (1902). *La science et l'hypothèse*. Quoted from the edition in 1992. Paris: Flammarion.

Poincaré, Henri (1907). 'Le hasard'. *La revue du mois* 3. In 1912, this paper is used as an introduction for the second edition of (Poincaré, 1896), and it is also included in Poincaré (1908, pp. 64–94, 257–76).

Poincaré, Henri (1908). *Science et méthode*. Paris: Flammarion.

Poincaré, Henri (1916–54). *Œuvres*. Paris: Gauthier-Villars.

Poincaré, Henri and Gösta Mittag-Leffler (1999). *La Correspondance entre Henri Poincaré et Gösta Mittag-Leffler, avec en annexe les lettres échangées par Poincaré avec Fredholm, Gyldén et Phragmén*. Edited by P. Nabonnand. Basel: Birkhaüser.

Quetelet, Adolphe (1835). *Sur l'homme et le développement de ses facultés, ou Essai de physique sociale*. Paris: Bachelier.

Rapport de la Commission du Prix offert par S.M. le roi Oscar ii de Suède (1889). *Œuvres* de Poincaré (Poincaré, 1916–54), t. 11, pp. 286–9.

Roque, Tatiana (2001). *Ensaio sobre a gênese das idéas matemáticas: exemplos da theoria dos sistemas dinâmicos*. Ph.D. Thesis, Universidade Federal do Rio de Janeiro, COPPE, Brazil.

Seneta, Eugene, Karen H. Parshall, and François Jongmans (2001). 'Nineteenth-century developments in geometric probability: J. J. Sylvester, M. W. Crofton, J.-E. Barbier, and J. Bertrand'. *Archive for history of exact sciences* 55, pp. 501–24.

Shafer, Glenn and Vladimir Vovk (2004). 'The sources of Kolmogorov's *Grundbegriffe*'. *Statistical science* 21, pp. 70–98. http://www.probabilityandfinance.com (working paper).

Sheynin, O. B. (1991). 'H. Poincaré's work on probability'. *Archive for history of exact sciences* 42, pp. 139–71.

Stigler, Stephen M. (1986). *The history of statistics. The measurement of uncertainty before 1900*. Cambridge, MA: The Belknap Press of Harvard University Press.

Von Kries, Johannes (1886). *Die Principien des Wahrscheinlichkeits-Rechnung*. Freiburg: Mohr.

Von Plato, Jan (1983). 'The method of arbitrary functions'. *The British Journal for the Philosophy of Science* 34, pp. 37–47.

Von Plato, Jan (1994). *Creating modern probability. Its mathematics, physics and philosophy in historical perspective*. Cambridge: Cambridge University Press.

7

Generality and structures in functional analysis: the influence of Stefan Banach

FRÉDÉRIC JAËCK

7.1 Introduction and setting the problem

The Polish mathematician Stefan Banach is mostly known, often far beyond the strict mathematical community, through his recognized book on linear operations (Banach, 1932), as well as through the spaces named after him. The latter, now called Banach spaces or, in modern terms, complete normed vector spaces (on $\mathbb{R}$ or $\mathbb{C}$), constitute a fundamental and omnipresent object in the mathematics of the second half of the twentieth century.

In our paper, centered on the concept of generality, we propose initiating a systematic study of Banach's work. Starting with his doctoral thesis, we shall analyze Banach's specific appropriation of earlier works and the line of thought he derives from his own reading.

His doctoral dissertation was presented at Lwów University (in Poland at that time), under Professor Lomnici's supervision, and was published in one of the very first issues of *Fundamenta Mathematicae*[1] two years later (Banach, 1922).

The article, written in French, bore a title that indicated its goal in Banach's view: *On operations on abstract spaces and their application to integral equations* (*Sur les opérations dans les ensembles abstraits et leur application aux équations intégrales*). Historians of mathematics often refer to this publication as the first occurrence of the Banach space structure in mathematics. We shall question this assertion, one of our main tools being to distinguish different types of generality in distinct publications by Banach.

---

[1] The journal was founded in 1920 by Zygmunt Janiszewski, Stefan Mazurkiewicz, and Wacław Sierpiński and is freely accessible at the Polish Virtual Library of Science: http://matwbn.icm.edu.pl/.

The Oxford Handbook of Generality in Mathematics and the Sciences. First Edition. Karine Chemla, Renaud Chorlay and David Rabouin. © Oxford University Press 2016. Publishing in 2016 by Oxford University Press.

The birth of Banach spaces can be placed within the general historic context of the emergence of structures in analysis. It constitutes a major and characteristic part of an evolution that can be traced across several domains in mathematics at the turn of the century.[2]

Focusing on Banach's doctoral thesis, we shall see that at this stage his ideas are motivated by three major objectives:

a) The intention, announced in the introduction of Banach's article, to deal with some collections of functions of various natures from a unique and undifferentiated point of view.

b) The project, expressed in the title of the article, to deal with operations defined on abstract spaces.

c) The motivation to obtain some new results for integral equations problems, this goal first appearing, once again, in the title of the paper.

Those three goals will be shown to be essential elements, and will constitute in a sense what determines the geometry of Banach's text. They will embody the different forms of generality that Banach's 1922 article generates.

Our study will be divided into two main parts. The first one analyzes in detail the two stages in the process by which Banach elaborated a new framework for functional analysis where structures were bound to play a central role. We shall focus on the work of reinterpretation and appropriation of previously unconnected material, and we shall determine some crucial ingredients in the emergence of the new theory. In this part we shall also argue that Banach spaces were not actually born in this first paper, although the axioms enunciated by Banach match the ones we now use to define those spaces. In fact, we will bring forward specific evidence that Banach's conception of such spaces as a general tool can only be definitely acknowledged later, in his 1932 book. A close examination of generality in the 1922 and 1932 publications will be essential for capturing a major change in this respect.

In a second part, we shall concentrate on what appears to be the core of Banach's 1922 paper and on the transformation into a general setting that it represents. The main achievements of this work, as well as all the essential features that bear witness to the birth of a new theory, are concentrated in the study of linear operations. Our goal is to show that we are here in the presence of a general theory developed by Banach in his doctoral dissertation: the theory of linear operations constitutes the pinnacle of a development in search of a high degree of efficiency and generality.

Banach's doctoral dissertation will be the first paper in a long series and at this early stage it is possible, maybe more than elsewhere, to capture in a mathematical discourse in the making and in the subtle interplay between some elements of the text, the emergence of various degrees of generality.

We shall use three main sets of sources to show the mechanisms involved into producing generality in Banach's 1922 paper:

---

[2] The reader can consult, for example, the following papers to obtain a general overview of the birth of modern analysis: Dorier (1996) and Dieudonné (1981).

1) Some texts written before Banach's time, by Peano, Pincherle, Fréchet, and others, all of whom worked on connected subjects. We shall establish some similarities with Banach's paper and thus highlight his specific and complex work of appropriation and selection.

   Comparison among several texts either reveals a precise similarity between two forms,[3] or shows the mark of a more subtle and manifold process. In particular, the mere formal correspondence between parts of two texts is not sufficient to give a faithful idea of the nature of the mathematical work, nor is it sufficient to discuss the emergence of new concepts. In order to give a precise account of the multiply-flavored expression of generality that arises in Banach's work it is crucial to explore the deep links and evolving interactions among several parts of the various texts, especially between definition-like material and the proofs of theorems. Our analysis will trace these interactions through the texts and will determine the type of generality that emerges in some specific situations.

   When this is the case, we shall bring forward the nature of the evolution, and elaborate on a distinction between various types in the expression of generality.

2) We shall also refer to Banach's 1932 book on linear operations.[4] It exposes a very polished and mature theory, the form of which is currently still completely acceptable. The reference to this book, although it was written some 12 years after Banach's doctoral dissertation will be essential in our discussion about the precise positioning of the birth of Banach spaces in a long historical process.

3) Finally, Banach's doctoral dissertation will constitute the central text of this study and a starting point for a new approach to his work and his wide influence.

7.2   Banach spaces

The notion of a space of functions is the result of a diverse movement the main aspects of which can be expressed in terms of generality. And Banach's work is from this point of view very specific, since it brings together many essential ideas and developments which shape an abstract and general conception in functional analysis.[5]

---

[3] In other words, some passages in two separate texts might be superposable, although their roles in the two contexts significantly differ.

[4] Banach (1932).

[5] The expression "functional analysis" deserves some comments here, since it was not quite a defined field at the time of Banach's dissertation publication. We use it here to refer to the branch of analysis that deals with sets of functions (as opposed to the study of such or such given function individually). The emergence of several domains stamped "analysis" at the turn of the century is deeply connected with the influential use of structures in the study of functional problems. This essential aspect is beyond the scope of the present paper, and a specific analysis will be presented somewhere else.

Banach's doctoral dissertation opens with a passage about methodology, in which generality appears as a central value in the chosen approach:[6]

> The present work will establish some theorems that are valid for several fields[7] of functions which I shall specify later on. But, in order not to have to prove them separately for each particular field, which would be tedious, I have chosen another way round: I first consider in a general way the sets of elements for which I postulate some properties. Then I deduce some theorems and I show that the postulates I had adopted were eventually true for each particular field of functions.

And this methodological introduction is immediately followed by a list of function sets (*champs fonctionnels*):[8]

> For the sake of simplicity, I have introduced the following notations for a few function fields:

The set of all continuous functions	(C)
The set of all summable functions (Lebesgue integrable)	(S)
The set of all rth power integrable functions (L)	(S^r)
The set of all bounded measurable functions	(M)
The set of all bounded Duhamel[9] functions	(D)

The set of all functions such that the $(p-1)$-derivative is absolutely continuous and the p-derivative is

continuous	$(C^p C)$
integrable (L)	$(C^p S)$
rth power integrable (L)	$(C^p S^r)$
bounded	$(C^p M)$
Duhamel	$(C^p D)$

Starting with this context, the first step taken by Banach consists in a process of abstraction involving two dimensions. On the one hand, Banach identifies and extracts some common characteristics from these collections of functions as well as from some already known situations (he refers to several mathematicians at the beginning of his work). On the other hand the selected material is reshaped with the intention of being

---

[6] Banach (1922: 134).
[7] Here the word "field" refers to collections of functions without reference to any algebraic structure. It is used by Banach in his dissertation in the same way as the word "set" (see next quote in our text). We shall discuss this in more detail later on.
[8] Banach (1922: 134).
[9] Duhamel functions are Lebesgue integrable functions f such that $\lim_{h \to 0} \frac{1}{h} \int_x^{x+h} f(t)\,dt = f(x)$.

efficient: it is meant to enable the mathematician to present in a unified and synthetic way some ideas that pertain to what he considers to be a coherent domain. In particular, the proofs of the main results will be organized in a specific way and will stress the role of some general tools that were identified and isolated.

7.2.1 Vector spaces

Banach's doctoral dissertation's first chapter opens with a list of axioms which he will add to throughout his paper.

The first sub-list defines what we now call a vector space. Yet, Banach himself does not mention the word "vector space" at any stage in the paper. In fact, the vocabulary he uses fluctuates throughout the article. Banach will in turn use the words "class" (*classe*), then "system" (*système*), or "field" (*champ*) to denote the collection of elements he deals with. It is not until his 1932 book that one can see a final choice of terminology for differentiating situations involving more or less structure.

For the reader's convenience, we shall nevertheless refer to these axioms using the expression "vector space axioms."

These axioms are introduced at the beginning of the first section in the following form:[10]

§1 Axioms and fundamental definitions. Let E be a class containing at least two elements, any arbitrary elements, which we will designate for example by $X, Y, Z, \ldots$

Given any three real numbers a, b, c, we define for E the two following operations:

1) the addition of the elements in E
    $X + Y, Y + Z, \ldots$
2) the multiplication of the elements in E by a real number
    $aX, bY, \ldots$

Let us admit that the following properties are fulfilled:
I_1. $X + Y$ is a well-defined element of the class E,
I_2. $X + Y = Y + X$,
I_3. $X + (Y + Z) = (X + Y) + Z$,
I_4. $X + Y = X + Z$ implies $Y = Z$,
I_5. There exists a specific element θ in the class E such that we always have $X + \theta = X$,
I_6. $a \cdot X$ is a well-defined element of the class E,
I_7. $a \cdot X = \theta$ is equivalent to $X = \theta$ or $a = 0$,
I_8. $a \neq 0$ and $a \cdot X = a \cdot Y$ implies $X = Y$,
I_9. $X \neq \theta$ and $a \cdot X = b \cdot X$ implies $a = b$,
I_{10}. $a \cdot (X + Y) = a \cdot X + a \cdot Y$,

---

[10] Banach (1922: 134–5).

I$_{11}$. $(a+b) \cdot X = a \cdot X + b \cdot X$,

I$_{12}$. $1 \cdot X = X$,

I$_{13}$. $a \cdot (b \cdot X) = (a \cdot b) \cdot X$.

Along with this, we introduce the following definitions:

(a) $-X = (-1) \cdot X$,

(b) $X - Y = X + (-1) \cdot Y$.

Although he lists these axioms, in his 1922 paper Banach does not consider the algebraic structure for itself. These statements are only properties shared by the collections of functions he considered in the introduction. In this respect, we shall see that the 1932 book differs considerably from this earlier paper. Here, these axioms are followed immediately by a short comment and a few examples. Then Banach goes on introducing a norm, and definitively settles his study in the new context:[11]

II. There exists an operation called norm (it will be denoted by $\|X\|$), defined on the field E, with counter-domain the set of real numbers, and satisfying the following conditions:

II$_1$. $\|X\| \geq 0$,

II$_2$. $\|X\| = 0$ is equivalent to $X = \theta$,

II$_3$. $\|a \cdot X\| = |a| \cdot \|X\|$,

II$_4$. $\|X + Y\| \leq \|X\| + \|Y\|$,

III. If 1°) $\{X_n\}$ is a sequence of elements in E, 2°) $\lim_{\substack{r \to \infty \\ p \to \infty}} \|X_r - X_p\| = 0$,

then there exists an element X such that:

$$\lim_{n \to \infty} \|X - X_n\| = 0.$$

Of course, it is possible to find some definitions that are very close to parts of these axioms in earlier texts, and one can read for example Jean-Luc Dorier's paper "A general outline of the genesis of vector space theory"[12] to get a view of the evolution of related ideas over a long period.

Giuseppe Peano's book, entitled *Geometric calculus* (Calcolo geometrico[13]), published in 1888, contains what is usually accepted as the first occurrence of a list of axioms defining what we now call a vector space.

At the end of this book, which deals with some specific geometrical situations involving vectors in the plane or in three-dimensional space, one can find the following axioms, which define what Peano calls a "linear system" (*sistema lineare*):[14]

---

[11] Banach (1922: 135–6).
[12] Dorier (1995).
[13] Peano (1888).
[14] Peano (1888: 141–2).

72. There exist systems of entities for which the following definitions hold:

1. The *equality* of two elements a and b of the system is defined, that is, there is a proposition, noted $a = b$, which expresses a condition that some but not all pairs of elements in the system may satisfy, and which satisfies the following logical equations:

$$(a = b) = (b = a) \qquad (a = b) \cap (b = c) < (a = c).$$

2. The *sum* of two entities a and b is defined, that is, there is an entity noted $a + b$, which belongs to the given system, and that satisfies the conditions:

$$(a = b) < (a + c = b + c), \; a + b = b + a, \; a + (b + c) = (a + b) + c,$$

and the common value of the two sides of the last equality will be denoted by $a + b + c$.

3. Given an entity a of the system, and a positive integer m, with ma we shall designate the sum of m entities all equal to a. It is easy to see that, given entities $a, b, \ldots$ of the system, as well as positive integers $m, n, \ldots$, we have:

$$(a = b) < (ma = mb); \; m(a + b) = ma + mb$$

$$(m + n)a = ma + na; \; m(na) = (mn)a; \; 1a = a.$$

We shall suppose that the expression ma is given a signification for any real number m, in such a way that the preceding equalities remain valid. The entity ma will be called product of the real number m by the entity a.

4. Finally we will suppose that there exists an entity in the system, which we will call the *null entity*, noted 0, such that, for each entity a, the product of the number 0 with the entity a always gives the entity 0:

$$0a = 0.$$

If we give to the expression $a - b$ the signification $a + (-1)b$, then we can deduce:

$$a - a = 0, \; a + 0 = a.$$

DEF. The systems of entities for which the definitions 1, 2, 3, 4 as well as the conditions imposed therein hold will be called *linear systems*.

This passage is followed by some examples, mainly geometrical ones, in dimension 1, 2, or 3, that have been developed and studied earlier in the book and that are cited here as special cases. Hence this early work by Peano indisputably develops a unifying point of view, encapsulating several known situations under the same definition. The introduction of this viewpoint constitutes one of the essential aspects of a kind of generality, which takes a specific form here. The axiomatic formulation, exposed at the end of Peano's book, was derived from a single and coherent view of a diversity of situations.

It is also remarkable that Peano already thought of the infinite dimensional case:[15] "A system may have infinitely many dimensions."

Yet no examples are given of infinite-dimensional spaces, and ideas of such situations might have been somewhat uncertain at that time. Anyway, this shows that the type of generality developed here is *open*: it goes beyond the strict unification of known cases, and presents itself as an open vision that wishes to point at some new and unknown situations, or even at some barely thinkable cases. We shall see a bit later how Banach himself uses a specific strategy to develop an efficient general point of view. Both approaches share some common ingredients: a unified formulation and a kind of *openness* that potentially covers unknown situations, while encompassing the usual ones.

Besides these first attributes characterizing the emergence of a kind of generality, Peano's text bears witness to the development of a reflexive point of view. The presence of reflexivity in a text constitutes, in our view, a necessary ingredient for acknowledging the emergence of a new concept or a new theory. Returning to Peano's book we notice that, just after his axioms, he develops a list of properties and starts a study of the newly defined linear systems (independence, dimension, etc.): this is an illustration of the reflexive activity that consists in exploring the new conceptual situation for itself. Highlighting the emergence of a similar reflexivity in Banach' publications will constitute a key point in our argumentation and this criterion will make it possible to establish a clear distinction among several situations.

Some 13 years later, in 1901, Pincherle and Amaldi[16] wrote a book of a similar flavor to what we analyzed in Peano's work, titled *The distributive operations and their applications to analysis* (*Le operazioni distributive e le loro applicazioni all'analisi*[17]), which opens with a first chapter titled "The general *n*-dimensional linear set." In fact, and regardless of the restriction contained in the title of the chapter, the authors first define linear spaces without any reference to dimension:[18]

> 2. We will admit first that, given two elements α and β in S one can say whether they are linked by a relation, called *equality*,[19] characterized by the following properties:
> I. If α is *equal* to β, then β is also equal to α.
> II. If α is *equal* to β, and β is *equal* to γ, then α is *equal* to γ.
> In order to express that α equals β, one shall write $\alpha = \beta$ ...

[15] Peano (1888: 143).
[16] Much of the content of the book had been previously published by Pincherle alone; nevertheless this book has two authors.
[17] Amaldi and Pincherle (1901).
[18] Amaldi and Pincherle (1901: 1–2).
[19] Words are italicized in Pincherle and Amaldi's original text.

3. Now we shall also admit that, given any two elements α and β in S, one can always determine a third element which will be denoted by $\alpha + \beta$.

The operation sending $\alpha 1$ and β to the third element is characterized by the following equalities:

I. $\alpha + \beta = \beta + \alpha$,

II. $(\alpha + \beta) + \gamma = \alpha + (\beta + \gamma)$,

Where $\alpha + (\beta + \gamma), (\alpha + \beta) + \gamma$ represent the elements deduced from α and $\beta + \gamma$, or from $\alpha + \beta$ and γ respectively, as $\alpha + \beta$ is deduced from α and β.

The *equal* elements $\alpha + (\beta + \gamma)$ and $(\alpha + \beta) + \gamma$ will be denoted by $\alpha + \beta + \gamma$ without further notice. The operation defined in this way will be called *addition* and $\alpha + \beta$ the *sum* of α and β.

The equalities I and II will be called *laws* or *properties* of the addition; the former will be called the *commutative law* and the latter the *associative law* for addition.

In any sum involving any number of elements, one can, in conformity with the laws I and II, group any of the elements or permute their order.

4. In the set S we shall suppose the existence of an element ω, which we shall call the *zero element*, or simply *zero*, which satisfies for any element α the equality $\alpha + \omega = \alpha$.

We shall ultimately designate the element ω with the symbol 0; but it will be important to distinguish this element from the zero of the set of all numbers.

Some intermediate notations then follow, making it possible to replace, for any integer n, the sum $\alpha + \cdots + \alpha$ by $n\alpha$, before outlining some extended distributivity properties.[20]

5. To provide some sort of generalization of the above property, we shall admit that, given a number a and any element α, there exists in S an element, noted $a\alpha$, and we shall call it the product of the element α by the number a. The operation through which $a\alpha$ is deduced from α will be called the *multiplication* of the element α by the number a.

This operation is characterized by the following properties. When a is a positive integer, $a\alpha$ will be the sum $\alpha + \cdots + \alpha$ of a elements all equal to α; for any given numbers a and b, the preceding properties will remain valid, that is:

I. $a(\alpha + \beta) = a\alpha + a\beta$,

II. $(a + b)\alpha = a\alpha + b\alpha$,

III. $a(b\alpha) = (ab)\alpha$,

To which one has to add the following:

IV. $0\alpha = 0$.

Finally, Pincherle and Amaldi name the new object "linear space" or "linear set" (*insieme o spazio lineare*):[21]

A set S that has the properties enunciated in paragraphs 2, 3, 4, and 5 will be called a *linear set* or *linear space*.

Then some (very few) examples follow:[22]

[20] Amaldi and Pincherle (1901: 3).
[21] Amaldi and Pincherle (1901: 4).
[22] Amaldi and Pincherle (1901: 4).

> A first example of linear system is given by the set of all (real or complex) numbers ...
> Another example will be very instructive for us, since it makes the consideration of
> general linear systems intuitive: ... the set of all vectors in the ordinary space.

Beyond the obvious similarities there are some important differences between Peano's work and the text produced by Amaldi and Pincherle. The fact that what we identified as the emergence of a structure first appears in the text at the very end of Peano's book, whereas it opens Pincherle and Amaldi's text, is not without importance. For Peano, the "linear systems" come at the end of an organized list of situations which are ordered in a nested progression: geometry of the line first, then the plane, then three-dimensional space. Linear systems come naturally as the next step in this list. No context ever becomes inappropriate because of the emergence of the next one and, as with a nested set of Matryoshka dolls, it is legitimate to choose the appropriate level and the best focus—plane, space, etc. for working in specific situations. Generality here is the expression of the fact that some results (for example in the plane) can be stated in an extended formal grammar (space or linear systems).

Starting from a different point of view, Pincherle and Amaldi begin their book with the definition of a linear system, then after a very short list of easy examples, they introduce determinants in a continued line of thought and use linear systems as a mere ingredient in their theory of distributive operations. In contrast to Peano's view, there is here no hierarchy in the situations studied by the authors when considering linear systems. Linear spaces do not appear as a concept encompassing a large spectrum of situations. They appear in a context that allows them to promote the idea that linear operations are a universal tool, adapted to many problems, especially in analysis. In particular, and this will be very different from Banach's strategy, they put forward very few properties of their spaces.

In fact, these historical documents show that contrasting and analyzing the strict formal contents of a list of axioms indeed sheds some light on the different points of view adopted by the authors. However, what we investigated just above shows that this is not enough, that the list of axioms and their differences or similarities do not give a complete account of the type of generality that arises in those texts.

Nevertheless, contrasting the texts from the perspective of their forms—I shall refer to this feature of the mathematical contents as "static," in contrast to its dynamic features, to be introduced below—leads one to notice some specificities of Banach's approach. For example, one can see that:

1) Banach carefully assumes that there are at least two elements in the classes he deals with. This rigorous remark allows him to avoid the trivial space (though Banach does not refer to this point here).

On their side, Pincherle and Amaldi use as many elements as their propositions require without any comment. It appears to be only a question of style here since one can find the same element several times in their propositions. Nevertheless one can compare a sentence like "given two elements α and β ..." in Banach's text or Pincherle and Amaldi's text.

2) Both Banach and Pincherle and Amaldi give some examples in order to illustrate their lists of axioms, and there is some overlap here.

For Banach:[23]

> One can think for example of the following systems: the Grassmann forms, the quaternions, the complex numbers, etc.

while Pincherle and Amaldi give the more restrictive collection of examples, referring mainly to real or complex numbers and to vectors "in the ordinary space."

In comparison, Peano's text, which was written earlier, does not embrace such a diversity of particular situations, but as we mentioned earlier, it nevertheless refers to the infinite-dimensional case.

Hence, we see here that it is possible to track down parts of several texts in earlier publications that match Banach's first list of axioms. Our claim is that this similarity is not enough to attribute the paternity of the concept to the author of the first extant text in which the axioms occur. The specificity of Banach's ideas in 1922 is due to two main factors. On the one hand, he brings about a shift in the point of view by straightaway extending his list of axioms, adding a new collection of propositions of a very different nature that define the norm. On the other hand, he uses the newly defined environment in his own way.

The emergence of any form of generality, or even the nature of generality cannot be accounted for by the mere list of axioms; generality has to be understood here as the expression of a shift of point of view that we shall continue to analyze now.

7.2.2 Topological aspects

Of course, we shall use the word "topology" not as an actors' category, but only as a taxonomic device: it will refer to several words in Banach's text, such as *norm, distance, convergence, continuity*, and related elements.

7.2.2.1 Norm

As we have already mentioned, in a second list of axioms Banach introduces the notion of norm. This list is immediately followed by an axiom dealing with completeness (although this word is not used by Banach who does not give a name to the property).

The notion of a norm, even when dealing with functions, is not new, and it is possible to find the following passage in an article published by F. Riesz in 1916:[24]

1. Definitions and theorems. In the following study we shall consider the set of all continuous functions on the segment $a \le x \le b$ ...

---

[23] Banach (1922: 135).
[24] Riesz (1916: 72).

We shall call the set under consideration a *function space*. Moreover, we shall call the *norm* of $f(x)$, and we shall write as $\|f\|$, the maximal value of $|f(x)|$; hence the quantity $\|f\|$ will generally be positive, and will vanish only when $f(x)$ is identically zero. Moreover, this norm satisfies the relations:

$$\|cf\| = |c| \cdot \|f\| \quad \text{and} \quad \|f_1 + f_2\| \le \|f_1\| + \|f_2\|$$

This passage is shaped in a very modern form, very close to the one we would use nowadays. Riesz introduces a norm, then defines the continuity of a linear transformation on what he calls a "function space" (*Funktionalraum*).[25]

   In this paper by Riesz, the scope of the definition, that is, the set of elements that fall under the definition, is totally determined in advance: the norm applies to continuous functions on the compact interval. At that time the set of all continuous functions on a closed interval had been studied for its own sake and was usually denoted by $C([a,b])$.

   Despite obvious similarities, Banach's point of view is very different since he does not explicitly list the objects that fall under his definitions: the function spaces cited at the beginning of the paper do not directly shape the list of axioms, and the norm is not specified, nor given *a priori* in terms of those function spaces. The vocabulary used here by Banach is characteristic: "Let us admit that ...there exists an operation called norm."[26]

   This gives rise to substantial differences between the two approaches. For Riesz, the norm is given and uniquely defined in the usual setting of continuous functions. Hence, what we call now completeness appears in his text as a useful property and is isolated as a tool and not as an axiom:[27]

> The term *distance* of two functions f_1 and f_2 will be understood as the norm of their difference $\|f_1 - f_2\| = \|f_2 - f_1\|$. It follows that the uniform convergence of a sequence $\{f_n\}$ to the limit function f is equivalent to the fact that the distance $\|f - f_n\|$ converges to zero. A necessary and sufficient condition for the uniform convergence of a sequence $\{f_n\}$ consists in the usual so-called principle of convergence $\|f_m - f_n\| \to 0$ when $m \to \infty$ and $n \to \infty$.

In particular, a sequence $\{f_n\}$ such that the distances $\|f_m - f_n\| (m \ne n)$ are non-vanishing, that is to say such that the lower bound is strictly positive, will never converge uniformly.

   As we saw previously, Banach introduces the corresponding property for any "sequence of elements" and without any reference to the nature of those elements. Banach even insists several times on the general nature of the elements using synonyms for "arbitrary" (*quelconque, arbitraire* in French).

   Of course, the reference to Riesz's work cannot be minimized in the analysis of the process leading to the shaping of a general context for the study of functional problems.

---

[25] Here the functions are defined on a compact interval $[a;b]$ with the supremum norm.
[26] Banach (1922: 135).
[27] Riesz (1916: 72).

The similarities show that at that time there was a need to develop a general point of view in analysis.

Moreover, one could argue from a historic point of view that, even if Banach's axioms refer to a vaster field than Riesz's, the space of uniformly bounded functions on the compact set [a; b] served as an archetype of the complete normed vector spaces, and that the notion of norm emerged by generalization from this instance. This is not the case, and Banach provides us with evidence that he did not model his introduction of a norm in general on that single example. Banach himself comments upon the fact that the context of continuous functions cannot be taken as a model:[28]

> The notion of a *function of a line*[29] was introduced by Mr. Volterra. Research on this subject has been carried out by Messrs. Fréchet, Hadamard, F. Riesz, Pincherle, Steinhaus, Weyl, Lebesgue, and many others. In their earlier works, they supposed that the domains and counter-domains were sets of continuous functions with derivatives of higher degree.[30] It was only Hilbert's work that brought some results that could be easily transferred into theorems about operations for which the domain and counter-domain were square integrable functions (L), even though it was about quadratic forms with infinitely many variables, and not about functions of a line.

Hence Banach wished to set his 1922 paper in a very general context, and the choice of vocabulary (*abstract set, sets of elements*) shows his intention to elaborate a framework that is not as limited as the one used by his predecessors. Yet, despite those characteristics which tend to define a very open structure, where those objects falling under the definitions and axioms are not fully identified, there are some restrictions that appear from the beginning of Banach's paper. In particular, as we have seen earlier, Banach enumerates a list of sets of functions for which he wants to establish new results. This situation has to be contrasted with Banach's 1932 book in which he would completely accept this open context for itself: the new exposition would no longer refer to an *a priori* given list of sets of functions and there would be a new step taken toward generality, an open generality freed from the initial restrictive context.

The notion of *openness* and its aptitude for describing a type of generality is hence first linked to the class of objects—identified or not, that potentially fall under a concept or a list of axioms. But further, *openness* also derives from the internal organization of the text, and from the dynamic chaining between some propositions. Contrasting with this point of view, Banach's dissertation and his book of 1932 will show some major differences concerning the type of generality emerging in each situation. Among other important features of Banach's work, the case of completeness, in particular its position and interaction with other ideas, is characteristic of this phenomenon as we shall continue to analyze.

---

[28] Banach (1922: 133).

[29] A *function of a line*, as introduced by Volterra (1887), is a function (or operation) with domain the set of all continuous functions on an interval, and with values in $\mathbb{R}$ or $\mathbb{C}$.

[30] Some ten years after Volterra's pioneering paper, Pincherle would study "functional operations" where the domain and counter-domain were functions (cf. Pincherle, 1897).

7.2.2.2   *Completeness: toward an open generality*

One aspect that affects the type of generality in Banach's work is the shaping of the text itself and the way he organizes and separates several ideas. In retrospect, one can see that the organization of the axioms in his 1932 book permitted him to study some independent properties of metric spaces. Reading the contents pages of the book clearly shows the new possibilities concerning the organization of the concepts and the possibility of bifurcating into new unexplored domains or of reinterpreting old theories from a new point of view. In his 1932 book Banach uses this organization of ideas fully, splitting his book into chapters that follow the construction of more and more complex structures: groups, general vector spaces, spaces of type (F),[31] normed spaces, and spaces of type (B).[32] In comparison, Banach's doctoral dissertation is less developed but the organization of the axioms is already present. There are no structures, and no sub-structures, but the three groups of axioms introduced are well differentiated and will make it possible to exactly point out what is necessary in the subsequent proofs.

In particular, the analysis of the role of completeness[33] in Banach's text and the contrast with what he does in his 1932 book lead one to reconsider the process involved in creating what we now call Banach spaces.

To understand the evolution of Banach's ideas, one has to remember that in 1922 one of his main objectives was to deal with linear operations and in particular to bring forth new ideas about the study of integral equations. Considering this, completeness appears to be dictated by the necessity of working on those equations.

More precisely, after the 1922 paper, it was essentially the necessity of using infinite series and linearity that seems to have motivated the use of completeness. Historically, series seemed to come up in unavoidable ways. A major breakthrough concerning integral equations described the solutions in terms of infinite sums and used their ability to approximate regular functions by polynomials (see, for example, Fréchet, 1910a, 1910b). Moreover, the methods used in Banach's paper for studying functional equations, which will give birth to the (now) well-known fixed point theorem for Lipschitz maps (II §2, theorem 6[34]) and will lead to the theorem on Fredholm equations (II §2, theorem 7[35]), are based on series of iterated values of a given function. The convergence of these series can be established thanks to completeness.

When considering Banach's use of completeness one can see that there are two conflicting aspects. On the one hand, the axioms concerning the norm are dispatched in two sets: the first one states the general definition of a norm, and the axioms are numbered

---

[31] (F) for Fréchet: complete metric spaces in current usage.
[32] (B) for Banach: complete normed spaces in current usage.
[33] Since Banach does not use the word "completeness," in our text it will refer to the property stated in Banach's axiom III.
[34] Banach (1922: 160).
[35] Banach (1922: 161)

II_1 to II_4, while the second set introduces completeness, and is numbered III. Hence, in conjunction with what we mentioned earlier, as present-day historians we are tempted to acknowledge that we are in the presence of an *open* situation: *openness* seems to manifest itself through possible bifurcation points which are emphasized in the structure of the text.

But, on the other hand, these bifurcations remain non actualized possibilities: in his dissertation, Banach does not deal with all the possible developments and definitely considers a unique context in which there is a norm, and a complete norm at that.

Moreover, the introduction of the norm is not itself dissociated from the vector space definition, and comes attached to it, in a coherent context. It is introduced by the sentence "Let us then admit that ..." (Banach, 1922: 135) followed by the norm axioms, generating a single flow of axioms.

As we shall see in Section 7.3, the discourse here is in fact oriented toward some purpose, namely the study of operations in a uniform presentation, a purpose that structures and gives a specific coherence to the new edifice. The *openness* that we see in the structure of the list of axioms and the tension between this openness and some prescribed horizon (namely, operations for the study of integral equations) both shape our interpretation of Banach's text.

At this point, when reading such an analysis, one could be tempted to acknowledge the usual statement that says that Banach spaces were born in this article. Indeed, one could put forward two points in favor of such a position:

1) a list of axioms that formally matches the list we now use to define a Banach space.
2) a coherent context in which the link among the axioms is emphasized (a single list, given at the beginning of Banach's paper).

On the contrary, we would like to argue that this is not enough to actually acknowledge the birth of Banach spaces in the 1922 paper. As a matter of fact, we shall prove in the remainder of our paper two facts which support such an argument. First, there is an essential parameter lacking, namely *reflexivity* (in the sense we mentioned earlier): the list of axioms here is never looked at for its own sake, and no independent theorem is derived solely from their statements. Every result derived from the axioms is presented as an almost pedagogic tool, meant to emphasize some parts of the reasoning or to simplify the proofs concerning the operations. Banach's main goal here is to clarify some situations concerning the study of operations and to simplify the diversity of the treatments of various sets of functions into a single coherent language. In his 1932 book, Banach will actually introduce a more reflexive point of view: he will name different spaces, as we mentioned previously, and will develop a specific chapter of the book dedicated to each space. Moreover, there appear some elements that constitute the premises of a geometry of Banach spaces with sections entitled "spaces of type (B)," "bases in spaces of type (B)," or "weak convergence in spaces of type (B)." Although the book is again orientated toward a general theory of linear operations, which appears to be Banach's main focus, there are specific parts of the text which are dedicated to the study of inner properties of spaces of type (B) and which develop a reflexive approach. For example, Theorem 13 on p.172 of the book, which is characteristic, reads:

> Given a separable space E of type (B) such that each norm bounded sequence of elements {x_i} of E contains a partially weak convergent sequence to an element in E, then the space E is equivalent to $\overline{\overline{E}}$ (the conjugate of $\overline{E}$[36]).

This reflexive point of view is even more perceptible in the last part of the book entitled "remarks," which was written primarily by S. Mazur. The formulations there show that spaces of type (B) or normed linear spaces constitute subjects of study by themselves: "Normed vector spaces have been treated independently ... by N. Wiener," "One can establish *for*[37] spaces of type (F) ...," etc.

We will use these remarks to address the important question of determining when Banach spaces were actually introduced and thus to make conclusions regarding the various types of generality that one meets in the various publications by Banach. But first we conclude this part, which is centered on topological aspects, with an analysis of the specific position given by Banach to some themes or elements connected with topology. Our purpose is to add weight to the argument that in the 1922 article there was no reflexivity in Banach's approach.

7.2.2.3 *Topology and geometry*

Once again, Banach does not really deal with topological aspects that could have been derived from the properties of the norm. After the introduction of the norm, Banach inserts a paragraph entitled "auxiliary theorems on the norms and the limits" (Banach, 1922: 136). This passage presents some facts that will be useful in his study of operations and isolates some lemmas concerning the manipulation of the norm.

Along those lines, one can find some theorems about the uniqueness of a limit (I §2 theorem 4[38]), about operations and limits (I §2 theorem 6[39]), and about series and convergence (I §2 theorem 7 and theorem 8[40]).

It is to be observed that Banach always points out very carefully the axioms involved at each stage of his proofs. In particular the proof of theorem 8 is characteristic:[41]

Theorem 8: If

1°) {X_n} is a sequence of elements,

2°) $\sum\limits_{n=1}^{\infty} \|X_n\|$ exists,

---

[36] $\overline{E}$, the "conjugate" of the space E as Banach calls it, is nowadays known as the space of all functionals on E.
[37] Our emphasis.
[38] Banach (1922: 137).
[39] Banach (1922: 138).
[40] Banach (1922: 138–9).
[41] Banach (1922: 138–9).

then the series $\sum\limits_{n=1}^{\infty} X_n$ is norm convergent.

The proof involves some partial sums $\sum\limits_{n=1}^{p} X_n$ and is written in Banach's particular style:

Proof. Let $S_r = \sum\limits_{n=1}^{r} X_n$. Whenever $p > q$, we have:

$$(4) \quad \left\| S_p - S_q \right\| = \left\| \sum_{n=q+1}^{p} X_n \right\| \leq \sum_{n=q+1}^{p} \left\| X_n \right\|.$$

But, by means of 2°):

$$\lim_{\substack{p \to \infty \\ q \to \infty}} \sum_{n=q+1}^{p} \left\| X_n \right\| = 0.$$

And hence inequality (4) gives:

$$\lim_{\substack{p \to \infty \\ q \to \infty}} \left\| S_p - S_q \right\| = 0.$$

Which proves, according to axiom III and definition 1, the convergence of the sequence $\{S_p\}$. But now, the convergence of the sequence $\{S_p\}$ is equivalent, in view of definition 2, to the convergence of the series $\sum\limits_{n=1}^{\infty} X_n$.

Finally, immediately after this result, Banach shows in theorem 9 of his paper that, under the assumptions in force in the article, the convergence of a sequence $\{X_n\}$ in norm implies that it is Cauchy.[42] This theorem in particular shows exactly where axiom III about completeness comes into play, while also validating part of the result without this assumption.[43] All the proofs refer to the axioms they rely on, which again contributes to the presentation of a text that could be interpreted as open, since it could in retrospect allow some inflection, some bifurcation dropping such or such hypothesis. But in fact the 1922 text remains strongly oriented toward a precise goal and leads to efficient theorems, following strictly the program enunciated by Banach in the introduction of his paper. Once again, what we could be tempted to analyze as an open situation is not actualized in this doctoral dissertation: as we shall see in the next section, it attempts to present the study of operations in such a way that every hypothesis in use is clearly identifiable.

After this passage on the convergence of series, Banach adds a paragraph titled "definitions and theorems on the sets" which makes use of a very geometric vocabulary:[44]

Definition 3. Let X_1 be an element of E, and r a positive real number. The set of all elements X that satisfy the inequality $\left\| X - X_1 \right\| \leq r$ will be called a *sphere*. The element X_1 will be called the *center* and r the *radius* of the sphere. We will denote it $K(X_1, r)$.

---

[42] Banach does not give a name to this property.
[43] One implication is a reformulation of axiom III, whereas the reverse statement does not use completeness and is obtained by a simple application of the triangular inequality.
[44] Banach (1922: 140).

(In this section, as well as in the sequel, we shall often use the word 'point' instead of 'element').

Since the title of the Banach paper refers to "abstract spaces," which is an expression that is due to Fréchet, it is important to analyze the framework adopted by Banach precisely. As a matter of fact, in his own doctoral dissertation and in several subsequent papers, Fréchet developed a notion of neighborhood that was meant to be largely emancipated from any geometrical background. His aim was to introduce topology in situations where geometrical tools were barely thinkable (sets of functions for example). In contrast, in 1922 Banach wants to deal with any set (of functions) so long as there is a norm allowing geometrical intuition. This may explain why there are no allusions to open sets[45] or neighborhoods as basic topological devices, but only to spheres which will constitute the main tools for dealing with the notions of proximity or continuity.

The reasoning developed by Banach aims to show that a complete norm guarantees a situation in agreement with geometrical intuition and does not present any pathological aspects. We are far from Fréchet's, Cantor's, or Hausdorff's discussions here.

We also notice that the properties of the norm are not studied for their own sake, but only in order to develop useful tools adapted to demonstrations using sequences. The theorems or propositions proved in this passage will be reused in the subsequent major theorems as efficient and basic tools to shorten proofs and make their logical progression clearer.

The following theorem is an eloquent example of this specific use of the tools Banach has isolated:[46]

Theorem 17. When an operation $F(X)$, defined on set E, is of the first Baire class with respect to this set, then $F(X)$ is pantachically continuous with respect to any perfect set.[47]

We have chosen this example here because it is an example of a non-trivial proposition where Banach's method and style are fully at play in the proof.

As a matter of fact, the proof following the theorem is very well structured: at every step, previously stated results or assumptions are clearly enunciated and referenced. No fewer than seven previous results or definitions are quoted by Banach in nearly two pages to justify the progression of the proof. This adds to the clarity of the development which is striking and characterizes Banach's style. Some technicalities are moved out of the main development while some important lines of thought and strategies are put forward. In particular among those key ideas, Banach uses mainly spheres and sequences of their centers to deal with what we retrospectively call topological aspects. Another idea widely used by Banach is the consideration of the set of all elements satisfying a given

---

[45] We do not say here that Banach does not use words like "open" or "closed" which belong to the usual vocabulary of topology. Banach mentions in some theorems that a set he deals with is closed. But his proofs in topology only use spheres and balls and never refer to general open sets.

[46] Banach (1922: 149).

[47] The reader who is interested in the definitions of the mathematical concepts involved herein can find the relevant definitions in Banach's paper which is nearly self-contained and scarcely refers to external sources (Banach, 1922: 149). We shall nevertheless continue to recall some notions when necessary in our study.

proposition. This, in particular, allows Banach to reinterpret some inequalities in terms of the geometry of the space and its topology. For example in the proof of the theorem cited above, Banach fixes an arbitrary ε and considers "the set of all points that satisfy the inequality: $\left\| F_n(X) - F_{n+p}(X) \right\| \leq \varepsilon$." This technique is used in several places in Banach's paper and shows to be a very efficient tool. What could be analyzed as a purely analytic problem or as a problem of estimation of some quantity is read here in terms of geometry.

Another important feature of Banach's text of 1922 resides in paragraph 4 on the applications of some topological ideas to operations. First of all, this passage is situated in the long first chapter of the paper, and is clearly part of the general background developed by Banach in order to study linear operations. In this part Banach defines the continuity of an operation as what we now call sequential continuity,[48] and he definitively anchors his study in the new general setting we explored above. The definition of continuity is given in the context of what precedes it in the text, and strictly in terms of convergent sequences. Again, there is no escape here from this specific setting which structures Banach's line of thought and establishes the coherence the 1922 paper demonstrates.

What is also new here is that Banach considers not only operations one by one, but also some sequences of operations:[49]

Definition. We shall say that a sequence of operations $\{F_n(X)\}$, defined on a set A, converges in norm[50] to an operation $F(X)$ defined on A, when every X in A satisfies:

$$\lim_{n \to \infty} F_n(X) = F(X).$$

This makes it possible for Banach to define some operations which are (pointwise) limits of sequences of continuous operations, which will be said to be of the first Baire class.

These passages again show that there is no study of normed spaces for themselves, and that the 1922 paper is mainly dedicated to the study of linear operations. Even if Banach considers sequences of operations, their norms, their additions, or multiplications by a number, operations will always keep a special status in this text, very different from the elements of the spaces defined in the introduction of his paper. Operations have a specific place in the Banach text that we shall analyze in detail in Section 7.3.

---

[48] A function $f : E \to F$ is *sequentially continuous* if whenever a sequence (x_n) in E converges to a limit x, the sequence $(f(x_n))$ converges to $f(x)$. Thus sequentially continuous functions preserve sequential limits. Every continuous function is sequentially continuous. If E is a first-countable space, then the converse also holds: any function preserving sequential limits is continuous. In particular, if X is a metric space, sequential continuity and continuity are equivalent (this covers all the spaces studied by Banach or Fréchet that we mentioned in our study). When the space is not first-countable, sequential continuity might be strictly weaker than continuity.

[49] Banach (1922: 148).

[50] We now call this convergence "pointwise convergence" as opposed to the operator convergence induced by the norm of the operator.

7.2.3   The birth of Banach spaces in question

As we mentioned in the introduction of our paper, the birth of Banach spaces is often attributed to this seminal work by Banach. As a matter of fact, we saw that the list of axioms I to III in Banach's first chapter in 1922 could be superposed to the one we use nowadays to *define*[51] the corresponding spaces. Moreover, and this seems to support the common idea, we established through a historical analysis that so exact a list of axioms was produced for the first time in Banach's doctoral dissertation. Nevertheless, in light of the approach that we developed in this article and of prolonging some ideas we mentioned earlier in our study, we would like to argue that the thesis regarding the birth of Banach spaces has to be reconsidered, and that there is a capital difference between what we refer to as Banach spaces nowadays and what was involved in Banach's 1922 paper.

As we have said before, our analysis is premised upon the idea that in order to comprehend mathematical texts and concepts it is not sufficient to focus on and compare locally their individual propositions from a strictly formal point of view. Some inner features of the mathematical texts and the interplay among several parts of the discourse have to be taken into consideration. These are the "dynamic" features of the text, as opposed to the static ones introduced above. They appear to us as essential tools for interpreting a text.

In a context where new pieces of knowledge arise, the analysis of some specific attributes of the generality emerging in the process of creation allows us to get a clearer idea of the mathematician's activity. Our analysis of the attributes of generality as appearing in Banach's landmark publications of 1922 and 1932 makes it possible to draw a clear line of separation between several situations. We shall now use this feature to prove that Banach spaces, as the general concept we now know, were not born in his doctoral dissertation.

The first attribute of generality, which we have already found in several places in Banach's text is *openness*. Banach works on abstract sets (in his own terms "ensembles abstraits"). Yet, in his 1922 paper, the list of axioms I and II appears to be merely a tool meant for dealing with several given specific cases in a single uniform way. In fact, the only sets Banach will consider are the ones listed at the beginning of the article. They will appear again in the last part of the paper where he shows that they all share the right properties.

Hence, and although the axioms and the results of the first section appeared to cover a wide and open diversity of situations, this very general point of view is not completely adopted in 1922. The specific list of sets of functions (defined on a closed interval), placed at the beginning of the text and constituting the only topics dealt with in the final part of the text, define the boundaries of this early paper in a still limited field.

So generality is bounded by a given horizon, realized by known sets of functions endowed with their natural norms and carrying with them already settled properties.

---

[51] We shall establish a typology of definitions in mathematics in a coming work. The word "definition" and connected expressions like "axiom," "hypothesis," or other modes used to isolate some objects play a fundamental and complex role in the mathematical discourse. In particular it will be necessary to analyze the deep changes in vocabulary that appear between Banach's doctoral dissertation and his 1932 book.

The second essential aspect is *reflexivity* and its counterpart *separation*,[52] which is related to some *distancing effect*. They are both necessary for acknowledging the creation of a new mathematical object.

While analyzing Banach's 1922 text we saw that axioms stated in the list which opens the paper served as a natural background and as a new sort of formal grammar to be used in the proofs. As we already mentioned, nowhere in the text is there any single proposition concerning an object such as a space defined by the list of axioms. Yet, Banach attempts to isolate the axioms I to III and insists upon their use in the last section:[53]

> We did not give these axioms [IV–VI] at the beginning in order to better emphasize the consequences of the axioms I–III.

But, more than a retrospective look at what Banach himself could have considered as a new mathematical object, we think that this statement has to be understood as a useful strategy for making the proofs clearer and for emphasizing what they really involve. All the results in part 1 and 2, except the ones concerning operators, are used in the last part in order to give proofs concerning function spaces that are simplified and made uniform.

One can see that there is a gap of a very special nature between this early text and what is at stake in current evocation of Banach spaces: there was no determination of a new concept, no Banach space, and no proposition assuming such a concept. This situation contrasts sharply with what we find in the book of 1932, some 10 years later, where we see a real change in the mathematical discourse. In this book we can observe a new point of view and the emergence of spaces of type (B). The process of the introduction of those spaces naturally involves the seminal paper by Banach, but also, as we saw, some earlier works, as well as some rethinking of Banach's doctoral publication by Fréchet who first termed the new non-totally achieved spaces "spaces of type (B)." Finally, Banach will develop the new theory in his book of 1932 and he will discuss what holds or does not hold when completeness is or is not assumed.

In conclusion, we believe that there are no Banach spaces in the 1922 publication. Perhaps in relation to the fact that historians have been obsessed with questions of origin, they concentrated mainly on the part of Banach's doctoral dissertation that could be related to that concept. However, the paper undoubtedly presents for the first time in analysis another kind of general structure, as we shall now show.

7.3   Linear operations: the birth of a new theory

This section argues for the thesis that the central objects of Banach's 1922 paper are linear operations on complete normed spaces.[54] We shall prove here why this is the main

---

[52] The words *reflexivity* or *separation* might be misleading since they also frequently appear in the mathematical discourses we analyze. Nevertheless, we have decided to keep them since they refer to what could be a topology of the discourse and are well adapted to the description of the emergence of generality here. They will be illustrated and their sense will be clarified in what follows.

[53] Banach (1922: 164).

[54] For the reader's convenience, when axioms I–III are assumed to hold in a given situation, we shall refer to the situation with the anachronistic expression "complete normed spaces," which Banach does not use in 1922.

achievement of Banach's dissertation and we shall show how it constitutes the most advanced object of this text from the point of view of generality.

7.3.1   Continuous operations as a limit case

The first part of Banach's 1922 paper ends with a paragraph on operations. The link with what precedes is hence formally stated. Moreover, the central notion of continuity for an operation is defined here in terms of sequences, making this section dependent on the developments of the beginning of the first part.

   The link between operations and the topics developed in the first part of Banach's 1922 article is even clearer if one considers that operations are defined in the very first lines. Banach's thesis begins as follows:[55]

> An operation is a univocal relation $x\mathcal{R}y$, that is to say such that:
>
> $y\mathcal{R}x$ and $z\mathcal{R}x$ implies $y = z$ for every x, y, z.
>
> Every relation $x\mathcal{R}y$ has a counter-domain (the reserve of the ys) and a domain (the reserve of the xs) also called a *field*.[56] The functional operation, also known as the function of a line, is an operation such that the domain and counter-domain are sets of functions.

We shall come back to this definition soon, but in order to emphasize the organization of Banach's text at this stage it is important to recall how he introduces the notion of continuity:[57]

Definition. The operation $F(X)$ is continuous for a point X_0 with respect to a set A, when:

1°) $F(X)$ is defined for each point of the set A,

2°) X_0 belongs to A and is a point of accumulation in A,

3°) $\overline{\overline{\lim}}_{n\to\infty} F(X_n) = F(X_0)$ for any sequence $\{X_n\}$ in A such that $\lim_{n\to\infty} X_n = X_0$.[58]

Once again, we are far from Fréchet's attempts to define neighborhoods, and Banach states this definition in terms adapted to normed spaces, defining what we would now call sequential continuity.[59]

   We shall analyze several aspects of this definition in the next section, but we would like here to argue for the thesis that this last section of part I introduces some degree of reflexivity in Banach's discourse, which makes it somewhat special in this first part. As a matter of fact, although the majority of the results exposed here are used in the subsequent

---

[55] Banach (1922: 133).
[56] The word "field" in English might be ambiguous here. Banach uses the word "*champ*" in French which has to be differentiated from the French word "*corps*," which also translates into "field" in English. Here "field" in English or "*champ*" in French can be thought of as "collection" of elements.
[57] Banach (1922: 145).
[58] The symbol $\overline{\overline{\lim}}$ refers to the convergence in norm: $\overline{\overline{\lim}}_{n} u_n = u$ means here $\lim_{n} \|u_n - u\| = 0$.
[59] See note 48 on the types of convergence.

sections and hence appear at least partially as technical tools intended for a given purpose, some propositions do not completely fit into this category.

For example, the first theorem which states that a linear combination of continuous operations is continuous is not used in the remainder of the paper. This is also the case for theorem 16, recorded below, which is not cited at any point after its proof:[60]

Theorem 16. When

1°) the operation $F(X)$ is continuous with respect to the set A for all the points in A,
2°) the set A is closed,
3°) m and p are two non-negative numbers with $m \geq p$, then the set L of all points X
   in A such that the following inequality holds

$$m \geq F(X) \geq p$$

is either empty or closed.

Another example which shows the emergence of a reflexive view on operations is the introduction of sequences of continuous operations in the same section (as we have seen). Some theorems using the first Baire class[61] then follow, which give a specific point of view on collections of operations. It important to notice that Banach introduces his second section with the words "classes of operations" (*classes d'opérations*):[62]

§1. Among all classes of operations, the additive ones definitely deserve some special attention.

Hence operations have a very specific status and are sometimes treated as elements in a well-defined collection of objects to be studied through a uniform theory. And Banach does not study any specifically given or isolated operation: he considers operations in general, then presents a classification into sub-classes (operations, continuous operations, linear operations, etc.) and derives theorems that are uniformly valid for a given class.

As we will now show, the study of operations definitely constitutes the core of Banach's development in his doctoral dissertation. Two aspects will play an essential role in his approach: the first one is a property of the operations, namely linearity, while the other is a property of the domain of the operation, that is, completeness. We shall now analyze in more detail those two aspects.

7.3.2   Linearity

The case of linear operations, which is explored in the second part of the 1922 publication, seems to be of a special nature. As we shall see, Banach's approach in this domain is much more reflexive, and he definitively develops a new theory.

---

[60] Banach (1922: 147).
[61] Let us recall that an operation is of first Baire class if it is the pointwise limit of a sequence of continuous operations. One important result stated by Banach and used in several theorems says that any additive operation that is of first Baire class is also continuous.
[62] Banach (1922: 151).

But before we turn to analyzing this aspect, and before we show exactly what kind of generality will emerge from this part, it is necessary to briefly recall some earlier contributions dealing with linear operations, which undoubtedly influenced Banach's work.

Once again the formal confrontation with earlier texts will shed some light on the operations chosen by Banach and will show the specific elements he selected.

While introducing operations at the beginning of his article, Banach refers to some mathematicians who worked in this domain (see last quote of Section 7.2.2.1). The study of linear operations is hence related to a historical context that is well identified by Banach himself and that is definitely motivated by the study of functions of a line (*fonction de ligne*). As we have seen earlier, Banach provides a highly specific definition of an operation right at the beginning of the paper.

The term operation (*opération*) is not new though, and it was regularly used at that time (prior to 1922). Operations were often defined as correspondences: one can consult, for example, the 1901 book by Pincherle and Amaldi *Distributive operations and their applications to analysis*. At the beginning of the second chapter of this book, the authors define operations in the following way (we include some passages showing their choices of notations):[63]

> **22.** Among the elements of two systems S and S' there might be a correspondence (*corrispondenza*) which associates one or more elements of S to each element of S' ...
> It is possible that to one element in S there corresponds only one element in S' ...
> From now on, we shall use capitalized Latin letters to designate operations. If the operation A sends an element α in S to an element α' in S' we shall write:

$$A(\alpha) = \alpha'.$$

In his 1922 paper, Banach gets rid of the binary relation notation $\mathcal{R}$ he introduced in his definition, and immediately after adopts a stable prefixed form:[64]

> We will denote the operations by $F(X)$, $P(X)$, etc.

We remark here that Pincherle and Amaldi use a single letter to name their operations ("the operation A"), and that they make a distinction between the name of the operation and the name of the variable. This was not generally the case even in the articles published after 1901. One can cite here several examples of articles (Fréchet, 1904; Riesz, 1910; Banach, 1922, 1932) where the authors use the notations $f(X)$ or U_f to name their functions or operations, a generic element of the domain appearing here.

A detailed examination of Banach's introduction leads one to remark that he pushes to the forefront what he considers to be three inseparable elements: two sets, called the domain (the reserve of xs) and counter-domain (the set of all images), and a binary

---

[63] Amaldi and Pincherle (1901: 17).
[64] Banach (1922: 145).

relation that associates to any element in the domain a unique element of the counter-domain. Historically, this choice in the presentation is not new, and in their book, Pincherle and Amaldi introduce in the first chapter the systems noted S and S' at the same time as the correspondence.

Nevertheless, this point of view was again not uniformly accepted even after the publication of *Le operazioni distributive*. For example, in a series of three papers on linear operations written between 1904 and 1907, Fréchet starts from a point of view where the nature of the reserve of initial elements definitively does not share the front of the stage with the operation. Even his notation puts the name of the operation, and the elements it acts on, on two separate levels. Operations are meant to act on continuous functions only (which seems to be imposed by the necessity of describing functions in terms of converging series):[65]

> I. Let us first recall some definitions. We shall say that an *operation* is defined, if to each real function $f(x)$ which is continuous between two given numbers a and b there corresponds a finite determined real number U_f.

This first point of view then evolved along with his papers. In the second one, written in 1905, the end of the article is devoted to a commentary on the reserve of functions:[66]

> *Importance of the function field in which the operation is defined.*

Until now, we have supposed that the functional operations we were dealing with would associate a number U_f to each function $f(x)$ that was continuous between a and b, without investigating whether U_f was defined for some other functions.

But the words following this question do not leave any doubt: "This hypothesis [that is the fact that the functions should be continuous] is essential, as we shall see" (Fréchet, 1905: 138). And the end of the article finally deals with the only possibility left, which is to consider operations defined on regular functions.

The third article on the topic, published in 1907, manifests a clear-cut change in Fréchet's point of view. As a matter of fact, the last paper of the series opens directly on the definition of the sets that were to constitute the *a priori* given domain for the operations:[67]

> **Definition of the field.** Let us consider a field of functions, with variable x, and defined on the interval $(0, 2\pi)$. I will suppose that if two functions belong to this field, then so does their sum. To each function $f(x)$ of the field we can associate a well-defined number U_f. By means of this we will define an operation in the field of functions.

In conclusion, the contrast with those papers of Fréchet highlighted by these remarks shows that Banach introduced the operations in a very specific way. In fact, Banach's way

---

[65] Fréchet (1904: 493).
[66] Fréchet (1905: 138).
[67] Fréchet (1907: 433).

of dealing with operations aligns more closely to Pincherle and Amaldi's line of thought, and he insists on some aspects first ignored or left out by Fréchet.[68]

Concerning now the more specific subject of linearity (as we now call it), it is again helpful to compare Banach's text to Fréchet's approach. In 1905, Fréchet wrote:[69]

> Following M. Hadamard we shall call a linear operation any operation that satisfies the following two properties:
>
> 1°) it is distributive, that is, given any two functions f_1 and f_2 that are continuous between a and b, we always have:
>
> $$U_{f_1+f_2} = U_{f_1} + U_{f_2}$$
>
> 2°) it is continuous, that is U_{f_1} tends to U_{f_2} whenever the function f_1 tends to f_2 uniformly between a and b.

Here, continuity is part of the definition, and is expressed in a specific situation, namely the context of functions that are continuous on a closed interval $(a;b)$, with uniform convergence. The notations used in the definition above do not put on the same level the algebraic operations on f (addition and multiplication by a number) and that on U_f. It is noticeable that the choices made by different mathematicians are not uniform on this subject. For example, Riesz in 1914 uses a formulation close to Fréchet's (see Riesz, 1914).

In contrast with these texts, and a few years earlier, Pincherle and Amaldi used the expression "distributive operations" (*operazioni distributive*) and dissociated linearity from continuity:[70]

> If α and β are any elements of the linear space S, and A is an operation that can be applied to those elements, one says that A satisfies the distributive property when:
>
> $$A(\alpha+\beta) = A(\alpha) + A(\beta).$$
>
> As a consequence … if c is a rational number,[71]
>
> $$A(c\alpha) = c\,A(\alpha).$$

In turn, Banach adopts this type of formalism, but uses the specific word "additive" instead of "distributive" to qualify operations with such a property. By this choice he puts on the same level all the elements constituting an operation (two sets with addition and multiplication by a number, and a correspondence) and he can then explore the possible properties of such an object:[72]

---

[68] The passages of Fréchet's work above show that he started his study of operations almost from scratch and the evolution of his point of view developed in those three papers is very subtle. One can read, for example, Jaëck (2010) for an analysis of these papers by Fréchet.

[69] Fréchet (1905: 493).

[70] Amaldi and Pincherle (1901: 25).

[71] We shall analyze in a few lines the passage from rational to real numbers.

[72] Banach (1922: 151).

Definition. We shall say that an operation $F(X)$ is *additive* when for each X and Y:

$$F(X+Y) = F(X) + F(Y).$$

It is obvious that $F(\theta) = 0$, since

$$F(\theta) = F(\theta + \theta) = 2F(\theta).$$

It is easy to prove that

$$F\left(\frac{p}{q}X\right) = \frac{p}{q} \cdot F(X)$$

where p and q are two integers and $q \neq 0$.

The difference between the two words "additive" and "distributive" is important here from the point of view of generality. The use of the word "distributive" for Pincherle and Amaldi or Fréchet is probably copied from its use for operations on numbers. This has an influence on the understanding of the two + signs and the role of the operation itself in a proposition such as: $A(\alpha + \beta) = A(\alpha) + A(\beta)$. Using the word "distributive" in such a situation, the author emphasizes the similarity with well-known properties of the usual addition, and, more than a mnemonic device, it keeps the operation A within a particular framework: the operation A lives here in the undifferentiated world of algebraic objects.[73] The words *operation* and *distributive* combine here in accordance with the laws of algebra. As with Banach's doctoral dissertation, there is an evolution in the understanding of operations and they suddenly acquire a new status. In his paper, Banach defines *additive* operations. He writes propositions such as $F(x + y) = F(x) + F(y)$. Such a notation correlates with his comments regarding the fact that the two + signs might live in two separate spaces. Moreover, what is crucial from the point of view of developing a new general concept here is the position given to the "operation" (we mean F) itself. The parallel with the usual laws for the addition is not mentioned at all and the operations, as we saw before, acquire an independent status and are now studied for their own sake.

To conclude this section, we would like to point out that the word *operation* and its use in several contexts (addition and multiplication of numbers versus linear operations or binary relations for example) and the diversity of notations show that the mathematicians were struggling to shape a new general way of thinking about some usual and new objects in a coherent way. In particular, among many aspects, it would be interesting to compare the distinction between Fréchet and Banach's points of view. Our study brought to light, on the one side, the emergence of the concept of module,[74] and on the other side, the

---

[73] We see here that algebra and algebraic computation acquire an increasingly general status: algebraic operations serve as an archetype for situations where calculi have to be performed.

[74] The parallel with the structure of module was suggested by K. Chemla (personal discussion).

progression toward a general theory of operators, as opposed to operations. There is a striking neighborhood of several ideas and forms here that deserves deeper analysis.

7.3.3   A theory of linear operations on complete normed spaces

We now turn to the analysis of what appears to be one of Banach's main achievements in his 1922 paper. Banach divides his second chapter, devoted to linear operations, into two distinct parts.

The first part deals with linear operations in a way that reveals a reflexive point of view: the only assumptions needed to obtain the results are linearity and continuity and operations are considered for their own sake.

Banach's first theorem shows that boundedness on a sphere is a sufficient condition for continuity:[75]

Theorem 1. An additive operation $F(X)$, which is bounded in a sphere K, is continuous at each point of the field E.

Then some properties concerning sequences of operations and the conservation of continuity in the case of pointwise convergence are established:[76]

Theorem 5. If 1°) $\{F_n(X)\}$ is a sequence of additive continuous operations,
　　　　　2°) $F(X)$ is an additive operation,
　　　　　3°) $\lim\limits_{n\to\infty} F_n(X) = F(X)$,

then we have:
　　　　　1°) the operation $F(X)$ is continuous,
　　　　　2°) there exists a number $M > 0$ such that for every n and every X we have:

$$\|F_n(X)\| \le M\|X\|.$$

Except for theorem 3 and lemma 3 in chapter 2, many of the results in this part are not used anywhere else in the text. One can thus assume that they are not meant as tools but as an independent exploration of linear operations.

In the second part, which differs from the first part with respect to the type of generality involved, Banach develops specific results which are, at least in their form, best adapted to the study of integral equations. This part does not use the preceding one, and only refers to theorem 8 in Chapter I, which proved that under the assumption of completeness the convergence of $\sum X_n$ always implies the convergence of the series $\sum X_n$. This part contains

---

[75] Banach (1922: 151).
[76] Banach (1922: 157).

two major results that we shall now record. Although none of these developments are explicitly related to integral equations in Banach's text, their form leaves no ambiguity and even in Banach's time they would certainly have been understood as dealing directly with integral equations.

The first result concerns general operations which satisfy a Lipschitz condition (no linearity assumed). This theorem is now known as the Banach fixed point theorem:[77]

Theorem 6. Let us suppose that
1°) $U(X)$ is a continuous operation in E, with counter-domain contained in E_1 and
2°) there exists a number $0 < M < 1$ such that for every X' and X'' we have

$$\|U(X') - U(X'')\| \le M \cdot \|X' - X''\|.$$

Then there exists an element X such that $X = U(X)$.

Of course completeness is essential here. Moreover, one can observe the return of the non-geometric vocabulary "element" instead of "point" despite Banach's remark in the first chapter. Strictly speaking, this theorem is stated by Banach under the extra hypothesis that the operation is additive, although this does not appear in the statement itself. In fact, at the beginning of the chapter, Banach writes: "We only consider in this chapter additive operations defined everywhere on the set E" and there is here a slight inconsistency. However, the word "additive" does not appear in the statement itself. Further, the proof does not use the property and does not make use of any theorem using additivity. What is more, the hypothesis is repeated in the statement of every theorem in this part, except for theorem 6. Hence this fixed point theorem has a very special place in this article since it is totally self-contained and will not be used in subsequent theorems. In fact this theorem directly addresses a well-known problem in differential equations, namely the Picard–Lindelöf problem as we now call it. Its specific analysis and the role it plays in the study of differential equations goes beyond the scope of our paper.

What is central here, and is well illustrated by the theorem immediately above, is the role played by the hypotheses. It reveals Banach's strategy in the construction of a theory of operations. Banach begins with a very general point of view, a mere correspondence, and he then attaches to the correspondence properties (additivity, continuity, completeness). Such a presentation allows Banach to specify for each proposition the exact properties used in the proof and hence to write the most general theorem, generality being determined here by the proof and the tools or assumptions it strictly requires.

In the same chapter, a second essential theorem addresses a celebrated problem, namely Fredholm-type equations. Contrasting with the previous one, here additivity comes back into play (one can observe that all additive operations are denoted by F whereas Banach uses the letter U when no linearity is assumed). Here again, the result given by Banach

---

[77] Banach (1922: 160).

does not leave any doubt about one of the purposes of his study. This theorem interprets Fredholm equations and is stated in the most useful form:[78]

Theorem 7. Let $X + \alpha \cdot F(X) = Y$ be an equation where Y is given and X is the unknown element.

Let us suppose that:

1°) $F(X)$ is an additive continuous operation in the field E with counter-domain included in E,

2°) M is the least number that satisfies the following inequality: $\|F(X)\| \le M \|X\|$,

3°) α is a given real number;

Then for all Y and for all h such that $|hM| < 1$ the above equation has a solution. Moreover, this one can be written:

$$X = Y + \sum_{n=1}^{\infty} (-1)^n \cdot h^n \cdot F^{(n)}(Y),$$

where the operations $F^n(Y)$ are determined for each n by the following equalities:

$$F^1(Y) = F(Y) \quad and \quad F^n(Y) = F(F^{n-1})(Y).$$

In both cases mentioned above, although the forms are adapted to well-known problems in integral equation theory, the results are stated as dealing with operations in a very general way. These two theorems, in particular the first one, have a form that goes far beyond the strict context of integral equations proposed by Banach in his title. All of the developments about operations show that these constitute the most sophisticated and general point of view of this paper, which develops some elements of an independent theory of operations on a complete normed vector space.

Nevertheless, the reference to integral equations cannot be ignored in this work, and shapes the type of propositions and the generality they bring. The study of the influence of Banach's work in the analysis of integral equations is beyond the scope of this paper and certainly needs some particular attention as we have already mentioned. Yet, the theme of integral equations enters the dynamic movement which is central in our study and influences the type of generality that emerges in the text. It cannot be disconnected from the global process leading to the emergence of the theory of linear operations.

Precisely speaking, the final part of Banach's paper, which is specifically devoted to the sets of functions listed in the introduction, and more generally the references to the integral equation problems, alter the text in a specific way and share in the emergence of a particular type of generality as follows.

At first sight, the necessity of dealing with the integral equations stated in the title of Banach's paper could appear as preventing Banach from developing a very general

---

[78] Banach (1922: 161).

setting from the outset. We have already invoked some of the general results relating to integral equations, which were partly shaped by this necessity. But when looking carefully at Banach's text, one can see that the references to a specific domain or purpose (integral equations, continuous functions) is far from being a limitation in terms of generality as concerns the specific domain of operations. On the contrary, it is a fundamental element in the shaping of the new theory.

First, the ground material, that is the list of function sets, was obviously determined by this theme: the sets of functions quoted here are the ones usually involved in the study of classical integral equations, with the exception of square integrable functions which are new candidates. Then linearity came from similar considerations: it was already isolated as an important aspect of the theory of integral equations by Fréchet and others.

But, more than just complying with the necessities of dealing with a given context, the hard work of the final part (in particular showing that all the spaces verify the axioms) guarantees, in a sense, the coherence of the axioms chosen by Banach. The "axioms" and "postulates"[79] proposed and used by Banach are here confronted by a given context and the necessity of dealing uniformly with a wide collection of functions and more generally with some given problems. And the long final part of Banach's paper establishes the coherence of the axioms and of the new theory of linear operations with regard to these preset objectives.

In conclusion I suggest that the reference to the known list of functions or to the usual integral equation problems does not limit Banach's enterprise to define a new theory of linear operators, but on the contrary that it is used to prove its solidity and its efficiency to generate a new coherent point of view.

7.4 Conclusion

Banach's contributions stimulated and gave rise to some essential developments (operator theory, geometry of Banach spaces and many more). His doctoral dissertation gives very valuable access to the work of the mathematician and allows one to determine how he proceeded toward a new and very general conception, encompassing many previous approaches that did not grant such a prominent role to generality. This study, centered on some chosen aspects of Banach's text, showed that he initiated a new focalization, deported from the established point of view and directed toward a more general point of view.

Two major aspects interplayed in this progress. First, Banach made a very selective choice of some already known but unconnected material: a vector space, a complete norm and linearity are for the first time seen as a coherent and unique ground material for a theory. But we have also seen that this summoning of formal elements was not the only essential aspect at stake here. It appears that the study of the inner shaping of the mathematical discourse and the interplay between several parts of the text were necessary to account for the dynamics involved in the emergence of such a complex and efficient generality. Of course, this dual aspect played a fundamental role in Banach's work, but

---

[79] The terms "axiom," "postulate," as well as "definition" appear to be key words in the emergence and expression of generality. Those concepts and their specific roles need to be analyzed separately.

more generally in the emergence of modern functional analysis where structures would play a prominent role.

Acknowledgements

The author would like to thank the referees for their constructive remarks, Rebekah Arana who helped a lot improve the English of the text, and Karine Chemla for essential discussions on the present subject.

···

REFERENCES

Amaldi, H., & Pincherle, S. (1901). *Le operazioni distributive e le loro applicazioni all' analisi.* Bologna: Zanichelli.

Banach, S. (1922). Sur les opérations dans les ensembles abstraits et leur application aux équations intégrales. *Fundamenta mathematicae* (3): 133–81.

Banach, S. (1932). *Théorie des opérations linéaires* (Vol. VIII). Warsaw: Mathematisches Seminar der Univ. Warschau (Monografje Matematyczne I).

Dieudonné, J. (1981). *History of functional analysis* (Vol. 49). Amsterdam: North-Holland Publishing Co.

Dorier, J.-L. (1995). A general outline of the genesis of vector space theory. *Historia Mathematica,* 22: 227–61.

Dorier, J.-L. (1996). Genèse des premiers espaces vectoriels de fonctions. *Revue d'histoire des mathématiques,* 2: 265–307.

Fréchet, M. (1904). Sur les opérations linéaires. *Transactions of the American Mathematical Society,* 5(4): 493–9.

Fréchet, M. (1905). Sur les opérations linéaires. *Transactions of the American Mathematical Society,* 6(2): 134–40.

Fréchet, M. (1907). Sur les opérations linéaires. *Transactions of the American Mathematical Society,* 8(4): 433–46.

Fréchet, M. (1910a). Les dimensions d'un ensemble abstrait. *Mathematische Annalen,* 68(2): 145–68.

Fréchet, M. (1910b). Sur les fonctionnelles continues. *Annales scientifiques de l'ENS,* 27(3): 193–216.

Jaëck, F. (2010). Éléments structurels en analyse fonctionnelle : trois notes de Fréchet sur les opérations linéaires. *Archive for History of Exact Sciences,* 64(4): 461–83.

Peano, G. (1888). *Calcolo geometrico secondo l'Ausdehnungslehre di H. Grassmann preceduto dalle operazioni della logica deduttiva.* (Vol. III). Torino: Torino Bocca.

Pincherle, S. (1897). Mémoire sur le calcul fonctionnel distributif. *Mathematische Annalen,* 49: 325–82.

Riesz, F. (1910). Sur les Opérations fonctionnelles linéaires. *Comptes Rendus de l'Académie des Sciences,* 149: 974–77.

Riesz, F. (1914). Démonstration nouvelle d'un théorème concernant les opérations fonctionnelles linéaires. *Annales de l'Ecole Normale,* 31(3): 9–14.

Riesz, F. (1916). Über lineare Funktionalgleichungen. *Acta Mathematica,* 41: 71–98.

Volterra. (1887). Sopra la funzioni che dipendono da altre funzioni. *Rend. R. Accademia dei Lincei,* 3(4): 97–105, 141–6, and 153–8.

Section II.2

A diachronic approach: continuity and reinterpretation

8

How general are genera? The genus in systematic zoology

YVES CAMBEFORT

8.1 Introduction

In systematic zoology and botany, animals and plants are classified and named according to their species, genera, and higher categories (family, order, etc.).[1] Linguistic relationships between the words "genus" and "general, generality" might have played a role in some intuitive meaning of the genus, and older naturalists sometimes attached great importance to genus category. Zoology and botany share a number of traits, but the former deals with human beings, in addition to animals, and part of its aim has been to study the differentiation of our species from its animal relatives. Zoology and botany used to be associated (together with mineralogy) in "natural history," with the old meaning of the word "history"—Greek *(h)istoria*—which means "research," "information," "inquiry," etc. The scientific status of natural history has been questioned. Zoology, botany, and sometimes even biology have been denied the possession of general laws, since they are founded on particular cases:

> Perhaps the only thing which can be said with any certainty is that there is no law about the value of searching for generalities in Biology, only pragmatics. But even though we can't expect to find any laws governing the search for generalities in Biology, some rough, pragmatic, guidelines might be very useful indeed.[2]

In fact, the use of species, genus, and higher categories of classification (family, order, class, kingdom ...) can be considered not only as pragmatics, but as a particular expression

---

[1] See Appendix 1 for a summary on animal classification.
[2] Fox Keller (2006).

The Oxford Handbook of Generality in Mathematics and the Sciences. First Edition. Karine Chemla, Renaud Chorlay and David Rabouin. © Oxford University Press 2016. Publishing in 2016 by Oxford University Press.

of a sort of generality proper to zoology, or to natural history as a whole, aimed to account for the living world, in order to better understand it. It is clear that plants and especially animals are so numerous that it is necessary to name and sort them just to be sure of what and which is referred to. As it was obviously not possible to name individual animals, the concepts of species and genus were introduced to arrange together animals of the same apparent kinds. The word "apparent" refers more correctly to species, because the word "species" comes from a stem connoting vision, look, in-*spec*-tion: any animal species contains individuals which resemble each other more than they do individuals of other species. In contrast, the word "genus" connotes *gen*-eration and *gen*-ealogy: any animal genus contains individuals which (are supposed to) descend from each other or from a common ancestor. In the present paper, I shall examine how the genus category was perceived and conceived in zoology (with occasional references to botany), in reference to species on the one hand and to higher categories on the other hand, from the antique notions of Plato and Aristotle, through its systematic conception by Linnaeus, to its evolutionary definition by Darwin, and up to the diversifications of the present time. In the late twentieth and early twenty-first century, various concepts of the genus reflect differences in epistemological cultures according to biological specialties (zoology, anthropology, paleontology ...), especially in the case of our own species, which is often referred to as "the human genus."

8.2   From antiquity to seventeenth century

8.2.1   Plato and Aristotle

Plato used the words *genos* (plural *genê*) and *eidos* (plural *eidê*) to sort a variety of things and beings. *Genos* connotes both gender and generation, lineage, and descent; it has been Latinized as *genus*, and directly passed from Latin to English. *Eidos* is related to the old root *(w)id*, Latin *uidere*, which—in a way analogous to *spec*—connotes vision, aspect, and even concept, as exemplified in Plato's *idea* (almost the same word as *eidos*); the word is given various English translations: "species," "kind," "form" ... Plato also used the scheme of a "downward" and binary division of beings: an ideal group (a *genos*) could be divided into two subgroups (two *eidê*); each subgroup could be considered then as a *genos* and divided again into two *eidê*, etc. The division into two parts (or "dichotomy") was mandatory: *Tertium non datur* ("there is no third term"), as expressly stated by medieval scholastics.

Aristotle criticized this system, although he used the words and concepts of *genos* and *eidos*, especially in the treatises of logic which composed his *Organon*.[3] *Genos*, together with "property," "definition," and "accident," was one of the four ways a predicate might be attributed to a subject ("differentia" was added to this list by later commentators). Aristotle gave a definition of *genos* in respect with *eidos*, and he obviously considered that

---

[3] Pellegrin (2007a).

genos was "a whole," a sort of higher rank, in contrast with *eidos*, which was "a part," a sort of lower rank:

> A *genos* is what is predicated in the category of essence of a number of things exhibiting differences in species (*eidos*)....The question, "Is one thing in the same *genos* as another or in a different one?" is also a generic question; for a question of that kind as well falls under the same branch of inquiry as the *genos*: for having argued that "animal" is the *genos* of man, and likewise also of ox, we shall have argued that they are in the same *genos*; whereas if we show that it is the *genos* of the one but not of the other, we shall have argued that these things are not in the same *genos*.[4]

Genera were divided into—or according to—*eidê*, which in the most complete Aristotelian conception were contrary to each other, as for example the two species "health" and "illness" in the *genos* "condition of the body;" but there could be intermediates, as in the *genos* "color," where the two *eidê* "black" and "white" were considered opposites, with all other colors between them.[5] In his zoological treatises, Aristotle did not accept Plato's ideas nor dichotomies: for him, animal kinds were not ideal but real; they were those recognized by people, not defined by philosophers, and were never based on dichotomies:

> Some writers propose to reach the definitions of the ultimate forms of animal life by bipartite division. But this method is always difficult, and often impracticable. Sometimes the final *differentia* of the subdivision is sufficient by itself, and the antecedent *differentiae* are mere surplusage. Thus in the series footed, two-footed, cleft-footed, the last term is all-expressive by itself, and to append the higher terms is only an idle iteration. Again it is not permissible to break up a *genos*, birds for instance, by putting its members under different bifurcations, as is done in dichotomies where some birds are ranked with animals of the water, and others placed in a different *genos*. Birds and fishes happen to be named, while other groups have no popular names; for instance, the animals that we may call sanguineous and bloodless are not known popularly by any designations. If such *genê* are not to be broken up, the method of dichotomy cannot be employed, for it necessarily involves such breaking up and dislocation.[6]
>
> ...
>
> The method then that we must adopt is to attempt to recognize the *genê*, following the indications afforded by the instincts of mankind, which led them for instance to form the *genos* of birds and the *genos* of fishes, each of which groups combines a multitude of *eidê*, and is not defined by a single character as in dichotomy. The method of dichotomy is either impossible (for it would put a single group under different divisions or contrary groups under the same division), or it only furnishes a single ultimate *differentia* for each *eidos*, which either alone or with its series of antecedents has to constitute the ultimate *eidos*.[7]

---

[4] *Topics* I 5 (translation W. A. Pickard-Cambridge on the internet, at the address http://classics.mit.edu/Aristotle/topics.1.i.html consulted on September 20, 2014), modified.

[5] Pellegrin (2007b); see also Cho (2003).

[6] *Parts of Animals* I 2, § 642b, translation by William Ogle on the internet at the address http://classics.mit.edu/Aristotle/parts_animals.1.i.html consulted on 20 September 2014 (modified according to Aristote, 1956).

[7] *Ibidem*, § 644b, (*idem*).

As we can see from these excerpts, Aristotelian zoological classification was of the "downward" type like Plato's one: a large group (a *genos*) of animals was first intuitively recognized ("by the instinct of mankind") and not from any preconceived (Platonic) idea, for example, birds or fishes; it was then divided into subgroups (*eidê*), usually more than two (contrary to dichotomy), and sometimes "a multitude." The system was neither rigid nor strict: like in colors, the most general Aristotelian rule could be lessened or diverted; but it was relative: there were no clear-cut concepts of *genos* or *eidos*, which were defined in relation with each other.

As far as animals were concerned, the Greek words *genos* and *eidos* have been translated into Latin respectively as *genus* and *species*. For example, Pliny the Elder reported that the *genus* fish comprised 74 *species*. In spite of the fact that Aristotle's chief aim was not to classify animals but rather to differentiate them, he nevertheless introduced some implicit classification, including a very important first division of animals (sanguineous vs. bloodless), which had a very long use in western zoology (up to the introduction of the dichotomy vertebrates/invertebrates by Lamarck in 1797).

In the Aristotelian conception as it was retained in the Middle Ages and early modern times, the theoretical categories "genus" and "species" designated relative categories, the same category being a genus when compared to those categories which it comprised, and a species when compared to another, more general category. Even more than a category, the "genus" of an object (inanimate or animate) was its "essence," and the "essence" of a species consisted of its genus plus its *differentiae*. Aristotelian categories were strictly conformed to logic; in spite of the fact that neither animals nor plants seem to conform to any logical system already known, Aristotelian genus and species were nevertheless retained to designate groups of organisms, probably assuming that they were compliant with logic not yet discovered. These "logical" principles were applied by older naturalists to animals and plants, and ultimately reworked by Linnaeus into a system which has persisted up to now (see Section 8.3.1).

8.2.2   Sixteenth- and seventeenth-century naturalists

Between antiquity and the Renaissance, almost no new data were added to the corpus accumulated in antiquity. Botany was considered more important than zoology because plants were the most important sources of *Materia medica*. For this reason, botany experienced an earlier development than zoology.[8] In his book *De Plantis* (1583), the Italian botanist Andrea Cesalpino introduced the concept of "species" almost in its modern meaning, but he still used downward classification and dichotomies.

Sixteenth-century zoologists were numerous, but their science rarely exceeded a rather elementary knowledge of the more obvious kinds of animals, which they did not know how to classify.[9] In the biggest zoological monograph of the century, Conrad Gessner's five volumes of *Historia animalium* (1551–87), animals were alphabetized "in order to make

---

[8] Magnin-Gonze (2004).
[9] Delaunay (1962) and Pinon (1995, 2000).

easier the use of the work." Ulisse Aldrovandi attempted a dichotomous sorting according to characters, but these were deceptively heterogeneous, as in birds (1599): "with a strong beak;" "good singers;" "aquatic;" etc. This was no improvement, but rather a caricature of previous authors' classifications. Aldrovandi partially abandoned dichotomy, especially in his volume on insects (1602): his greater divisions were dichotomous, but the ultimate, and sometimes also the penultimate, were multifarious (i.e., they comprised usually more than two "genera").

In the following century, improvements of zoological knowledge were achieved by the British naturalist John Ray. Originally a botanist, he published important books on zoology, where he retained the downward model of classification (great animal "genera" were recognized at first glance, and then divided into smaller kinds); he is acknowledged for having carried out one of the first attempts to define animal species in their more or less modern meaning (just as Cesalpino did for plants), even if a clear difference was not yet made between category and character:

> Specific identity of bull and cow, of man and woman, came from the fact they were born from the same parents, often the same mother.... A particular form preserves its species in perpetuity (*speciem suam perpetuo servat*), and never a species was born from another one's seed.[10]

In the same way, Ray's concept of the genus retained its Aristotelian, relative character, especially in animals: he recognized the genus quadruped, as well as the genus dog, the genus hare, etc. He noted that older animal classification has "sinned" by reducing the *vivipara* and *ovipara* to the same genus, in spite of the fact that "they differ by essential and generic features" (*quae notis essentialibus & genericis differunt*). As in Aristotelian and scholastic categories, the words "genus" and "essence" were more or less synonymous. Thanks to extensive information, and in spite of his conservatism, Ray was one of the most important zoological sources of Linnaeus.

A generation later, the French botanist Joseph Pitton de Tournefort (1656–1708) was one of the first who explored distant countries in order to bring home new plants and animals. Ray and Tournefort's time was a turning point: naturalists' aim was no longer to recognize and classify species and genera which had been known since antiquity, but, in addition, naturalists had to make known new, "exotic" organisms that explorers had brought back from all over the world. Contrary to Ray, Tournefort wrote only botanical treatises, especially his *Éléments de botanique* (1694), which was translated into Latin and published under the title *Institutiones rei herbariae* in 1700. In this book, which was soon famous all over Europe, Tournefort gave the first firm principles of systematics of living organisms, and Linnaeus acknowledged him the "honor" of the "invention" of genera: for the first time, the relative, Aristotelian concept of *genos* was abandoned. Tournefort's genus category was given a clear-cut, precise meaning, and his 698 genera were all considered at the same level (rank) of classification. Tournefort is also acknowledged for the invention of binomial nomenclature, a system which was to be later generalized by Linnaeus (see

---

[10]   Ray, *Synopsis methodica Animalium Quadrupedum* 1693, quoted by Carus (1880: 340–3).

below): each plant was named by a genus name consisting of one word, followed by a specific *differentia* under the form of a few words or sometimes just one word.

8.3   Eighteenth century

8.3.1   Linnaeus

The Swedish naturalist Carl (von) Linné, or Linnaeus (1707–1778), is famous for his generalization to botany and zoology of a binomial nomenclature which is still in use today, as well as for a highly influential "system" of plant and animal classification, whose basic principles have been retained up to now.[11] As far as animals are concerned, Linnaeus's most important works are the zoological sections of his *Systema Naturae*, especially in the tenth (1758) and twelfth (1766–7) editions. Linnaeus was a physician, like most naturalists of his time, and he received the basic botanical and zoological training implied by his medical studies; in addition, he had a personal interest in plants, as well as in insects (especially butterflies). Eighteenth-century Sweden was one of those European countries where cabinets of natural history were most appreciated. Both King Frederik Adolph and Queen Louisa Ulrica each had a large cabinet of zoology. Linnaeus had access to these two extensive collections, and he published catalogs of both. The catalog of the King's cabinet, published in 1754, was the first work where Linnaeus applied binomial nomenclature to animals, as he did soon after in the tenth edition of *Systema Naturae* (1758).

For our particular purpose, Linnaeus is important for his theoretical writings on taxonomic categories, and especially for the key position he had always attached to the genus, even if his ideas had been first conceived for plants. The *Genera plantarum* of 1737 and the *Philosophia botanica* of 1751 were two important texts where he exposed his method or system (he did not make a difference between these two words). Although these texts dealt especially with botany, some animal examples were used too. Since he established his systems on the two seemingly Aristotelian categories of genus and species, he has been and still is thought of having been influenced by scholasticism and "essentialism." The debate was quite active in the years 1950's; but it has recently been reopened with a series of papers published on the occasion of Linnaeus's tercentenary.[12] His system was almost complete from the beginning; one of its clearest and most explicit summaries was given in the *Ratio operis* of the *Genera plantarum* (1737):[13]

> 6. There are as many genera as there are common, proximate attributes of different species, as they were created in the beginning. This is confirmed by revelations, discoveries, and observations. Thus: *Genera and Species are all natural.*

---

[11] Hoquet (2005).

[12] For some historians, Linnaeus was essentialist (from the classic papers of Cain, 1956, 1958, to more recent ones, e.g., Stamos, 2005); for others, he was not (more numerous papers, e.g., Winsor, 2001, and Müller-Wille, 2007).

[13] See Appendix 2 for quotations in original languages.

Indeed, it is not allowed to join the horse and the pig under one genus, even if both species were one-hoofed; nor is it allowed to distinguish the goat, the reindeer, and the elk by genus, even if they differed by the shape of the horns. Therefore, we have to study the limits of genera with attentive and diligent observation, since it is very difficult to determine them *a priori*, even though this work takes effort. 'For should the genera be confused, everything must be confused' (Cesalpino).

...

8.... Everyone capable of such work tried to make it useful and to build systems; all with the same inclination and with the same aim; but with unequal success. Because only a few knew the fundamental rules, which, if not observed by the builders, would cause the most splendid building to be ruined with the first tempest. Boerhaave's[14] *Institutiones medicae*, aphorism 31: *Teachers* 'are to proceed from generalities to particulars, while explaining discoveries; while *Inventors*, to the contrary, have to pass from particulars to generalities.' For some have assumed various parts as systematic principle, and with it, they have descended according to laws of division from classes to orders all the way down to species. And by these hypothetical and arbitrary principles they broke and tore apart the natural, non-arbitrary genera, and did violence to nature.... Because, as they say, if [these species] cannot be joined by class, they can be joined still less by genus. But they do not observe that they themselves constructed the classes, but the Creator Himself made the genera. Hence so many false genera! So many controversies among authors! So many bad names! So much confusion! ...

9. ... The method that is to be preferred over the rest is the one that leads to the genera by the more certain and trouble-free path, and the one that is the most universal. For I believe there is hardly anyone born with such a memory that he could retain the genera without a system. The method must therefore lead the way; for orders are subaltern classes. And no one will deny that it is easier to distinguish a few genera than all at once. I do not deny, however, that natural classes can be given as well as natural genera. And I do not deny that a natural method will be much preferred to ours and all methods invented. But I laugh at all natural methods hitherto proclaimed. And provoked to my defence, I venture to affirm that not a single class given so far, in any system, is natural, as long as such genera and such characters that are currently used are serving under them. It is easy to refer the greatest part of known genera to their natural classes, but more difficult to do this for the rest. And it is not possible to hope that our age will be able to see any natural system, nor perhaps will posterity. Nevertheless, we are striving to know the species; so meanwhile artificial and substitute classes have to be assumed.

10. Having assumed natural genera, two things are required to keep them pure and well inculcated, namely that true species, not others, are reduced to their genera ...; and that each genus is circumscribed by true limits and terms, which we call generic characters.[15]

This remarkable text exposes Linnaeus's ideas on the genus as a keystone of his "philosophy." Genera and species really exist: they are not made by naturalists, but they have been created by God in the beginning. Therefore, the purpose of naturalists is not to create but merely to discover them, especially genera, which are the most important part of the

---

[14] Hermann Boerhaave (1686–1738), Professor of Botany and Medicine at Leiden, helped Linnaeus for the publication of the first edition of *Systema Naturae* (Leiden, 1735).

[15] English translation by Müller-Wille & Reeds (2007: 565–7). See original Latin text below.

system, species being only subordinates to genera. Since genera were made by God, and therefore "natural," they are separated from each other by clear and precise discontinuities, and the only purpose of a naturalist is to discover their limits and boundaries. For this purpose, he must use inspection, *a priori* intuition, and flair. A naturalist must keep in his mind a much larger number of genera than a layman (if possible all available genera), and it is for this purpose that names must be attached to genera; because "If you don't know the names, the knowledge of things perishes also."[16] By this famous formula, Linnaeus expressed unequivocally his position in the controversy of nominalism vs. realism which has divided Western philosophy since the middle ages. Not only the names, but also the "mental pictures" of genera must be kept in mind in order to recognize them at once, and a good system must help a naturalist's memory: this retrieval role was the other purpose of Linnaean system.[17] If a naturalist does not recognize a genus at once, or if he wants a confirmation, he may use available generic characters as well, but always keeping in mind another famous Linnaean formula: "Characters do not make a genus, but a genus makes characters."[18] Only from the *facies* (or *habitus*) of a genus, or from well-chosen characters, a naturalist must distinguish the limits of genera, which in most cases are clear-cut and obvious. Common sense too is useful: even if horse and pig would have the same form of hoof, which naturalist would classify them into the same genus? On the contrary, goats, reindeers, and elks clearly belong to the same genus even if their horns are different. Being in possession of genera, the naturalist must then follow two complementary ways: a downward one, that is, from genera to species (as a teacher); an upward one, that is, from genera to higher categories (as an inventor). A naturalist should take great care not to start from supposed natural, higher categories, for example, orders and classes, since "orders are subaltern classes," and "they themselves [naturalists] constructed the classes, but the Creator Himself made the genera." On the contrary, he is allowed to use an upward process in the feed-back from species to genera, in the condition that only "true species" were used, and "reduced to their genera." This is a condemnation both of the older, arbitrary definition of higher categories, and of the one-way downward process, which had led to so many controversies, mistakes, and wrong names.

In his first writings, Linnaeus still adhered to a more or less scholastic logical code, and he called the whole animal and vegetable kingdoms each a *summum genus*, whose species were classes of animals or plants. After 1735, he abandoned this use and restricted the word "genus" to a hierarchical level immediately above the species. If it were possible to identify *essential* characters, he said, one would need only those essential characters. However, as no appropriate method exists to identify them, one must use also *artificial* or factitious characters, which "distinguish a genus from all other ones in an artificial order." Finally, a genus "natural definition gathers every possible generic feature; and it therefore includes Essential and Artificial."[19] Linnaeus often omitted to mention variations

---

[16] "*Nomina si nescis, perit et cognitio rerum*" (*Philosophia botanica*, 1751, § 210).
[17] Cain (1958).
[18] "*Scias Characterem non constituere Genus, sed Genus Characterem*" (*Philosophia botanica*, 1751, § 169).
[19] *Philosophia botanica*, 1751, § 186–90.

presented by aberrant species, as far as, for him, any species "evidently" belongs to a genus which has been given *a priori*.

In the *Systema Naturae*, Linnaeus numbered genera in a continuous series from the beginning to the end; for example, in the *Regnum animale* of the tenth edition (1758), genera were numbered from 1: *Homo* to 312: *Volvox*. On the contrary, species were numbered within each separate genus (genus 1: *Homo*, species 1-2;[20] genus 2: *Simia*, species 1-21; genus 3: *Lemur*, species 1-3; etc., to genus 312: *Volvox*, species 1-2). Also from the tenth edition on, only one word ('specific difference') was added to the genus name, and this pair [Genus + species] made the "binomen" which has become characteristic of Linnaean binomial nomenclature. For this reason, the tenth edition of *Systema Naturae* (1758) is still considered by most zoologists as the starting point of animal binomial nomenclature.[21]

8.3.2 Buffon

Buffon is often opposed to Linnaeus: the latter carved Nature into clear-cut classes, when the former affirmed that Nature was a continuum. In fact, his opposition with Linnaeus was introduced by Buffon chiefly for polemic reasons; also, it has sometimes been considered as another step in the always renewed controversy of nominalism vs. realism.[22] As early as 1745, Buffon wrote to a correspondent that "the method of Monsieur Linnaeus is the less sensible and the most monstrous of all." He accused Linnaeus to mix very different animals in the same class, and to use in most cases only one character to separate his genera. Already in the first volume of his *Histoire naturelle* (1749), he harshly criticized Linnaeus's zoological system and insisted on the fact that Nature was continuous and did not know our categories:

> Nature walks through unknown gradations, and therefore cannot submit herself entirely to our divisions, since she passes from one species to another species, and often from one genus to another genus, through imperceptible nuances; so much so that there are many medium species and half-partitioned objects which one does not know where to place, and which necessarily disturb the project of a general system.[23]

The idea of intermediate species was later that of Charles Bonnet:

> There always exist, between two closely related classes or genera, *median* Productions, which seem to belong neither to the one nor to the other, and join them: the polyp chains vegetable to animal; the flying squirrel links bird to quadruped; the monkey unites quadruped and man.[24]

---

[20] The second species, *Homo troglodytes*, was an artificial combination of orangutan and chimpanzee.

[21] International Commission of Zoological Nomenclature (1999).

[22] Buffon (1954) and Hoquet (2003, 2007).

[23] Buffon (1749: 13).

[24] "Il est toujours entre deux classes ou entre deux genres voisins, des Productions *moyennes*, qui semblent n'appartenir pas plus à l'un qu'à l'autre, & les lier. Le Polype enchaîne le Végétal à l'Animal. L'Écureuil-volant unit l'Oiseau au Quadrupède. Le Singe touche au Quadrupède et à l'Homme." (Bonnet, 1781: 35).

Buffon opposed Linnaeus to older authors, especially to Aristotle, whom he acknowledged as the best zoologist ever; he explained that Aristotle took a great care to recognize species as individuals and give them "proper names," instead of classifying them into vague and inform genera, to which "common names" were applied (i.e., by Linnaeus). Yet, when he began to publish his *Natural history*, he knew almost nothing about Zoology, and his appreciation at that time was more that of a layman than of a zoologist. He used a highly subjective order, starting with those animals which were most familiar to him as well to his supposed readers (who apparently were country gentlemen, like himself): the first animal he dealt with was the horse (which was also the first species dealt with by Ray in 1693). Gradually, Buffon would get acquainted with animals, would refine his zoological knowledge, and would come closer to Linnaean conceptions. Later in his work, he was obliged to introduce some sorting and classification of the many species he was dealing with, monkeys and especially birds being the clearest cases:

> Instead of dealing with birds one by one, that is to say by separate species, I shall reunite some into a same genus, however without confusing and renouncing to get them distinguished whenever they can be.[25]

It is usual to remark that Buffon's dealing of such groups as monkeys and birds was "systematical," even if he took a great care to avoid any use of Linnaean categories, speaking for example of "family" when Linnaeus defined a genus, etc. In his *Nomenclature des Singes* ("Nomenclature of Apes"), for example, he began by explaining that he himself was obliged to recognize in the group of monkeys some "families," to which he gave "generic names," as, for example, "ape" (*singe*) and "baboon" (*babouin*): note that he took great care to use French instead of Latin names, and without a capital letter. But certain species could not fit into these families/genera. For example, between the *singe* and the *babouin*, he says, there is a species which one does not know where to place: the *magot* (Barbary ape). Later on, Buffon will mention another such intermediate species, between the genera *babouin* and *guenon*: the *maimon* (pig-tailed macaque). His conclusions were from a higher perspective and gave more insight into his conception of genera:

> If, from such a large picture of similarities through which the Universe presents itself as if it were only one family, we pass to that of differences, where each species demands an isolated place and must have his own portrait, we shall recognize that, except for a few major species, as the elephant, the rhinoceros, the hippopotamus, the tiger, the lion, which must receive each its own frame, every other seems to join with its neighbours and form groups of degraded similarities: those genera that our Nomenclators have presented by a lace of figures, some of them hold by feet, others by teeth, by horns, by hair, and by other even weaker connections. And even those whose forms appear to us the most perfect, that is to say, the closest to ours: the apes, present themselves together, and need rather attentive eyes to be distinguished from each other, because it is less to form than to size

---

[25] "Au lieu de traiter les oiseaux un à un, c'est-à-dire par espèces distinctes et séparées, je les réunirai plusieurs ensemble sous un même genre, sans cependant les confondre et renoncer à les distinguer lorsqu'elles pourront l'être." (*Histoire naturelle des oiseaux*, tome I, 1770, p. xx.)

that the privilege of isolated species is attached, and that man itself, although unique in its species, infinitely different from those of animals, being only of a rather small size, is less isolated and has more neighbours than larger animals. It will be seen in the history of the orangutan, that if attention would be paid to face only, this animal might be regarded as the first ape, or the last man, because with the exception of soul, it lacks nothing we have, and because it differs less from man as far as body is concerned, than it differs from other animals which are given the name of ape.[26]

Buffon's opposition to Linnaeus is less clear-cut than would appear at first sight: both have changed opinion much during their careers, and it is somehow reductionist to arrest them at any moment of their reflection. Moreover, the methodological proposition according to which Linnaeus was a rationalist philosopher of the Aristotelian school and Buffon the prototype of an inductivist naturalist could be more realistic if reversed![27] The former opinion has been suggested by those who claimed that only Linnaeus's systematics was clearly scientific, and that, on the contrary, Buffon's *Natural History* was affected by a regrettable "naturalist" character which placed it far from authentic science. In fact, more than being representative of mediaeval nominalism or realism, Linnaeus and Buffon both tried to solve the even older problem of continuity vs. discontinuity: Linnaeus divided nature into clear-cut classes, for a purpose of classification, whereas Buffon, trying to describe it, affirmed that it was continuous and indivisible, and that natural bodies differentiated with each other only by imperceptible nuances. But when Buffon had to deal with complex groups, such as, for example, monkeys or birds, he had no other choice but to divide them into classes. When he resigned himself to practice this sort of classification, his kind of systematics was of the downward model: as he did not really know animals (i.e., species), he started from larger groups (e.g., monkeys or birds, in the most obvious cases), and then divided them into subgroups down to the species he had to deal with. By this procedure, he more or less intuitively conformed to Aristotle's zoological principles, and in fact Aristotle was Buffon's great model. On the contrary, Buffon never accepted Linnaeus's concept of the genus, and always carefully avoided to use something that could be perceived as similar or analogous to it.

8.4 Lamarck, Cuvier, and other nineteenth-century zoologists

In the 1780s, disciples of Linnaeus and Buffon tried to unite the two men's conceptions into one: Buffon's chapters were arranged according to Linnaeus's genera in a series of new editions of *Natural History*, complete or abridged, which were all published during the nineteenth century. Some editions were completed by sequels ("*Suites à Buffon*") dealing with all animals (and even vegetables), species which were not included in the original work, and which of course were sorted into Linnaean genera.[28] As original

---

[26] *Nomenclature des singes* (Buffon, 1766: 29–30).
[27] Barsanti (2005).
[28] For a list of these editions see Buffon (1954: 522–30).

Buffon's volumes were generally no longer in use, it seems that an agreement was gradually established according to which he himself had used from the beginning such kind of Linnaean system!

In the same period, downward sorting was abandoned in favor of an upward one. This was especially the case in the study of lower animals (sponges, jellyfish, polyps, various worms, etc.): they were poorly known creatures, with few or no subgroups intuitively perceived or conceived, even by zoologists. Both Cuvier—from 1795—and Lamarck (who was originally not a zoologist and was not knowledgeable in animals)—from 1801—tried to sort the mass of unknown, mostly marine animals from the Paris *Muséum* collections, into species, genera, and higher categories. That their observations brought them to very different conclusions is another story: as far as our subject is concerned, the important thing is that they introduced and established upward classification, starting from species, classifying them into genera, families,[29] orders, and classes. In spite of this change, nineteenth-century naturalists retained most Linnaean ideas, especially that genera had some natural or "essential" characters. For example, this is clear when Lamarck said:

> One gives the name of *genus* to clusters of races, called *species*, brought together by consideration of their relationships, and making as many small series limited by characters which are arbitrarily chosen in order to circumscribe them. When a *genus* is well made, all the races or species it comprises resemble each other by most essential and numerous characters, must be naturally sorted besides each other, and do not differentiate from each other except by characters less important but sufficient to distinguish them. Therefore, well-made genera are in fact small *families*, i.e., true portions of Nature's very order.[30]

Many younger naturalists, however, thought that genera did not exist *per se*, that a genus had no "essential" qualities, nor had it been "created," and that it was a mere artificial combination made for convenience; but they acknowledged as well that any genus must be adapted to its role in classification. Cuvier for example insisted on the importance of using multiple characters, especially those he considered more significant than others. The aim was to achieve a classification as "natural" as possible (this aim being also that of the botanists of the time). Cuvier went into further, more technical details:

> There is scarcely a single being which has a simple character, or can be recognized by one single feature of its conformation; a union of several of these traits is almost always required to distinguish one being from those which resemble it in some, but not in all its peculiarities, or which have some of them combined with others of which the first is destitute. The more numerous the beings to be distinguished, the greater should be the number of traits; so that to distinguish an individual being from all others, it would be necessary, without some concise method, to enter into a complete description of its character. It is to avoid this inconvenience, that divisions and subdivisions have been invented. A certain number only of neighboring beings are compared with each other,

---

[29] The family rank was not used by Linnaeus; introduced in botany by Michel Adanson in 1763 (Magnin-Gonze, 2004), it was used for the first time in zoology by the entomologist Latreille (1797).
[30] Lamarck (1809: 32).

and their characters need only express their differences, which, by the supposition itself, are the least part of their conformation. Such a combination of beings is termed a *genus*.[31]

But many zoologists still persisted by the middle of nineteenth century to find "essential" qualities in their genera, according to the scholastic tradition, and to associate them with nominalistic and typological considerations:

> When a genus is subdivided into other genera, the original name should be retained for that portion of it which exhibits in the greatest degree its essential characters as at first defined. Authors frequently indicate this by selecting one species as a fixed point of reference, which they term the 'type' of the genus.[32]

Other zoologists, still more anecdotal, invented curious systems, based on a sort of numerology, using the numbers three, four, or five as bases. The most famous is "quinarianism," which explained that everything in nature was based on the number five: all zoological groups and subgroups were arranged by five along "circles," adjacent circles touching themselves. In a given family (or tribe), there were precisely five genera—but a variable number of species: again, only genera, not species, were to conform to the "essence" of the system!

8.5 Darwin and the evolutionary concept of the genus

Charles Darwin's theory of natural evolution completely upset the preceding conceptions of biology, including those of the genus. He was an experienced naturalist, with extensive knowledge of plants and animals. He undertook in October 1846 (aged 37) a revision of a peculiar zoological group: cirripeds, which first were classified as mollusks, then established as a separate class, before they turned out to be recognized as highly modified crustaceans.[33] Darwin devoted exactly eight years of his life (till October 1854) to this project, and published it under the form of a two volume monograph. He later recognized that this work had been useful for his redaction of the *Origin of species*, however adding that he wondered whether the benefit was worth the investment of time it has demanded. In this text, Darwin revealed himself to be very aware of the basically practical role of genera, not hesitating to reject a particular genus, even perfectly valid, because he does not find it appropriate: for him, an equilibrium was necessary between the number of species and that of genera. In addition, just as other zoologists—and especially entomologists—of his time, he recommended that the number of genera should not be too much multiplied:

> In my opinion, this inordinate multiplication of genera destroys the main advantages of classifications [i.e., their role of retrieving system].[34]

---

[31] Cuvier 1817: 8–9; 1834: 4).
[32] Strickland (1842), quoted by Stamos (2005: 82); see also Farber (1976).
[33] In Darwin's time, crustaceans were still included in insects.
[34] Darwin (1851: 216).

He devoted a shorter time to the redaction of his *Origin of species*, which was published five years later (November 1859). Explanations about systematics were included in Chapter 13, where Darwin introduced for the first time a genealogical, evolutionary meaning of classification, basing his explanations on Fig. 8.1.

In the diagram each letter on the uppermost line may represent a genus including several species; and all the genera on this line form together one class, for all have descended from one ancient but unseen parent, and, consequently, have inherited something in common. But the three genera on the left hand have, on this same principle, much in common, and form a sub-family, distinct from that including the next two genera on the right hand, which diverged from a common parent at the fifth stage of descent. These five genera have also much, though less, in common; and they form a family distinct from that including the three genera still further to the right hand, which diverged at a still earlier period. And all these genera, descended from (A), form an order distinct from the genera descended from (I). So that we here have many species descended from a single progenitor grouped into genera; and the genera are included in, or subordinate to, sub-families, families, and orders, all united into one class. Thus, the grand fact in natural history of the subordination of group under group, which, from its familiarity, does not always sufficiently strike us, is in my judgement fully explained.

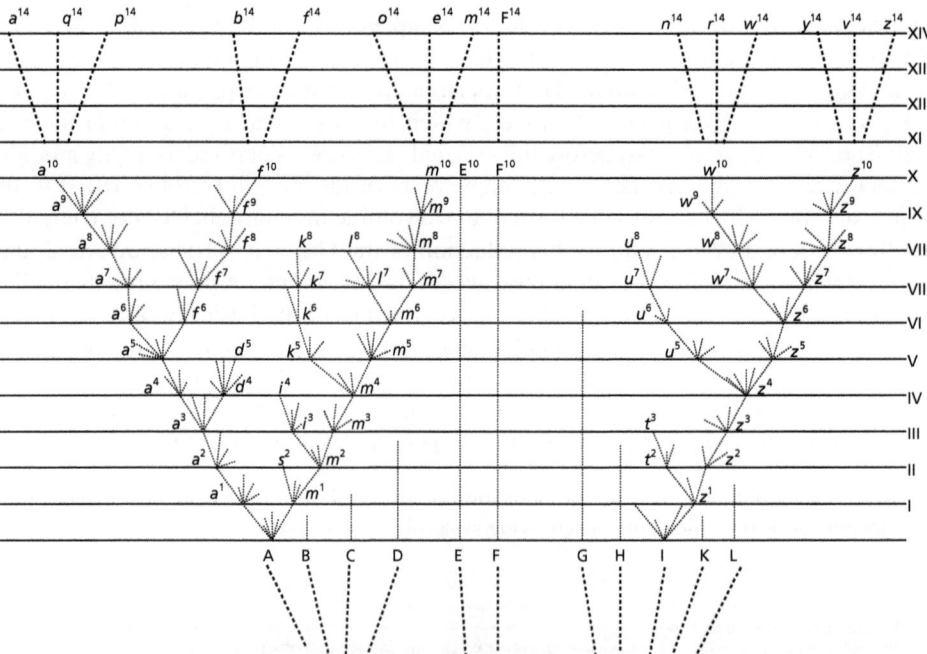

Figure 8.1 *Darwin's diagram of descent from various genera (redrawn from Darwin, 1859).*

Naturalists try to arrange the species, genera, and families in each class, on what is called the Natural System. But what is meant by this system? Some authors look at it merely as a scheme for arranging together those living objects which are most alike, and for separating those which are most unlike; or as an artificial means for enunciating, as briefly as possible, general propositions—that is, by one sentence to give the characters common, for instance, to all mammals, by another those common to all Carnivora, by another those common to the dog-genus, and then by adding a single sentence, a full description is given of each kind of dog. The ingenuity and utility of this system are indisputable. But many naturalists think that something more is meant by the Natural System; they believe that it reveals the plan of the Creator; but unless it be specified whether order in time or space, or what else is meant by the plan of the Creator, it seems to me that nothing is thus added to our knowledge. Such expressions as that famous one of Linnaeus, and which we often meet with in a more or less concealed form, that the characters do not make the genus, but that the genus gives the characters, seem to imply that something more is included in our classification, than mere resemblance. I believe that something more is included; and that propinquity of descent—the only known cause of the similarity of organic beings—is the bond, hidden as it is by various degrees of modification, which is partially revealed to us by our classifications....

Finally, with respect to the comparative value of the various groups of species, such as orders, sub-orders, families, sub-families, and genera, they seem to be, at least at present, almost arbitrary ... But I must explain my meaning more fully. I believe that the *arrangement* of the groups within each class, in due subordination and relation to the other groups, must be strictly genealogical in order to be natural; but that the *amount* of difference in the several branches or groups, though allied in the same degree in blood to their common progenitor, may differ greatly, being due to the different degrees of modification which they have undergone....Thus, on the view which I hold, the natural system is genealogical in its arrangement, like a pedigree; but the degrees of modification which the different groups have undergone, have to be expressed by ranking them under different so-called genera, sub-families, families, sections, orders, and classes.[35]

This was sort of a revolution, the most drastic rupture in the whole history of classification: from that time on, classifications were no longer artificial, subjective systems the only purpose of which was to retrieve kinds and sorts of living beings; but classification reflected—or at least tried to reflect—the history of these living beings. In the second paragraph, Darwin discussed the relative importance of arrangement and hierarchical ranking. For him, both operations were completely different: if ranking was a pure matter of convention, arrangement was strictly determined by the descent of the group under study.

8.6 Twentieth-century concepts of the genus

The last century has elaborated a greater number of systematic (and therefore of generic) concepts than the previous epochs. I shall mention three of these concepts, in a somehow

---

[35] Darwin (1859: 412–14, 419–22).

chronological order; they all have Darwinian evolution theory as both philosophical and scientific basis.

8.6.1 The "evolutionary" or gradist concept

In spite of illuminating principles introduced by Darwin, zoological and botanical systematics have hesitated during almost a century before finding ways which would enable convenient applications of Darwinian theory, one should rather say "neo-Darwinian," since original evolution theory was modified soon after 1859, including by Darwin himself. This process of transformation found a first culmination in the so-called "synthetic evolution theory," whose four principal advocates have been Julian Huxley (1887–1975), Theodosius Dobzhansky (1900–1975), George Gaylord Simpson (1902–1984), and Ernst Mayr (1904–2005); it is to their synthesis that the term "neo-Darwinism" has precisely been applied. Ernst Mayr was especially influential in the systematic part of the synthesis, a subject to which he devoted a number of books that have become classics.[36] He was anxious to conciliate Darwinian theory and traditional zoology, and it is interesting to observe the way he dealt with history to give his ideas the best possible presentation and allure. In his three last books, published between the ages of 93 and 100, he made every effort to convince benevolent readers that his system was the most "evolutionary" and the most "Darwinian" of all (he applied the two epithets only to his system, in spite of the fact that adversary ones were equivalent in these respects).[37] Instead of evolutionary, or Darwinian, Mayr's system has been more often referred to as "gradist," in opposition to another system termed "cladist" (see Section 8.6.2). These words both refer to evolutionary trees, which have been produced in great number since Darwin published the first one in 1859. In a genealogical tree (and in a classification), gradists attach a greater importance to clusters based on similitude, which are called "grades" (Latin *gradus*, a grade, a step, a stage), rather than to those based on strict phylogenetic criteria, or "clades" (Greek *klados*, a branch, a twig). Gradists indeed use evolutionary trees; but they also attach some importance to traditional sorting of taxa, taking care not to disturb too much the established classifications; they insist that commonsense characters, based on ecology, ethology, sometimes on intuitive or popular knowledge, etc., must be considered as well. This pragmatic position is well exemplified in the following Mayr quotation:

> The best definition of a genus seems to be one based on the honest admission of the subjective nature of this unit, and it may not be possible to say much more than that the genus, to be a convenient category in taxonomy, must in general be neither too large nor too small. A tentative definition might be: 'A genus is a systematic unit including one species or a group of species of presumably common phylogenetic origin, separated by a decided gap from other similar groups. It is to be postulated for practical reasons that the size of the gap shall be in inverse ratio to the size of the group.' The second sentence

---

[36] Mayr (1942, 1963, 1969).
[37] Mayr (1997: 136–42); see also Mayr (2001, 2004).

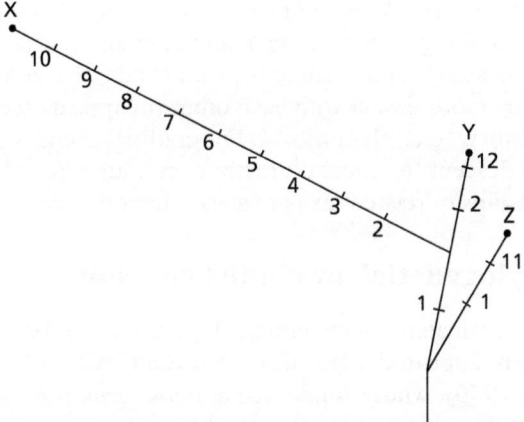

Figure 8.2 *An example of gradist taxonomy.*

of this definition requires an explanation. We have groups, as has been stated repeatedly, that are exceedingly variable and others that are very uniform. If the same standards of difference were applied, we might have to put every species of one family into monotypic genera and all the species of another family into a single genus. Genera with 500, 1000, or even 2000 species are very inconvenient, and any excuse for breaking them up into smaller units should be good enough. In birds of paradise, on the other hand, morphological differences between the species are so well developed that not only all the species, with two exceptions, have been put into different genera, but that even representative species in some genera have been separated generically on what seems superficially very good reasons. The genus would become completely synonymous with the species and would lose its function as category of convenience, if we were to recognize genera that are based merely on certain morphological differences.

It is obvious from this discussion that the delimitation of the genus is, to a considerable extent, a matter of judgment, and that this judgment in turn depends on wide experience and on some intangibles. It has happened frequently that some older taxonomists has divided a family into let us say eight genera, using what seemed entirely superficial characters. When the complete anatomy and life history of the species became known, it turned out that the original generic arrangement was quite correct.[38]

In a later text, Mayr gave a precise example of the gradist position. He considered the following diagram (see Fig. 8.2), a supposed phylogenetic tree of three supposed taxa: X, Y, and Z.[39]

This example could be identified as a true case: X could be the human species, Y the chimpanzee(s), and Z the gorilla. Numbers 1–12 represent derived characters. Gradist taxonomists would separate the taxon X (e.g., at the genus level) from the group

[38] Mayr (1942: 283–4).
[39] Mayr (1986).

(Y + Z), because X differs from Y by eight derived characters (3–10), and share only one (2) with Y; they would also group together Y and Z in another taxon (also supposedly at the genus level), because Y and Z share the same "adaptive zone," or "niche,"[40] that is, they are supposedly more similar with each other (in appearance, but also in ecology, ethology, "natural history," etc.) than with X. For gradists, overall similarity prevails over a closer proximity in descent (or, more correctly, over a supposed closer proximity, since genealogical relationships of related taxa are always hypothetical).

8.6.2 The "phylogenetic" or cladist concept

Gradist systematics dominated botany and zoology up to the 1970s. However, another system had previously appeared: cladistics. Its inventor was a German entomologist, Willi Hennig (1913–1976), whose fundamental book *Grundzüge einer Theorie der phylogenetischen Systematik*, published in 1950, had been largely ignored until an English translation was published in 1966.[41] Since this date, the Hennigian system has dominated systematics and taxonomy, under the various terms of "phylogeny" (a metaphorical use of the word, since the Hennigian system is no more phylogenetic than gradism is evolutionary), "cladistics" (which refers to clades, above), or "cladification" (used especially by gradists, with a possible nuance of pejorative connotation). Basically, cladistics is a dichotomous classification system in which only derived states of characters ("apomorphies") are taken into consideration; on the other hand, supposedly primitive states of characters ("plesiomorphies") are considered uninformative. For example, if we consider again the diagram in Fig. 8.2, a cladist will identify the group (X + Y) as a taxon, for example, a genus, because X and Y share one derived state of character (2). The genus (X + Y) will then be separated from genus Z, in spite of the fact that X is very different from Y (and Y very similar to Z) in appearance, "adaptive zone," or "niche," etc. In cladistic terms, a taxon such as (Y + Z) is called a "paraphyly," or paraphyletic group, that is, a group of species that share a common ancestor but does not include all the descendants of the common ancestor, and is radically rejected.

8.6.3 The biological concepts

In addition of concepts based on fossils, or on dead specimens preserved in collections and museums, other concepts are based on characters observable only on living individuals. These concepts are well known in the case of a given species, considered as a pool of genes which are supposed to circulate freely: all members of a species actually or potentially interbreed; in addition, some of them are capable of fertile interbreeding with members of other species.[42] Products, which are called "hybrids," are either fertile or sterile. For example, interbreeding between the two species lion and tiger produces

---

[40] Mayr (1963: 78, 593, etc.) considered the two terms "adaptive zone" and "niche" as synonyms.

[41] Hennig (1966).

[42] Interbreeding with other species is difficult and therefore infrequent under natural conditions, but sometimes easy and common with captive animals.

fertile offspring; but the product of the two species ass and horse (which is called mule) is sterile. From this basis, a biological definition of the genus has been introduced by some zoologists: all species capable of producing fertile hybrids must be classified into the same genus; but zoologists hesitate when hybrids are not or only partly fertile.[43] Be that as it may, such biological definition of the genus can be used only in those few cases where the taxa in question have been observed or reared. But the vast majority of animal species are still known only from museum specimens, and therefore it is impossible to verify whether or not they comply with the requirements of a supposed biological definition.

8.7 Controversies in the early twenty-first century

In the same time as living species as well as zoologists have decreased in number, theoretical biology has taken a great importance and produced a large quantity of works published in more or less specialized journals. In the same way as biological research of the twentieth century took place within the boundaries of Darwinian evolution theory, almost all systematicians and taxonomists of our time situate themselves within the Hennigian system. Mayr's efforts have been unsuccessful: most of his "evolutionary" system disappeared when he died in 2005, aged 101.

But the Hennigian system has also been altered. In its original conception, it only aimed to classify living organisms. Diagrams Hennig obtained from analyses of character matrices always referred to living species, because he studied insects, and living insects are vastly more numerous and better known than fossils. Hennig's direct followers, including "vertebrologists," also studied and cladified almost exclusively living species. But nowadays, basic Hennigian methodology of a step-by-step hand analysis has been universally replaced by computer treatments of very large matrices, which make use of a vast number of characters, either molecular (for living species) or drawn from fossil specimens (for paleontological studies). The result is that cladistics of today is only remotely related to original Hennigian methodology.[44] In addition, although Hennig's methodology was especially designed for classification, it is heavily used today to reconstruct phylogenies, a perversion of the original meaning of the word "phylogenetic" in Hennig's opinion. In other words, so-called "pattern cladistics" (which was aimed to make phylogenetically grounded, more exact classifications than previously), has been replaced by "process cladistics" (aimed to reconstruct pure phylogenies, i.e., scenarios of zoological groups history and evolution). It is not possible to enter here into details or to examine a series of cases. Only two examples which refer to our subject will be dealt with: the controversy of classification vs. cladistics, and the recurrent problem of the "human genus."

---

[43] See especially the books and papers of Alain Dubois, for example, 1988, 2004, etc.
[44] Krell (2005).

8.7.1   Cladification versus classification

When true classifications are abandoned in favor of pure and simple "phylogenies," a series of new problems emerge, because phylogenies are not classifications: the latter indeed have philosophical importance,[45] but classifications mostly were and still are used for practical purposes, which may be significant in certain domains. Let us take an example in mosquitoes of the tribe Aedini (subfamily Culicinae, family Culicidae, order Diptera, class Insecta). The future tribe was originally represented by a single species described and named *Aedes cinereus* by the German entomologist Johann Wilhelm Meigen in 1818. Less than two centuries later, in 2004, the tribe Aedini comprised 11 genera (the genus *Aedes* itself being divided into 44 subgenera), and some 1240 species widespread all over the world. Many species are important vectors of pathogen organisms, especially viruses: for example, the recently (re-)appeared *Chikungunya* and *Zika* are transmitted by a species of the genus *Aedes*. In 2004, a team of entomologists published a cladistic analysis of Aedini, where 32 subgenera of *Aedes* were elevated to generic status, bringing the total number of genera in the tribe from 11 to 43.[46] This result was indeed interesting, even important from a scientific or philosophical point of view; but it provoked a complete upheaval in the tribe Aedini. Another reputed specialist of the group replied with a paper expressing serious reservations about such a cladification, which did not take into account the true role of a classification, retrieval of information:

> To facilitate communication and information exchange among professional groups interested in vector-borne diseases, it is essential that a stable nomenclature be maintained. For the Culicidae, easily identifiable genera based on morphology are an asset. Major changes in generic concept, the elevation of 32 subgenera within *Aedes* to generic status, and changes in hundreds of species names proposed in a recent article demand consideration by all parties interested in mosquito-borne diseases. The entire approach to Aedini systematics of [Reinert et al.] was flawed by an inordinate fear of paraphyletic taxa or Paraphylyphobia, and their inability to distinguish between classification and cladistic analysis.[47]

This example invokes practical rather than purely "scientific," that is, theoretical criteria, and demonstrates than animal systematics can still be a source of disagreement between different epistemological cultures, as close as they might appear in this example.

8.7.2   Discordant ideas about the genus *Homo*

Linnaeus included the human species in the *Regnum animale* of his *Systema Naturae*, into the genus *Homo* and under the binomial name *Homo sapiens*.[48] The genus comprised

---

[45] Dupré (2001).
[46] Reinert et al. (2004).
[47] Savage (2005: 923).
[48] Linnaeus (1758, 1: 20).

another species: *Homo troglodytes*, with a commentary in the text *"Homo sylvestris Orang Outang"* and a footnote:

> There is no doubt that the *troglodytes* species is completely distinct from *Homo sapiens*, neither of our genus, nor our blood, although very similar in size.[49]

It is interesting to notice in this excerpt that Linnaeus made a distinction between a systematic or technical sense of the world "genus" and the ordinary meaning of the word. His species *Homo troglodytes* seems to have been an artificial combination of apes from Africa (chimpanzees) and Asia (true orangutan); another Linnaean name, *Simia satyrus*,[50] probably also referred to this combination. Buffon evoked "the great and unique family of our human genus"[51] (a formula which might have been partly anti-Linnaean); he too considered that man belonged to animals, but only under a "material" point of view, and he recognized that the orangutan "might be regarded as the first ape, or the last man," as we have seen above.

In the late twentieth and early twenty-first century, there are still diverse, even discordant ideas about the systematics of our species and genus. It should be reminded once more that a genus is a pure matter of convention, conviction, or convenience. In this perspective, different kinds of conceptions can be identified. For example, according to the relative size they give to their genera, zoologists have been sorted into "lumpers" and "splitters": the former give preference to similarities over differences, and therefore cluster in the same, broad genus series of species sometimes rather different; the "splitters," on the contrary, prefer to insist on differences and to make almost as many (small) genera as there are species.[52] This difference in the genus conception might be correlated with epistemological cultures: we shall here consider briefly the contrasted opinions of two groups of scholars: paleontologists and anthropologists.

One of the characteristics of paleontologists is their extreme tendency to splitting. For example, a recent synthesis recognized two living genera (*Pan* and *Homo*), and not less than six fossil genera (*Ardipithecus, Australopithecus, Kenyanthropus, Orrorin, Paranthropus,* and *Sahelanthropus*) in a "subfamily Homininae."[53] The authors explained their conceptions in the following words:

> We will deal with evolutionary history at a unique level of taxonomic detail, that of the species and genus. The species category is the lowest taxonomic level commonly used, and genera are composed of one, or more, species. For a group to qualify for the rank of genus, the taxa within it are generally taken to be both adaptively homogeneous and members of the same clade. To comply with the latter requirement, the genus must contain all the descendants of a common ancestor and its members must be confined to that

---

[49] *"Speciem Troglodytae ab Homine sapiente distinctissimam, nec nostri generis illam nec sanguinis esse, statura quamvis simillimam, dubium non est."* (*Op. cit.*: 24).
[50] *Op. cit.*: 25.
[51] *"La grande et unique famille de notre genre humain"* (Buffon, 1766: 311).
[52] Mayr (1942: 280–3).
[53] Wood & Constantino (2004).

clade. Species that are "adaptively similar" but belong to different clades do not qualify for the rank of genus.[54]

Remarkably, they refer both to cladistics and to a vague "adaptive similarity" which reminds us of Simpson's (1961) "adaptive zone" and Mayr's gradism (above). In the following paragraph, they invoke a "taxonomic philosophy," which they seem to consider different when human species is concerned from the prevailing situation in other animals. This "philosophic" point has been studied in a recent paper explaining that a possible, intuitive human-nonhuman distinction may have contributed to the idea that human evolution is somehow exceptional, resulting in a true inflation of genera in paleoanthropology.[55]

On the contrary, anthropologists working on living beings are not reluctant to consider that human species is not different in its ontology from other animal species, and that it should be included in the same genus as its closest relatives: African chimpanzees. Because the Linnaean genus name *Homo* has priority over *Pan*,[56] man cannot be termed a "third chimpanzee" (as some authors did a few years ago), but chimpanzees must be recognized as other members of our own genus, *Homo*, as Linnaeus himself did in the tenth edition of his *Systema Naturae*, however incorrectly and reluctantly (see above). This opinion, typical of the anthropologists "lumper" style, has been most clearly and completely presented in a recent paper:

> The accumulating DNA evidence provides an objective non-anthropocentric view of the place of humans in evolution. We humans appear as only slightly remodeled chimpanzee-like apes. This is apparent when the DNA evidence is translated into a phylogenetic classification based on principles first envisioned by Darwin and elaborated on by Hennig. The paramount principle is that each taxon should represent a clade in which all species in the clade share a more recent common ancestor with one another than with any species from any other taxon. The second basic principle, a corollary of the first, is that the hierarchical groupings of lower ranked taxa into higher-ranked taxa (e.g., species into genera, genera into families) should describe the degrees of phylogenetic relationships among taxa, i.e., among clades.[57]

In conformity with their arguments, these authors classified the human species together with the two chimpanzee into a same genus, however with a reservation: the two chimpanzees were classified into a particular subgenus, *Pan*, separated from the "true" human (sub-)genus *Homo*;[58] all hominid fossil species (split into six genera by paleontologists, above) were explicitly lumped together into the subgenus *Homo*.[59]

8.8   Conclusion: the *PhyloCode* and the end of genera

Today, most botanists and zoologists continue to use the genus category. For example, it is still used in what has become the major journal of taxonomic zoology, available both

---

[54]   *Op. cit.*: 518.
[55]   De Cruz & De Smedt (2007).
[56]   The genus *Pan* was created and named by Oken in 1816 to accommodate the African chimpanzee(s).
[57]   Wildman et al. (2003: 7181).
[58]   When a genus is divided into subgenera, the subgenus comprising the type species of the genus keeps the genus name, for example, *Homo (Homo) sapiens*.
[59]   Wildman et al. (2003: 7182 and Table 1).

on the net and on paper: *Zootaxa*.[60] However, the most radical cladists have arrived to the conclusion that Linnaean ranking—and especially the genus level—was no longer adapted to cladistic systematics, and they have decided to introduce a new, rank-free system called the *PhyloCode*. As they explain:

> The *PhyloCode* is a formal set of rules governing phylogenetic nomenclature. It is designed to name the parts of the tree of life by explicit reference to phylogeny. The *PhyloCode* will go into operation in a few years, but the exact date has not yet been determined. It is designed so that it may be used concurrently with the existing codes based on rank-based nomenclature. We anticipate that many people whose research concerns phylogeny will find phylogenetic nomenclature advantageous.[61]

Only a few zoologists have already described and named animals using the *PhyloCode* rules, which are based only on the mention of specific epithet, without mention of a supposed "genus." For example, here is the case of a recently described marine mollusk species:

> The species can be placed confidently in the clade Discodorididae, but not in any of its subclades (traditionally taxa of genus rank). The unique, epithet-based name of the species is '*aliciae* Dayrat, 2005.' The combination Discodorididae *aliciae* may also be used, once the unique, epithet-based name has been cited. Discodorididae *aliciae* is an example of how a new species of Discodorididae could be named in the context of phylogenetic nomenclature. I argue that epithet-based species names and their combinations with clade addresses should be very appealing to people who think phylogenetically. I also discuss two advantages of such combinations: first, they should be more stable than Linnaean binomials, which often change for arbitrary (e.g. non-phylogenetic) reasons; second, they should help taxonomists avoid creating multiple names for the same species.[62]

However, most authors insist that the Linnaean system of *Genus*-plus-*species* is millions of taxa in advance of the *PhyloCode*, and that, in spite of its shortcomings, it is still more convenient than the latter.

··

APPENDIX 1

Animal classification

Linnaeus established a system of classification and naming of plants and animals which has persisted up to now, and in which each species is designated by a "binomen" consisting of two words: a generic name and a specific name or epithet, for example, *Canis familiaris* (*Canis* is the genus name, *familiaris* is the specific name or epithet; a specific epithet has no meaning if its genus is not mentioned). Species and genera are arranged into higher categories, which in turn could be used for a more general comprehension of the animal world (Table A.I).

The words "species," "genus," "family," "order," "class," and "phylum" designate the principal categories of classification (there are other, additional ones, especially in insects, like superfamily,

---

[60] See http://www.mapress.com/zootaxa/
[61] From the Society for Systematic Biologists site (http://systbio.org/).
[62] Dayrat (2005: 216).

Table A.I *Example of Linnaean classification of a particular species: the domestic dog (Canis familiaris) [Family and Phylum ranks were introduced in the nineteenth century, as well as intermediate ranks which have not been mentioned here (subgenus, subfamily, etc.)]*

Species *familiaris* (the dog in the most restricted meaning)
Genus *Canis* (the dog, the wolf, and related species)
[Family Canidae (dogs *sensu lato*, foxes, and related genera)]
Order Carnivora (dogs, cats, and related families)
Class Mammalia (mammals)
[Phylum Vertebrata (vertebrates)]
Kingdom Animalia (animals)

subfamily, tribe, subgenus ...). In technical terms, these categories are called "taxa" (singular "taxon"), derived from "taxonomy," French *taxonomie*, a word coined by the Swiss botanist Augustin-Pyramus de Candolle in his *Théorie élémentaire de la botanique* (1813). The "species" taxon comprises only individuals, whereas all the other taxa—from the genus upward—comprise individuals already sorted and clustered into lower taxa. Thus, the genus comprises species, the family comprises genera, the order comprises families, the class comprises orders, and the phylum comprises classes. It is generally thought that the species itself—or at least each particular species— has some reality since it comprises real "objects" (living beings); but although species has been the object of intense debates and controversies among scientists and philosophers, it has not yet been possible to clearly define or understand it.[63]

APPENDIX 2

Original Latin and French texts of long quotations[64]

A.2.1   Linnaeus: *Ratio operis* of the *Genera plantarum* (1737)[65]

6. *Genera* autem tot sunt, quot attributa communia proxima distinctarum specierum, secundum quae in primordio creta fuere: confirmant haec revelata, inventa, observata. Hinc *Omnia Genera & Species naturalia sunt.*

Non enim licet conjungere sub eodem genere Equum & Suem: licet ambae monungulae essent; nec Capreolum, Rangiferum & Alcen genere distinguere, licet cornuum figurâ differant. Generum itaque limites attenta & sedula observatione inquirere debemus, cum a priori difficilius determinentur, licet hoc opus hic esset labor, nam *Confusis generibus omnia confundi necesse est.* Caesalp.

8.... Mox in usum vertere, mox Systema struere allaborarunt omnes hisce laboribus apti; omnes quidem eodem animo, & in eundem finem; at non omnes eodem cum successu. Paucis quippe

---

[63] Lherminier & Solignac (2005).

[64] Short Latin and French quotations have been given in footnotes. Original Greek texts (Plato, Aristotle) are not given.

[65] The quotation is extracted from the 2nd edition (Linnaeus, 1742: ii–vi).

notus fuit Canon fundamentalis, quem si non observant aediles, illico ruit prima oborta tempestate splendidissimum quamvis aedificium: BOERH. Inst. 31. DOCENTI *procedendum a generalibus ad singularia quaeque, dum inventa explicat ; ut* INVENTORI, *contra, a singularibus ad generalia eundum fuit.* Assumserunt enim Varii diversas partes [Fructificationis] pro principio Systematico, & cum eo secundum divisionis leges a Classibus per Ordines descenderunt ad Species usque, & hypotheticis ac arbitrariis his principiis fregerunt & dilacerarunt naturalia, nec arbitraria [6] genera; naturaeque vim intulerunt ...

Quod si, dicunt, non classe conjungi possunt, multo minus genera; sed non observant se construxisse Classes qualescunque, ipsum verò Creatorem Genera. Hinc tot falsa genera ! tot controversiae inter Authores ! tot mala nomina ! tanta confusio !

9.... Parum refert, qua methodo potestis pervenire ad genera facillimâ viâ. Istae reliquis preferenda est, quae certiori tramite securius ad ea ducit; quaeque maxime universalis est; vix enim credo ullos facili adeo natos esse memoria, ut, absque Systemate, genera tenere queant. Methodus itaque erit viae dux ; Ordines enim sunt Classes subalternae: & paucae distinguere genera facilius esse, quam omnia, nullus inficias ibit. Non quidem nego aeque dari Classes naturales ac Genera naturalia ; non nego quin methodus naturalis & nostrae & omnium inventarum methodis longe praeferri deberet, sed rideo omnes methodos naturales hactenus exclamatas, & in me defensionem provocatus suscipio, quod nulla, ne unica classis, hactenus data, in ullo systemate, naturalis sit, quamdiu ista genera istique characteres, qui jam sunt, sub iisdem militant. Facile est plus quam dimidiam notorum generum partem ad suas classes amandare, at eo difficilius reliqua. Nec sperare fas est, quod nostra aetas Systema quoddam Naturale videre queat, & vix seri Nepotes. Attamen plantas nosse studemus ; ideoque interim Artificiales assumendae sunt Classes & Succedaneae.

10. Assumtis Generibus naturalibus [6, 7] ad ea tenenda pura & inculcata duo requiruntur, ut scilicet verae Species, nec aliae, ad sua genera reducantur; utque Genera singula veris circumscribantur limitibus & terminis, quos *Characteres* vocamus *genericos.*

A.2.2 Buffon

A.2.2.1 *Histoire Naturelle, tome 1er (1749: 13)*

Mais la Nature marche par des gradations inconnues, & par conséquent elle ne peut pas se prêter totalement à ces divisions, puisqu'elle passe d'une espèce à une autre espèce, & souvent d'un genre à un autre genre, par des nuances imperceptibles ; de sorte qu'il se trouve un grand nombre d'espèces moyennes & d'objets mi-partis qu'on ne sait où placer, & qui dérangent nécessairement le projet du système général.

A.2.2.2 *"Nomenclature des Singes," Histoire naturelle, tome 14e (1766: 29–30)*

Si de ce grand tableau des ressemblances dans lequel l'Univers vivant se présente, comme ne faisant qu'une même famille, nous passons à celui des différences, où chaque espèce réclame une place isolée et doit avoir son portrait à part, on reconnaîtra qu'à l'exception de quelques espèces majeures, telles que l'éléphant, le rhinocéros, l'hippopotame, le tigre, le lion, qui doivent avoir leur cadre, tous les autres semblent se réunir avec leurs voisins et former des groupes de similitudes dégradées, des genres que nos Nomenclateurs ont présentés par un lacis de figures dont les unes se tiennent par les pieds, les autres par les dents, par les cornes, par le poil et par d'autres rapports encore plus petits. Et ceux même dont la forme nous paraît la plus parfaite, c'est-à-dire la plus approchante de la nôtre, les singes, se présentent ensemble et demandent déjà des yeux attentifs pour être distingués les uns des autres, parce que c'est moins à la forme qu'à la grandeur qu'est attaché le privilége de l'espèce isolée, et que l'homme lui-même quoique d'espèce unique, infiniment différente de toutes celles des animaux, n'étant que d'une grandeur médiocre, est moins isolé et a plus de voisins que les grands animaux. On verra dans l'histoire de l'orang-outang, que si l'on ne faisait attention qu'à la figure on pourrait également regarder cet animal comme le premier des singes ou le dernier des hommes, parce qu'à l'exception de l'âme, il ne lui manque rien de tout ce que nous avons, et parce qu'il diffère moins de l'homme pour le corps, qu'il ne diffère des autres animaux auxquels on a donné le même nom de singe.

A.2.3   Lamarck: *Philosophie zoologique* (1809: 32)

On donne le nom de *genre* à des réunions de races, dites espèces, rapprochées d'après la considération de leurs rapports, et constituant autant de petites séries limitées par des caractères que l'on choisit arbitrairement pour les circonscrire.

Lorsqu'un *genre* est bien fait, toutes les races ou espèces qu'il comprend, se ressemblent par les caractères les plus essentiels et les plus nombreux, doivent être rangées naturellement les unes à côté des autres, et ne diffèrent entre elles que par des caractères de moindre importance mais qui suffisent pour les distinguer. Ainsi, les genres bien faits sont véritablement de petites *familles*, c'est-à-dire de véritables portions de l'ordre même de la nature.

A.2.4   Cuvier: *Le Règne animal distribué d'après son organisation* (1817: 8–9).

Presque aucun être n'a de caractère simple, ou ne peut être reconnu seulement par un des traits de sa conformation ; il faut presque toujours la réunion de plusieurs de ces traits pour distinguer un être des êtres voisins qui en ont bien aussi quelques-uns, mais qui ne les ont pas tous, ou les ont combinés avec d'autres qui manquent au premier être ; et, plus les êtres que l'on a à distinguer sont nombreux, plus il faut accumuler de traits ; en sorte que, pour distinguer de tous les autres un être pris isolément, il faut faire entrer dans son caractère sa description complète. C'est pour éviter cet inconvénient que les divisions et subdivisions ont été inventées. L'on compare ensemble seulement un certain nombre d'êtres voisins, et leurs caractères n'ont besoin que d'exprimer leurs différences qui, par la supposition même, ne sont que la moindre partie de leur conformation. Une telle réunion s'appelle un *genre*.

....

REFERENCES

Aristote (1956) *Les parties des animaux.* Text established and translated by Pierre Louis. Paris: Société d'édition "Les Belles Lettres."

Barsanti, Giulio (2005) 'Linné et Buffon: deux visions différentes de la nature et de l'histoire naturelle', in *Les fondements de la botanique: Linné et la classification des plantes*, ed. Thierry Hoquet. Paris: Vuibert, 103–29.

Bonnet, Charles (1781) *Contemplation de la nature*. Neuchâtel: Samuel Fauche.

Buffon, Georges-Louis Leclerc de (1954) *Œuvres philosophiques*, ed. Jean Piveteau. Paris: Presses universitaires de France.

Buffon, Georges-Louis Leclerc de, and Daubenton, Louis Jean Marie (1749; 1766) *Histoire naturelle, générale et particulière, avec la description du cabinet du roi. Tome Premier; Tome 14e.* Paris: Imprimerie royale.

Cain, Arthur J. (1956) 'The genus in evolutionary taxonomy', *Systematic Zoology* 5: 97–109.

Cain, Arthur J. (1958) 'Logic and memory in Linnaeus's system of taxonomy', *Proceedings of the Linnaean Society of London* **169**: 144–163.

Carus, Victor (1880) *Histoire de la zoologie depuis l'antiquité jusqu'au XIXe siècle.* Paris: Baillière [French translation of the original German edition of 1872].

Cho, Dae-Ho (2003) *Ousia und Eidos in der Metaphysik und Biologie des Aristoteles.* Stuttgart: F. Steiner.

Cuvier, Georges (1817) *Le Règne animal distribué d'après son organisation, pour servir de base à l'Histoire naturelle des animaux, et d'introduction à l'Anatomie comparée.* Tome I. Paris: Deterville.

Cuvier, Georges (1834) *Animal kingdom: arranged according to its organization,* translated from the French and abridged for the use of students by H. McMurtrie. London: Orr and Smith.

Darwin, Charles (1851) *A monograph of the sub-class Cirripedia. Pars I.* London: The Ray Society.

Darwin, Charles (1859) *On the origin of species by means of natural selection, or the preservation of favoured races in the struggle for life.* London: John Murray.

Dayrat, Benoît (2005) 'Advantages of naming species under the *PhyloCode*: an example of how a new species of Discodorididae (Mollusca, Gastropoda, Euthyneura, Nudibranchia, Doridina) may be named', *Marine Biology Research* 1: 216–32.

De Cruz, Helen, and De Smedt, Johan (2007) 'The role of intuitive ontologies in scientific understanding—the case of human evolution', *Biology and Philosophy* 22: 351–68.

Delaunay, Paul (1962) *La zoologie au seizième siècle.* Paris: Hermann.

Dubois, Alain (1988) 'The genus in zoology: a contribution to the theory of evolutionary systematics', *Mémoires du Muséum national d'histoire naturelle, Série A, Zoologie* 140: 1–124.

Dubois, Alain (2004) 'Developmental pathway, hybridizability, and generic taxonomy in Amphibians', *Alytes* 22: 38–52.

Dupré, John (2001) 'In defence of classification', *Studies in History and Philosophy of Biology and Biomedical Sciences* 32: 203–19.

Farber, Paul Lawrence (1976) 'The type-concept in zoology during the first half of the nineteenth century', *Journal of the History of Biology* 9: 93–119.

Fox Keller, Evelyn (2006) 'Systems biology and the search for general laws', Seminar REHSEIS, Paris, December 4. Paris: REHSEIS.

Hennig, Willi (1966) *Phylogenetic systematics.* Urbana: University of Illinois Press.

Hoquet, Thierry (2003) 'La comparaison des espèces: ordre et méthode dans l'*Histoire naturelle* de Buffon', *Corpus* 43: 355–416.

Hoquet, Thierry (ed.) (2005) *Les fondements de la botanique: Linné et la classification des plantes.* Paris: Vuibert.

Hoquet, Thierry (2007), 'Buffon: from natural history to the history of nature?', *Biological Theory* 2: 413–19.

International Commission of Zoological Nomenclature (1999) *International code of zoological nomenclature.* Fourth edition. London: International Trust for Zoological Nomenclature.

Krell, Frank-Thorsten (2005) 'A Hennigian monument on vertebrate phylogeny', *Systematics and Biodiversity* 3: 339–41.

Lamarck, Jean-Baptiste de (1809) *Philosophie zoologique.* Paris: Dentu.

Latreille, Pierre-André (1797) *Précis des caractères génériques des insectes, disposés dans un ordre naturel.* Brive: Imprimerie de F. Bourdeaux, an 5 de la R[épublique].

Lherminier, Philippe, and Solignac, Michel (2005) *De l'espèce*. Paris: Syllepse.

Linnaeus [Linné, Carl von] (1742) *Genera plantarum eorumque characteres naturales secundum numerum, figuram, situm, & proportionem omnium fructificationis partium. Editio secunda aucta et emendata*. Lugduni Batavorum [Leiden]: apud Conradum Wishoff et Georg. Jac. Wishoff.

Linnaeus [Linné, Carl von] (1751) *Philosophia botanica in qua explicantur fundamenta botanica cum definitionibus partium, exemplis terminorum, observationibus rariorum*. Stockholmiae [Stockholm]: Godofr. Kiesewetter.

Linnaeus [Linné, Carl von] (1758) *Systema Naturae. Edition decima, reformata. Tomus I (Animalia)*. Holmiae [Stockholm]: Laurentius Salvius.

Magnin-Gonze, Joëlle (2004) *Histoire de la Botanique*. Paris: Delachaux & Niestlé.

Mayr, Ernst (1942) *Systematics and the origin of species from the viewpoint of a zoologist*. Cambidge (Mass.): Harvard University Press.

Mayr, Ernst (1963) *Animal species and evolution*. Cambridge (Mass.): Harvard University Press.

Mayr, Ernst (1969) *Principles of systematic zoology*. New York: McGraw Hill.

Mayr, Ernst (1986) 'La systématique évolutionniste et les quatre étapes du processus de classification', in *L'ordre et la diversité du vivant: quel statut scientifique pour les classifications biologiques?*, ed. Pascal Tassy. Paris: Fayard, 143–60.

Mayr, Ernst (1997) *This is biology: the science of the living world*. Cambridge (Mass.): Harvard University Press.

Mayr, Ernst (2001) *What evolution is*. Jackson (Tn.): BasicBooks.

Mayr, Ernst (2004) *What makes biology unique? Considerations on the autonomy of a scientific discipline*. Cambridge: Cambridge University Press.

Müller-Wille, Staffan (2007) 'Collection and collation: theory and practice of Linnaean botany', *Studies in History and Philosophy of Biological and Biomedical Sciences* 38: 541–62.

Müller-Wille, Staffan, and Reedes, Karen (2007) 'A translation of Carl Linnaeus's introduction to *Genera plantarum* (1737)', *Studies in History and Philosophy of Biological and Biomedical Sciences* 38: 563–72.

Pellegrin, Pierre (2007a) 'Introduction générale à l'*Organon*', in Aristotle *Catégories. Sur l'interprétation*. Paris: Flammarion, 7–49.

Pellegrin, Pierre (2007b) *Dictionnaire Aristote*. Paris: Ellipses.

Pinon, Laurent (1995) *Livres de Zoologie de la Renaissance: Une anthologie (1450–1700)*. Paris: Klincksieck.

Pinon, Laurent (2000) 'Les livres de zoologie de la Renaissance: objets de mémoire et instruments d'observation (1460–1605)', Thèse Université de Tours.

Ray, John (1693) *Synopsis methodica animalium quadrupedum et serpentini generis*. London: S. Smith and B. Walford.

Reinert, J.F., Harbach, R.E., and Kitching, I.J. (2004) 'Phylogeny and classification of Aedini (Diptera: Culicidae)', *Zoological Journal of the Linnaean Society* 142: 289–368.

Savage, Harry M. (2005) 'Classification of mosquitoes in tribe Aedini (Diptera: Culicidae): P araphylyphobia, and classification versus cladistic analysis', *Journal of Medical Entomology* 46: 923–7.

Simpson, George G. (1961) *Principles of animal taxonomy*. New York: Columbia University Press.

Stamos, David N. (2005) 'Pre-Darwinian taxonomy and essentialism: a reply to Mary Winsor', *Biology and Philosophy* 20: 79–96.

Tournefort, Joseph Piton de (1700) *Institutiones rei herbariæ. Editio altera, Gallicâ longè auctior*. Paris: Imprimerie royale.

Wildman, Derek E., Uddin, Monica, Liu, Guozhen, Grossman, Lawrence I., and Goodman, Morris (2003) 'Implications of natural selection in shaping 99.4% nonsynonymous DNA identity between humans and chimpanzees: enlarging genus *Homo*', *Proceedings of the National Academy of Sciences of the U.S.A.* **100**: 7181–8.

Winsor, Mary P. (2001) 'Cain on Linnaeus: the scientist-historian as unanalysed entity', *Studies in History and Philosophy of Biological and Biomedical Sciences* **32**: 239–54.

Wood, Bernard, and Constantino, Paul (2004) 'Human origins: life at the top of the tree', in *Assembing the tree of life*, ed. Joel Cracraft and Michael J. Donoghue. Oxford: Oxford University Press, 517–35.

9

Homology: an expression of generality in the life sciences

STÉPHANE SCHMITT

9.1 Introduction

The concept of homology, in its current biological sense, is used to refer to two or more organs, parts, or molecules (proteins, genes ...) that belong to different species and yet show so much similarity in their structure, regardless of their function, that they are interpreted as coming from the same organ, part, or molecule in a common ancestor. One example is the arm of man, when compared with the forelimb of other vertebrates (forefoot of horse, wing of bird or bat, pectoral fin of fish ...): the general morphology (number, mutual disposition of bones, muscles, etc.) is the same, but there are differences in the relative size that explain the differences of function, so that one says that all these limbs are homologous. In turn, similarity of function, regardless of morphology and evolutionary origin, corresponds to the notion of "analogy": for example, an insect wing is analogous (but not homologous) to a bird wing. Today, even though evolutionary biologists (particularly cladists) attempt to replace these concepts by more precise notions, such as synapomorphy or symplesiomorphy, homology and analogy are still commonly used in most fields of life sciences. When a molecular biologist is working on mouse or *Drosophila* genes in order to understand a biological phenomenon in the human being, he or she is using (even if implicitly) the concept of homology at the level of the gene.

   Given the current definition of homology, it seems to be closely linked to evolutionary theory. However, its history began long before, and it gradually emerged in contexts where knowledge or understanding of evolutionary theory was marginal (before 1850), or did not exist at all, in the modern sense (before 1750).[1] From antiquity, there existed a more

---

[1] On the history of comparative anatomy, see, in particular Russell (1916), Cole (1944), and Schmitt (2006).

The Oxford Handbook of Generality in Mathematics and the Sciences. First Edition. Karine Chemla, Renaud Chorlay and David Rabouin. © Oxford University Press 2016. Publishing in 2016 by Oxford University Press.

or less vague concept of homology (even if it was not so named), in order to designate relations between structures in different species bearing similarities that were not (or not only) functional. Throughout its long history, it expressed a quest for generality in the understanding of animal anatomy by attempting to interpret a diversity of forms as result-ing from modifications of a single "primitive" structure. But the meaning of this quest, and the practices associated with it, deeply changed with the different theoretical context of the life sciences. It was, thus, an element of continuity in the history of biology, and a central actor in the emergence of some developments, particularly evolutionary theory.

9.2 Homology in pre-transformist comparative anatomy

Homology is closely connected with the concept of "unity of plan," namely, the idea that a general structure would be common to all animals, or, at least, to large groups such as quadrupeds, and of which the individual animals or species would be variations. Homology concerns single organs or structures, unity of plan entire organisms. Both ideas are very old. For example, we can find some passages in Aristotle's biological treatises where the philosopher distinguishes between organs with a same function and of a same nature (we would say today "analogous" and "homologous," in a wide sense), and those with a same function but of a different nature (i.e., "analogous" but not "homologous").[2] More generally, during late antiquity and the Middle Ages, it was often accepted that man and animals (at least higher animals such as quadrupeds) shared the same plan of organization and had corresponding organs.

Generality in human/animal structures was, thus, merely stated, with no further charac-terization. It was sometimes implicitly admitted: this enabled physicians to extrapolate to human beings anatomical knowledge they had gotten from animals. It could also be used in a symbolic way: for example, in a Christian context, anatomical similarities were interpreted as resulting from the unity of God's design and used as a rhetorical commonplace.[3]

These approaches to homology were common up to the sixteenth century. In the Renaissance, new considerations appeared. In 1555, French naturalist Pierre Belon (1517–1564) published a treatise on birds in which he represented, side by side, a human skeleton and a bird skeleton, with the same letters indicating corresponding bones (see Fig. 9.1).

This plate was often considered the starting point of comparative anatomy, not only because it emphasizes a general similarity between human and avian skeletons by giving to the latter an unnatural, more human attitude, but because it shows a precise relation between single, precise parts of them (Crié, 1882). This opinion was, however, chal-lenged by Delaunay (1962)[4] and Foucault (1966). Foucault thought that the status of this

---

[2] See especially Aristotle (1964–9), *Historia animalium*: 486a–487a.

[3] See, for example, Lactantius (1974), *De opificio Dei*, chap. 5–7.

[4] According to Delaunay (1962: 189), Belon went not really farther than Aristotle in the comparative method.

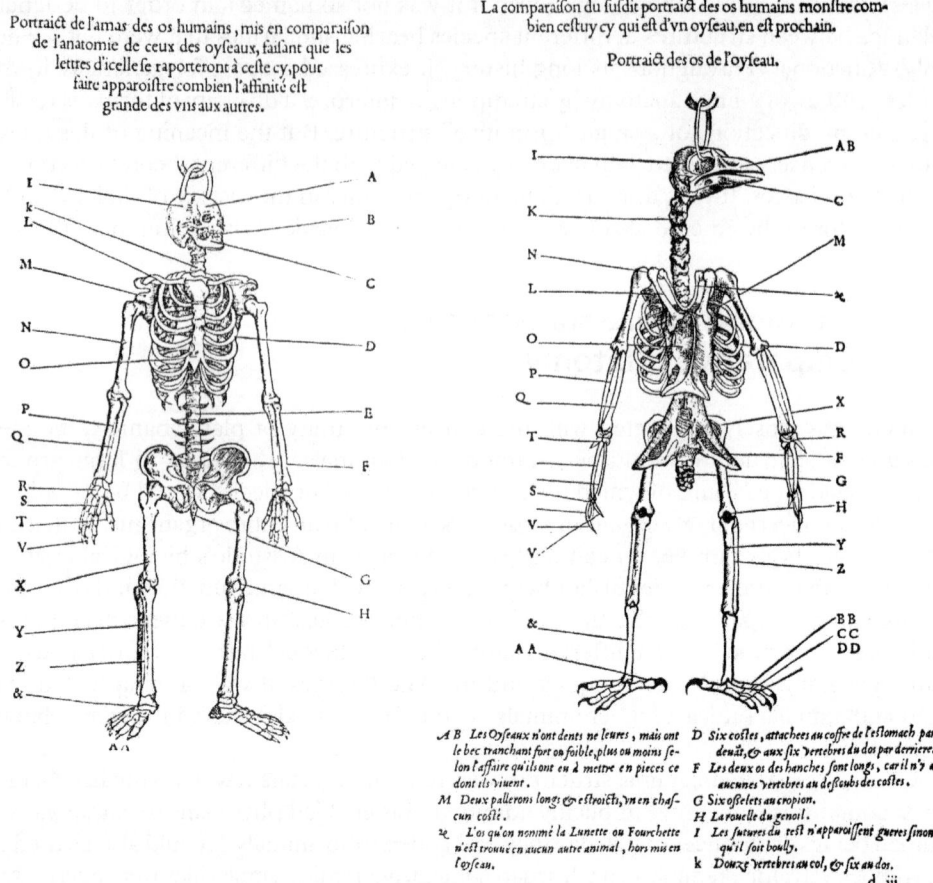

Portraict de l'amas des os humains , mis en comparaison de l'anatomie de ceux des oyseaux, faisant que les lettres d'icelle se raporteront à ceste cy, pour faire apparoistre combien l'affinité est grande des vns aux autres.

La comparaison du susdit portraict des os humains monstre combien cestuy cy qui est d'vn oyseau, en est prochain.

Portraict des os de l'oyseau.

A B Les Oyseaux n'ont dents ne leures , mais ont le bec tranchant fort ou foible, plus ou moins selon l'affaire qu'ils ont en à mettre en pieces ce dont ils viuent .
M Deux palerons longs & estroicts, vn en chascun costé .
ʒ L'os qu'on nomme la Lunette ou Fourchette n'est trouué en aucun autre animal , hors mis en l'oyseau.

D Six costes , attachees au coffre de l'estomach par deuât, & aux six vertebres du dos par derriere.
F Les deux os des hanches sont longs , car il n'y a aucunes vertebres au dessoubs des costes .
G Six osselets au cropion.
H La rouelle du genoil .
I Les sutures du test n'apparoissent gueres sinon qu'il soit boully.
k Douzes vertebres au col, & six au dos.

d iii

Figure 9.1 *Pierre Belon's representation of a human, compared with a bird skeleton (Belon, 1555: 40–1). The legend is: "Representation of the set of human bones, compared to the anatomy of those of birds, so that the letters of the latter are related to those of the former, in order to show how great the similarity between them is." (left); "The comparison of the preceding representation of human bones with that of a bird, which follows, demonstrates how similar they are."*

comparison in Belon's work was the same as the status of the relations between plants, minerals, parts of the human body, etc., in traditional alchemy and astrology (Foucault, 1966: 38). Certainly, these considerations are isolated in Belon's books, and if the correspondences between bones are rather precisely established, he does not try to invent a concept or to coin a word to designate this kind of anatomical relationship. But it cannot be denied that Belon introduced, at least implicitly, a new program concerning anatomical generality, consisting of finding exact relationships between structures of different species. In turn, this idea of a precise correspondence between structures enabled one to

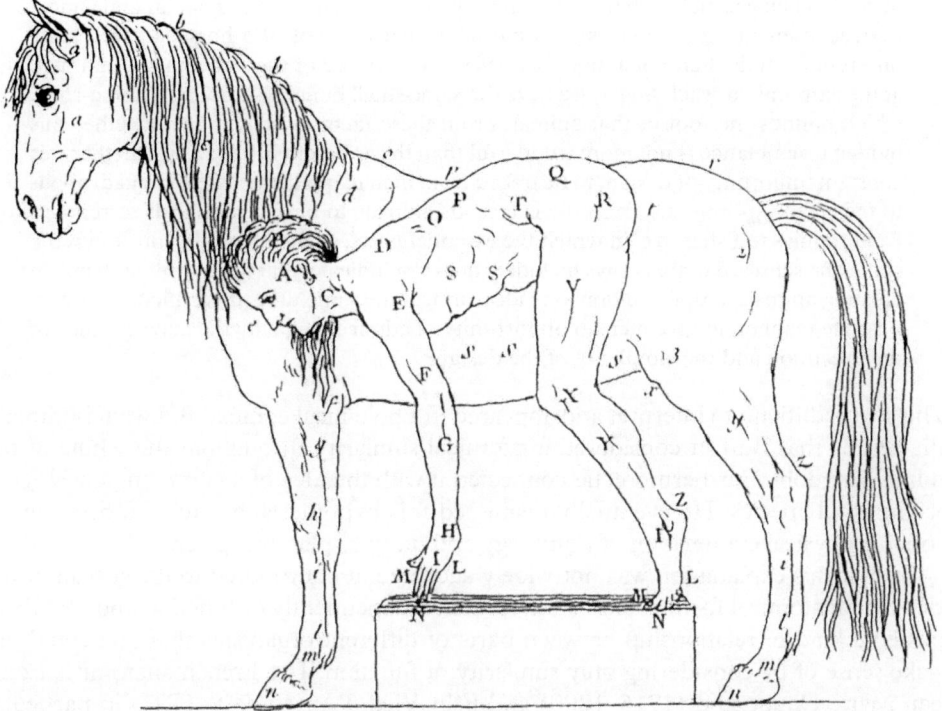

Figure 9.2 *Comparison of man and horse according to De Garsault (1741: plate II, fig. B).*

conduct anatomical observations on animals and man since, if a structure was found in one species, it was reasonable to research it in other species.

Such considerations became rather common, especially from the late seventeenth century (see Fig. 9.2). But a deeper investigation about the meaning, or the causes, of similarity between organs or organisms began emerging during the eighteenth century. We find an example of this exploration in Buffon's text on the ass:

> If, from the immense number of animated beings which people the universe, we select a single animal, or even the human body, as a standard, and compare all other organized beings with it, we shall find that each enjoys an independent existence, and that the whole are distinguished by an almost infinite variety of gradations. There exists, at the same time, a primitive and general design, which may be traced to a great distance, and whose degradations are still slower than those of figure or other external relations: For, not to mention the organs of digestion, of circulation, or of generation, without which animals could neither subsist nor reproduce, there is, even among the parts that contribute most to variety in external form, such an amazing resemblance as necessarily conveys the idea of an original plan upon which the whole has been conceived and executed. When, for example, the parts constituting the body of a horse, which seems to differ so widely from that of man, are compared in detail with the human frame,

instead of being struck with the difference, we are astonished at the singular and almost perfect resemblance ... Let us next consider, that the foot of a horse, so seemingly different from the hand of a man, is, however, composed of the same bones, and that, at the extremity of each finger, we have the same small bone, resembling a horse-shoe, which bounds the foot of that animal. From these facts we may judge, whether this hidden resemblance is not more wonderful than the apparent differences; whether this constant uniformity of design, to be traced from men to quadrupeds, from quadrupeds to the cetaceous animals, from the cetaceous animals to birds, from birds to reptiles, from reptiles to fishes, &c. in which the essential parts, as the heart, the intestines, the spine, the senses, &c. are always included, does not indicate, that the Supreme Being, in creating animals, employed only one idea, and, at the same time, diversified it in every possible manner, to give men an opportunity of admiring equally the magnificence of the execution and the simplicity of the design?[5]

This text is difficult to interpret and appeared in a polemical context. But what is important here is that Buffon considered anatomical similarity throughout the whole of the animal kingdom. Furthermore, he connected it with the idea of a common genealogical origin of all species. He eventually dismissed this hypothesis, but for the first time, a possible physical explanation of homology and unity of plan was given.

Even if this explanation was not widely accepted, it contributed to the spread of the idea that anatomical form could be interpreted independently of function and that there existed a kind of relationship between parts of different organisms that one could not make sense of by considering only similarity of function. The French anatomists Louis Jean Marie Daubenton (1716–1800) and Félix Vicq d'Azyr (1748–1794) in particular developed this theme, which became central in the life sciences around 1800. At that time, anatomists and naturalists, particularly in France and Germany, tried to clarify the definition of such morphological relations and to establish criteria enabling others to recognize them.

The term "homology" first appeared in life sciences at that time, but the terminology was not fixed until the mid-nineteenth century. The concept, however, was commonly used from the first decades of the century. Étienne Geoffroy Saint-Hilaire (1772–1844) was one of its most famous supporters. He was convinced that all animals were built according to a single plan (*théorie des analogues*) and, as a consequence, that it was possible to find equivalents of every organ of a species in all other species. More importantly, he tried to establish a clear criterion for recognizing such correspondence between species and introduced the principle of connections: for him, an organ in one species was homologous to one in another species if it had similar connections with neighboring organs known to be homologous. Thus, even if two organs had completely different functions and if their forms were not similar at first sight, it was possible to demonstrate their homology (i.e., the essential similarity of their forms). This criterion led to the realization of the program of research already suggested by Belon nearly three centuries earlier. There was, then, a positive approach of generality in comparative anatomy.

[5] Buffon (1753, vol. 4: 379–81; transl. Smellie, 1781, vol. 3: 399–401).

An important aspect for Geoffroy Saint-Hilaire was that unity of plan concerned the whole animal kingdom, and not only a single group such as vertebrates. For example, according to him, insects and other arthropods shared the organization of vertebrates, apart from the fact that they lived inside their vertebral column and that the mutual position of nervous system and digestive tube was inverted. In 1830, he tried to demonstrate that the organization of mollusks was essentially the same as that of vertebrates and insects, but his rival, Georges Cuvier (1769–1832), was strongly opposed to these views and attacked them energetically.

Cuvier had a very different view of animal anatomy. The key concept, in his thought, was the organism. According to him, an animal was a whole, each part being closely connected to the others in a functional way, to the advantage of the whole. Each zoological group was thus characterized by a specific plan of organization that worked perfectly and could not be related to another one. He acknowledged four such groups (he named them *embranchements*): vertebrates, articulates (i.e., arthropods and annelids), radiates, and mollusks and, thus, could not accept Geoffroy's attempts to unify at least three of them (vertebrates, articulates, and mollusks).

In a sense, the Geoffroy–Cuvier debate was about how to conceive of generality in animal anatomy, and also how much scope to give it. Geoffroy considered primarily morphological similarities and thought it was possible to find a unique plan for the whole animal kingdom, whereas Cuvier favored function and, thus, thought that anatomical similarity existed within a given embranchment, but not between different embranchments.

This debate created a considerable "media sensation" (Appel, 1987). It is important to emphasize that the central point was not transformism. Geoffroy Saint-Hilaire himself never tried to explain the unity of organization of *all* animals by a common genealogical origin. Certainly, elaborate evolutionary theories did already exist at that time (cf. Lamarck, etc.), but anatomical thought mostly developed itself in a fixist context, or, at least, with no direct reference to the possibility of species transformation. Therefore, the concept of homology was not directly connected to the idea of evolution. It was mostly interpreted as the expression of a general "type" (or "archetype") whose exact nature varied according to the different authors.

The English anatomist and paleontologist Richard Owen (1804–1892) tried to make this notion precise. According to him, the vertebrate "archetype" was a Platonic idea, of which the real species were modified projections, and he represented its skeleton on a plate, next to the skeletons of each class of vertebrate.

Owen also contributed to the attempt to fix the fluctuating terminology and gave the following definition for the adjective "homologous" (or "homologue") in 1843: "Homologue ... The same organ in different animals under every variety of form and functions."[6] This definition gradually spread. It is very close to the current definition, except that it was not connected in any way with any evolutionary theory. The debates on anatomical generality, up to the second half of the nineteenth century, took place at an ideal, not physical or historical level.

---

[6] Owen (1843: 374, 379). See also Owen (1848).

9.3 Homology and the evolutionary theory: a physical interpretation of generality

When Charles Darwin published the first edition of *On the origin of species* in 1859, transformation of species was not a new idea (in contrast to the idea of natural selection), but this publication contributed to a large-scale dispersion of this theory in the life sciences. In relatively few years, most biologists accepted it and re-interpreted the data in their own field in the light of transformism. Darwin himself was aware of the advantages his theory could have for anatomical analysis and the concept of homology: he devoted one chapter to these questions in his book. He claimed that transformism could offer a satisfactory theory for all the homologies anatomists had found in the preceding decades, which could then be explained, not ideally, but physically, by a common genealogical origin of the considered organs or parts: different species had similar organs since they had common ancestors. Conversely, all these structural similarities were evidence in favor of evolution.

Thus, all concepts connected to structural generality in animals that had been introduced in a fixist context could be re-interpreted in the new evolutionary frame. Homology was no longer an ideal relation, but a historical one. In 1870, Edwin Ray Lankester (1847–1929), a disciple of Darwin, gave a new definition of this concept: "Without doubt the majority of evolutionists would agree that by asserting an organ A in an animal α to be homologous with an organ B in an animal β, they mean that in some common ancestor κ the organs A and B were represented by an organ C, and that α and β have inherited their organs A and B from κ."[7] In the same vein, Owen's archetype could be considered as a common ancestor: it was no longer a Platonic idea, but acquired a physical reality. Indeed, in spite of the considerable conceptual change caused by evolutionary theory regarding the interpretation of generality among living beings, many concepts could paradoxically be re-used.

This continuity is particularly remarkable if we consider the anatomists' practices and methods of investigation. Before and after 1859, they were looking for homologies between different species, and for that they used the same methods (i.e., anatomical observations on adult or embryonic forms) and the same criteria (e.g., principle of connections). The particular questions (e.g., "Are vertebrates' eyes homologous to arthropods' eyes?," "Is there a homology between this bone of a fish and that bone of a mammal?," etc.) did not change. Surely, their underlying meaning was not the same, since ideal homology was replaced by genealogical relations, and the aim of research program in comparative anatomy was, then, to understand (or reconstruct) the history of species; but the realization of this program, that is, the daily work of anatomists, was not altered by this change. This was probably an important factor in the relative quickness of the shift from idealistic to transformist morphology in the 1860s,[8] as if the most important thing was the quest for generality, whatever meaning this generality could have.

---

[7] Lankester (1870: 36).
[8] See Coleman (1976), Bowler (1989), Richards (1992), and Schmitt (2004: part 2).

Deeper transformations in life sciences occurred only some decades later, in particular with the generalization of the experimental approach in such fields as embryology, and with the emergence of new biological disciplines such as genetics. But these new fields were also able to integrate the concept of homology, as we can see in the case of genetics.

9.4   The new meanings of homology in genetics

Genetics emerged as an autonomous field in the first decades of the twentieth century. Its position regarding homology was, at first, rather negative, and geneticists rejected this concept inherited from morphology and often considered obsolete and unclear. A good example of this attitude is given by the Scottish geneticist Francis Albert Eley Crew (1886–1973), who wrote in 1925:

> The feather of the bird is not necessarily the modified scale of a reptile but may be a distinctly different characterization based upon an entirely different genotype. A certain genotype results in a certain characterization—scales—; mutation—alteration—in this genotype results in a new genotype and thus leads to another characterization. The old genotype is transformed into the new but the old characterization is not transformed, it disappears and is replaced. Scales and feathers are not homologous structures—homology attempts to establish a similarity in origin and nature of structures seemingly different and is based on the assumption that during the course of evolution structures have undergone transformation yet remain fundamentally the same. In fact this conception of homologous structures cannot be accommodated by the chromosome hypothesis until it can be experimentally demonstrated that the genes themselves can pass through a process of gradual modification.[9]

Thus, what was important for Crew was the understanding of the proximal, genetic causes leading to the formation of an organ. For him, chromosomes (and, then, genes) were (discontinuously) modified in the course of evolution, so that it was meaningless to compare an organ in one species (controlled by certain genes) with another organ in another species (controlled by different genes): a feather was only a feather, and not a modified scale. Generality could not be conceived in absence of immediate causality and material continuity.

The geneticist point of view was also expressed by James Edwin Duerden (1869–1937):

> Following the older methods of comparative morphology, we can maintain that the relationship of scale and feather affords evidence that in the course of evolution the reptilian—scale has grown upwards into a filament, and by a complicated system of incisions of the epidermis, due to ingrowths of the dermis, the filament has frayed out, and given rise to the many structural divisions of the feather—shaft, barbs, and barbules. This is the view that has hitherto been largely accepted ... But now, says the Mendelian, a feather is a new structure; it is *sui generis*: it is an epidermal mutation, the result of a

---

[9] Crew (1925: 152). The genotype is the set of genes of an organism.

separate germinal change; its origin is quite apart from that of scales ... The delight of the morphologist in the tracing of a corresponding structure through all its many transitional stages, from its one extreme to the other, is but a delusion if a genetic relationship is the underlying idea (Duerden, 1923–4, quoted in Boyden, 1935: 449).

Homology as it was conceived by evolutionary comparative anatomy, thus, seemed to be very far from the scope and approaches of the first geneticists. But some of them soon found different ways to re-use this concept by adapting it to the new field. As early as 1920, Alexander Weinstein introduced the notion of "homologous genes" to designate genes of different species for which mutations had about the same effect and with similar position on the chromosomes (Weinstein, 1920). This definition was similar to the definition of homology by anatomists, but at the molecular level. Homology of genes, just as homology of organs, expressed the genes essential similarity (resulting from a common evolutionary origin). Similar criteria could be used to identify homologies between organs or between genes, since Weinstein considered the relation of each gene with its neighbors, just as Geoffroy Saint-Hilaire had introduced the principle of connections. The quest for biological generality was thus able to shift its conceptual tools from the macroscopic to the microscopic scale.

But genetics was also able to acknowledge homology on the scale of organs using genes as a new kind of criteria. Some authors began, in the 1930s, to try to find evidence for homologies by analyzing the genes that are involved during the development of the different organs. In 1935, Alan Boyden suggested that, since homology results from common descent, it was a "genetic phenomenon." Therefore, according to him, "knowledge of gene action should therefore be able to illuminate and explain homology, and might even provide more precise criteria by which it might be recognized." Since "the structure of any particular organ is determined by the interaction of many if not by all of the genes presents in the particular individual ..., **however** discontinuous the changes in individual genes may be the fact is that through the interaction of many genes the organ is still essentially and fundamentally similar," so that "the determination of taxonomic relationships and of homology could then be based on methods of testing identity of genes" (Boyden, 1935: 448–51).

Homology at the microscopic (genetic) level was thus connected to homology at the macroscopic (anatomical) level. This link was not only a theoretical one, but it suggested new experimental approaches to find homologies and, then, new programs of research. This view was challenged in 1938 by Gavin De Beer (1899–1972), who stressed that "the interesting paradox remains that, while continuity of homologous structures implies affinity between organisms in phylogeny, it does not necessarily imply similarity of genetic factor or of ontogenetic processes in the production of homologous structures" (De Beer, 1938: 71). But Boyden's idea remained, and it is behind many studies in evolutionary developmental biology today, a field of which a major aim is to find phylogenetic relations between structures of different species by analyzing and comparing gene expression in these species.

Homology thus represented an important factor of continuity in life sciences during the last 250 years and succeeded in being redefined and reused in different theoretical contexts, in different disciplines and at different levels, in spite of the important changes undergone by biology (emergence of new theories such as evolution, of new fields such as genetics ...). The reason for this is probably that it is one of the major concepts that fulfil the quest for unifying principles and generality in this field.

··

REFERENCES

Appel, T. A. (1987) *The Cuvier–Geoffroy debate. French biology in the decades before Darwin.* New York: Oxford University Press.

Aristotle (1964–9) *Histoire des animaux*, translated and edited by Pierre Louis. Paris: Belles Lettres, 3 vols.

Belon, P. (1555) *Histoire de la nature des oyseaux, avec leurs descriptions et naïfs portraicts retirez du naturel.* Paris: Cavellat.

Bowler, P. J. (1989) *The non-Darwinian revolution: reinterpreting a historical myth.* Baltimore: John Hopkins University Press.

Boyden, A. (1935) 'Genetics and homology', *Quarterly Review of Biology* 10: 448–51.

Buffon, G.-L. (1753) *Histoire naturelle, générale et particulière, avec la description du Cabinet du Roi. Tome quatrième.* Paris, Imprimerie royale; transl. William Smellie (1781) *Natural history, general and particular.* Edinburgh: William Creech.

Cole, F. J. (1944) *A history of comparative anatomy from Aristotle to the eighteenth century.* London: Macmillan.

Coleman, W. (1976) 'Morphology between type concept and descent theory', *Journal of the History of Medicine* 31: 149–75.

Crew, F. A. (1925) *Animal genetics. An introduction to the science of animal breeding.* Edinburgh: Oliver and Boyd.

Crié, L. (1882) 'P. Belon du Mans et l'anatomie comparée', *Revue scientifique*, 3ᵉ série, 3ᵉ année, 16: 481–5.

De Beer, G. (1938) 'Embryology and evolution', in *Evolution. Essays on aspects of evolutionary biology, presented to Professor E. S. Goodrich on his Seventieth Birthday*, ed. De Beer, G. Oxford: Clarendon Press, 57–78.

De Garsault, F.-A. (1741) *Le Nouveau parfait Maréchal, ou connaissance générale et universelle du cheval.* Paris: Despilly.

Delaunay, P. (1962) *La Zoologie au seizième siècle.* Paris: Hermann.

Duerden, J. E. (1923–24) 'Methods of evolution', *Science Progress* 18: 556–64.

Foucault, M. (1966) *Les mots et les choses. Une archéologie des sciences humaines.* Paris: Gallimard.

Lactantius (1974) *L'ouvrage du Dieu Créateur*, French transl. by Michel Perrin. Paris: Cerf, 2 vols.

Lankester, E. R. (1870) 'On the use of the term Homology in modern Zoology, and the distinction between Homogenetic and Homoplastic agreements', *The Annals and Magazine of Natural History* 6: 34–43.

Owen, R. (1843) *Lectures on the comparative anatomy and physiology of the vertebrate animals, delivered at the Royal College of Surgeons of England, in 1843.* London: Longman, Brown, Green, and Longmans.

Owen, R. (1848) *On the archetype and homologies of the vertebrate skeleton.* London: John van Voorst.

Richards, R. (1992) *The meaning of evolution: the morphological construction and ideological reconstruction of Darwin's Theory.* Chicago: University of Chicago Press.

Russell, E. S. (1916) *Form and function. A contribution to the history of animal morphology.* London: Murray.

Schmitt, S. (2004) *Les Parties répétées. Histoire d'une question anatomique.* Paris: Éditions du Muséum National d'Histoire Naturelle.

Schmitt, S. (2006) *Aux origines de la biologie moderne. L'anatomie comparée d'Aristote à la théorie de l'évolution.* Paris: Belin.

Weinstein, A. (1920) 'Homologous genes and linear linkage in *Drosophila viridis*', *Proceedings of the National Academy of Sciences* 6: 625–39.

Section II.3

Circulation between epistemological cultures

10

The role of *genericity* in the history of dynamical systems theory

TATIANA ROQUE

10.1 Introduction

In mathematics, general descriptions of mathematical beings in a given universe often engage a tension between the generality of the description and the relevance of the properties chosen to be described. In the case under study, mathematicians' aim, in their attempt to classify or, better, to provide a general description of dynamical systems, was to find a compromise between the two poles: generality and relevance. The notion of *genericity*, along with its different mathematical definitions, plays a fundamental role in the history of dynamical systems theory.

In his memoir "Sur les courbes définies par une équation différentielle," published in four parts, from 1881 to 1886, and known as the starting point of the study of dynamical systems, Poincaré analyzed the behavior of curves that are solutions for certain types of differential equations.[1] He succeeded in classifying them by focusing on singular points, described the trajectories' behavior in important particular cases and provided new methods that proved to be extremely useful. Some historians have already analyzed these works: Christian Gilain in the context of research on differential equations in Poincaré's time,[2] Anne Robadey from the perspective of generality in a different sense from the one exposed here,[3] and myself, in relation with stability issues.[4]

Despite its fundamental importance, this study was far from describing the classes of differential equations themselves. Over 20 years after this first breakthrough, during

---

[1] Poincaré (1881, 1882, 1885, 1886).
[2] Gilain (1977).
[3] Robadey (2006).
[4] Roque (2011).

The Oxford Handbook of Generality in Mathematics and the Sciences. First Edition. Karine Chemla, Renaud Chorlay and David Rabouin. © Oxford University Press 2016. Publishing in 2016 by Oxford University Press.

the 1908 International Congress of Mathematicians in Rome, Poincaré announced his prospects for future research on differential equations as follows:

> Much has already been done for linear differential equations, and one only needs to perfect what was started. However, concerning nonlinear differential equations, developments have been too modest. The hope of integrating with known functions was forsaken long ago; we therefore need to study in and of themselves the functions defined by differential equations, starting with an attempt to systematically classify them. A study of the growth mode in the vicinity of singular points will most probably provide the first elements for such a classification, but we will not be satisfied until some group of transformations has been found (such as the Cremona transformations), playing—vis-à-vis the differential equations—the same role as that played by the group of bi-rational transformations in the case of the algebraic curves. Then, we would be able to place in a single class all the equations derived from one of them by transformations. We would have the analogy with an existing theory to guide us: that of bi-rational transformations and the genus of an algebraic curve.[5]

The above quotation clearly demonstrates that Poincaré's wish was that this classification could have the same general character as that of the algebraic curves classified according to their genus. We need not go into the details of the analogy that Poincaré points to in this quote, but we ought to underline three elements that will play a part in this chapter.

First, Poincaré points to a shift in terms of objects and investigative means. In elementary cases, one is accustomed to solving a differential equation by exhibiting a formula built from elementary functions (such as exponentiation or arctangent). Usually, a single formula describes a family of solutions, since it contains arbitrary constants. However, in all but the most elementary cases, this family fails to encompass all solutions: "singular" solutions are, more often than not, left out. In more complex situations, the functions that are solutions of a given differential equation usually cannot be expressed by simple formulas, hence the very meaning of what it means to solve a differential equation has to be altered. In his early work on the qualitative study of differential equations (1881–6), Poincaré showed that a new viewpoint allowed for relevant information to be gained even when explicit formulas are beyond reach. This new and qualitative viewpoint relies on two significant shifts. First, Poincaré chose to focus on the *curves* that are defined by the solutions, hence a shift from the analytic to the geometric. Second, the theory does not deal with individual solutions (seen as functions or as curves), but with the system of *all* solutions. In particular, singular solutions must be taken into account from the beginning since they provide the first elementary pieces of information on the basis of which the set of solutions can be investigated as a geometric object.

Second, when it comes to studying the sets of solutions to all differential equations of a given type, some classification principle has to be devised. One needs to find a way of capturing technically the idea that two different sets of solutions are equivalent from the qualitative viewpoint. If that could be achieved, all equivalent sets of solutions could

---

[5] Poincaré (1908: 177).

be "sorted into a single class," leaving two tasks for the mathematician: identifying the relevant geometric properties that tell to which class a given differential equation belongs; and enumerating the different classes thus obtained. To serve this purpose, Poincaré suggests that groups should be brought into the picture: two differential equations (or two sets of solution curves) would be considered as equivalent if and only if they could be transformed one into the other by a transformation from a given group. In this framework, finding the right equivalence relation boils down to finding the right transformation group. More often than not, there are several possible groups to choose from, and hence several (possible) equivalence relations: the larger the group, the coarser the equivalence relation and the cruder the classification.[6] This is one of the aspects for which the analogy with the theory of algebraic curves is enlightening. A fine-grained classification of algebraic curves relies on the group of algebraic isomorphisms. Using the larger group of birational transformations entails a loss of information (curves which are algebraically inequivalent can be birationally equivalent), but this is the price to be paid for a convenient classification: algebraic curves (more precisely smooth, complex, projective algebraic curves) are birationally equivalent if and only if they have equal genus; the genus being a numerical invariant which can easily be computed, the coarser classification proves tractable. This coarse classification can be seen as a first step on the way to the finer classification.

Third, in this quotation, Poincaré presents a rather loose analogy between the theory of algebraic curves and that of systems of curves defined by differential equations. He merely suggests that a very general strategy, which had been successfully used in the case of algebraic curves, should be used in another theory. In this chapter, we will see several cases in which proof strategies *and concepts* were introduced in a given theoretical context by mathematicians who explicitly relied on an analogy with another theoretical context. This phenomenon is fundamental for the understanding of the relationship between the various concepts of *genericity* that we will encounter.

In this study, we will focus on the second half of the twentieth century, hence on a much later phase of the development of the qualitative study of differential equations. The theory would be re-christened the "theory of dynamical systems." Sets of solutions would be tackled in the language of function sets, or function spaces. The progressive discovery of the complexity of systems of solutions would lead to several classification attempts, based on different groups; on different notions of equivalence or "sameness" or "proximity;"

---

[6] For the reader who is not familiar with these ideas, a geometric example can be given (in the spirit of F. Klein's Erlangen program). In elementary plane geometry, one can consider three different groups, the group of translations (T), the group of direct isometries (I), and the group of similitudes (S). The second one is larger than the first one, since it also contains rotations; S is larger than I since it also contains homotheties (scalings). If only T is taken into account, plane geometric objects fall into a multitude of inequivalent classes; for instance, not all (infinite, straight) lines are equivalent, two lines being equivalent (up to a translation) if and only if they are parallel. If I is taken into account, all lines become equivalent and make up a single class; parallelism stops making sense. Yet I is enough to classify more complex geometric objects: for instance, two circles are equivalent relative to I if and only if they have the same radius (the "radius" thus forming a complete set of invariants). Moving on to S, all circles become equivalent and the notion of radius becomes meaningless. In Euclidean geometry (school variety), S is the standard group: the theorems proven for circles of radius 1 hold for circles of radius 2 all the same, since in the wording of theorems only ratios of lengths are referred to.

on different "suggesting sciences" (differential topology, then measure theory). All these elements fit the general scheme outlined by Poincaré in the above quotation.

However, one fundamental element is not related to this description. As we shall see, mathematicians would soon set themselves the more tractable task of classifying "almost all" solutions. To implement this strategy, two concepts must be defined: that which characterizes the subset of solutions whose description is tractable; and that which gives a technical meaning to "almost all." Our historical narrative will document the use of several pairs of concepts. From a more epistemological viewpoint, the core issue is that of the *interdependence* of the two concepts, and of the *compromises* that must be made: the subset picked out by the first concept must be small enough to exclude cases which are beyond our descriptive capacities, but large enough to be considered "generic," that is, to satisfy the "almost all" requirement. This dialectic can be thought of as a *productive tension* in the development of the theory of dynamical systems; a tension which the historical development of the concept of *genericity* helps one to understand.

10.2   The influence of singularity theory on the first definitions of genericity

At first, the qualitative study of differential equations Poincaré proposed was explored by few mathematicians,[7] the best known examples being Hadamard and Birkhoff. Dynamical systems, as mathematical objects, were defined in 1927 by Birkhoff,[8] and his definition already implied a qualitative view.[9] Nonetheless, it was only in the fifties that the theory gained a new impulse, mainly with the classification efforts which will be analyzed throughout this chapter.

The establishment of dynamical systems as a theory is related to classification goals, in close relationship with what was to be later called singularity theory of differentiable mappings. One of the major aims of the theory was to classify mappings by the type of their singularities. The classification of algebraic curves, mentioned by Poincaré, succeeds in explaining the type of *all* such curves by their genera. In singularity theory, mathematicians also tried to classify *all* objects of a given universe, for instance mappings from R^n to R. However, they only obtained a classification of *almost all* objects of this universe. In dynamical systems theory, as the curves have much more complex behaviors, the maximum one could expect to obtain is a classification of *almost all* objects of a given universe. So, prior to classification, we must know if the classifiable objects are *almost all* objects in a universe.

---

[7] This means that few mathematicians explored dynamical systems with the general perspective discussed here. Roque (2015) shows that a number of astronomers and mathematicians of Poincaré's time employed some of the tools he proposed, as periodic solutions.

[8] Birkhoff (1927).

[9] For a discussion about qualitative implications in the definition of a dynamical system, see Roque (2007) and Roque (2011).

The aim of that which would later be called singularity theory was to obtain a picture of the kind of singularities that an arbitrary function from R^n to R^p can have. Since it is impossible to characterize all functions in this way, the strategy used was to determine the kind of singularities a certain prototype function can have for each universe (obtained for fixed values of n and p), and then show that any function in such a universe can be approximated by the prototype. This means that we classify the prototypes and show that any being in the universe can be approximated by one prototype. It was René Thom who expressed this last property of prototypes in terms of genericity. In the search for generic properties, the question is to establish that *almost all* (and not necessarily *all*) beings are classifiable.

Beginning in 1925,[10] Marston Morse studied the singularities of functions from a manifold $M \subset R^n$ to R. He found that any function of this type can be approximated by a function having only isolated singular points.[11] A smooth real-valued function on a manifold M was later called a "Morse function" if it had no degenerate critical points[12]. Nowadays, the most important result of Morse theory says that almost all functions (in C^2 topology) are Morse functions. Besides, as Morse functions have only one type of singularity, which may be easily characterized, it is possible to obtain a satisfactory description of almost all functions in the universe of real-valued functions.

In a paper published in 1955, on the singularities of mappings from R^2 to R^2, Hassler Whitney defined an "excellent" mapping as being a mapping for which all singular points are "folds" (analogous to singularities of Morse functions) and "cusps," which are well characterized types of singularity.[13] Then, he showed that, arbitrarily close to any mapping of the plane into the plane, there is an excellent mapping. This means that any mapping of that type can be approximated by a mapping having only well-characterized types of singularities (see Fig. 10.1).

The following year, in an article presented in the Bourbaki Seminar,[14] Thom proposed studying, with the same methods as Morse and Whitney, the mappings from R^n to R^p and characterizing them by the type of their singularities. It was thus necessary to analyze mappings that can have well-determined singularities and that can approximate an arbitrary function in the space of functions from R^n to R^p. The question is to find classifiable prototypes and to show that any function can be approximated by one of them. So, the properties of the prototypes would be called generic. But we still do not know exactly what this means.

Thus, Thom defines a generic property P of a mapping of class C^m from R^n to R^p as a property verifiable by all functions belonging to the space of functions of this type, except for a "thin" subset of that space. This fact can be viewed as an extension to differentiable

[10] Morse (1925).

[11] With the Hessian being non-zero at each.

[12] A point p in a manifold M is a critical point if there is a local coordinate system $(x_1,...,x_n)$ about p such that $\nabla f(p) = 0$. Such a critical point is non-degenerate if and only if the $n \times n$ matrix of second partial derivatives, called the Hessian of f at p, is non-singular, that is, its determinant does not vanish.

[13] Whitney (1955).

[14] Thom (1956a).

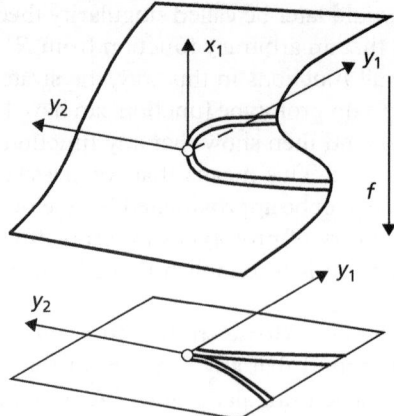

Figure 10.1 *The Whitney cusp drawn by Vladimir I. Arnold (Arnold, 1992: 4). This kind of singularity arises when a surface like the one in the figure is projected onto a plane.*

structures of the notion of genericity used in algebraic geometry. In this theory, a generic property was already defined as a property that can be satisfied for all points of a space, except for the points of a thin sub-manifold of that particular space. In the spring of 1952, while spending a year in Princeton University, Thom had, in his own words, "a striking conversation" with Claude Chevalley who suggested introducing the notion of genericity used by the Italian geometers to the world of differential structures.[15]

In a subsequent article, published a few months after the Bourbaki one, Thom explicitly notes the parallelism between the definition of genericity presented in his first article and the one employed in algebraic geometry, with the difference that, in the context of differentiable mappings, the functions studied constitute a functional space that is a Baire space and the exceptional thin sub-manifold is replaced by a closed subspace without interior points.[16] But Thom was not completely satisfied with this terminology. Despite this fact, the adjective "generic" has been disseminated with the meaning that generic transformations are those that can approximate any transformation. This notion of "approximation" had a precise definition. A generic property was then known as a property that is satisfied by the elements that constitute an open and dense subspace of the domain, which is the complement of a closed subspace without interior points.

A successful classification program in the theory of singularities of differentiable functions should be able to define classes of functions with the following characteristics: (1) each class is sufficiently particular to be geometrically well described by the type of its singularities; and (2) such classifiable functions are sufficiently general to include "almost all" functions in the sense that they constitute an open and dense subspace of the domain of all functions.[17]

---

[15] Thom (1989: 200).

[16] "Une propriété (P) des applications *f*: $R^n \to R^p$, définie localement en tout point de l'espace source sera dite *générique*, si l'ensemble des *f* qui ne présentent pas la propriété (P) en au moins un point d'un compact K de R^n forme dans $L(R^n R^p : r)$ un ensemble *fermé rare* (fermé sans point intérieur)" (Thom, 1956b: 52). See also Thom (1956c).

[17] "Pour tout couple d'entiers *n, p* [concernant les applications de R^n dans R^p] (...) on se propose de décrire et classifier un certain ensemble de singularités (S), tel que les applications qui présentent des singularités du type (S) et uniquement de celles-là forment um ensemble 'générique'" (Thom, 1956c: 59).

For differentiable functions, these two conditions express the compromise mathematicians were searching for, between the relevance of the properties under study (given by the type of singularities) and the generality of these properties or its genericity (the size of the subset of entities satisfying these properties). The type of the singularities of a function successfully captures its qualitative aspect and, if the two conditions cited above are fulfilled, we can conclude that almost all functions are classifiable by their singularities. The importance of this conclusion justifies the search for classifiable functions that form a generic subset in the domain of all functions. A similar goal would motivate the development of dynamical systems theory in the fifties and the sixties.

10.3 Globally understanding dynamical systems through the notions of structural stability and genericity

During its renewal at the end of the fifties, the style that the theory of dynamical systems acquired was greatly influenced by the program of classifying singularities of real functions, and vice-versa. In his doctoral thesis, David Aubin shows how the practices of what he calls "applied topologists"—working at IHES (*Institut des Hautes Études Scientifiques*) under Thom's leadership and having inherited from the Bourbakists the interest in structures—influenced the style acquired by qualitative dynamics, notably in the pioneering works of Peixoto and Smale.[18] An understanding of the way this style was transmitted involves, as we will see later, the direct collaboration among the three mathematicians, not only at the IHES but also at the Brazilian IMPA (*Instituto de Matemática Pura e Aplicada*).

The Brazilian mathematician Mauricio Peixoto was convinced that the main goal of the mathematics of his times was to classify mathematical objects, with emphasis on their structures, and by means of equivalence relations between them.[19] He thought it would be a useful challenge to express the theory of differential equations in a set-theoretic language. From his point of view, the suggestion given in Poincaré's quotation (transcribed in the introduction) had to be fulfilled with notions extracted from set theory.

Poincaré's and Birkhoff's work was certainly the point of departure for such a study, but in order to express their theory in a set theoretical basis it was still necessary to introduce two new elements:[20]

A) A space of differential equations, or dynamical systems, possessing a topological structure.

B) A notion of qualitative equivalence between two differential equations (analogous to Cremona transformations as was claimed by Poincaré).

Both requirements were fulfilled, primarily in two articles: the first written in 1958 and published in 1959 and the second published in 1962.[21] Peixoto defines the space of dynamical systems by considering a dynamical system as a point of a Banach space,

---

[18] Aubin (1998).
[19] Peixoto (2000).
[20] Peixoto (1987).
[21] See Peixoto (1959a) and Peixoto (1962), respectively.

and proposes that an equivalence relation between two systems in this space should be a homeomorphism, transforming trajectories of one system into trajectories of the other. This last definition is inspired by the work of Andronov and Pontryagin.

In 1937, these two Soviet mathematicians had published a paper called "Systèmes grossiers,"[22] in which they studied dynamical systems defined in a two-dimensional space and proposed that the trajectories of two systems should be considered equivalent if they could be transformed into one another by means of a transformation that is a homeomorphism (with the additional condition that this transformation is close to the identity transformation[23]). A system is called "grossier" (which means "coarse" and can also be translated as "robust") if its trajectories remain qualitatively similar after a perturbation in the definition, and the homeomorphism is precisely the transformation considered in order to maintain trajectories that are "qualitatively similar."

The importance of the coarseness property consists in the role it plays in modeling physical systems. If a system is not coarse, or robust, its fundamental properties are easily lost after a small perturbation. As mathematical models are just idealizations of physical realities, we cannot avoid perturbations in the definition of the mathematical system. Thus, a coarse system is a good candidate to serve as a model for a physical situation.

Around 1950, as Dahan-Dalmedico showed,[24] "grossier" was renamed "structurally stable," following a suggestion of Lefschetz. By this time, there were some researchers working on the subject in Princeton, and that led Peixoto to join them in 1957.

In 1952, a mathematician of Lefschetz's team, De Baggis, managed to provide the demonstrations that were lacking in Andronov and Pontryagin's article and alleviated some of the requirements they had recognized as necessary. He opens his article by claiming that:

> In the study of nonlinear problems it is difficult for the mathematician to find rich classifications of nonlinear systems which are sufficiently homogeneous in their properties to yield an interesting theory.[25]

He explicitly refers, in this quotation, to the productive tension we claim is present in the efforts undertaken, during the fifties and sixties, by mathematicians trying to obtain some kind of classification of dynamical systems.

The difficulty of this research direction is due to the enormous variety of behaviors of trajectories of dynamical systems in the nonlinear case. In order to provide general descriptions it is necessary to impose some restrictions that make it possible to choose a subset of interesting systems in the universe of nonlinear systems. But what are the interesting properties that can be used to restrict this enormous universe?

---

[22] Andronov and Pontryagin (1937).
[23] In Peixoto (1962), Peixoto demonstrated this requirement was not necessary.
[24] Dahan-Dalmedico (1994). See also Roque (2008).
[25] De Baggis (1952: 37).

Issues of physical relevance can provide a clue to the right concept of structural stability. From the standpoint of applied mathematics, a supplementary advantage of such a concept is that it would solve the problem of legitimacy of modeling raised by the Soviet mathematicians because, even if we know a model is never exact, it can always be approximated by a structurally stable one, which means it can be approximated by another model, similar to the original, that does not lose its main properties after perturbation.

Andronov and Pontryagin attempted to give a mathematical description of two-dimensional structurally stable systems, proposing that the essential features of these dynamical systems that are able to be preserved under small perturbations are: (1) it has a finite number of singularities, all of which are simple; and (2) no trajectory of the system goes from a saddle point (that is a type of singularity) to another saddle point. These features express the topological character of the set of trajectories of a dynamical system defined on a two dimensional manifold. In his article of 1959,[26] Peixoto showed that structurally stable systems, having Andronov and Pontryagin's features, form an open and dense subset in the space of all systems defined on a sphere. It is the first general result in the theory of dynamical systems.[27] Even though it only holds for two-dimensional systems defined on a specific surface, Peixoto's theorem is a general result, since it succeeds in describing the relevant features of *almost all* two-dimensional dynamical systems.

In the following years, the result was reformulated in the mode of singularity theory. Smale was introduced to Peixoto in the autumn of 1958, during his postdoctoral studies at Princeton. Less than two years later, at the beginning of 1960, the American mathematician went to Rio de Janeiro. In 1961, Thom also visited Rio and, after becoming familiar with similar problems from singularity theory, Peixoto reformulated his result in his paper of 1962,[28] using the term "generic" for the first time in the context of dynamical systems theory.

Peixoto starts with the following assertion: the fact that structurally stable systems form an open and dense subset of the space of all systems means they are "generic." It is something significant "because these systems are precisely the ones that exhibit the simplest possible features being amenable to classification."[29] Thus, Peixoto manages to show that structurally stable systems defined on a compact bidimensional orientable surface are generic. But how are these systems precisely the ones that are classifiable?

We know the special features of structurally stable two-dimensional systems, given in conditions (1) and (2) of Andronov and Pontryagin. If these systems are generic, we can hope to classify them by their topologically distinct types, for instance, by their number of distinct singularities, even if it can be a difficult task.[30] That is why we can say

---

[26] Peixoto (1959a).

[27] In the paper of 1962 he extended this result to orientable two-dimensional manifolds that are compact and differentiable.

[28] Peixoto (1962).

[29] Peixoto (1962: 101).

[30] Later on, Smale classified the topological distinct types of structurally stable systems on a two-dimensional compact differentiable manifold. He shows there are only a finite number of topologically distinct types of structurally stable systems having a given number of singularities and closed orbits.

that the goal of demonstrating that structurally stable systems are generic is to establish conditions that make obtaining a classification result possible. Thus we propose calling it a *pre-classification* program.

But we remind ourselves that this problem was solved only for dynamical systems defined on specific two-dimensional spaces, and that in general the models dealt with have more dimensions. What happens in larger dimensions?

In the history of dynamical systems theory, in order to get to a pre-classification result for larger dimensions, the notions of stability and genericity had to be redefined using different mathematical tools. First of all, it was necessary to extend the definition of structurally stable systems for any dimension, which was done by Peixoto in the same year (1959) using the language of functional spaces.[31]

On the basis of this new definition, mathematicians need to find ways of describing as explicitly as possible those systems which are structurally stable in any dimension. In other words, mathematicians need to generalize conditions (1) and (2) of the characterization of Andronov and Pontryagin.

The work of the American mathematician Steve Smale proved to be of fundamental importance for providing a solution to the problem. As we have said, he met Peixoto in 1958 and the encounter, according to Smale, "sparked my interest in structural stability."[32] That is why he went to Rio de Janeiro to visit IMPA in 1960.

The generalization of the conditions characterizing structurally stable dynamical systems for dimensions greater than two is inspired by the work of Morse, which attests to the influence of the singularity theory of differentiable mappings. Morse was responsible for associating the number of singularities of a function to the topology of the manifold on which the function is defined. This motivated Smale to obtain a similar result in 1960, as he related the number of periodic trajectories of a dynamical system to the homology of the manifold on which the system is defined.[33] Inspired by this result, he proposed, in the same article, to study systems with a finite number of periodic trajectories, and in which all trajectories approach periodic trajectories. As a consequence, these periodic trajectories are the only attractors of the system. In addition, he showed the stable and unstable

---

[31] The definition chosen in Peixoto (1959b) can be summarized as follows:

Let $X = (X_1,...,X_n), n \geq 2$ be a differential system $\dfrac{dx_i}{dt} = X_i(x_1,...,x_n), i = 1,...,n$ of class C^1 defined in the unit ball $B^n, x_1^2 + \cdots + x_n^2 \leq 1$. The system X is said to be *structurally stable* in B^n if:

(i) the vector field of X has no contact with the boundary S^{n-1} of B^n and, say, always points inward;
(ii) there exists $\delta > 0$ such that, whenever a system $Y = (Y_1,...,Y_n)$ satisfies

$$\rho(X,Y) = \max_{B^n} (\sum_{i=1}^{n} |X_i - Y_i| + \sum_{i,j=1}^{n} |\partial X_i / \partial x_j - \partial Y_i / \partial x_j|) < \delta,$$

we can find a homeomorphism T of B^n onto itself mapping trajectories of X onto trajectories of Y.

[32] Smale (1990: 45).
[33] Smale (1960).

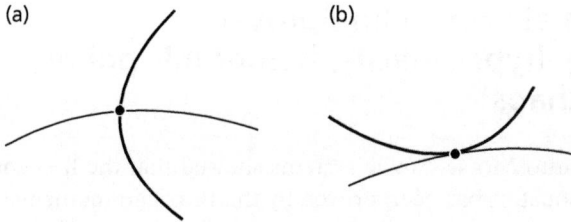

Figure 10.2 *(a) A local image of transversality. We have a transversal intersection whenever we can locally find an image like this one. (b) A non-transversal intersection.*

trajectories associated with those periodic trajectories must intercept transversally.[34] Systems with these characteristics generalize the Andronov and Pontryagin conditions for higher dimensions and were called "Morse–Smale systems" by René Thom.

What we have here is the prototype for classifiable systems in higher dimensions that has been searched. The finiteness of the number of periodic trajectories satisfies the demand that attractors must be easily described for the purpose of classification. The demand that unstable and stable trajectories must intercept transversally[35] is fundamental to establish robustness and genericity, something already suggested by Thom's famous transversality theorem.[36] It is not difficult to imagine that non-transversal intersections can be easily undone by very small perturbations. So, the qualitative aspect of trajectories in the presence of an intersection of this kind changes considerably after small perturbations. In Fig. 10.2, we show, respectively, a transversal and a non-transversal intersection of two curves in R^n. We can note that even if one perturbs the curves in Fig. 10.2(a), they remain transversal, whereas, in Fig. 10.2(b), a small perturbation can turn the intersection into a transversal one.

Yet, what remained to be shown is that Morse–Smale systems are open and dense, and therefore generic in the space of systems. This is the essential part of a famous conjecture proposed by Smale in 1960. If it were true, the classification program would be fulfilled, because it actually suffices to classify *almost all* systems by their number of attractors. Morse–Smale systems are classifiable systems and, if they are generic, it is possible to classify dynamical systems, in the sense that: (1) a system characterized by Morse–Smale properties is robust (its qualitative essential features are not destroyed by small perturbations); (2) any arbitrary system can be approached by a Morse–Smale, and, therefore, they are generic (dense).

---

[34] The stable and unstable trajectories are those neighboring the periodic one that respectively approach it or recede from it as time grows. The fact that they intercept transversally means this intersection occurs in a point with non-collinear tangents.

[35] The transversality property is satisfied only by these stable and unstable trajectories, which tend to periodic trajectories with increasing or decreasing time.

[36] Thom (1956b).

10.4   The Smale conjecture proven wrong: hyperbolicity, homoclinic points, and "chaos"

A deeper inquiry into Morse–Smale systems showed that the first condition holds, that is, that they are robust. It has been proven by the Brazilian mathematician Jacob Palis.[37] Unfortunately, they are not dense in the space of systems. Smale's conjecture is therefore false. In fact, a glance at some previous works of Poincaré, Birkhoff, or Hadamard would suggest that systems with a finite number of periodic points cannot be dense in the space of systems. Smale admitted that he did not know these works, and that is why he ended up being in error. However, beyond this first disappointment, the subsequent developments provide a very good example of how a wrong conjecture can become productive in mathematical research.

In his work on celestial mechanics, Poincaré had already noticed the possibility of what he called "doubly asymptotical points" or "homoclinic points."[38] They would be periodic points of a three-dimensional system's dynamics, whose vicinity contains other trajectories with a very complex behavior. Poincaré studied the behavior of these trajectories by means of their intersections with a two-dimensional section, which associates to one point the point of the next return of the trajectory to the section. Thus, the study of the original dynamics of this now called "Poincaré section" is reduced to an analysis of a two-dimensional map defined on the section. The intersection of a periodic trajectory with this section is a fixed point of this map, and the other trajectories in its neighborhood also intersect the section in points (not fixed).[39] There is a special kind of fixed point, also studied by Poincaré, now called "hyperbolic," for which the dynamics in its neighborhood can be characterized by means of a stable manifold and an unstable manifold. The stable manifold is the set of all points that approach the fixed point under iteration of the map (shown by the almost vertical line in Fig. 10.3),[40] and the unstable manifold is the set of all points that approach the fixed point under iteration of the inverse map (shown by the almost horizontal line in Fig. 10.3). In two dimensions, a fixed point of this type is a saddle point (the point p in Fig. 10.3).

If we start with a small ball of initial points centered on a saddle point and iterate the map, the ball will be stretched and squashed along the line of the unstable manifold, and the opposite occurs along the line of the stable manifold. We have a homoclinic point if the stable and the unstable invariant manifolds, from this same fixed point, intersect again.

---

[37] At this time Palis was a doctoral student working with Smale and this was the subject of his thesis; see Palis (1969).

[38] These names and the description of the dynamics associated with this kind of systems were introduced, respectively, in Poincaré (1890) and Poincaré (1892–9). For a history of this discovery made by Poincaré, see Barrow-Green (1997) and Anderson (1994).

[39] In Roque (2007) and Roque (2011), this method of Poincaré is explained in relationship with the definition of a dynamical system and with the problem of stability.

[40] This figure and the next one were taken from the article "Unstable periodic orbit" from the site Scholarpedia, at the address: http://www.scholarpedia.org/article/Unstable_periodic_orbits.

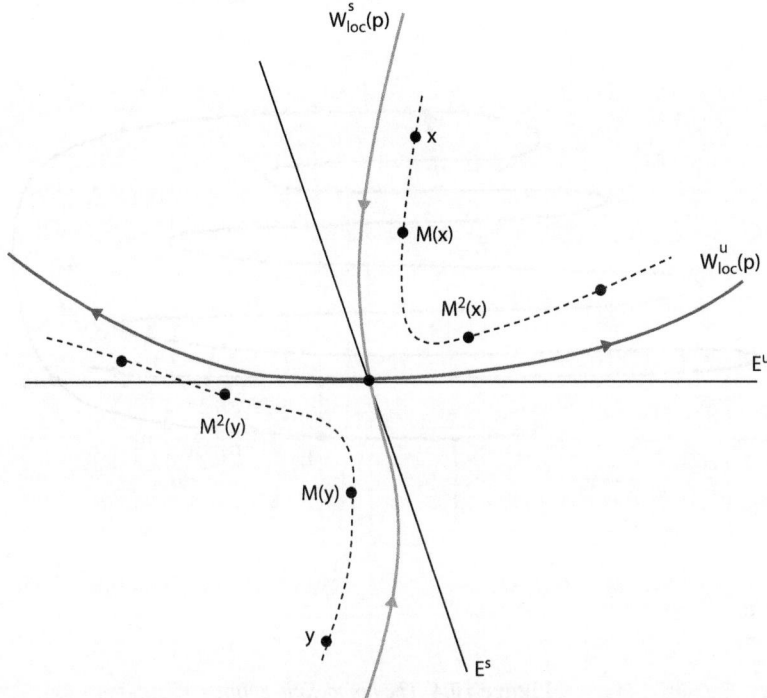

Figure 10.3 *Unstable and stable manifolds near a saddle point p are named respectively $W_{loc}^{u}(p)$ and $W_{loc}^{s}(p)$. The map M gives the iterations of each point. The iterates of x and y near the stable manifold will get closer together and move toward the saddle while trajectories near the unstable manifold will diverge in time.*

On the Poincaré section, it gives rise to a very complex figure of trajectories. Figure 10.4 shows the successive intersections between the stable and the unstable manifolds.

Birkhoff had shown that in the neighborhood of a homoclinic point there are infinitely many others.[41] The dynamic associated with a homoclinic point is not easy to destroy, as Norman Levinson already knew in 1949.[42] Smale tells how he received a letter from him suggesting that the conjecture on the genericity of systems with a finite number of periodic points should be false.[43] If Morse–Smale systems were dense, in a neighborhood of any system it would be possible to find a Morse–Smale system, which does not happen in the neighborhood of a system containing a homoclinic point. Indeed, in the presence of a homoclinic point it is possible to verify the phenomenon of infiniteness

[41] See Birkhoff (1920, 1935, [1927] 1966).
[42] Levinson (1949).
[43] Smale (1998).

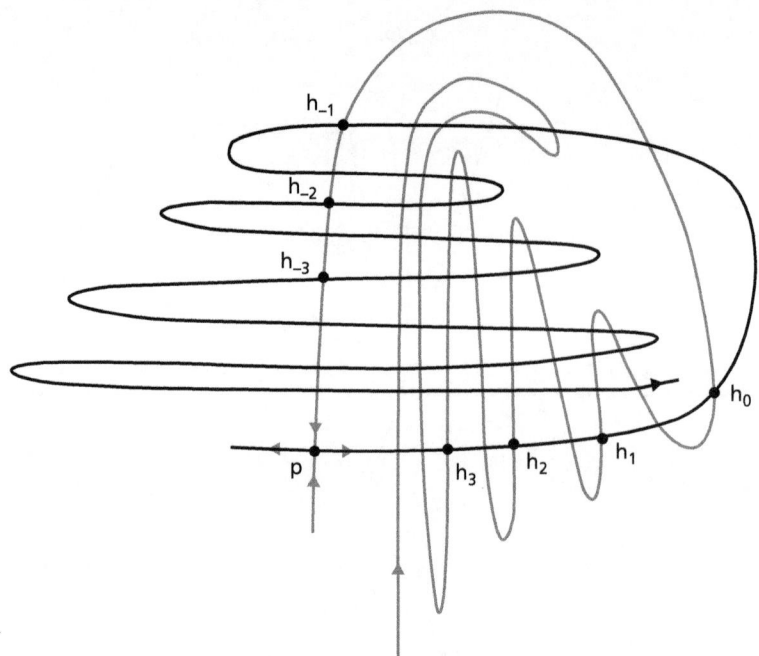

Figure 10.4 *The homoclinic point.*

of periodic points and to affirm its persistence (in particular, that this is a structurally stable phenomenon).

Smale easily admitted his error and immediately proposed a model to describe the set found by Levinson geometrically.[44] This study gives birth to his "horseshoe" (in Fig. 10.5)[45] which became very famous afterward as the prototype of chaotic behavior, associated mainly with its sensitivity to initial conditions.[46]

In a horseshoe, the cross section of the final structure corresponds to a Cantor set.[47] This is one of its most interesting properties. Even if the complexity associated with a homoclinic intersection was already known, the model proposed by Smale allowed one to grasp the mechanism that produced its essential properties. In Section 10.5.1 we explore further the interesting phenomena with which these Cantor sets are related.

---

[44] Smale (1963).

[45] I thank the Brazilian mathematician Maria José Pacifico for having offered me this beautiful image of a horseshoe.

[46] Alain Chenciner (2007) comments on how unfortunate is the designation of "chaos" in this case. As we will see later, the structure of trajectories in hyperbolic systems is very well understood, and has no reason to be called "chaotic."

[47] Just to give an example, we explain how to obtain the Cantor ternary set. This set is created by repeatedly deleting the open middle thirds of a set of line segments. One starts by deleting the open middle third (1/3, 2/3) from the interval [0, 1], leaving two line segments: [0, 1/3] ∪ [2/3, 1]. Next, the open middle third of each of these remaining segments is deleted, leaving four line segments: [0, 1/9] ∪ [2/9, 1/3] ∪ [2/3, 7/9] ∪ [8/9, 1]. This process is continued ad infinitum. The Cantor ternary set contains all points in the interval [0, 1] that are not deleted at any step in this infinite process.

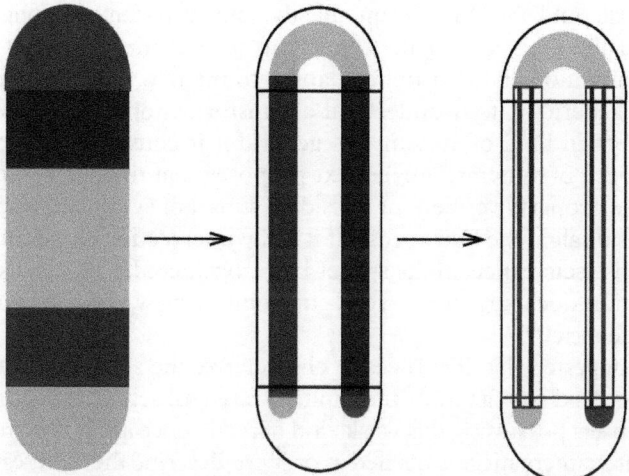

Figure 10.5 *The Smale horseshoe map consists of a sequence of operations on the rectangular middle section in the first figure. Expand vertically by more than a factor of two, then compress horizontally by more than a factor of two. Finally, fold the resulting figure and fit it back onto the rectangle, overlapping at the top and bottom, and not quite reaching the ends to the left and right (and with a gap in the middle), as illustrated above. Repeat the same process indefinitely.*

In an article called "Finding horseshoes in the beaches of Rio"—actually a note presented for the birthday of the Brazilian National Center of Scientific Development (CNPQ) in 1996—Smale asserts that the work on the horseshoe is the byproduct of the opportunity of being at that moment (in IMPA) at the confluence point of three traditions: the field in which Levinson (but also Cartwright-Littlewood) worked; the traditional contributions of Poincaré and Birkhoff; and the writings about structural stability of Soviet mathematicians that he had come to know from Peixoto.

In an inspiring example of how a wrong conjecture can produce novelty in mathematical practice, Smale uses the counter-example of the horseshoe in his favor and proposes to incorporate this phenomenon into a new dynamics prototype that should renew the classification on another basis. The role that periodic trajectories played in the Morse–Smale prototype should be played now by hyperbolic sets (of which the horseshoe is a particular case).[48] If hyperbolic sets were proved to be structurally stable, they would be good candidates to constitute the new prototype of dynamics, which received the unfortunate denomination of "Axiom A."[49]

---

[48] Intuitively speaking, all points in a hyperbolic set are locally similar to a saddle point. For a more technical explanation, a compact smooth manifold M is considered and $f: M \to M$ is a diffeomorphism, with $Df: TM \to TM$ being the differential of f. An invariant subset of M is said to be hyperbolic if the tangent bundle in this subset admits a splitting into a sum of two subbundles, called the stable bundle and the unstable bundle. In the stable bundle we have a contraction and in the unstable bundle, an expansion. We refer here to systems for which the entire manifold M is hyperbolic.

[49] In Palis (1997), Jacob Palis notices the lack of creativity in the inventions of denominations in those times and gives, in particular, the example of this designation of Axiom A systems.

In 1967, Smale published "Differentiable dynamical systems," a long paper in which he summarized the advances of the theory up to that moment.[50] At that point, the mathematical definitions needed to be adapted to the new classification program, and the concept of genericity itself underwent a transformation. In the first version of the conjecture, a certain kind of system was generic if it constituted an open and dense subset in the space of systems. Smale next proposes that the name "generic" must be associated with a property verified for a residual subset of systems. A subset of the space of systems is residual, in the Baire sense, if it is the intersection of a denumerable family of open and dense sets. Since, in the spaces here considered, a residual set is also dense, this new definition does not imply, from our point of view, a significant change in the conception of genericity.

Smale also suggested that it suffices to characterize the subsets of trajectories whose dynamics do not lead too far from their initial state (subsets of trajectories that do not escape). The subsets possessing this weak kind of recurrence are called "non-wandering." They are all that mathematicians needed in order to describe the "interesting properties" of dynamical systems, since, in this domain, they are always considering phenomena that recur in some sense, and not the transient ones. In the new classification prototype, the non-wandering sets must be hyperbolic.

One more thing Smale did not know at the time was that in the Soviet Union, mathematicians, like Dmitri Victorovich Anosov, a student of Pontryagin, were already working to understand the structure of hyperbolic sets, in particular the one deriving from Hadamard's example of geodesics on surfaces of negative curvature, later called Anosov diffeomorphisms.[51]

The definition of a new kind of system, based on homoclinic intersections was very proficuous to the research, since it can be associated with the existence of infinitely many periodic points. The work Anosov did on geodesic flows on surfaces of negative curvature showed that these are hyperbolic. Furthermore, this is a persistent behavior, something that could have already contradicted the first conjecture on the genericity of Morse–Smale systems.

Poincaré had suggested, in his writings on celestial mechanics, that periodic trajectories were the only breach[52] through which we can penetrate in the space of trajectories in higher dimensions, in particular the trajectories in three-dimensional space, which he was studying at that time.[53] Smale's first conjecture asserted that periodic trajectories were effectively a good breach to penetrate the space of trajectories of a system of any dimension, which are far from being describable. Any system should be well understood if it could be approximated by a system with a finite number of periodic sets that attract all trajectories. But this program failed, and the next attempt was to test whether hyperbolic systems could be the breaches needed to describe the great complexity of

---

[50] Smale (1967).
[51] Anosov (1967)
[52] The term in French is "brèche."
[53] Poincaré (1892–9).

dynamics. If hyperbolicity were a generic property of dynamical systems, it would be a very useful notion for furnishing a global theory, since hyperbolic systems can be very well understood (to such an extent that they can even be said to be "integrable").[54] But, once again, this is not true.

During the seventies, many examples were given revealing the existence of entire domains of dynamics covered with non-hyperbolic phenomena, which led to perplexity in the mathematical community.[55] At this point, we must mention the influence of computational research—Lorenz and Hénon attractors are good examples as they exhibit a "persistent" non-hyperbolic dynamics. Besides those persistent non-hyperbolic attractors, there are other kinds of phenomena such as duplication of periods, coexistence of infinite pits in which the dynamics tend to disappear, and so forth.[56] These are all "persistent" non-hyperbolic phenomena and Jacob Palis named them the "dark realm of dynamics."[57] We will see in the next section what "persistent" means in this context.

The perplexity that struck the researchers' community in the seventies, generating a number of incomplete and challenging works, converged with the fact that, about this same moment, Western scientists became acquainted with the works of the Soviet school, leading the research in a new direction. These discoveries raised two new questions:

1) Would it be possible to find phenomena capable of generating non-hyperbolicity?

2) Are topological notions appropriate? Perhaps these notions, used in theory up to this time—such as structural stability or density as the definition for genericity[58]—were not appropriate for describing the behavior of the majority of dynamical systems.

Considerable efforts were made to penetrate the complement of the hyperbolic world—the "dark realm" of dynamics. Afterward, the research in the field took the two questions mentioned above as a point of departure, following two main directions: 1) the study of bifurcations of homoclinic tangencies (described in Section 10.5.1); and 2) ergodic theory (see Section 10.5.2). The latter theory uses probabilistic tools in a general study of dynamical systems, following the suggestion of the Soviet mathematicians. In the next topic we will briefly mention the relation these problems have to the issue of genericity.

[54] For a discussion of this different meanings of integrability, see Chenciner (2007).

[55] Palis (1997).

[56] Palis recalls the "supreme humiliation" of mathematicians in the seventies because the major part of those examples came from physics. The tendency of referees in mathematical journals, in this time, was to refuse articles from experimentalists that exhibited these kinds of behavior. Only Ruelle was more receptive (Palis, 1997).

[57] Palis (1997, 2000).

[58] The Soviets realized this fact much earlier and used probabilistic notions instead, even in their study of hyperbolic systems. But their work started to have an influence on topological Western researchers by 1977, particularly during a meeting at Warsaw.

10.5   The changing definitions of genericity in the attempts to describe the "dark realm of dynamics"

10.5.1   Bifurcation of homoclinic tangencies

Many examples exhibited during the sixties and the seventies suggested that complex non-hyperbolic dynamics arise in relation to an interesting phenomenon called the "unfolding of a homoclinic tangency."

Transversal intersections between stable and unstable manifolds, associated with periodic points, became an essential property for ascertaining that a system is robust, because non-transversal intersections, as shown in Fig. 10.2(b), can be easily destroyed. We have seen that transversality is associated with hyperbolicity. So, it is not very difficult to imagine that, if we want to understand non-hyperbolic behavior, we must study those cases in which the homoclinic points are *not* transversal intersections between stable and unstable manifolds.

Suppose p is a periodic point, if in the homoclinic point q the stable and the unstable manifold do not intersect transversally but are tangents to each other, as in Fig. 10.6, we call this last point a "homoclinic tangency."

In the neighborhood of a homoclinic intersection there must be infinitely more points of the same type, produced by the intersections of the stable and unstable manifolds in the vicinity of the first one. So, a homoclinic point is part of a family of points of the same kind. It had already been proven that in the hyperbolic case, when the homoclinic point is a transversal intersection, there is a horseshoe in the vicinity of this point. In his geometrical model, as we cited above, Smale had already showed that a horseshoe is the product of two Cantor sets.

In the universe of dynamical systems defined on two-dimensional manifolds, as mathematicians want to study the trajectories of dynamical systems in the neighborhood of one system containing a homoclinic tangency, they define a family of diffeomorphisms f_μ such that for $\mu = 0$ this tangency is exhibited. In Fig. 10.6, we can see a picture of what happens for f_0. When this system is perturbed, the value of μ changes and the homoclinic

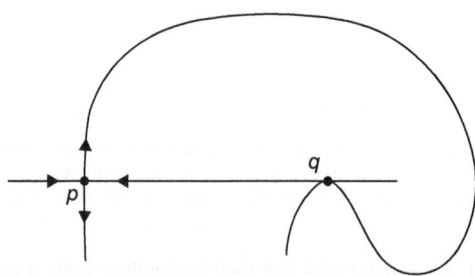

Figure 10.6 *Homoclinic tangency.*

tangency is unfolded. The system thus obtained can maintain the homoclinic tangency or can change it into a transversal homoclinic intersection. We can do an exercise to imagine the two curves in Fig. 10.6 are being perturbed—they can thus intersect in the same way or even transversally. As a consequence, the qualitative nature of the system can undergo an important change when homoclinic tangencies are unfolded. That is why this is called a "bifurcation point."

In this context, genericity has to do with the following question: are tangencies a persistent phenomenon or can they be easily destroyed by small perturbations? What happens when we perturb (or unfold) a homoclinic tangency: "how many" phenomena arising after the perturbation are hyperbolic and "how many" are non-hyperbolic?

It was through the seminal works of Sheldon Newhouse[59] that hyperbolicity was shown to be not dense in the space of a special kind of dynamical systems.[60] The underlying mechanism here was the presence of a homoclinic tangency leading to the phenomenon called nowadays "Newhouse phenomena," that is, non-negligible subsets of dynamical systems displaying infinitely many periodic attractors.

During the 1980s, mathematicians already knew some examples of non-hyperbolic behavior that functioned as prototypes.[61] Some of them expected that a non-hyperbolic system containing a prototype could be approximated by another system containing a homoclinic tangency. If that happens, it would be possible to say that the unfolding of the tangency would give birth to the non-hyperbolic phenomena. There were many works of Jacob Palis, Floris Takens, Carlos Gustavo Moreira, Jean-Christophe Yoccoz, Sheldon Newhouse, and others in this direction.[62] They tried to understand the appearance of complex non-hyperbolic behavior after the bifurcation of homoclinic tangencies.

Roughly speaking, mathematicians working in this field try to determine, in the space of parameters μ, of the dynamical systems f_μ that are being unfolded, the size of the subsets that corresponds to transversal or tangent intersections. This is the main concern in the bifurcation theory of homoclinic tangencies: to understand complex phenomena in the complement of the hyperbolic world by means of homoclinic bifurcations.

Here, once again, we can see an attempt to draw a general picture of the major non-hyperbolic phenomena that can appear in dynamics. Yet, the goal is no longer to "classify" them, but to discover the phenomenon that triggers non-hyperbolicity. A system with a homoclinic tangency would indicate how to penetrate into the dark realm of dynamics.

In order to accomplish this project, it was necessary to describe the prevailing phenomena in dynamics after unfolding a homoclinic tangency. We have said earlier that non-hyperbolic phenomena are "persistent," but this notion had to be defined. Mathematicians thus proposed mathematical refinements related to the definition of genericity in order to express the complexities of this new universe. In their research, they used some numerical invariants of Cantor sets to understand exactly what "persistence" means.

---

[59] Newhouse (1970, 1974, 1979).
[60] However, let us point out that in the C^1 topology it is still open.
[61] The Lorenz attractor, which characterized the famous "butterfly effect," is one of these prototypes.
[62] For a complete reference, see Palis and Takens (1993).

In this paragraph and the one that follows, we try to describe, as briefly and intuitively as possible, the mathematics involved in this direction of research. Suppose we have a family of dynamical systems defined by f_μ such that for $\mu = 0$ the system exhibits a homoclinic tangency. We want to know, near the bifurcation point .., how big in the parameter space (in the space of μ) is the set of values that correspond to hyperbolic or non-hyperbolic dynamical systems. The answer to this question has to do with different types of Cantor sets appearing in the space of parameters μ.

A system containing a tangency corresponds to a parameter μ_0 belonging to a Cantor set. Since a Cantor set is full of holes, we could imagine that, with a small perturbation of the dynamics, it is possible to make the tangency disappear. But this is not the case. In order to study the unfolding of homoclinic tangencies, mathematicians had to impose conditions on Cantor sets that concern some invariants describing whether a set of this type is "small" or "large." So, the possibility of destroying a tangency depends on some numerical properties that make one type of Cantor set different from another. Even if a Cantor set has no interior points, it is possible to define a property of "thickness,"[63] that determines how it behaves with respect to perturbations. When a system is perturbed, the homoclinic intersections may or may not be persistent and the exact meaning of "persistence" here refers to the previously mentioned numeral properties of Cantor sets.

A notion of genericity emerges combining the notion of denseness—already in use in the hyperbolic context—with other notions concerning dynamically defined properties of Cantor sets. The main proposition in the theory asserts that if there is a homoclinic tangency between the stable and the unstable manifold of some periodic point and the related Cantor sets are sufficiently thick, the tangencies are persistent.[64] In some sense, this result attests that when thickness is big enough, hyperbolicity cannot be generic.

Hopefully the brief explanations furnished thus far indicate how the concept of genericity necessarily changes in the development of research. As in the topological phase of dynamical systems theory, prevailing in the West up to the 1960s, the persistence of tangencies is mathematically expressed in terms of denseness. Nevertheless, denseness is studied by the authors cited in this section in relation with invariant numerical measures of Cantor sets.

10.5.2   Ergodic theory

Ergodic theory develops a probabilistic (i.e., measure-theoretic) approach in the study of determinist dynamical systems. Soviet mathematicians, such as Andrei Kolmogorov, had already introduced the idea of measure to describe the genericity of some kinds of

---

[63] The notion of *thickness* was introduced exactly to show that homoclinic tangencies are persistent. The precise meaning of this statement is that we can find a neighborhood U in the domain of dynamical systems treated here (C^2 diffeomorphisms of a two-manifold) in a way that for each $f \in U$, there is a tangency between the stable and the unstable manifolds associated to a fixed point. Otherwise, this tangency can be obtained by a small perturbation of f (the diffeomorphisms which have a homoclinic tangency are dense in U). The persistence of homoclinic tangencies can be defined by this denseness property.

[64] For more mathematical details, see Palis and Takens (1993: chap. 6).

systems. In the context of systems we are treating here, hyperbolic systems, the theory was proposed by Yacov Sinai, David Ruelle, and Robert Bowen. If we have a measure invariant by the dynamics, it is possible to describe "typical" properties of systems in the sense of the measure. This means that we do not describe the behavior of all the trajectories, but only that of "almost all" trajectories in a subset of trajectories with positive measure. What kind of properties can be studied in this way?

In the study of dynamical systems, sensitivity to initial conditions became a key notion, meaning that two motions departing from very close initial conditions can considerably diverge in the future. The systems presenting this behavior were called "chaotic." In order to understand the behavior of trajectories, mathematicians first divide the space into cells and then study how often the trajectories pass into each cell (this implies that we do not consider individual trajectories anymore and that we do not need to know exactly where a trajectory is inside a cell). It is thus possible to analyze the dynamics by taking averages that constitute a probabilistic evaluation of deterministic dynamics.

The reason for the appearance of statistical laws is the instability of trajectories. There is such a variety of trajectories' behaviors that the average aspect of trajectories in general tends to be stable.[65] As an example of a general characteristic, there is the fraction of time that a trajectory spends in a given cell. The number of cells can get larger and we become interested in the time averages of the trajectories in each cell. This is a good example of a relevant statistical property that can be studied in deterministic systems.

Based on this property, it is possible to redefine the concept of stability in statistical terms. The topological definition of structural stability points out the qualitative (or topological) nature of the set of trajectories in a dynamical system. But in actual experiments, each time we observe the state of a system it is actually a different system that is being considered. Explained mathematically, suppose we have a system defined by means of a function f. When we study the iterations of x_n, this leads to an x_{n+1} which is not exactly $f(x_n)$, because there is an aleatory noise. Thus after each iteration, we are considering a slightly different system.

We say that a system is stochastically stable if time averages are not affected by this noise. The notion of topological equivalence employed in the definition of structural stability gives too much importance to the structure of trajectories and, consequently, to their pathologies. From the viewpoint of stochastic research, structural stability does not eliminate a satisfactory number of pathologies and takes into account properties that are not sufficiently relevant. Furthermore, it is a typical property of hyperbolic systems. In order to study non-hyperbolic systems, it can be more appropriate to use a stochastic definition of stability.

Once stochastic stability is defined, the possibility of classification depends on the genericity of well-characterized stochastically stable systems. This characterization can be obtained with a defined system having a finite number of "good" attractors whose attraction basin covers "almost all" the ambient space, what means that the basin's[66] measure must be equal to the measure of the ambient space. Systems of this kind are

---

[65] Sinai (1992).
[66] For each attractor, its *basin of attraction* is the set of initial conditions leading to long-time behavior that approaches that attractor.

significant in the description of the global aspect of dynamics since attractors characterize the asymptotical behavior of the system.

"Good" attractors are those with good measures, that is, physically relevant measures, and whose dynamics in attraction basins are stochastically stable. These can be non-hyperbolical attractors, but if they are finite, it is possible to classify dynamics by the number and nature of those attractors. Yet, this possibility relies on the genericity of such well-characterized stochastically stable systems.

Again, in the same style as the pre-classification program set by Peixoto and Smale, mathematicians in the 1980s proposed a conjecture regarding the genericity of stochastically stable systems. We note the central role that the research of IMPA, in Rio de Janeiro, played in this proposal, Jacob Palis being is a key figure in the general shape it acquired.[67] The meaning of genericity employed here is still translated by the denseness of stable systems, since it is hard to define a measure in the infinite dimensional space of all systems.

At the International Mathematical Congress of 1954, held in Amsterdam, Kolmogorov gave a closing lecture with the title "The general theory of dynamical systems and classical mechanics."[68] This event played an important role in the development of what is now called Kolmogorov–Arnold–Moser (or KAM) theory. In his lecture, Kolmogorov discusses the occurrence of some special kinds of motions in the space of trajectories of Hamiltonian dynamical systems. He proposes to investigate the persistence of properties that are known for integrable systems after small, non-integrable, perturbations of the equations describing the system. But once again, what does persistence mean?

The issue at stake is to state that if certain conditions are fulfilled, after a sufficiently small perturbation of the integrable system, the space of motions is "mostly" filled by deformations of the invariant tori whose existence is known for the unperturbed system. "Mostly" here means that the complement of the set where it occurs has small measure.[69] Starting with Kolmogorov's version, the theorem was improved upon by V.I. Arnold and J. Moser in the 1960s, producing the result known nowadays as KAM theory.

It is not within the limits of our proposal to analyze the history or the implications of Kolmogorov's theorem. We mention it only to show that the notion of "mostly" here becomes mathematically precise when a measure concept is employed.[70] A phenomenon is persistent if it is frequent, that is, when the set of systems in which it is displayed has a big measure in the set of all systems.

In Section 10.1, we explained how the description of "almost all" systems in the universe of systems was translated in a topological language. In the study of bifurcations of homoclinic tangencies, some measure concepts were introduced. But before Western mathematics had gotten acquainted with the works developed in the Soviet Union,

---

[67] See Palis (2000) for more details.

[68] Kolmogorov (1954).

[69] The main question in KAM theory is to show that, generically, after small perturbations of integrable systems, the union of quasi-periodic tori has positive measure. The statement applies to many concrete models of classical mechanics, as in some versions of the n-body problem.

[70] We must recall that the measure concept is in straight relation with probability, one of the major fields of Kolmogorov's research.

mathematicians such as Kolmogorov and Sinai had already suggested the inconvenience of topological notions to express the mathematical meaning of "almost all."

The complement of a set of positive Lebesgue measure can be an open and dense set. Thus, a property that is topologically generic, fulfilled by an open and dense subset of systems, can be absent of all elements in a set of positive measure,[71] which indicates that it is not generic in the measure-theoretic sense. So, a generic property should be defined as a property holding for a subset of "big" measure in the set of systems. A notion of genericity based on measures would be stronger than the notion based on denseness that is also used in the general study of dynamical systems. The only reservation with respect to this proposal is that it is hard to define an observable measure (a measure having a physical sense, such as Lebesgue measure) in the infinite dimensional space of all systems.

In the context of ergodic theory, generic properties must be related to statistical and probabilistic notions. But sometimes, it is necessary to use denseness as a definition for genericity, and in these cases, the mathematical tools with which genericity is assessed differ from those used to express the properties whose genericity is investigated.

10.6 Concluding remarks

We gave a very brief and panoramic report of different branches in dynamical systems theory, with special emphasis on the transformation undergone by key concepts, such as genericity. Our main goal was to furnish evidence for the thesis that mathematics very often develops under a tension in the search for a good compromise between a general understanding of mathematical objects and the properties making these objects relevant.

In the beginning, the definition of genericity was related to classification purposes, since not all systems are classifiable, and those that are classifiable have to be generic. The hitch is that both the concepts of "generic" and "classifiable" have to be defined. In an initial phase, a property was called generic if the subset of systems satisfying this property was dense in the space of all systems of a kind. As to the characterization of classifiable systems, it underwent quite a few changes. The topological aspect of the set of trajectories used by Peixoto and Smale, inspired by Andronov and Pontryagin's works, was too restrictive and it was only appropriate for systems defined on two-dimensional manifolds. Even structural stability revealed itself as a property of hyperbolic systems, which are not very general themselves, even if they are very well understood. Bifurcation theory strived for an analysis of the mechanism that triggers non-hyperbolicity, noting that general results depend on numerical invariants of Cantor sets—some of which already existed and others had to be defined to advance the theory.

---

[71] A variant of the Cantor set furnishes an example of a nowhere dense set with positive measure. Remove from [0,1] all dyadic fractions of the form $a/2^n$ in lowest terms for positive integers a and n and the intervals around them $[a/2^n - 1/2^{2n+1}, a/2^n + 1/2^{2n+1}]$; since for each n this removes intervals adding up to at most $1/2^{n+1}$, the nowhere dense set remaining after all such intervals have been removed has measure of at least 1/2 (in fact just over 0.535... because of overlaps) and so in a sense represents the greatest part of the ambient space [0,1]. Generalizing this method, one can construct in the unit interval nowhere dense sets of any measure less than 1.

Finally, topological characterization turned out to be too restrictive, and only appropriate for the particular case of hyperbolic systems. Properties of a different nature, observed in the same systems, had to be expressed by statistical tools. From these tools, new stochastic definitions were proposed, which proved to be more useful for a general description of all systems, including non-hyperbolic systems. In turn, the advent of this new, probabilistic, notion of genericity triggered the search for new characterizations of what can be considered a good definition for "almost all" systems. From a certain perspective, the denseness definition is not as good as other ones using measure concepts. But, on the other hand, it is difficult to define a measure on the set mathematicians were working with.

Is there a definite answer to what is the best notion of genericity? That is not our point. As the complexity of the possible set of trajectories defined by a dynamical system increases, new notions are invented to deal with unexpected behaviors. When the research turned from bidimensional to more general systems, the hope of classifying all systems vanished, but the search for a general theory still motivated mathematicians. The stages of the development of the theory analyzed here furnish a good example of how the compromise between the generality of the description and the relevance of the studied properties entails a very productive tension in actual mathematical research.

REFERENCES

Anderson, K. G. (1994) 'Poincaré's discovery of homoclinic points', *Archive for History of Exact Sciences* 48(2): 133–47.

Andronov, A., and Pontryagin, L. (1937) 'Systèmes grossiers', *Doklady Akademi Nauk SSSR* 14(5): 247–50.

Anosov, D. V. (1967) 'Geodesic flows on closed Riemannian manifolds with negative curvature', *Proc. Steklov Inst.* 90: 1–235. Translated in 1969 by the American Mathematical Society.

Arnold, V. I. (1992) *Catastrophe theory.* Berlin/Heidelberg: Springer-Verlag.

Aubin, D. (1998) *A cultural history of catastrophes and chaos: around the Institut des Hautes Études Scientifiques, France.* Ph.D. Dissertation, Princeton University, USA.

Barrow-Green, J. (1997) *Poincaré and the three body problem.* Providence, RI: American Mathematical Society, London Mathematical Society.

Birkhoff, G. D. (1920) 'Surface transformations and their dynamical applications', *Acta Mathematica* 43: 1–119.

Birkhoff, G. D. (1935) 'Nouvelles recherches sur les systèmes dynamiques', *Memoriae Pontifical Academia Scientia Novi Lyncaei* 1(3): 65–216.

Birkhoff, G. D. ([1927] 1966) *Dynamical systems.* Providence, RI: American Mathematical Society.

Chenciner, A. (2007) 'De la mécanique céleste à la théorie des systèmes dynamiques, aller et retour: Poincaré et la géométrisation de l'espace des phases', in *Chaos et systèmes dynamiques: éléments pour une épistémologie,* eds. S. Franceschelli, M. Paty, and T. Roque. Paris: Hermann, 11–36.

Dahan-Dalmedico, A. (1994) 'La renaissance des systèmes dynamiques aux États-Unis après la Deuxième Guerre Mondiale: l'action de Solomon Lefschetz', *Rendiconti dei circolo matematico di Palermo (II)* 34: 133–66.

De Baggis, H. F. (1952) 'Dynamical systems with stable structures', in *Contributions to the theory of nonlinear oscillations*, Vol. 2, Lefschetz, S (org.). Princeton, NJ: Princeton University Press, 37–59.

Gilain, C. (1977) *La théorie géométrique des équations différentielles de Poincaré et l'histoire de l'analyse.* Thèse de doctorat, Université de Paris 7, France.

Kolmogorov, A. N. (1954) 'The general theory of dynamical systems and classical mechanics', in *Proceedings of the International Congress of Mathematicians*, Vol. 1. Amsterdam: North Holland, 315–33 [in Russian]. Reprinted in *International Mathematical Congress in Amsterdam, 1954 (Plenary Lectures)*. Moscow: Fizmatgiz, 1961, 187–208. English translation as Appendix in R.H. Abraham and J.E. Marsden, *Foundations of mechanics*, 2nd ed. Benjamin/Cummings, 1978, 741–57.

Lefschetz, S. (1952) *Contributions to the theory of nonlinear oscillations*, Vol. 2. Annals of Mathematics Studies. Princeton, NJ: Princeton University Press.

Levinson, N. (1949) 'A second order differential equation with singular solutions', *Annals of Mathematics* 50: 127–53.

Morse, M. (1925) 'Relations between the critical points of a function of n independent variables', *Transactions of the American Mathematical Society* 27: 345–96.

Newhouse, S. (1970) 'Non-density of axiom A(a) on S^2', *Proceedings of Symposia in Pure Mathematics—American Mathematical Society* 14: 191–202.

Newhouse, S. (1974) 'Diffeomorphism with infinitely many sinks', *Topology* 13: 9–18.

Newhouse, S. (1979) 'The abundance of wild hyperbolic sets and non-smooth stable sets for diffeomorphisms', *Publications mathématiques de l'Institut des Hautes Études Scientifiques* 50: 101–51.

Palis, J. (1969) 'On Morse–Smale dynamical systems', *Topology* 8: 385–405.

Palis , J. (1997) 'Interview with Tatiana Roque at the Instituto de Matemática Pura e Aplicada', Rio de Janeiro, Brazil.

Palis, J. (2000) 'A global view of dynamics and a conjecture on the denseness of finitude of attractors', *Astérisque* 261: 339–51.

Palis, J., and Takens, F. (1993) *Hyperbolicity and sensitive chaotic dynamics at homoclinic bifurcations.* Cambridge: Cambridge University Press.

Peixoto, M. (1959a) 'On structural stability', *Annals of Mathematics* 69: 199–222.

Peixoto, M. (1959b) 'Some examples on n-dimensional structural stability', *Proceedings of the National Academy of Sciences of the United States of America* 45: 633–6.

Peixoto, M. (1962) 'Structural stability on two-dimensional manifolds', *Topology* 1: 101–20.

Peixoto, M. (1987) 'Acceptance speech for the TWAS 1986 award in mathematics', in *The future of science in China and the Third World*, eds. A. Faruqui and M. Hassan. Beijing: World Scientific, 600–14.

Peixoto, M. (2000) 'Interview with Tatiana Roque at the Instituto de Matemática Pura e Aplicada', Rio de Janeiro, Brazil.

Poincaré, H. (1881) 'Mémoire sur les courbes définies par une équation différentielle (1re partie)', *Journal de Mathématiques (3e série)* 7: 375–422. Reprinted in Poincaré (1951–6, t. I: 3–43).

Poincaré, H. (1882) 'Mémoire sur les courbes définies par une équation différentielle' (2ᵉ partie)', *Journal de Mathématiques (3e série)* 8: 251–96. Reprinted in Poincaré (1951–6, t. I: 44–84).

Poincaré, H. (1885) 'Mémoire sur les courbes définies par une équation différentielle' (3ᵉ partie)', *Journal de Mathématiques (4e série)* 1: 167–244. Reprinted in Poincaré (1951–6, t. I: 90–158).

Poincaré, H. (1886) 'Mémoire sur les courbes définies par une équation différentielle' (4ᵉ partie)', *Journal de Mathématiques (4e série)* 2: 151–217. Reprinted in Poincaré (1951–6, t. I: 167–222).

Poincaré, H. (1890) 'Sur le problème des trois corps et les équations de la dynamique', *Acta Mathematica* 13: 1–270. Reprinted in Poincaré (1951–6, t. 7: 262–479).

Poincaré, H. (1892–9) *Méthodes nouvelles de la mécanique céleste.* Paris: Gauthier-Villars.

Poincaré, H. (1908) 'L'avenir des mathématiques', *Atti del IV Congresso internazionali dei matematici de Roma* 1: 167–82.

Poincaré, H. (1951–6) *Œuvres d'Henri Poincaré.* Paris: Gauthier-Villars.

Robadey, A. (2006) Différentes modalités de travail sur le général dans les recherches de Poincaré sur les systèmes dynamiques. Thèse de doctorat, Université de Paris 7, France.

Roque, T. (2007) 'Les enjeux du qualitatif dans la définition d'un système dynamique', in *Chaos et systèmes dynamiques: éléments pour une épistémologie,* eds. S. Franceschelli, M. Paty, and T. Roque. Paris: Hermann, 37–66.

Roque, T. (2008) 'De Andronov a Peixoto: a noção de estabilidade estrutural e as primeiras motivações da escola brasileira de Sistemas Dinâmicos', *Revista Brasileira de História da Matemática* 7(14): 233–46.

Roque, T. (2011) 'Stability of trajectories from Poincaré to Birkhoff: approaching a qualitative definition', *Archive for History of Exact Sciences* 65: 295–342.

Roque, T. (2015) 'L'originalité de Poincaré en mécanique céleste: pratique des solutions périodiques dans un réseau de textes', *Revue d'Histoire des Mathématiques* 21: 31–105.

Sinai, Y. G. (1992) 'L'aléatoire du non-aléatoire', in *Chaos et déterminisme,* eds. A. Dahan-Dalmedico, J.-L. Chabert, and K. Chemla. Paris: Seuil, 68–87.

Smale, S. (1960) 'Morse inequalities for a dynamical system', *Bulletin of the American Mathematical Society* 66: 43–9.

Smale, S. (1963) 'A structurally stable differential homeomorphism with an infinite number of periodic solutions', *Report on the Symposium on Non-linear Oscillations of Kiev.* 365–6.

Smale, S. (1967) 'Differentiable dynamical systems', *Bulletin of the American Mathematical Society* 73: 747–817. Reprinted in Smale (1980).

Smale, S. (1980) *The mathematics of time: essays on dynamical systems, economic process and related topics.* New York: Springer-Verlag.

Smale, S. (1990) 'The story of the higher dimensional Poincaré conjecture (what actually happened on the beaches of Rio)', *The Mathematical Intelligencer* 12(2): 44–51.

Smale, S. (1998) 'Finding a horseshoe on the beaches of Rio', *The Mathematical Intelligencer* 20(1): 39–44.

Thom, R. (1956a) 'Les singularités des applications différentiables', *Séminaire Bourbaki* 3 (1954–1956), Exp. n. 134: 357–69.

Thom, R. (1956b) 'Les singularités des applications différentiables', *Annales de l'Institut Fourier* 6: 43–87.

Thom, R. (1956b) 'Un lemme sur les applications différentiables', *Boletin de la Sociedad Matematica Mexicana (II)* 1: 59–71.

Thom, R. (1989) 'Problèmes rencontrés dans mon parcours mathématique: un bilan', *Publications mathématiques de l'I.H.É.S.* 70: 199–214.

Whitney, H. (1955) 'On singularities of mappings of Euclidean spaces. Mappings of the plane into the plane', *Annals of Mathematics* 62(3): 379–91.

Part III

Practices of generality

Section III.1

Scientists at work

11

Leibnizian analysis, canonical objects, and generalization

EMILY R. GROSHOLZ

11.1 Introduction

Leibniz's notion of analysis can be understood as an art of both discovery and justification in mathematics that aims for generalization rather than abstraction, and explanation rather than formal proof. This characterization may seem odd to those whose view of Leibniz as the champion of formal proof was shaped by Bertrand Russell and Louis Couturat, and the later twentieth century philosophers who interpreted Leibniz under their influence. However, an attentive reading of Leibniz's own practice as a mathematician supports my claim, as do many of his philosophical reflections on that practice. In this essay, I review aspects of his investigation of transcendental curves, focusing on the catenary, where analysis is the search for conditions of solvability of problems, and then go back to some of his pronouncements on analysis as the search for conditions of intelligibility. Finally, I bring his notion of analysis into relation with discussions by contemporary philosophers, including Carlo Cellucci, Herbert Breger, and Nancy Cartwright.

11.2 Leibniz on the catenary or *la chainette*

When Leibniz investigates a novel transcendental curve, he treats it as a paradigm in order to exhibit procedures or the algorithms that can be elicited from them. His working out of problems typically exhibits intermediary steps in a process of reasoning that contributes to the meaning of the final result and indicates how it might be extended. The exhibition of the meaning of procedures and the correctness of algorithms in terms of paradigmatic problems and canonical objects for Leibniz typically involves the combination of distinct modes of representation, including figures that exhibit spatial articulation,

The Oxford Handbook of Generality in Mathematics and the Sciences. First Edition. Karine Chemla, Renaud Chorlay and David Rabouin. © Oxford University Press 2016. Publishing in 2016 by Oxford University Press.

and descriptions of how to reason both upward and downward. It is a progressive and pedagogical search for the reasons that underlie general procedures and the constitution of objects, a search for deeper as well as broader understanding.

For Leibniz, the key to a curve's intelligibility is its hybrid nature, the way it allows us to explore numerical patterns and natural forms as well as geometrical patterns; he was as keen a student of Wallis, Huygens, and Cavalieri as he was of Descartes. These patterns are variously explored by counting and by calculation, by observation and tracing, and by applications of the language of ratios and proportions on the one hand and the new algebra on the other. To think them all together in the way that interests Leibniz requires the calculus of infinites as an *ars inveniendi*. The excellence of a characteristic for Leibniz lies in its ability to reveal conditions of intelligibility: for a transcendental curve, those conditions are arithmetical, geometrical, mechanical, and algebraic. What Leibniz discovers is that this "thinking-together" of number patterns, natural forms, and figures, where his powerful and original insights into analogies pertaining to curves considered as hybrids can emerge, rebounds upon the very algebra that allows the thinking-together and changes it. The addition of the new operators d and ∫, the introduction of variables as exponents (which changes the meaning of the variables), and the entertaining of polynomials with an infinite number of terms, are all examples of this. Indeed, the names of certain canonical transcendental curves (log, sin, sinh, etc.) become part of the standard vocabulary of algebra and analysis.

This habit of generalization is evident throughout Volume I of the VII series (*Mathematische Schriften*) of Leibniz's works in the Berlin Akademie-Verlag edition, devoted to the period 1672–1676. As M. Parmentier admirably displays in his translation and edition *Naissance du calcul différentiel, 26 articles des Acta Eruditorum*, the papers in the *Acta Eruditorum* taken together constitute a record of Leibniz's discovery and presentation of the infinitesimal calculus (Leibniz, 1989). They can be read not just as the exposition of a new method, but as the investigation of a family of related canonical items, that is, algebraic and transcendental curves. In these pages, sequences of numbers alternate with geometrical diagrams accompanied by ratios and proportions, and with arrays of derivations carried out in Cartesian algebra augmented by new concepts and symbols. For example, "De vera proportione circuli ad quadratum circumscriptum in numeris rationalibus expressa,"[1] which treats the ancient problem of the squaring of the circle (see Fig. 11.1), moves through a consideration of the series

$$\pi/4 = 1 - 1/3 + 1/5 - 1/7 + 1/9...$$

to a number line designed to exhibit the finite limit of an infinite sum. Various features of infinite sums are set forth, and then the result is generalized from the case of the circle to that of the hyperbola, whose regularities are discussed in turn. The numerical meditation culminates in a diagram that illustrates the reduction: in a circle with an inscribed square, one vertex of the square is the point of intersection of two

---

[1] *Acta Eruditorum* Feb. 1682: Leibniz (1962, V: 118–22).

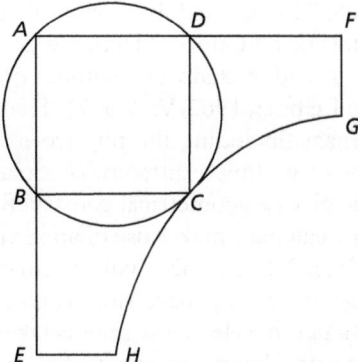

Figure 11.1 *De vera proportione circuli ad quadratrum circumscriptum in numeris rationalibus expressa.*

perpendicular asymptotes of one branch of a hyperbola whose point of inflection C intersects the opposing vertex of the square. The diagram also illustrates the fact that the integral of the hyperbola is the logarithm. Integration takes us from the domain of algebraic functions to that of transcendental functions; this means both that the operation of integration extends its own domain of application (and so is more difficult to formalize than differentiation), and that it brings the algebraic and transcendental into rational relation.

During the 1690s, Leibniz investigates mathematics in relation to mechanics, deepening his command of the meaning and uses of differential equations, transcendental curves and infinite series. In this section I discuss one of these curves, the catenary. In the "Tentamen Anagogicum," Leibniz discusses his understanding of variational problems, fundamental to physics since all states of equilibrium and all motions occurring in nature are distinguished by certain minimal properties; his new calculus is designed to express such problems and the things they concern. The catenary is one such item; indeed, for Leibniz its most important property is the way it expresses an extremum, or, as Leibniz puts it in the "Tentamen Anagogicum," the way it exhibits a determination by final causes that exist as conditions of intelligibility for nature. And indeed the catenary, and its surface of rotation, the catenoid (which is a minimal surface, along with the helicoid), are found throughout nature; their study in various contexts is pursued by physicists, chemists, and biologists (Leibniz, 1978 VII: 270–9).

The differential equation, as Leibniz and the Bernoullis discussed it, expresses the mechanical conditions which give rise to the curve: in modernized terms, they are $dy/dx = ws/H$, where ws is the weight of s feet of chain at w pounds per foot, and H is the horizontal tension pulling on the chain. Bernoulli solves the differential equation by reducing the problem to the quadrature of a hyperbola, which at the same time explains why the catenary can be used to calculate logarithms (Bernoulli, 1742, III: 494). The solution to the differential equation proves to be a curve of fundamental importance in purely mathematical terms, the hyperbolic function $y = a \cosh x/a$ or simply $y = \cosh x$ if a is chosen equal to 1.

In "De linea in quam flexile," Leibniz exhibits his solution to the differential equation in different, geometrical terms; he announces "Here is a geometrical construction of the curve, without the aid of any thread or chain, and without presupposing any quadrature" (*Acta Eruditorum* June 1691; Leibniz, 1962, V: 243–7). That is, he acknowledges various means for defining the catenary, including the physico-mechanical means of hanging a chain and the novel means of writing a differential equation; but in order to explain the nature of the catenary he gives a geometrical construction of it (see Fig. 11.2). The point-wise construction of the catenary makes use of an auxiliary curve labeled by points 3ξ, 2ξ, 1ξ, A (origin), $1(\xi)$, $2(\xi)$, $3(\xi)$.... This auxiliary curve, which associates an arithmetical progression with a geometrical progression, is constructed as a series of mean proportionals, starting from a pair of selected segments taken as standing in a given ratio D : K; it is the exponential curve. Having constructed e^x, Leibniz then constructs every point y of the catenary curve to be $1/2(e^x + e^{-x})$ or $\cosh x$. "From here, taking ON and O(N) as equal, we raise on N and (N) the segments NC and (N)(C) equal to half the sum of Nξ and (N)(ξ), then C and (C) will be *points of the catenary* FCA(C)L, of which we can then determine geometrically as many points as we wish." (*Acta Eruditorum* June 1691; Leibniz, 1962, V: 243–7).

Leibniz then shows that this curve has the physical features it is supposed to have (its center of gravity hangs lower than any other like configuration) as well as the interesting properties that the straight line OB is equal to the curved segment of the catenary AC, and the rectangle OAR to the curved area AONCA. He also shows how to find the center of gravity of any segment of the catenary and any area under the

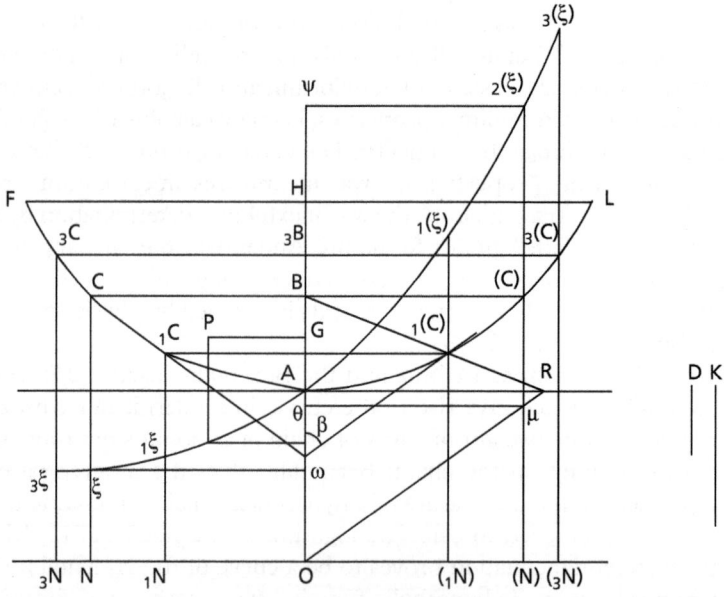

Figure 11.2 *The catenary.*

curve delimited by various straight lines, and how to compute the area and volume of solids engendered by its rotation. It also turns out to be the evolute of the tractrix, another transcendental curve of great interest to Leibniz; thus it is intimately related to the hyperbola, the logarithmic and exponential functions, the hyperbolic cosine and sine functions, and the tractrix; and, of course, to the catenoid and so also to other minimal surfaces.

An important difference between Descartes and Leibniz here is that Leibniz regards the mechanical genesis of these curves not as detracting from their intelligibility, but as constituting a further condition of intelligibility for them. As new analogies are discovered between one domain and another, new conditions of intelligibility are required to account for the intelligibility of the hybrids that arise as new correlations are forged. The analytic search for conditions of intelligibility of things that are given as unified yet problematic (like the catenary) is clearly quite different from the search for a small, fixed set of axioms in an axiomatization. The catenary is intelligible because of the way it aids in the calculation of logarithms; and embodies the function that we call cosh, from whose shape we can "read off" its rational relation to both the exponential function and the hyperbola; and expresses an equilibrium state in nature; and displays a kind of duality with the tractrix; and whatever deep and interesting aspect we discover next.

Generally, we can say that the things of mathematics, especially the items that are fundamental because they are canonical, become more meaningful with time, as they find new uses and contexts. Thus the conditions of their intelligibility may expand, often in surprising ways. When the differential equation of the catenary is "fitted out" by a geometrical curve or an equilibrium state in rational mechanics (to use Nancy Cartwright's term), the combination of mathematical representations allows us both to solve problems and to refer successfully, that is, to discover new truths.

11.3   Leibniz on analysis

Leibniz's method of analysis is a rational search for grounds of thinkability, and moves from something that is complex but unified to something simple and explanatory. In general, more than one way of proceeding from the complex to the simple is available, and what may play the role of the "simple" requires reflection, and may be revised. The complex is not, as it is on the reductive account of Descartes, exhaustively defined by means of the simple; rather, for Leibniz, the simple is defined in terms of the complex as a kind of degree zero, different from but standing in rational relation to it. The simple and the complex, while heterogeneous, are held together in virtue of Leibniz's principle of continuity. Leibnizian analysis is not the unpacking of merely concatenated simples; mere concatenation, supposing there is such a thing, rarely appears in the constitution of complexes and when it does it is thoroughly problematized, as when the truths of arithmetic are led back to the primitive notion of the unit. Thus Leibniz leads the labyrinth of the continuum back to the point, the phenomenon of motion to rest, *vis viva* to *vis mortua*, perception to *petites perceptions*, the complex forms of biology to rudimentary monads, and the projects of law and science to rational, self-conscious monads. This family of methods related by analogy is fundamentally opposed to the method of Descartes, for the analysis it pursues is not reductionist.

Leibniz explains his principle of continuity in a public letter to Christian Wolff, written in response to a controversy over the reality of certain mathematical items sparked by Guido Grandi; it was published in the Supplementa to the *Acta Eruditorum* in 1713 under the title "Epistola ad V. Cl. Christianum Wolfium, Professorem Matheseos Halensem, circa Scientiam Infiniti" (A. E. Supplementa 1713, V, sec. 6; Leibniz, 1962, V: 382–7). Toward the end, he presents a diagram and concludes, "All this accords with the *Law of Continuity*, which I first proposed in the *Nouvelles de la République des Lettres* of Bayle, applied to the laws of movement. It entails that with respect to continuous things, one can treat an external extremum as if it were internal [*ut in continuis extremum exclusivum tractari posit ut inclusivum*], so that the last case or instance, even if it is of a nature completely different, is subsumed under the general law governing the others."[2] He cites as illustration the relation of rest to motion and of the point to the line: rest can be treated as if it were evanescent motion and the point as if it were an evanescent line, an infinitely small line. Indeed, Leibniz gives as another formulation of the principle of continuity the claim that "the equation is an infinitesimally small inequality" (see Leibniz, 1962, VII: 25, for example).

For Leibniz, simplicity is the cardinal feature of the canonical; canonical items prove themselves fundamental to a discipline or inquiry. The discovery of canonical items is the outcome of philosophically and mathematically reflective analysis that typically results in a hypothesis that may need to be reconsidered; moreover, the ways in which canonical things may be associated also provides the occasion for further philosophical reflection and revision. Leibnizian analysis is thus not a trivial spelling out of what has already been thought in the terms involved in a claim, as Kant would have it, nor is it the extracting of an empty form that universally determines the truth of any contents inserted into it. It is not reductionist, and cannot be reversed into a Cartesian process of retrieving the complex from the simples. It is, rather, a method that works variously but analogously in many fields of human endeavor, a philosophical project that searches for the rational conditions underlying manifold complexity, but which can never come to a definitive conclusion.[3]

Leibnizian analysis lends itself to generalization in mathematics. Generalization starts from a set of solved problems, and asks how successful procedures may be extended to new problems and how their success may be explained. Problems involve problematic items, mathematical things that are intelligible but not wholly understood: a right triangle may be well defined, but we are still far from understanding the relation among its legs and hypotenuse. We may understand the relation among its legs and hypotenuse, but we are still far from understanding how a family of right triangles inscribed in a circle can define the transcendental functions sine and cosine. We may express the relation among its legs and hypotenuse in an equation, but we are still far from understanding the conditions under which whole number solutions to that equation may or may not exist. In the course of mathematical history, certain items (like the right triangle and the circle) prove to be canonical; canonicity is not an intrinsic quality that we discern by a sixth sense, but

² 'Réplique à l'abbé D.C. sous forme de lettre à Bayle,' Feb. 1687; Leibniz (1978, III: 45).
³ See Grosholz (1991, Ch. 1) and Grosholz and Yakira (1998, Ch. 1).

a feature of items that we discover in mathematical practice. Transcendental curves like sine, cosine, and logarithm, and those discussed above, the catenary and tractrix, were not canonical for the Greeks, who were aware of just a few transcendental curves, as it were by accident. They became canonical only in the seventeenth century, once their important geometrical and mechanical properties were discerned, and their relations to each other and to the conic sections were studied.

Leibniz characterizes his own mathematical practice vis-à-vis Descartes as a generalizing search for the conditions of intelligibility of canonical items, and the conditions of solvability of the problems in which those items are involved. In the introductory paragraph of his essay "De la chainette," he writes, "The ordinary analysis of Viète and Descartes consist[ed] in the reduction of problems to equations and to curves of a certain degree ... M. Descartes, in order to maintain the universality and sufficiency of his method, contrived for that purpose to exclude from geometry all the problems and all the curves that one couldn't treat by that method, under the pretext that they were only mechanics." This kind of exclusion, however, cuts off the process of generalization artificially, and Leibniz criticizes it.

> But since these problems and curves can indeed be constructed or imagined by means of certain exact [tracing] motions, and since they have important properties and since nature makes frequent use of them, one might say that he [Descartes] committed an error in doing this, rather like that with which we reproach some of the Greeks, who restricted themselves to constructions by ruler and compass, as if all the rest were mechanics (*Journal des Sçavans* 1692; Leibniz, 1962, V: 258–63).

By contrast, Leibniz is inspired by the cogency and urgency of the excluded problems and curves to look for ways of expressing them in useful form and discovering conditions of solvability for them; he calls them "transcendental" problems and curves, because they go beyond ordinary algebra.

> This is what he [Leibniz] calls *the analysis of infinites*, which is entirely different from the geometry of indivisibles of Cavalieri, or Mr. Wallis' arithmetic of infinites. For the geometry of Cavalieri, which is by the way very restricted, is attached to figures, where it seeks the sums of ordinates; and Mr. Wallis, in order to facilitate research, gives us by induction the sums of certain sequences of numbers: by contrast, the new analysis of infinites doesn't focus on figures or numbers, but rather on general magnitudes, as ordinary algebra does. But it [the new analysis] reveals a new algorithm, that is, a new way to add, subtract, multiply, divide and extract roots, appropriate to incomparable quantities, that is, to those which are infinitely big or infinitely small in comparison with others. It employs equations involving finite as well as infinite quantities, and among those that are finite, allows equations with exponents that are unknowns, or rather, instead of powers and roots, it makes use of a novel appropriation of variable magnitudes, which is variation itself, indicated by certain characters, and which consists in differences, or in the differences of differences of certain degrees, to which the sums are reciprocal, as roots are to powers (*Journal des Sçavans* 1692; Leibniz, 1962, V: 258–63).

Analysis for Leibniz is thus the search for the conditions of solvability of problems and the conditions of intelligibility of the things involved in them. Mathematical things insofar as they exist are intelligible unities, but they are also problematic because they do not wear the conditions of their intelligibility on their faces. These conditions must rather be sought in reflection that pursues both the inconstancy and the orderliness they present: things require explanation. Analysis is the search for the reasons or causes, the requisites, which are necessary for a thing to be what it is. This means, not to unpack a concept as if it were a concatenation of concept-parts nor to impose externally a set of relations on a thing as if it were a mere placeholder, but to distinguish or develop its content. The orderliness of a thing, expressed in its relations (both logical and analogical) to other things, is so to speak internal to it: things always have content. Thus for Leibniz, there are no surds or mere, mute givens: whatever exists has reasons or causes and therefore a *logos*.

For Leibniz, analysis guides any attempt to move from perceptual experience to science, or from science to philosophy; and the intelligibility it pursues is objective, in the sense of being independent of the accidents of the empirical world, and in particular of human subjectivity. It is not a question of searching for the "necessary conditions of the possibility of human experience," that is, Kant's transcendental conditions, but for the conditions of the intelligibility of existing things. Though mathematical items do not exist in the way that material or perceived objects exist, their existence is called for in order to explain the intelligibility of the latter. Their formal unity answers to the (sometimes material and perceptual) unity or selfsameness of things, their extent and multiplicity to the differentiation and otherness of things, and their relations by analogy to relations discovered in the world.

Leibniz writes in a letter of 1702 to Sophia Charlotte, Queen of Prussia, that only the intelligible allows us to account for the force of demonstration, and to furnish rational explanations, explanations that do not merely correlate or describe but exhibit the reasons for things being as they are, adding "Intelligible truth is independent of the truth or existence of sensible and material things" (Leibniz, 1978, VI: 499–508). To be intelligible is to be a possibility for thought, a possibility that may or may not ever be actualized. For Leibniz, of course, no mathematical thing is part of the created world; and yet mathematical things are not nothing. Intelligible things are neither mental acts nor extrinsically or "barely" given objects (no object exists without a reason or without intrinsic connection to other things); they are what is possible for thought independent of human subjectivity or the constitution of the world. And thought always outruns us: intelligible things are not contained by Cartesian or Lockean or Kantian strictures. To be intelligible is also to exhibit rational consistency, the absence of contradiction (for contradiction drives entities out of the realm of the possible); and to be expressive of other things, which is another way of saying that relations are intrinsic, and different kinds of things exist in different ways and yet as rationally related to each other.

11.4   Modern adumbrations of Leibnizian analysis

Analysis as the search for conditions of intelligibility entails a metaphysics of differentiation, of heterogeneous things existing in different ways. The notion of intelligibility (as opposed to truth), like that of the sublime (as opposed to the beautiful), confesses that knowledge lags behind being, and that in mathematics what we encounter transcends us and so prevents any complete determination, as Jules Vuillemin observed in a meditation on Gödel's incompleteness theorems (Vuillemin, 1997). Leibnizian analysis admits that when we know, we may confront something highly infinitary and in a sense inhuman that both lends itself to knowledge and yet always outstrips us. All the same, analysis as the investigation of conditions of intelligibility avoids the traps of irrationalism: one might say it constitutes a rationalism of the sublime. The conditions of intelligibility of the human world then prove to be different in kind from that world, and both being and knowledge are differentiated. Leibniz, moreover, never falters in his faith that the infinitesimal, finite, and infinite can be held together in rational relation; indeed, the principle of continuity is the expression of this confidence.

The analysis of intelligible items, the explanation of problematic facts by plausible explanatory hypotheses, and the convergence of disparate traditions in the service of problem-solving take place within a mathematical tradition that continually reformulates and reorganizes its results in order to transmit them. Mathematics textbooks sometimes present material in axiomatic form: we are used to the axiomatic presentation of Euclid's *Elements*, and of the highly abstract, highly axiomatized, diagram-free presentations of the twentieth century Bourbaki school and its offspring. However, there are other ways to analyze collections of problems and procedures and to teach the methods of mathematics. Many modern textbooks use an exposition that prizes generalization over abstraction, and the exploration of the meaning of procedures and algorithms in terms of paradigmatic problems that may be used to exhibit their correctness, clarify their domain of application, and indicate how that domain may be extended. Generalizing analysis is concerned with explaining *why* a result is correct and not just establishing that it *is* correct. Generalization as an ideal has not been wholly overshadowed by the ideal of abstraction and axiomatization in the twentieth century. On the contrary, the two ideals appear to co-exist dialectically.

If we consult, for example, the earlier textbooks cited in the Introduction to the *Topologie* of Paul Alexandroff and Heinz Hopf (Alexandroff and Hopf, 1935) we see that topology in 1935 had become divided into two distinct research programs. Pursuing one avenue, Oswald Veblen and Solomon Lefschetz write presentations of algebraic geometry and differential geometry, the study of algebraic varieties and differentiable manifolds by topological methods. Lefschetz's presentation is generally historical; he shows the student where and why various problems arose, and how they have been addressed (Lefschetz, 1924, 1929). Veblen explores the foundations of the field and offers an axiomatization of differential geometry (Veblen and Whitehead, 1932). The other avenue is explored by Maurice Fréchet and Casimir Kuratowski, who use topology to investigate the abstract function spaces (infinite-dimensional spaces) and infinitary point-sets that arise in real and complex analysis. Fréchet's presentation (Fréchet, 1928), like that of Lefschetz, is historical, while Kuratowski concentrates on an axiomatization of topology. Kuratowski writes at the beginning of his textbook, "the methods of reasoning that I use in this volume belong to set theory; the methods of combinatorial topology (homology, Betti groups, etc.) in general don't intervene in the questions treated here" (Kuratowski, 1933). The authors whose approach is historical—Lefschetz on the one hand and Fréchet on the other—present the analysis of canonical objects in a process of generalization. The authors who aim at logical systematization—Veblen on the one hand and Kuratowski on the other—present axiomatizations that synthesize the whole domain by relating principles, rules, and definitions inferentially in a process of abstraction. Both approaches, in tandem, are needed to bring up students who will do research in topology.

Carlo Cellucci give a particularly lucid account of the process of Leibnizian analysis in his essay, "The growth of mathematical knowledge: an open world view" (Cellucci, 2000), upon which he elaborates in Chapters 8 and 9 of his book *Filosofia e matematica* (Cellucci, 2002) He writes:

> [The analytic method] consists in solving a problem by reducing it to another one, which is provisionally assumed as a hypothesis and shown to be adequate to solve that problem. Such a hypothesis in turn generates a problem to be solved in much the same way, that is, introducing a new hypothesis, and so on. Thus solving mathematical problems appears as a potentially infinite process that consists in introducing more and more general hypotheses and considering their consequences. Every new hypothesis establishes new connections between the given problem and other areas of mathematical or non-mathematical knowledge, thus paving the way for a higher level of abstraction. From this viewpoint the growth of mathematical knowledge essentially consists in looking for new hypotheses to solve specific mathematical problems and establishing new connections between concepts involved in such problems and concepts in other areas" (Cellucci, 2000: 162).

The solution of a problem establishes a pathway through mathematics; but it is often, and perhaps typically, not a pathway through an axiomatized system, since it proceeds by means of provisional hypotheses (which are found in the process of proof generation) and often moves sideways into other domains by means of correlations.

It is a striking feature of the growth of mathematical knowledge that an important problem often leads the mathematician outside the domain in which it arose. Cellucci observes that these pathways "do not depend on permanent axioms but on provisional hypotheses that may be changed in the course of proof and are found by a trial-and-error process" and that "they involve constant interactions with other systems, where dialogue is essential because one cannot generally expect that solving a problem concerning a particular mathematical field will require only concepts and methods of that field" (Cellucci, 2000: 162).

In his essay "Tacit knowledge and mathematical progress" (Breger, 2000), Herbert Breger describes a process of mathematical analysis which has two stages. The first is the rise of a specific "know-how" or knowledge at the meta-level as mathematicians come to know their way around certain objects and the problems associated with them, a know-how which at first they may not be able to articulate; it remains tacit, or partially expressed. The second step turns tacit knowledge at the meta-level into general principles or new abstract objects along with theorems that govern them. But for mathematicians, such generalization is not an end in itself; it is pursued only if it reorganizes knowledge in deep and novel ways, explaining the know-how and suggesting ways to generate new problems. Thus, the investigation of particular things (e.g., transcendental curves like the catenary) leads to a partially formulated sense of what methods work to solve what kinds of problems (e.g., methods of differentiation, integration, and the solution of differential equations). This consolidation leads in turn to presentations of "a method that works for every particular curve" and thence to a method whose validity for every curve can be proved, and a formal theory. Detlef Laugwitz (2000) adds that the virtue of formal theory is precisely that it makes explicit what is tacit in a family of problem-solutions or methods, so that these procedures can be taught; otherwise great mathematicians take their inventive know-how with them to the grave.

But Breger warns that not all know-how can be made explicit and translated into an abstract theory. The relation of an axiomatized system to a thinking person, the know-how we employ in interpreting and applying its symbolic knowledge, the relations among axiomatized systems, even the source of the axioms themselves: all these things remain unstated in an axiomatized system. (Cellucci reminds us that Gödel's incompleteness results really mean that "any particular formal system is intrinsically provisional, subject to the eventual need to go beyond it" (Breger, 2000: 159).) Moreover, every advance in mathematical knowledge generates tacit knowledge: new tacit knowledge at the meta-level that will one day demand articulation, and older, more concrete levels of knowledge that, because they have been abstracted away from, have lost their articulation. The more concrete levels at one time constituted evidence for the more abstract levels, and yet they are often erased and cannot be read off the formal system. So there are always at least two dimensions of the tacit.

The accounts of both Cellucci and Breger reveal that the outcome of mathematical progress is not simply abstraction, even when abstract concepts are used to organize earlier results. Philosophers of mathematics often fail to reflect sufficiently on what they mean by the terms "abstract" and "concrete." In *The dappled world: a study of the boundaries of science*, Nancy Cartwright's meditation on why we must reflect more deeply on the relationship between the abstract and the concrete bears directly on this point (Cartwright, 1999: Ch. 2) She uses her argument to illuminate episodes in modern physics, yet clearly what she says bears on philosophy of mathematics. She argues, "First, a concept that is abstract relative to another more concrete set of descriptions never applies unless one of the more concrete descriptions also applies. These are the descriptions that can be used to "fit out" the abstract description on any given occasion. Second, satisfying the associated concrete description that applies on a particular occasion is what satisfying the abstract description consists in on that occasion" (Cartwright, 1999: 49). In other words, abstract descriptions can only be used to say true things if they are combined with concrete descriptions that fix their reference in any given situation. This insight holds for mathematics as well as for physics. When abstract schemata are applied in mathematics, their successful application typically depends on the mathematician's ability to find a useful concrete description for the occasion, which will mediate between complex mathematical reality and the general theory.

This does not, however, entail that abstract concepts are no more than collections of concrete concepts. Cartwright argues, "The meaning of an abstract concept depends to a large extent on its relations to other equally abstract concepts and cannot be given exclusively in terms of the more concrete concepts that fit it out from occasion to occasion"

(Cartwright, 1999: 40). The more abstract description of the situation adds important information that cannot be unpacked from any or even many of the concrete descriptions that might supplement it, or from our awareness of the thing or things successfully denoted. Nor is the relation between the abstract description and the concrete descriptions the same as the Aristotelian relation between genus and species, where the genus-concept is arrived at by subtracting content from the species-concept. Likewise, the more concrete descriptions have meanings of their own that are to a large extent independent of the meaning of any given abstract term they fall under; we cannot deduce the content of the concrete descriptions by specifying a few parameters or by plugging in constants for variables in the abstract description.

Here is another way of putting the insight. Abstract terms may be used to say something true when they are combined with more concrete locutions in different situations that help us to fix their reference. And there is no complete sum of such concrete locutions that would be equivalent to the original term. Conversely, as Leibniz argues, concrete terms can be used to say something true only when they are combined with more abstract locutions that express the conditions of intelligibility of the thing denoted, the formal causes that make the thing what it is and so make its resemblance to other things possible. And there is no complete sum of the conditions of intelligibility of a thing (Grosholz and Yakira, 1998: 56–72). We cannot totalize concrete terms to produce an abstract term and we cannot totalize abstract terms to produce a concrete term that names a thing; and furthermore we cannot write meaningfully and truthfully without distinguishing as well as combining concrete and abstract terms. So we are left with an essential ambiguity that results from the logical slippage that *must* obtain between the concrete and the abstract. There is an inhomogeneity that cannot be abolished, which obtains between the abstract terms that exhibit and organize the intelligibility of things, and the concrete terms that exhibit how our understanding bears on things that exist in the many ways that things exist. We need to use language that both exhibits the "what" of the discourse, and identifies the formal causes, the "why" of the things investigated. To do so, we need to use different modes of representation in tandem, or to use the same mode of representation in different ways, to use it ambiguously.

11.5   Coda

Breger's examples of the articulation of tacit knowledge in the development of mathematical theories reminds us that bringing objects at higher levels of abstraction into focus entails the forgetting or obscuring by formal systems of the canonical objects that exist at more concrete levels, levels which cannot be retrieved from the formal systems that forget them. Students who study function spaces lack instruction in the peculiarities of various families of curves and differential equations; and this information cannot be read out of the formalisms they study, but must be sought in the textbooks of an earlier era. Mathematical analysis often leads from the study of the conditions of solvability of problems about more concrete objects, to the elaboration of general methods, to the construction of formal systems that both delimit and generalize the scope of a method, processes which lead not only to new problems but to new, more abstract objects. Fermat's methods, for example, led from the study of numbers to the study of polynomials; and Galois's way of systematizing the study of the roots of polynomials led to group theory. The different levels of analysis are rationally related: indeed, I want to say that the integers serve as conditions of the intelligibility of polynomials, as the latter serve as conditions of intelligibility for group theory; yet the former cannot be retrieved from the latter. At the heart of mathematical analysis lies the solution of problems, even though problem-solving leads to methods and systems; methods and systems arise out of families of problems about related objects. And yet methods and systems often forget the original problems and objects from which they arose, their own conditions of intelligibility which in fact they cannot reduce; and so do the philosophers that contemplate them.

Andrew Wiles' proof of Fermat's last theorem establishes a "perfect" correlation between modular forms and "semi-stable" elliptic curves that we might call a reduction, but the correlation is established in the service of problem solving, and lies outside the bounds of any formal theory. It does not collapse the distinction between modular forms (which belong to complex analysis) and elliptic curves (which belong to algebraic number theory), but rather raises the question why they should share the same structure: it calls for explanation (Wiles, 1995). So too does Leibniz's inauguration of the domain of higher analysis, which investigates families of transcendental curves by means of differential equations and infinite series.[4] Typically, the mathematician posits a fragmentary formal, syntactic linkage between two heterogeneous domains; that link draws the two domains into a novel proximity, the link is extended and deepened, and the domains are altered and extended because of it. This procedure is useful in problem solving because mathematical domains tend to give rise to problems that can be articulated within that domain, but not solved there. To bring to bear another domain, with distinct but analogous ways of framing and solving problems, and to render it pertinent to the problem at hand by establishing relations of reduction between the domains, can be a very effective strategy. The problem, and the objects involved in it, can then be regarded from different perspectives simultaneously, and the resources for solving the problem are thereby enriched. The syntactic connectedness and the semantic heterogeneity are clearly both needed for this strategy to work. This heterogeneity corresponds to that located by Breger, between less abstract and more abstract stages held in rational relation by processes of analysis that move from the study of relatively more concrete objects to methods to systems or procedures.

One can react to the transcendence or sublimity or inhumanity of mathematical things by turning against the situation. One can become a sophist, content to shift appearances at the first level of the divided line, or a materialist, content to remain with the sensible at the second level, or an empiricist, who hopes that the third level of the line is just an abstractive or constructive extension of the second. And these are natural choices, to try to shelter oneself within the ambit of the human, the computable, and the visible. But philosophers who try to rest within what can be encompassed by finitary construction and perception in the end somehow never rest easy, and in any case never do justice to mathematics.

---

[4] Mancosu (1999). Paolo Mancosu here gives an instructive discussion of mathematical explanation.

They are plagued by difficulties (revealed by *reductio* arguments), unanswerable questions (revealed by burden of proof arguments), and by the way even the constructions and perceptions to which they cling seem willy-nilly to point beyond themselves. The other option, besides fleeing reason altogether, is to engage in analysis, that is, to recognize the top half of the divided line and try to find a philosophical speech about it, which acknowledges that it is real, that it stands in rational relation to us, and that its reality is different from ours. Even the finitary parts of mathematics involve the infinitary and even the visible world-system involves the invisible, as conditions of intelligibility.

The project of philosophical analysis is also self-reflective: it puts into question the relationship of the human mathematician (or philosopher) to the things of mathematics. We must not only try to explain the way in which knowledge unifies but also the way in which it holds things apart. Human beings, and their perception and constructions, are finitary and fleeting; but the things of mathematics are infinitary and eternal. If we try to reduce mathematics to merely human terms, to perceptions or finitary constructions, we avoid rather than confront the question of our rational relation to it, and the differences involved in that relation. In a sense, such confrontation—which sets us beside the working mathematician—is an exercise of our freedom, something about us which is after all infinitary and eternal (Vuillemin, 1997). If we pay attention to how differentiation and rational relatedness accompany each other, then not only the systematic unity of mathematics becomes salient but also the way in which systematic unities change over time as mathematicians explore the analogical likenesses of unlike things, or conversely try to articulate and conserve the intrinsic features of things apart from pressures exerted upon them by impinging analogies, or finally even try to consider likenesses in themselves, as novel methods and systems precipitate new items. In other words, the history of mathematics becomes pertinent to the philosophy of mathematics.

Logic, especially the expressive and subtle instrument of modern predicate logic, helps philosophers to examine mathematical systems considered as fixed and stable; but history provides evidence to help us understand how and why mathematics changes, and why it always moves beyond systematization. More generally, history is pertinent to the question of how mathematics brings very different kinds of things into determinate but revisable rational relation, and how this bears upon our philosophical understanding of rationality itself. Mathematics inspires us because it is at once inhuman and intelligible; it outstrips our finitary powers of construction, our perception, even our logic, at every turn, and nevertheless guides us because it stands in rational relation to us. The things of mathematics are problematic and yet intelligible, severely constraining what we can say about them; however, because they are so determinate, they render the little that we do manage to say about them necessary. Any finite thing in mathematics is at the same time an expression of the infinite, and the infinitary things that occur in our mathematics find finitary expression. This tension between the infinite and the finite in all of mathematics insures that our knowledge, despite its precision, must remain incomplete. Mathematics also stands at the crossroads of history and logic: essential as logic is to the articulation of relations among mathematical items, the very constitution of a problem in mathematics is historical, since problems constitute the boundary between the known and the yet to be discovered. We cannot explain the articulation of mathematical knowledge into problems and theorems without reference to both logic and history.

The philosophical reconstruction of mathematical practice is thus a delicate task, which requires scholarly familiarity with the detail of the mathematics of a given period, as well as the imaginative ability to go beyond the perspective of the mathematicians of that period and a respectful sense of their rationality. The problem with logic is its penchant for totalization and its intolerance of history; the outcome of mathematical progress is not always, and perhaps only rarely, an axiomatized system, where solved problems recast as theorems follow deductively from a set of special axioms, logical principles, and definitions. Careful study of the history of mathematics, even twentieth century mathematics, may discover that mathematicians pursue generality as often as they pursue abstraction, and sometimes prefer deeper understanding to formal proof. An axiomatic system is not the only model of theoretical unity, and deduction from first principles is not the only model for the location and justification of mathematical results.

··

REFERENCES

Alexandroff, P. and Hopf, H. (1935) *Topologie*. Berlin: Springer Verlag.

Bernoulli, J. (1742) *Opera Johannis Bernoulli*. Edited by G. Cramer. Geneva.

Breger, H. (2000) 'Tacit knowledge and mathematical progress', in *The growth of mathematical knowledge*, eds. E. Grosholz and H. Breger. Dordrecht: Kluwer, 221–30.

Cartwright, N. (1999) *The dappled world*. Cambridge: Cambridge University Press.

Cellucci, C. (2000) 'The growth of mathematical knowledge: an open world view' in *The growth of mathematical knowledge*, eds. E. Grosholz and H. Breger. Dordrecht: Kluwer, 153–76.

Cellucci, C. (2002) *Filosofia e matematica*. Rome: Editori Laterza.

Fréchet, M. (1928) *Les espaces abstraits et leur théorie considérée comme introduction à l'analyse générale*. Paris: Gauthier-Villars.

Grosholz, E. (1991) *Cartesian method and the problem of reduction*. Oxford: Clarendon Press.

Grosholz, E. and Yakira, E. (1998) 'Leibniz's Science of the Rational', *Studia Leibnitiana* Sonderheft 26. Stuttgart: Steiner Verlag.

Kuratowski, C. (1933) *Topologie I*. Warsaw/Lvov.

Laugwitz, D. (2000) 'Controversies about numbers and functions', in *The growth of mathematical knowledge*, eds. E. Grosholz and H. Breger. Dordrecht: Kluwer, 177–98.

Lefschetz, S. (1924) *L'analysis situs et la géométrie algébrique*. Paris: Gauthier-Villars.

Lefschetz, S. (1929) *Géométrie sur les surfaces et les variétés algébriques*. Paris: Gauthier-Villars.

Leibniz, G. W. (1962) *Mathematischen Schriften*, 7 vols. Edited by C. I. Gerhardt. Hildesheim: Georg Olms.

Leibniz, G. W. (1978) *Die Philosophischen Schriften*, 7 vols. Edited by C. I. Gerhardt. Hildesheim: Georg Olms.

Leibniz, G. W. (1989) *La Naissance du calcul différentiel, 26 articles des* Acta Eruditorum. Translated by M. Parmentier. Paris: Vrin.

Mancosu, P. (1999) *Philosophy of mathematics and mathematical practice in the seventeenth century*. Oxford: Oxford University Press.

Veblen, O. and Whitehead, J. H. C. (1932) *The foundations of differential geometry.* Cambridge: Cambridge University Press.

Vuillemin, J. (1997) 'La question de savoir s'il existe des réalités mathématiques a-t-elle un sens?', *Philosophia Scientiae* 2/2: 275–312.

Wiles, A. (1995) 'Modular elliptic curves and Fermat's Last Theorem', *Annals of Mathematics* 141/3: 443–551.

12

Models, structure, and generality in Clerk Maxwell's theory of electromagnetism

OLIVIER DARRIGOL

12.1 Introduction

In *La théorie physique* of 1906, Pierre Duhem opposed the "factory" of English physics to "the tidy and peaceful abode of deductive reason." He reproached his most eminent British colleagues, William Thomson and James Clerk Maxwell, with accumulating complex and disparate models at the expense of rigor and generality. This caricature has affected Maxwellian historiography, which often harbors the idea that Maxwell's taste for mechanical pictures ran against abstract generality.[1]

In the present essay, I defend the opposite thesis, namely: Maxwell aimed at general structures through his models, illustrations, formal analogies, and scientific metaphors. To this end, I analyze the gradual development of his electromagnetic theory, from his extension of an analogy invented by William Thomson to the great *Treatise on electricity and magnetism* of 1873. I also discuss a few texts in which Maxwell expounds his conception of physical theories, of their construction, and of their relation to mathematics.

12.2 An analogy by William Thomson (1842)

In 1842, the sixteen-year old William Thomson published a brief article in which he exploited an analogy between Fourier's theory of heat propagation and Poisson's theory of the equilibrium of electricity. According to the latter, the electric potential

$$\phi(\mathbf{r}) = \frac{1}{4\pi} \int \frac{\rho(\mathbf{r}')}{|(\mathbf{r} - \mathbf{r}')|} \mathrm{d}^3 r'$$

---

[1] Duhem (1914: 101).

The Oxford Handbook of Generality in Mathematics and the Sciences. First Edition. Karine Chemla, Renaud Chorlay and David Rabouin. © Oxford University Press 2016. Publishing in 2016 by Oxford University Press.

created by a charge distribution ρ satisfies the equation

$$\Delta\phi + \rho = 0.$$

According to Fourier's theory, the steady flow of heat in an unlimited homogenous medium of unit specific heat and unit thermal conductivity satisfies the similar relation

$$\Delta\theta + s = 0$$

between the temperature θ and the density s of the heat sources included in the medium. In the latter system, it is physically evident that the heat flux $\int -\nabla\theta \cdot d\mathbf{S}$ across a surface enclosing all sources must be equal to the total quantity of heat $\int s \, d^3 r$ emitted by the sources per unit time. By analogy, Thomson deduced the theorem according to which the integral $\int \mathbf{E} \cdot d\mathbf{S}$ of the electric force $\mathbf{E} = -\nabla\phi$ (that would act on a unit point charge) over a surface enclosing all charges must be equal to the total charge. He also established similar theorems by similar means.[2]

Through this recourse to physical intuition in the demonstration of a mathematical theorem, Thomson contradicted the idea of an abstract generality of mathematics. Yet he did not generate a physical kind of generality, because he shrank from defending a profound similarity between heat propagation and the mode of action of electricity. His interest in Faraday's ideas only began three years later, after he had understood that the properties of Faraday's electric and magnetic lines of force were very similar to those of the lines of heat flow in Fourier's theory.[3]

12.3   Maxwell on Faraday's lines of force (1855)

Some ten years later, James Clerk Maxwell saw in Thomson's analogical reasoning an opportunity for giving a precise geometrical expression to Faraday's ideas. Instead of taking heat propagation as a model, Maxwell considered the steady flow of an incompressible fluid through a resisting, porous medium in which the current is proportional to the pressure gradient. The latter relation plays the same role as the proportionality between heat flux and temperature gradient in Fourier's theory—to such an extent that one may wonder why Maxwell did not content himself with Thomson's thermal model. The reason probably is that the ideal character of the new model was more apparent and that it remained within the more primitive framework of mechanics.[4]

Thanks to adequate notions of tubes of flow and isobaric surfaces, Maxwell obtained a simple, rigorous, geometric description of his resisted flows, from which he drew a few useful theorems. For example, the pressure within an isobaric closed surface must be a constant if this surface does not contain any source. In order to show that, Maxwell reasoned that when there is no source, a thin tube of flow passing through a point of the

---

[2] Thomson (1842) and Wise (1981). These theorems were in part known to Green, Gauss, and Chasles.
[3] Smith and Wise (1989).
[4] Everitt (1975: 87–93).

internal volume would either form a loop within this volume or connect to points of the isobaric surface; since in the porous medium a pressure gradient must exist along any tube of flow, there cannot be any flow nor any pressure gradient in the internal volume.[5]

Maxwell related his ideal flows to three kinds of phenomena: steady electric currents in conductors, the electrostatics of linear dielectrics, and the magnetostatics of linear media. To the current of the ideal fluid corresponded, respectively, the electric current ($\mathbf{j}$), the "electric quantity" ($\mathbf{D}$), and the "magnetic quantity" ($\mathbf{B}$). To the sign-reversed pressure gradient corresponded, respectively, the "electric intensity" ($\mathbf{E}$), the same vector, and the "magnetic intensity" ($\mathbf{H}$). In Maxwell's later notation, to the proportionality between current and pressure gradient corresponded the three relations:

$$\mathbf{j} = \sigma\mathbf{E}, \ \mathbf{D} = \varepsilon\mathbf{E}, \ \text{and} \ \mathbf{B} = \mu\mathbf{H},$$

wherein σ is the electric conductivity, ε the dielectric permittivity, and μ the magnetic permeability. On the one hand, these notions corresponded to concepts already introduced by Faraday for experimental reasons. On the other, they permitted a precise mathematical theory of the three relevant classes of phenomena. To give only one example of this fecundity, the electric realization of the aforementioned theorem yields the uniformity of the electric potential within any equipotential closed surface that does not contain any charge.[6]

Maxwell explained the philosophy of his method in the following terms:

> The first process ... in the effectual study of [electrical] science, must be one of simplification to a form in which the mind can grasp them. The results of this simplification may take the form of a purely mathematical formula or of a physical hypothesis. In the first case we entirely lose sight of the phenomena to be explained; and though we may trace out the consequences of given laws, we can never obtain more extended views of the connexions of the subject. If, on the other hand, we adopt a physical hypothesis, we see the phenomena only through a medium, and are liable to that blindness to facts and rashness in assumption which a partial explanation encourages. We must therefore discover some method of investigation which allows the mind at every step to lay hold of a clear physical conception, without being committed to any theory founded on the physical science from which that conception is borrowed, so that it is neither drawn aside from the subject in pursuit of analytical subtleties, nor carried beyond the truth by a favorite hypothesis.—In order to obtain physical ideas without adopting a physical theory we must make ourselves familiar with the existence of physical analogies.

Thus, Maxwell regarded his method of physical analogy as a *via media* between the Charybdis of mathematical formalism and the Scylla of physical hypotheses. His aim was to grasp a formal analogy between two domains of physics without taking the risk of confusing the nature of the phenomena. Regarding the imaginary flows on which he based his memoir, he wrote:

> By referring everything to the purely geometrical idea of the motion of an imaginary fluid, I hope to attain generality and precision, and to avoid the dangers arising from a premature theory professing to explain the cause of the phenomena.

[5] Maxwell (1856).
[6] Wise (1979).

Maxwell believed he had succeeded in conciliating generality and ontological neutrality for the following reason: a single model, the steady flow of an incompressible fluid through a porous medium, served to illustrate three distinct classes of phenomena. Hence it was clear, on the one hand, that the model served as the paradigm of general structure belonging to at least these three classes of phenomena. On the other hand, the same model could evidently not represent the deeper physical nature of each of the three classes at the same time.[7]

Maxwell here defended a conception of generality similar to that detected by other historians in quite different contexts. In her studies of ancient Chinese mathematics, Karine Chemla has shown that the common practice of expressing algorithms through a typical concrete application did not imply the ignorance of their generality. In her study of Poincaré's seminal works on topology, Anne Robadey found that Poincaré frequently preferred to express general relations through paradigmatic examples and let his reader imagine abstract statements of more explicit generality.[8]

In Maxwell's memoir, however, the hydrodynamic model was not the only way to point to a general structure. Maxwell encountered the following difficulty: his model was only pertinent for electrokinetic, electrostatic, and magnetostatic phenomena taken separately. As soon as electromagnetic phenomena came into play, this model was no longer of any help. In particular, electromagnetic induction according to Faraday called for a concept of "electro-tonic" state of which Maxwell knew no mechanical counterpart. Facing this temporary failure of the illustrative approach, Maxwell retreated to mathematical formalism:[9]

> The idea of the electro-tonic state, however, has not yet presented itself to my mind in such a form that its nature and properties may be clearly explained without reference to mere symbols, and therefore I propose in the following investigation to use symbols freely, and to take for granted the ordinary mathematical operations.

Accordingly, Maxwell introduced the Latin letters a, b, c for the three Cartesian components of a "quantity" analogous to the current of an ideal fluid, and the three Greek letters α, β, γ for those of an "intensity" analogous to a force. The indices 1 and 2 specify the magnetic or electric nature of these vectors; the index 0 corresponds to an electro-tonic magnitude. For the sake of transparency, I here use Maxwell's later terminology in which $\mathbf{E}$ and $\mathbf{H}$ denote the electric and magnetic "forces," $\mathbf{D}$ and $\mathbf{B}$ the electric and magnetic "fluxes." From the Amperean equivalence between a current loop and a double magnetic sheet, Maxwell deduced the relation

$$\int \mathbf{H} \cdot d\mathbf{l} = \int \mathbf{j} \cdot d\mathbf{S}$$

between the circulation of the force $\mathbf{H}$ along a closed curve and the flux of the electric current $\mathbf{j}$ across a surface delimited by this curve, as well as its infinitesimal counterpart

$$\nabla \times \mathbf{H} = \mathbf{j}.$$

In a general manner, the operation $\nabla \times$ relates a flux to a force, since by definition a force naturally occurs in linear integrals (which give the work done) and a flux naturally

[7] Maxwell (1856: 156, 159).
[8] Chemla (2005); Robadey (2004).
[9] Maxwell (1856: 187).

occurs in a surface integral. Next, Maxwell expressed Faraday's law of electromagnetic induction in the form

$$\int \mathbf{E} \cdot d\mathbf{l} = -\frac{d}{dt} \int \mathbf{B} \cdot d\mathbf{S}.$$

With the flux **B** he associated the force **A** such that

$$\mathbf{B} = \nabla \times \mathbf{A},$$

which allowed him to rewrite the law of induction in the form

$$\mathbf{E} = -\frac{\partial \mathbf{A}}{\partial t},$$

in conformity with Faraday's intuition according to which the electromotive force of induction (**E**) is the temporal variation of the electro-tonic state (**A**).

Maxwell thus exploited the symbolic expression of his distinction between intensity and quantity (alias force/flux) in order to generate the equations that capture Faraday's empirical laws. This distinction kept playing an important role in his electromagnetic theory. Even though he did not dwell on its hydrodynamic origins in his subsequent writings, it would be wrong to believe that the words of flux and force became mere residues of an obsolete mode of illustration. On the contrary, Maxwell kept emphasizing the importance of scientific metaphors and refused to separate mathematical symbols from simple intuitive illustrations of their functioning, no matter how incompatible these illustrations could be with the global theory in which the symbols were used.

12.4 Maxwell's honeycomb (1861–2)

A few years elapsed before Maxwell discovered a satisfactory model for all known phenomena of electricity, magnetism, and their interactions. The aim of this model no longer was to clarify and sharpen Faraday's ideas; it rather was to demonstrate the mechanical essence of electromagnetic phenomena. The model consisted in an array of hexagonal cells separated by a layer of idle wheels—hence Duhem's name of "honeycomb." The cells' rotation corresponded to the magnetic field; the circulation of the idle wheels to the electric current. Elastic deformations of the cells corresponded to an electric field; their variation implied an additional shift of the idle wheels, thus causing the "displacement current" which entered Maxwell's final system of equations.[10]

Although Maxwell firmly believed in the reality of certain aspects of this model, he promptly remarked that the details of the suggested mechanism were too artificial to be plausible:

> The conception of a particle having its motion connected with that of a vortex by perfect rolling contact may appear somewhat awkward. I do not bring it forward as a mode of connexion existing in nature, or even as that which I would willingly assent to as an electrical hypothesis. It is, however, a mode of connexion which is mechanically conceivable, and

---

[10] Maxwell (1861–2). See also Siegel (1991).

easily investigated, and it serves to bring out the actual mechanical connexions between the known electro-magnetic phenomena; so that I venture to say that anyone who understands the provisional and temporary character of this hypothesis, will find himself rather helped than hindered by it in his search after the true interpretation of the phenomena.

As Maxwell more briefly explained in a contemporary letter to his friend Peter Guthrie Tait, "The nature of this mechanism is to the true mechanism what an orrery is to the solar system." In his *Treatise* of 1873, Maxwell underlined that an infinite number of distinct mechanisms were able to produce the same connections between two parts of a mechanical system.[11]

In sum, Maxwell obtained his celebrated system of equations and the concomitant unification of electricity, magnetism, and optics through a model which he himself judged largely arbitrary. In order to consolidate his results, he had to provide a proof that they did not depend on the details of the model. In other words, he needed to identify a general structure shared by all the possible mechanical models of the electromagnetic field. He reached this goal three years later, in his memoir entitled "The dynamical theory of the electromagnetic field."[12]

12.5 The Lagrangian approach

In an electrodynamics system, the empirically controllable quantities are the currents and the positions of the conductors in which they circulate. In conformance with the general assumption of the mechanical nature of physical systems, Maxwell assumed that these quantities completely determined the motion of a mostly hidden mechanism. Consequently, for a given value of impressed electromotive and mechanical forces these quantities evolve according to Lagrange's equations. These equations only depend on the expressions of the kinetic and potential energies of the system as functions of the observable quantities, which can be determined empirically.

Maxwell thus replaced the choice of a specific model of the underlying mechanism with the broader requirement of the Lagrangian structure of the fundamental equations. This requirement, together with Faraday's assumptions about the electric current and with the empirically known expressions of electric and magnetic energies, led to the same field equations as his model of 1862. It did so in a more economical and less speculative manner, since any consideration of the true mechanism chosen by nature was thus short-circuited. Only the possibility of a mechanism mattered, not its knowledge. Maxwell privileged this point of view in his *Treatise* of 1873, and so did three of his most eminent readers: Hermann Helmholtz, Henri Poincaré, and Hendrik Lorentz. The modern requirement of the Lagrangian form of the fundamental equations of physics is a direct descendant of this new style of theoretical physics.[13]

---

[11] Maxwell (1861–2: 486); Maxwell to Tait, 23 December 1867, in Harman (1990–2002, vol. 2: 337); Maxwell (1873: §531).

[12] Maxwell (1865).

[13] Moyer (1977); Stein (1981); Buchwald (1985: 20–3).

It would nonetheless be excessive to believe that Maxwell accorded to the Lagrangian structure the same status as it now has in field theories. He kept hoping that a mechanical model of the ether, simpler and better than the honeycomb of 1862, would someday enlighten his theory. Moreover, he did not regard the Lagrangian form of the fundamental equations of a physical theory as a purely formal requirement. Along with his friends Thomson and Tait, he interpreted Lagrange's equations as an extension of Newton's second law in which generalized forces were equated to the temporal derivatives of generalized momenta. He concretely defined the generalized forces through the work produced during an infinitesimal change of configuration, and the generalized momenta as the impulses needed to suddenly start the motion of the system from rest.[14]

Thus, Maxwell's striving for structural generality did not eliminate concrete reference. We earlier saw that his distinction between force and flux kept bearing the trace of the hydrodynamic metaphor that engendered it. Similarly, Maxwell accompanied his requirement of a Lagrangian structure with metaphors borrowed from the energetic theory of machines. The most famous of these metaphors is that of the belfry, which Maxwell included in a review of Thomson and Tait's *Treatise on natural philosophy*.[15]

Granted that the angles of the bells are completely determined by the vertical shift of the driving cords of the belfry, a learned clerk could in principle predict the spontaneous motion of these cords from the initial value of their shifts and velocities. To this end, he would only have to determine, by a preliminary experiment, the value of the impulses $p_i(q, \dot{q})$ necessary to communicate to the cords any given sequence of velocities $\dot{q}_i$ for any given sequence of values of the shifts q_i. In the general case, if the forces applied to the cords have the values f_i, the motion is then given by Lagrange's equations

$$f_i = \frac{\mathrm{d}}{\mathrm{d}t} p_i(q, \dot{q}).$$

It is therefore possible to determine this motion without knowing the hidden mechanism of the belfry, just as it is possible to determine the evolution of an electrodynamic system without knowing the hidden mechanism of the ether.[16]

12.6 Maxwell's ulterior reflections

All along the elaboration of his electromagnetic theory, Maxwell was guided by general structural requirements that were inspired by partial and temporary models. The *Treatise on electricity and magnetism* contains a systematic exposition of these requirements. Through this voluminous and dense treatise, Maxwell not only aimed at gathering contemporary experimental and theoretical knowledge but also at illustrating a new style of theory construction. This style implied two basic ingredients: the

---

[14] Smith and Wise (1989: 390–5).
[15] Maxwell (1879: 783–4).
[16] For the sake of simplicity, I have neglected the potential energy.

classification of physico-mathematical quantities, and the Lagrangian structure. In a long "Preliminary on the measurement of quantities," Maxwell arranged the various quantities according to their dimension (Fourier), to their continuous or discontinuous character, to the scalar/vector distinction (Hamilton), to the force/flux distinction which he had himself invented, and to topological properties. In a separate chapter devoted to connected systems, he propounded a proof that the energy principle by itself implied the Lagrangian structure which he subsequently applied to electromagnetism.[17]

That Maxwell treated the classification of physical quantities and the Lagrangian structure in different parts of his treatise clearly shows that he did not place them on the same footing. The reason for this separation is not difficult to guess: in the first case, it was a matter of partial structures controlling the insertion of the quantities in particular sectors of the theory; whereas in the second case it was a matter of a structure globally imposed on the theory.

We may now compare the kinds of generality implied in these two kinds of requirements. In the case of the force/flux distinction, Maxwell meant to identify the common structure that made various partial theories (electrokinetics, electrostatics, and magnetostatics) analogous to the same hydrodynamic model. In the case of the Lagrangian structure, Maxwell associated a single theory (the global electromagnetic theory) with an infinite class of mechanical models that were all able to mimic its laws; and he meant to identify the structure shared by all these model representations.

While he was composing his treatise, Maxwell gave some thoughts to the merits of the kind of classification and illustration that he had developed. These thoughts can be found in two conferences which he gave in 1870, one for the London Mathematical Society and the other for the British Association for the Advancement of Science. In front of the mathematicians, he emphasized the economy of time that resulted from the "mathematical classification of physical quantities":

> It is evident that all analogies of this kind depend on principles of a more fundamental nature; and that, if we had a true mathematical classification of quantities, we should be able at once to detect the analogy between any systems of quantities presented to us and other systems of quantities in known sciences, so that we should lose no time in availing ourselves of the mathematical labours of those who have already solved problems essentially the same.

As examples of such classifications, he gave those he included in the *Treatise*. He also introduced the terms *convergence, curl,* and *concentration* for the operators $-\nabla\cdot$, $\nabla\times$, and $-\nabla^2$ formed from the gradient operator ∇ (nabla). And he drew field archetypes for which these quantities had a local extremum. In general, Maxwell did not introduce a symbol without accompanying it with a simple geometrical or mechanical illustration. While he emphasized the benefits that physics drew from the mathematical classification of

---

[17] Maxwell (1873: §§ 1–26, 553–68). Cf. Harman (1987). Maxwell's proof that the Lagrangian structure derives from the energy principle is flawed.

quantities, he also reminded his audience that mathematics could benefit from associated physical contents.[18]

The latter idea of the symbiotic development of mathematics and physics provides the central theme of the conference that Maxwell pronounced in front of the mathematical and physical sections of the British Association:

> If the skill of the mathematician has enabled the experimentalist to see that the quantities which he has measured are connected by necessary relations, the discoveries of physics have revealed to the mathematician new forms of quantities which he could never have imagined for himself.

Maxwell meant that physics borrowed from mathematics the arithmetic needed for the measurement of quantities while the classification of the various kinds of quantities and the mathematics associated with this classification proceeded from physical ideas.[19]

More broadly, Maxwell distinguished between two kinds of "clear" thinking:

> The human mind is seldom satisfied, and is certainly never exercising its highest functions, when it is doing the work of a calculating machine. What the man of science, whether he is a mathematician or a physical enquirer, aims at is, to acquire and develop clear ideas of the things he deals with. For this purpose, he is willing to enter on long calculations, and to be for a season a calculating machine, if he can only at last make his ideas clearer.—But if he finds that clear ideas are not to be obtained by means of processes the steps of which he is sure to forget before he has reached the conclusion, it is much better that he should turn to another method, and try to understand the subject by means of well-chosen illustrations derived from subjects with which he is more familiar.

Formalist and algorithmic clarity had to give way to the clarity of familiar pictures when the complexity of the subject defied computational methods. Formal analogies or illustrations not only permitted the more efficient and pleasant teaching of a subject, they also enriched our knowledge of the two compared domains.[20]

Maxwell's preference for the illustrative approach is evident in his researches. Unlike most of today's physicists, he did not regard the equations that he wrote in his memoirs as mere concatenations of symbols according to rigid rules. To each equation or to each part of an equation he associated images and processes, most often geometrical or mechanical. Instead of algebraically combining a few equations in order to generate new laws, he mentally combined the images associated with these equations. Even though this way of thinking generated Maxwell's most shining successes, it is not without dangers. The images associated with two differential equations may be incompatible, in which case their combination leads to aberrant conclusions.[21]

---

[18] Maxwell (1870b: 258).
[19] Maxwell (1870a: 218).
[20] Maxwell (1870a: 219).
[21] Achard (2003).

To cite only one example of the latter pitfall, in his memoir of 1865 on the dynamical theory of the electromagnetic field, Maxwell illustrated the equation $\mathbf{D} = \varepsilon\mathbf{E}$ (relating the "polarization" $\mathbf{D}$ to the "force" $\mathbf{E}$) by microscopic shifts of electricity within the molecules of the medium. This picture, borrowed from Poisson's and Mossotti's theories of magnets and dielectrics, implied the macroscopic electric charge

$$\rho = -\nabla \cdot \mathbf{D}$$

in a medium of heterogeneous polarization. In addition, Maxwell admitted Faraday's idea that the source of the magnetic field must be the total electric current $\mathbf{J}$ made of the displacement current $\partial\mathbf{D} / \partial t$ and of the conduction current $\mathbf{j}$. Through the relation

$$\nabla \times \mathbf{H} = \mathbf{J}$$

this assumption leads to

$$0 = \nabla \cdot \mathbf{J} = \frac{\partial}{\partial t}\nabla \cdot \mathbf{D} + \nabla \cdot \mathbf{j} = -\frac{\partial\rho}{\partial t} + \nabla \cdot \mathbf{j},$$

whereas the conservation of electricity requires

$$\frac{\partial\rho}{\partial t} + \nabla \cdot \mathbf{j} = 0.$$

As Maxwell did not combine the equations in a purely algebraic manner, he did not see this contradiction in this memoir of 1865. He solved it much later, in his *Treatise* of 1873.[22]

Even though Maxwell favored the illustrative approach in his own researches and in his teaching, he acknowledged the existence of two kinds of minds, somewhat like Duhem's French and English physicists:

> There are men who, when any relation or law, however complex, is put before them in a symbolical form, can grasp its full meaning as a relation among abstract quantities. Such men sometimes treat with indifference the further statement that quantities actually exist in nature which fulfill this relation. The mental image of the concrete reality seems rather to disturb than to assist their contemplations.—But the great majority of mankind are utterly unable, without long training, to retain in their minds the unembodied symbols of the pure mathematician, so that, if science is ever to become popular, and yet remain scientific, it must be by a profound study and a copious application of those principles of the mathematical classification of quantities which, as we have seen, lie at the root of every truly scientific illustration.

Accordingly, Maxwell recommended a balanced pedagogy, one that would satisfy the two kinds of minds:[23]

---

[22] Maxwell (1865, 1873: §64).
[23] Maxwell (1870a: 219–20).

For the sake of persons of these different types, scientific truth should be presented in different forms, and should be regarded as equally scientific, whether it appears in the robust form and the vivid colouring of a physical illustration, or in the tenuity and paleness of a symbolical expression.

12.7 Conclusions

Maxwell's pedagogical precepts reflect the ambiguity inherent in his use of formal analogies, illustrations, or scientific metaphors. On the one hand, his artistic sensitivity and his impregnation with Scottish Common Sense philosophy induced him to define the intelligibility of phenomena through their representation by geometrical or mechanical constructs.[24] On the other hand, he was completely aware of the arbitrary and particular character of these representations and he sought to extract from them universal structures expressible in mathematical symbols. Three forms of generality resulted. In the first form, a specially clear and simple model played the role of a common paradigm for a substructure frequently encountered in theoretical physics. In the second form, specific notation and adequate symbolic rules permitted a direct imposition of these substructures. In the third form, a general principle such as the energy principle or the principle of least action served to express the possibility of a mechanical model of the global theory without explicating this model.

Some of Maxwell's readers, Helmholtz and Poincaré notoriously, saw in his use of models a springboard toward a more modern theoretical physics based on very general structural requirements. This is what Poincaré called "the physics of principles," the principles being energy conservation, least action, equality of action and reaction, etc. Maxwell, for himself, did not wish to relegate mechanical images to the attic of science. In his eyes, a lively, clear, and intuitive physics needed a constant recourse to images. Generality was better conceived as the focal point of a beam of sensible specificities.[25]

Acknowledgments

I thank Karine Chemla and Renaud Chorlay for their fruitful comments on an early version of this text.

....

REFERENCES

In the following, *MSP* refers to Maxwell, J. C. (1890) *The scientific papers*, 2 vols. Cambridge: Cambridge University Press.

Achard, F. (2003) Argumentation et thèmes méthodologiques dans les publications théoriques de James Clerk Maxwell. Thèse de doctorat, EHESS, Paris, France.

[24] Cat (2001).
[25] Poincaré (1904).

Buchwald, J. (1985) *From Maxwell to microphysics: Aspects of microphysics in the last quarter of the nineteenth century.* Chicago: Chicago University Press.

Cat, J. (2001) 'On understanding: Maxwell on the methods of illustration and scientific metaphor', *Studies in history and philosophy of modern physics* 32: 395–441.

Chemla, K. (2005) 'Geometrical figures and generality in ancient China and beyond: Liu Hui and Zhao Shuang, Plato and Thabit ibn Qurra', *Science in context* 18: 123–66.

Duhem, P. (1914) *La Théorie physique. Son objet et sa structure*, 2nd ed. Paris: Rivière.

Everitt, C. W. F. (1975) *James Clerk Maxwell. Physicist and natural philosopher.* New York: Scribner.

Harman, P. (1987) 'Mathematics and reality in Maxwell's dynamical physics', in R. Kargon and P. Achinstein (eds.), *Kelvin's Baltimore lectures and modern theoretical physics: Historical and philosophical perspectives.* Cambridge: Cambridge University Press, 267–97.

Harman, P. (ed.) (1990–2002) *The scientific letters and papers of James Clerk Maxwell*, 3 vols. Cambridge: Cambridge University Press.

Maxwell, J. C. (1856) 'On Faraday's lines of force,' *Cambridge Philosophical Society, Transactions*, also in *MSP* 1: 155–229.

Maxwell, J. C. (1861–2) 'On physical lines of force,' *Philosophical magazine*, also in *MSP* 1: 451–513.

Maxwell, J. C. (1865) 'A dynamical theory of the electromagnetic field,' *Royal Society of London, Philosophical transactions*, also in *MSP* 1: 586–597.

Maxwell, J. C. (1870a) 'Address to the mathematical and physical sections of the British Association,' *British Association report*, also in *MSP* 2: 215–29.

Maxwell, J. C. (1870b) 'Remarks on the mathematical classification of mathematical quantities,' *Mathematical Society of London, Proceedings*, also in *MSP* 2: 257–66.

Maxwell, J. C. (1873) *A treatise on electricity and magnetism.* Oxford: Oxford University Press.

Maxwell, J. C. (1879) 'Thomson and Tait's natural philosophy,' *Nature*, also in *MSP* 2: 776–85.

Moyer, D. (1877) 'Energy, dynamics, hidden machinery: Rankine, Thomson, Tait, and Maxwell', *Studies in history and philosophy of science* 8: 251–68.

Poincaré, H. (1904) 'L'état actuel et l'avenir de la physique mathématique' (Saint-Louis conference), *Bulletin des sciences mathématiques* 28: 302–24.

Robadey, A. (2004) 'Exploration d'un mode d'écriture de la généralité: L'article de Poincaré sur les lignes géodésiques des surfaces convexes (1905)', *Revue d'histoire des mathématiques* 10: 257–318.

Siegel, D. (1991) *Innovation in Maxwell's electromagnetic theory: Molecular vortices, displacement current, and light.* Cambridge: Cambridge University Press.

Smith, C., and Wise, N. (1989) *Energy and empire: A biographical study of Lord Kelvin.* Cambridge: Cambridge University Press.

Stein, H. (1981) 'Subtler forms of matter in the period following Maxwell', in G. Cantor and M. J. S. Hodge (eds.), *Conceptions of ether theories, 1740–1900.* Cambridge: Cambridge University Press, 309–40.

Thomson, W. (1842) 'On the uniform motion of heat in homogeneous solid bodies, and its connection with the mathematical theory of electricity', *The Cambridge mathematical journal*, also in *Reprint of papers on electrostatics and magnetism.* London: MacMillan, 1872, 1–14.

Wise, N. (1979), 'The mutual embrace of electricity and magnetism', *Science* 203: 1310–18.

Wise, N. (1981) 'The flow analogy to electricity and magnetism—Part I: William Thomson's reformulation of action at a distance', *Archive for the history of exact sciences* 25: 19–70.

Section III.2

A diachronic approach: continuity and contrasts

13

Biological generality: general anatomy from Xavier Bichat to Louis Ranvier

JEAN-GAËL BARBARA

> *Avis au lecteur, Cet ouvrage est très différent de ceux qui ont paru sur le même sujet.*
> *Les uns remplis de préceptes communs, rebutent par leur longueur : les autres bornés*
> *à de simples catalogues ...*
>
> Le Cuisinier Gascon, Amsterdam, 1741.[1]

13.1 Introduction

Epistemological studies on generality in the life sciences have primarily focused on the concept of natural law and the generality of theories. Contemporary philosophers often conclude that no such laws or theories, nor natural kinds,[2] or even generalizations[3] exist in biology. Similarly, Richard Burian and coworkers have suggested that generalizing had its dangers in the philosophy of science. Consequently, these authors have argued that the philosophy of science should study how biological knowledge developed locally. They advocate a descriptive epistemological approach[4] within precisely defined historical and geographical contexts. Even so, we may ask why and how generality was occasionally seen to possess great importance. In keeping with these views, studies focusing on the latter questions are more likely to bring answers if they focus on specific historical milieus. This is precisely the way, in this chapter, that we suggest addressing the following issues: How was generality built and expressed locally? What did it mean?

---

[1] "Note to readers, this work is far different from those published on the same subject. Some full of common precepts discourage by their length: others are limited to simple tables ..." Louis-Auguste de Bourbon (1700–1755), Prince de Dombes, expresses a common idea in the eighteenth century, a middle way between the art of dealing with details ("simple catalogues") and the aim of highlighting general principles ("préceptes communs"), in agreement with the spirit of analysis (Bourbon, 1741). All translations by the author.

[2] I. Hacking, 25 April 2006 Lecture at the *Collège de France*. See the paper by Y. Cambefort in this volume.

[3] Burian et al. (1996).

[4] Burian et al. (1996: 25).

The Oxford Handbook of Generality in Mathematics and the Sciences. First Edition. Karine Chemla, Renaud Chorlay and David Rabouin. © Oxford University Press 2016. Publishing in 2016 by Oxford University Press.

The discipline of *anatomie générale* was founded in France in 1800 by Xavier Bichat (1771–1802). Shortly after his death, it was taken as a model throughout Europe for the study of medicine, human anatomy, and pathology. Bichat's anatomy was later developed in the 1870s in France by Louis Ranvier (1835–1922) at the *Collège de France,* by means of microscopy. Contemporary debates in science studies suggest paying attention to local epistemic cultures and usually deny general biological laws. This chapter echoes the first suggestion while rejecting the second; it is intended to study Bichat's and Ranvier's interests in generality as an actor's category. We hope that such an approach can renew discussions on the role of generality in the life sciences.

Bichat's concern regarding generality was not a simple consequence of his vitalism. In fact, Bichat claimed not to rely on a unique and mysterious vital principle, but rather to study particular and complex properties of living matter. Nevertheless, his approach focused on the observation of more or less strict experimental regularities in biological phenomena and observations of structures, an attitude which we will try to account for. Bichat's work is of interest for understanding which kind of concept of generality gained favor in the life sciences at the start of the nineteenth century, and why that happened. Moreover, it sheds light on its present-day status in experimental biology.

The later career of Louis Ranvier, who held the first chair in *Anatomie générale* in France, will inform us about the ways that generality was searched for at the microscopic level and its significance in the discovery of real and minute biological objects. With Ranvier, we will get closer to our understanding of the value of generality in biology today. We will study the context in which Ranvier came to value generality as a heuristic category and how this led him to discover new microscopic objects. We will concentrate in particular on the role of cellular theory and Claude Bernard's experimental physiology.

These case studies will allow us to analyze the changes in meanings and expressions of generality in Bichat and Ranvier. In both cases, generality was at the heart of a new disciplinary approach. The search for generality promoted the search for general facts—common traits encountered in multiple observations, thought of as being part of one ensemble, defined by a criterion—and their use to define the objects of study. The works of both Bichat and Ranvier involved the two disciplines of anatomy and physiology. The sense of generality in biological sciences changed in this cross fertilization. Their studies aimed to correlate facts common to anatomy and physiology and to search for criteria which did not rely solely on either one of them. Consequently, their method could define biological objects and categories by spatially overlapping anatomical and physiological traits.

13.2   Bichat's *anatomie générale*

Xavier Bichat is usually considered the founder of general anatomy. This young surgeon had a private course in anatomy and edited, after the death of his master Pierre Joseph Desault (1744–1795), his courses in surgery. Bichat died prematurely at age 31, after publishing two anatomical books, *Traité des membranes*[5] and the *Anatomie générale*,[6] which had profound impacts on the teaching and future directions of research in human anatomy and medicine.

His work was based on a new concept borrowed from the English school of surgery, the concept of "tissue." The study of such tissues, namely "histology," now relies on subtle microscopic observations and selective staining procedures that were not available in Bichat's time. Rather, Bichat made use of physiology and pathology to make distinctions between anatomically similar tissues.

Bichat's method can be most easily understood by using the example of the first type of tissue in his first category of membranous tissues, mucous membranes.[7] A first approach was to search for general traits defining an ensemble of anatomical parts as a specific tissue. This search was guided by *a priori* criteria which could be changed after study. Bichat's criterion for mucous membranes defined them as those inside cavities in continuity with the skin (mouth and nose, for example). Using another criterion of spatial continuity, Bichat divided mucous membranes into two ensembles, whose membranes were in continuity. Accordingly, a first ensemble lay in the interior of nose, mouth, pharynx, larynx, esophagus, stomach, intestine, and anus, while a second is in the urethra, ureter, kidneys, and prostate or vagina. At a physiological and pathological level, a common pathological property of the first kind of membrane seemed apparent when, during a cold, inflammation of the mucous membrane spread from nose to throat, or vice versa, possibly to sinus or bronchi. In case of stomach acid reflux, inflammation can invade the esophagus to larynx and ears. This example shows how a priori criteria were used to define kinds of tissues, and how tissue categories were defined a posteriori, by reference not only to anatomical, but also to physiological and pathological properties. This approach enabled Bichat to build a single category for membranes of nose, mouth, pharynx, larynx, esophagus, stomach, intestine, and anus, which is referred to as a single mucous tissue.[8] The novelty of this approach relied on defining anatomical parts of the body using non anatomical properties, and the discovery of converging criteria from different disciplines, in the partition of the human body, for defining single types of tissues.

Consequently, the discipline that Bichat established is usually described as a "physiological anatomy."[9] Defining general anatomical categories as types of tissues (general anatomy), Bichat attempted to describe their organization within organs in a new manner (descriptive anatomy), as well as to understand their functional properties and functions (physiology). His systematic use of the concept of tissue revolutionized anatomy, descriptive anatomy, physiology, pathology, and medicine. After his death, Bichat's method was recognized as the future direction to follow in all of these fields. Its main feature was the expression of the general and the particular in a new physiological perspective, following

---

[5] Bichat (1799).

[6] Bichat (1801).

[7] Bichat (1799: art. II).

[8] Bichat will later make other, finer, distinctions in his tissue categories and his system of tissues will be later criticized for its overwhelming complexity.

[9] See, for example, Flourens (1858: 244–7).

the work of Haller (1708–1777), in correlating anatomy and pathology. This required a diversity of practical procedures, that is, experimental physiology performed on animals, human dissections of healthy and diseased bodies, and the inspection of sick persons, which Bichat skillfully used within a single perspective. In all these studies, Bichat was focused on general facts in order to compare each of them to others obtained with different procedures from other disciplines. This explains how and why generality as he practiced it was central to his approach.

13.2.1 Bringing anatomy and physiology closer together

We shall now analyze Bichat's work within a historical perspective, in order to highlight the contexts in which he came to study anatomy, surgery, and medicine conjointly in this new way. However, before addressing this issue, let us make some preliminary remarks on the way generality had been used as a category for comparing knowledge coming from different disciplines. In the history of biology, generality has often been associated with explanation and causality. The central goal of Bichat was not only a new description of the human anatomy. For medical practice, he also wished to understand the function of each type of tissue and the causes of pathological lesions. Such relations between generality and causation appear in ancient Greek physics and medicine. In his *Physics*, Aristotle claimed to look for general causes within general facts, and particular causes within the particular.[10] The study of general facts was considered a path to discover the general causes involved.

In eighteenth- and nineteenth-century anatomy and physiology, general anatomical facts were extracted from observations and considered as possible causes of the function of organs in an Aristotelian perspective.[11] As late as the nineteenth century, Rudolf Virchow expressed a similar opinion, when he asserted that new general anatomical observations had to be developed prior to making any physiological discovery.[12] General anatomical categories were thus frequently sought in order to explain function.[13] This is what, for instance, Georges Canguilhem (1904–1995) stressed, in his comment on Galen's approach in *De usu partium*:

> Anatomy describes organs, physiology explains their functions. How can one claim to deduce physiological rules from anatomical techniques? In fact, any form of physiology understood in this way amounted more or less to a discourse on the utility and use of parts of organisms.[14]

---

[10] Aristotle (*Physics*, Book II, part 3).
[11] Debru (1996: 28).
[12] Virchow (1861).
[13] Hall (1968).
[14] The discourse on the use of parts of animals refers to Galen's project to deduce the function of organs from the knowledge of their structures. "L'anatomie est la description des organes, la physiologie est l'explication de leurs fonctions. Comment prétendre déduire des techniques de la première les règles de la seconde ? En fait, toute physiologie ainsi entendue revenait plus ou moins ... à un discours sur l'utilité et l'usage des parties de l'organisme" (Canguilhem, 1968a: 227).

By contrast, Claude Bernard's experimental physiology was built against such a principle.[15] However, even if, before Claude Bernard, physiology was already empirical and often experimental, on a theoretical level it provided merely a discussion of general anatomical facts to explain the functions of organs. Haller referred to it as *Anatomia animata*. Although generality was central in discussions linking anatomy and physiology, the relations between these fields were teleological and polarized from anatomy to physiology, never relying systematically on experimental physiology or pathology.

In the eighteenth century, physiologists did not only seek to explain function in terms of general structures, but they also began to propose physiological principles—vitalist or mechanist—to account for life. The vitalist approach defined functional properties, for which no causal explanation was sought. For example, irritability was defined as the property of muscles to contract, without inquiring into its origin. Haller avoided any discussion of the causes of physiological properties, referring to Newton's lack of discussion of the origin of gravitational force. *Sensibility*, the power of a stimulated tissue to alert an animal, and *irritability*, the power of inducing contraction in a stimulated tissue, were considered two general physiological properties to be studied. Experimental physiology established these concepts as properties mapped onto organs throughout the body. These researches invented a generality of physiological facts and subsumed them in broad categories of vital properties, which Bichat adopted.

Such physiological studies did not try to account for vital properties in terms of underlying anatomical structures and mechanisms, but rather aimed to understand how they contributed to specific functions. However, such studies were not made without reference to anatomy. Why, then, were general anatomical facts of interest in such an approach?

Bichat answered this question in a specific way. He was among the leading anatomists adopting Haller's style of physiological research. His personal interest was in establishing correlations between the general physiological properties described by Haller and general anatomical descriptions in order to define tissues as overlapping categories between those of anatomy and physiology. As we have sketched above, since, for instance, inflammation of the mucous membranes of the urethra never occurred after a cold, for him it belonged to a different tissue category than that of the membrane of the respiratory and digestive tracks. Therefore, he correlated anatomical, physiological, and pathological properties to define categories of tissues. All such observations converged on precisely determined surfaces of the body to define a tissue as a single and general object. Bichat considered tissues as anatomical entities "carrying"[16] specific properties. His goal was, on the one hand, to describe functional differences in apparently similar anatomical objects which could account for specific physiological properties and pathologies, and on the other hand, to search for functional similarities between anatomical objects with different textures, colors, or tastes. Bichat conducted all possible confrontations between classical anatomy and medicine, by relying on the senses and new experimental physiology. He wished to arbitrate between structure and function, to discover a middle way of defining tissues.

---

[15] Bernard advocated that functions should be discovered by experimental physiology alone, without prior reference to structures. Accordingly, physiology should, in his view, be independent from anatomy.

[16] Bichat's term.

Locating physiological and pathological properties on anatomical maps allowed him to localize them onto particular territories, defined as tissues. Although Bichat adopted Hallerian physiology, he no longer associated anatomy and physiology in a unidirectional and causal relationship. Their connections relied on overlapping generalities in the organization of life, described only in terms of visible entities, where anatomy and physiology could compare, correlate, and combine their spatial partitions of the body.

13.2.2   The choice to focus on tissues within a historical perspective

Let us now examine the historical contexts in which Bichat decided to found his research on tissues. Bichat's approach is both original and representative of anatomy in the second half of the eighteenth century. As we have already suggested, his quest for generality is apparent in the identification of general anatomical entities and their spatial correlations with physiological properties, which led to the creation of new objects. The history of this approach has been studied by Othmar Keel,[17] who discovered that the concept of tissue was first described by students from John Hunter's school of surgery (1728–93), followed by Philippe Pinel (1745–1826), and later Bichat. Tissues were initially identified by their distinct sensitivity to inflammation, and thus spatially localized by the extension of physiopathological properties.[18] The attention paid to such properties is representative of Bichat's method, which he used systematically. However, Bichat distinguished himself as the only author to make use of general anatomy in this context. Most authors, including British scientists, agree on this fact. We may account for this feature of his approach by referring to his personal method of searching for generality. This search is characteristic of his dominant style of research, as he was the only one who required that a collection of specific practices define the topography of physiological and pathological properties, and demanded their spatial correlations with anatomical observations. Moreover, Bichat gave a theoretical content to the use of the tissue as a concept, and made it the centerpiece of his new method of anatomy. Practices from different disciplines, systematically developed conjointly, were for him a means of avoiding the limitations of anatomy alone and of defining real biological objects.

Bichat's starting point was anatomy studied by the senses. He inspected hundreds of cadavers and believed that the repetition of dissections led to clear and general ideas, since "in this matter, observation is all, as in most physical sciences…. Images last only when they are repeated: the first image is fleeting, the second confused and the third is often indistinct. Senses can teach us better than books."[19] For Bichat, general facts emerged from repeated observations and were those which most struck the mind and represented the common denominator of all the observations.

---

[17] Keel (1979, 1982).

[18] According to Flourens, the term *tissue* was taken by Bichat from Bordeu (Flourens, 1858: 235).

[19] "Ici, l'inspection est tout, comme dans la plupart des sciences physiques … Les images ne sont durables qu'autant qu'elles sont répétées : la première fuit ; la seconde est confuse ; souvent la troisième n'est pas distincte. Les sens, mieux que les livres, peuvent nous instruire" (Bichat, 1829: xxv).

Bichat emphasized the primary importance of the senses in observation, advising that one should "point out phenomena, and even often refrain from looking for the connection between them;" [Bichat added], "this is most often what we ought to."[20] But Bichat's own practice seemed to some of his contemporaries to contradict this claim. François Magendie (1783–1855) emphasized this point in a note to the 1827 edition of the *Traité des membranes*: "Principles given here by Bichat are wise and truly philosophical; we regret his active imagination led him away from them too often."[21] While Magendie's specific approach advocated against extracting general facts from observations, Bichat used "connections" and "analogies"[22] freely for grasping generality. Thus in Bichat's practice, constructing and deriving general facts relied on a psychological procedure, which actively drew on connections and analogies to reach a higher level of generality in a rational manner.

Thus, we come across two modes of creating generality in Bichat's method:

(1) One is based on spatial correlations between anatomical boundaries and the extension of pathophysiological properties, which leads to the definition of general tissues. Such tissues are general in the sense that they represent general entities within three sets of orders: anatomical, physiological, and pathological. They are also general since they can be found in different organs of the body. However, this latter form of generality derives from the former since the properties under consideration and their associations are general and consequently occur in different areas of the body.

(2) A second mode of achieving generality is present in each of these disciplines—anatomy, for example—when similar observations are abstracted by mental processing to build a single and general representation. In this case, this representation is what remains in memory or is actively built. Moreover, it has an educational virtue for teaching.

All these processes for achieving generality require comparing abstractions derived by focusing on similarities and differences. Bichat's way of observing evokes later practices of nineteenth-century anatomists, as described by Ranvier or Mathias Duval (1844–1907). In these practices, the search for similarities belongs to the domain of the senses, a domain of unconscious perception, from which generality emerges. It also belongs to the domain of the rational construction of analogies.

---

[20] "Indiquer les phénomènes, s'abstenir même souvent de rechercher la connexion qu'ils ont entre eux, c'est presque toujours ici ce que nous avons à faire" (Bichat, 1829: 116–17).

[21] "Les principes qu'émet ici Bichat sont très sages et vraiment philosophiques ; il est à regretter que son imagination active l'ait trop souvent conduit à s'en écarter" (Bichat, 1829: 117). These words show the weight that Magendie placed on facts, an emphasis later criticized by Claude Bernard and Louis Ranvier.

[22] "*Connexions*," "*rapprochements*," and "*analogies*." In the introduction of a new edition of Bichat's *Traité des membranes*, F. Magendie wrote "anyone will understand the shortcomings of his book if one observes the necessity he [Bichat] felt to connect similar facts, in ways that were often more unexpected and newer than true;" "son besoin de rapprochements, souvent plus nouveaux et inattendus que vrais, expliqueront à chacun les défauts de l'ouvrage ..." (Bichat, 1827: viii).

Were these methods specific to Bichat, or can one find such principles at play in other circles? Bichat's principles seem to have been shaped in opposition to the traditional anatomical teaching he received from his physician father and his master, Parisian surgeon of the Hôtel-Dieu, Pierre Joseph Desault. This education was based on memorizing large collections of particular facts. Desault was aware of this fact and criticized it. He therefore developed a pragmatic aspiration, common to many surgeons of his time, to rationalize the anatomical knowledge compiled over past centuries.[23] Bichat's biographer, Husson, noted how:

> [Desault] had long wished to collect in a regular and methodical framework all the discoveries he added to surgery; he wanted to transform his journal, removing all isolated facts, keeping only those which allowed general inductions; in a word, he wished to establish a code of surgical doctrine.[24]

Desault advocated an analytical surgical anatomy, in which each chapter would begin with general facts. Thus, generality was already praised in some circles of surgeons in order to organize and simplify the anatomical knowledge used in the art of surgery.

This search for generality reveals a trend in all areas of knowledge typical of the eighteenth century. While the progress of surgical knowledge, on the smallest parts of the body, attracted students to this discipline, this material could seem "full of scholarly minutiae, so dry that it discouraged young people who were to study the art of healing."[25] After Desault's premature death during the French Terror, Bichat continued his lectures on surgery, and published them,[26] with a synopsis for each chapter on the organization of general facts. Bichat was conscious of the necessity of assembling both general knowledge and relevant particular observations. This same approach is found in his studies on membranes, where Bichat noted that "this science lacks ... some of these general thoughts which begin the treatise of each organic system in our anatomical textbooks ..."[27] Bichat's teaching perspective was clear. He stressed that "method in the sciences is the

---

[23] Barbara (2008a).

[24] "Depuis longtemps Desault formait le projet de rassembler dans un cadre régulier et méthodique toutes les découvertes dont il avait enrichi la chirurgie ; il voulait refondre son journal, en retrancher tous les faits isolés, conserver ceux dont l'ensemble pût fournir des inductions générales ; en un mot, il voulait créer un code de doctrine chirurgicale" (Bichat, 1827: xviii).

[25] L'anatomie est "hérissée des minutes scolastiques, [elle rebute] trop souvent par sa sécheresse les jeunes gens destinés à l'étude de l'art de guérir" (Bichat, 1827: xxv). In the preface of the *Œuvres chirurgicales* of Desault, Bichat commented on the state of the teaching of anatomy: "The teaching of anatomy was isolated at the time within boundaries that contingency had drawn and that were maintained by usage. It was on the one hand characterized by an actual lack of details in descriptions, and, on the other, by a mass of superfluous and almost isolated facts. The former needed to be enlarged and the latter diminished. Anatomy ought to present, in a more methodical table, a better way to conceive all organs, and a more reliable guide to surgeons through a description of their relations, in less inaccurate terms."; "L'enseignement anatomique, alors enfermé dans des limites que le hasard avait posées & que l'habitude entretenait, offrait d'un côté une insuffisance réelle dans les détails de la description ; de l'autre, un amas superflu de faits presqu'isolés. Il fallait, en ajoutant aux uns, retrancher à l'autre ; présenter dans un tableau plus méthodique, un ensemble mieux conçu de nos organes, & donner sur-tout dans une histoire moins inexacte de leurs rapports, un guide plus fidèle aux chirurgiens" (Desault, 1798).

[26] Bichat (1827).

[27] Bichat (1827: 2).

link joining those who learn and those who demonstrate…. [Teaching methods] become the sharing of judgment which classifies, arranges, and coordinates this scattered and confused material."[28] In my view, Bichat's interest in generality may have initially been dictated by the need to teach and thereby rationalize anatomical knowledge so that it was correctly used in surgery and medicine. This trend occurred within a wider cultural context in the eighteenth century, whereby various fields of knowledge underwent similar reorganization. However, Bichat later imported this practice into research, taking other disciplines as a guide, such as philosophy, mathematics, and botany, where the question of the value of generality was under discussion.

Bichat intended to simplify anatomical knowledge with a concept of generality that would enable the building of a large framework that embraced a selection of interdependent facts.[29] This was achieved in line with a new philosophical and analytical method. He recommended relying on philosophy, the humanities,[30] physics, chemistry, and mathematics in particular, since in his view "mathematics … educates our spirit of method and analysis."[31] In this context, Bichat praised the geometrical spirit of Blaise Pascal (1623–1662), which "defines all terms and proves all statements."[32] Bichat makes use of specific criteria and principles to precisely define his categories. He also followed the anatomist Félix Vicq d'Azyr (1748–1794), who adopted anatomical nomenclature and the use of language as an analytical method,[33] in agreement with Fontenelle's ideas[34] on the spirit of analysis and the thoughts of Étienne Bonnot de Condillac (1714–1780) on language.[35] Bichat appreciated these ideas, asserting:

> Language influences the sciences to a certain point. Condillac says, 'there is a true method of analysis which guides us all the more safely that it is exact.' In the descriptive sciences, the perfection of language lies in attaching images to each term, tying memory to nomenclature, and describing many objects by means of a few words. Language ought to be … an abridgment of science itself.[36]

[28] "La méthode, dans les sciences est le lien qui attache celui qui apprend à celui qui démontre…. [Les méthodes d'enseignement] deviennent le partage du jugement, qui classe, arrange, coordonne ces matériaux confusément épars" (Bichat, 1829: vii).

[29] On studies devoted to membranes: "a science, subject of so numerous discourses, where what is to be removed exceeds plausibly what is to be added."; "une science où l'on a déjà tant écrit, et où ce qui est à retrancher surpasse sans doute ce qui reste à ajouter" (*Ibid.*, xiv).

[30] "les sciences humaines" including "les belles-lettres," "la morale," and "la philosophie universelle."

[31] "Nous aimons les sciences mathématiques, parce qu'elles forment l'esprit de méthode et d'analyse" (Bichat, 1798: ix).

[32] "définir tous les termes et à prouver toutes les propositions" (Pascal, 1985).

[33] Vicq d'Azyr commented: "Since the whole of language is an analysis, how important it is, in the study of the sciences, to improve the methods thanks to which diverse parts of a whole are dissociated, examined, known, named, compared, and united! For long, only geometers know how to use these procedures: physicists, naturalists have at last learnt to use them."; "Puisque tout le langage est une analyse, combien n'importe-t-il pas, dans l'étude des sciences, de perfectionner des méthodes à l'aide desquelles les diverses parties d'un tout sont séparées, examinées, connues, nommées, comparées et réunies ! Longtemps les seuls géomètres surent employer ces procédés utiles : les Physiciens et les naturalistes ont enfin appris à s'en servir" (Vicq d'Azyr, 1805, vol. 4: 210).

[34] Fontenelle (1708, vol. I: 17–18).

[35] Condillac (1780).

[36] "Le langage, dit-il, influe jusqu'à un certain point sur l'étude des sciences. 'Il est, dit Condillac, une véritable méthode analytique, qui nous dirige d'autant plus sûrement qu'elle est plus exacte.' Dans les sciences

Thus, the manner Bichat envisaged to express generality was closely associated with new ways of using language and analysis in many areas of science.

In addition, the eighteenth-century "spirit of analysis" also related to scientific practices, which Condillac described using the metaphor of the dressmaker able to take apart and reassemble a dress.[37] Bichat considered this spirit crucial in studies of general anatomy. Such an analysis was, in his mind, akin to that "practiced by an architect who, before building a house, attempts to know in detail the distinct materials he will have to use."[38] Thus, Bichat's rationality is not only expressed with language, but it also involves decomposing and reassembling parts in the body with the practical procedures of dissections, the latter being carried out with a superimposed and operation-associated descriptive terminology. In this respect, analysis allows practical anatomy and fictive anatomy, as often practiced in teaching, to be closely associated, while remaining distinct.[39]

Asking whether anatomy must name all details, Bichat warned: "anatomy has two pitfalls that must be equally feared: superfluous details … exaggerate precision on the one hand …, too narrow a framework [which] only allows us to catch a glimpse of the whole picture that it contains …"[40] Naming all details characterizes descriptive anatomy, in which generality is absent, whereas using only gross categories misses essential points. In Bichat's work, a middle way was to be found at two levels, between things and words, but also among anatomy, physiology, and pathology, in line with evolutions in eighteenth-century life sciences.

Concerning the second level, where a middle way was needed, the method of Bichat consisted of an analytical approach to both physiology and anatomy. Marie-Jean-Pierre Flourens (1794–1867) distinguished between Haller's *analyse physiologique* and Bichat's *analyse anatomique*. Bichat's method relied on a new analytical anatomy developed conjointly with physiology. Consequently, Bichat's *analyse anatomique* did not limit itself to favor general forms of objects, since "differences in form may only be accessory, and the same tissue sometimes shows different states…. Main differences must therefore also be derived from the organization of properties."[41] This principle was also justified by the

de description, attacher des images à chaque terme, enchaîner, pour ainsi dire, la mémoire à la nomenclature, exprimer beaucoup d'objets par un petit nombre de termes, voilà la perfection du langage. Il faudrait, si je puis m'exprimer ainsi, que le langage fût un abrégé de la science elle-même" (Bichat, 1829: xx–xxi).

[37] "[Dressmakers] will imagine how to take apart and reassemble again the dress you ask naturally. Thus, they know analysis as well as philosophers."; "[Les couturières] imagineront naturellement de défaire & de refaire la robe que vous demandez. Elles sçavent donc l'analyse aussi-bien que les philosophes" (Condillac, 1780 : 23).

[38] "[L'analyse] à laquelle se livre un architecte, qui, avant de construire une maison, cherche à connaître en détail tous les matériaux isolés qu'il a à employer" (Bichat, 1829: x).

[39] Foucault (1966).

[40] "prenons-y garde, répond Bichat, l'anatomie a deux écueils également à craindre : d'un côté les détails superflus … une précision exagérée … [de l'autre] un cadre trop étroit [qui] ne laisse qu'entrevoir le tableau qu'il renferme ; de même une méthode trop concise ne présente qu'à demi les objets qu'elle embrasse"(Bichat, 1829: xiii).

[41] "Les différences de formes peuvent n'être qu'accessoires, et le même tissu se montre quelquefois sous plusieurs états différens…. C'est donc de l'organisation des propriétés, que les principales différences doivent se tirer" (Bichat, 1799: lxxx–lxxxi). The *analyse anatomique* criticized those who put too much emphasis on differences in structures: "anatomists, struck by differences in the structure of organs, have forgotten that their distinctive membranes could be analogous; they neglected to establish relations between them and this is an essential lacuna"; "les anatomistes, frappés de la différence de structure des organes, ont oublié que leurs

over generality of observations emerging from analyses of form alone, as Bichat's critique showed Haller had done on membranes.[42] Bichat found the scope of Haller's generalizations on membranes too large,[43] since he himself described three types of membranes, where Haller, judging on a similar appearance, had only considered one. Bichat's middle way thus relied on associating disciplines to define the right level of generality. Physiology regulated the shaping of generality in anatomy and *vice versa*, by correlating specific criteria from each of these fields in the definition of tissues.

In doing so, Bichat aimed at founding new objects. He considered that putting together general facts from anatomy, physiology, and pathology was a means to define real objects. In procedural terms, Bichat's general anatomy begins by abstracting from physiological and pathological observations, and then relates the abstractions obtained to abstractions of anatomical structures. For Bichat's biographer, Husson, there is "more merit in anticipating differences in the organization of anatomical parts from the diversity of diseases, than in classifying disorders by anatomical knowledge of these same parts."[44] From this perspective, Bichat did not aim to construct a new classification of disorders, but he wished to order them according to a new anatomy. His method proceeded from knowledge on both healthy and sick anatomical organs and tissues to define general and real anatomical categories, which were impossible to define with anatomy, physiology, or pathology alone.

Establishing a specific web of complex relations between normal and pathological anatomy and physiology represented Bichat's own way of finding a middle ground. This explains why Bichat advocated the unification of anatomy and physiology. This thesis became a major idea which Auguste Comte (1798–1857) developed in his philosophy of biology.[45] Physiological properties were already central to Desault's surgical anatomy. Bichat regretted that other scientists, physicians in particular, considered these disciplines separately:

> Struck by this difference between the parts of a same science, physicians had drawn between them a line of demarcation which was established and respected with time. Cadavers belonged to the domain of anatomy; physiologists were concerned with phenomena attached to life, as if the studies of the former were not inextricably linked with the researches of the latter; as if the knowledge of an effect could be separated from that of the agent that produced it.... One could wonder here which of the two, anatomy or physiology, lost more from this long separation.[46]

membranes respectives pouvaient avoir de l'analogie ; ils ont négligé d'établir entre elles des rapprochements, et c'est là un vide essentiel" (Bichat, 1827: 2).

[42] It is also reminiscent of Bernard's later attack of anatomical deduction of organic functions.

[43] Bichat (1827: 3).

[44] "il y a plus de mérite à pressentir, d'après la diversité de nos maladies, la différence dans l'organisation de nos parties, qu'il n'y a de difficulté à classer nos affections d'après la connaissance parfaite de ces mêmes parties" (Bichat, 1827: xx–xxi).

[45] Comte (1830: Leçon 44).

[46] "Frappés de cette différence entre les parties d'une même science, les médecins avaient tiré entre elles une ligne de démarcation que l'habitude consacra et que le temps a respectée. Les dépouilles des morts furent le domaine de l'anatomiste ; le physiologiste eut en partage les phénomènes de la vie : comme si les travaux de l'un n'étaient pas immédiatement enchaînés aux recherches de l'autre ; comme si la connaissance de l'effet

Accordingly, Bichat's anatomical generality has multiple facets, taking advantage of structures, illness, and experiments which make use of desiccations, putrefactions, macerations, boiling, cooking, treating with acids or alkali, etc.[47] Furthermore, as we have already suggested, Bichat's emphasis on generality is motivated by his conviction that such an approach will bring out real entities. He asserts:

> The idea of considering thus abstractively the different simple tissues of our parts is not an imaginary conception; its basis is most real and will exert, I think, a most powerful influence on physiology, as well as on medical practices. Indeed, tissues never appear similar, whatever our point of view is. Nature delineated them, not science.[48]

The search for truth by means of generality relied on assembling homologous observations to provide natural divisions among things and on defining tissues as real objects.

13.3   Classifying tissues by combining different frameworks: another eighteenth century legacy

Bichat arrived at a certain generality through the use of a multidisciplinary strategy that aimed to define biological objects from complementary analyses. He himself perceived that this was another legacy of Desault. This point appears clearly in Bichat's own description of his master's work as:

> A vast framework split into secondary ones by salient lines. External form belongs to a first framework; structure to a second one; another one deals with properties; the last one is for functions: each is subdivided into several sections linked to each other without merging together and following each other without overlapping. Their union gives a general formula applicable to organs from all systems, every point being described and located, when previously omissions left empty spaces in previous descriptions, leaving to the reader following this sketch the exact knowledge to be remembered on each part.[49]

---

pouvait se séparer de celle de l'agent qui le produit.... On pourrait se demander ici laquelle, de l'anatomie ou de la physiologie, a le plus perdu à ce long isolement" (Bichat, 1829: vi).

[47] Bichat (1801: vi).

[48] "L'idée de considérer ainsi abstractivement les différens tissus simples de nos parties, n'est point une conception imaginaire ; elle repose sur les fondemens les plus réels, et je crois qu'elle aura sur la physiologie comme sur la pratique médicale, une puissante influence. En effet quel que soit le point de vue sous lequel on considère ces tissus, ils ne se ressemblent nullement. C'est la nature, et non la science, qui a tiré une ligne de démarcation entre eux" (Bichat, 1801: lxxx).

[49] "C'est un vaste cadre que des lignes saillantes séparent en plusieurs autres cadres secondaires. Dans l'un se range la conformation externe ; à l'autre appartient la structure ; un autre embrasse les propriétés ; le dernier est réservé aux usages : chacun se subdivise en plusieurs sections qui s'enchaînent sans se confondre et se succèdent sans empiéter sur leurs limites. De leur réunion naît une formule générale, applicable aux organes de tous les systèmes, offrant à chaque point de leur description, une place à occuper, indiquant ce qu'on omet par les vides qu'elle présente, & laissant à celui qui l'a parcouru, le tableau exact de tout ce qu'il faut apprendre sur chaque partie" (Desault, 1798: préface, 11).

Bichat claimed that the reality of tissues was grounded in the combination of such frameworks. It was his work to put Desault's ideas into practice in his definitions and classifications of tissues.

Bichat's studies led to complex classifications. His first study of membranes proposed categories of elementary tissues, which provided a partial classification of tissues. Aspects of Bichat's method also reflected new trends in eighteenth-century taxonomy. As a matter of fact, Bichat described his method as "natural":

> It must be precisely determined which membranes belong to the same class.... The group-ing of two membranes into a single class must rely only on the simultaneous identity of their external configuration, structure, vital properties, and functions. Only natural methods can lead us to useful results.[50]

How should we understand "natural method" here? Probably the term referred to discussions in botany and in natural sciences more generally. The idea of a natural method was the object of much discussion in the second half of the eighteenth century. Methodological advances in botany were strikingly similar to those in Bichat's anatomy. In both fields, there emerged the conception that Nature was a continuous ordering of things with relative affinities, defined from similarities and differences among them, and that it could be described by a single—natural—classification. Carl von Linné (1707–1778) suggested that elements and groups of elements should be delineated as a function of all fundamental affinities and distinctions between them.[51] Antoine Laurent de Jussieu (1748–1836) warned that anatomical characteristics should be used and that they should be weighted according to their relative importance and contribution to the function of organs.[52] Georges Cuvier (1769–1832) acknowledged this shift toward a more nat-ural method and influenced in that respect the botanist Augustin Pyramus de Candolle (1778–1841).[53] For the latter, equivalent distinctions should be made whatever main plant function was chosen.[54] For these authors, a natural method required studying various morphological and functional traits, if they were to define real categories of plant species which reflected the order of Nature. This approach is similar to that of Bichat, who also used as many characteristics as possible, both anatomical and physiological, to identify

---

[50] "Il faut donc fixer avec précision quelles membranes appartiennent à la même classe.... Ce n'est que sur l'identité simultanée de la conformation extérieure, de la structure, des propriétés vitales et des fonctions, que doit être fondée l'attribution de deux membranes à une même classe.... ce n'est que par les méthodes naturelles que nous pouvons être conduits ici à d'utiles résultats" (Bichat, 1827: 5–6).

[51] Larson (1968: 312–13).

[52] Jussieu (1789: 5–9).

[53] Lorch (1961: 284–5).

[54] "The author of a natural method is not free to choose [characteristics]; he follows rigorous principles in the observation of all organs, and in attributing to each of them a relative importance, which does not rely on the facility to see them, but to the part that the organs play in the life of beings."; "L'auteur d'une méthode naturelle n'a pas la liberté du choix [des caractères] ; il est conduit par des principes rigoureux à observer tous les organes, et à donner à chacun une importance relative, non à la facilité que nous avons de le voir, mais au rôle que cet organe joue dans la vie des êtres" (Candolle, 1819: 52–53); "... truly natural classes, established according to the one of the major functions of plants, are necessarily identical to those established upon another."; "... les classes vraiment naturelles, établies d'après une des grandes fonctions du végétal, sont néces-sairement les mêmes que celles qui sont établies sur l'autre" (Candolle, 1819: 79).

tissue classes. Bichat's method also relied implicitly on the equivalence of the distinctions that could be derived from anatomy, physiology, or pathology. Hence, the routes to generality that Bichat developed had affinities with the natural methods of botany. Bichat's conviction that his tissue classes represented real distinctions of Nature,[55] since they were founded on a sum of anatomical and physiological characteristics, in agreement with their function, is an echo of contemporary attitudes toward botanical systems. One may thus assume that Bichat was inspired by practices in botany when he defined his natural method. This method was thought of as natural because it was seen as the single and correct way to approach and define real categories, botanical taxa, or tissues.

Like botanical and zoological taxa, Bichat's objects remained theoretical. In his *Traité des membranes*, Bichat defined three classes of simple membranes: serous, fibrous, and mucous; and three complex mixed membranes: fibro-serous, sero-mucous, and fibro-mucous.[56] In the eyes of some of his contemporaries, his method showed some weaknesses. An astute choice of the characteristics observed would have been crucial for such a comparative approach, as it had been in botany. Magendie felt that choices made by Bichat led him into "foolish views and even mistakes."[57] Nonetheless, Bichat's conception of generality had an indisputable heuristic potential that enabled him to access new objects, with a combination of physiological and anatomical facts that were later ignored when anatomists, including Geoffroy Saint-Hillaire,[58] developed analogies based on form alone. Bichat's classifications were later much criticized—because of the multiplication of the tissues he defined—but his method remained used.

Other scientists followed quite parallel paths. In his *Encefalotomia nuova universale*,[59] Vicenzo Malacarne (1744–1816) had initiated a systematic, universal, and topographical study of the anatomical parts of the brain, using geometrical and geographical abstractions to highlight the stable configurations of elements. Malacarne used form, physiology, and pathology to classify fibers, membranes, and humors. This similarity of Malacarne's and Bichat's approaches can be traced back to the work of the pathologist Giovanni Battista Morgagni (1682–1771). They both finally came to consider tissues as general entities,[60] while both rejecting microscopy.[61] Malacarne's ideas can thus be perceived as another outcome of a reflection on generality, the results of which are convergent with Bichat's.

---

[55] Bichat (1801).

[56] Bichat (1801: 8).

[57] "… vues hasardées et même [les] erreurs" (Bichat, 1799: viii–x).

[58] See paper by S. Schmitt in this volume.

[59] Cherici (2005); Malacarne (1780).

[60] Cherici (2005).

[61] In Bichat's days, microscopy was tricky and often led to great variety in the descriptions of minute anatomical entities. Magendie attacked Bichat's opinion that each observer could see what he himself imagined (Bichat 1799: 35, 35-36n). However, Ranvier recognized the inadequacy of microscopical instruments at that time. According to him, they could have brought only confusion to Bichat's general anatomy: "Bichat was a thousand times right when he refused to use such defective instruments"; "Bichat a eu mille fois raison de ne pas vouloir se servir d'instruments aussi imparfaits [comme les microscopes de son époque]" (Ranvier, 1880: 4). Bichat's rejection of microscopy perhaps reflects his idea that general microscopic observations could not be gathered, if one used the instruments of his time, and that a diversity of particular observations could not help anatomy.

However, Bichat's work is more explicit regarding the ways that generality is searched for, expressed, and separated from mere speculations.

In conclusion, we see that Bichat lies at an epistemological crossroad, where the convergence of several mature disciplines (anatomy, physiology, and pathology) raised a new question regarding how these distinct bodies of knowledge could be combined. Bichat wished to compare and link together general facts that had been defined in each of these disciplines. The search for generality within each of these disciplines was previously felt to be necessary for teaching and for the light it cast on the question of causation. These were ancient questions requiring generality built on the basis of sense perception and actively created with analogies. However, the comparison of several orders of generality, each perceived from within a given discipline, raised the question regarding at which level generality was to be grasped within different disciplines. A balance was to be achieved between concrete and abstract facts to establish connections between general facts from different disciplines. Within disciplines, categories were to be defined at the right level of generality, because broad categories making connections easier would not be of great interest for defining generality across disciplines. The search for such a balance can be seen in the anatomists' interest in terminology as well as in the new method that was shaped in eighteenth-century botany in order to get closer to natural classifications.

In all these cases, the key issue was that of discovering reality as such, and not simply what was visible. Whereas different disciplines focus on particular aspects of reality, real objects are not readily accessible to our senses, but require different orders of analyses. Convergences between different disciplines are necessary to create a novel rationality leading to the definition of real objects.

Thus, Bichat embodies an epistemological model in the way in which he brought anatomy, physiology, and pathology closer to each other. His great emphasis on generality provided him with the means to create a new science. However, Bichat remained closer to anatomy. Therefore his work leads to the characterization of anatomy in a novel manner, with using the concept of tissue. Nevertheless, general anatomy allowed new researches in other disciplines and made possible closer connections between them.

In order to perceive the future of Bichat's epistemological approach, it is possible to draw parallels between Bichat's concept of tissue and the foundation of the neuron as a biological object in the twentieth century.[62] The construction of the neuron from an ensemble of concepts taken from different disciplines (histology, electrophysiology, pharmacology) required the correlation of both structural and functional data. Thus, different and partial visions, deriving from multiple techniques, encouraged the convergence of incomplete descriptions of the neuron into a single biological object. Twentieth-century neuroscientists were guided by homologies between concepts of the nerve cell, led as they were by the conviction that different structural and functional techniques clarified distinct facets of a general object. Thus, we now accept that the generality of a biological object derives from homologies among general facts from various disciplines. In the eighteenth century, the generality of Bichat's tissue was also achieved by combining disciplines.

[62] Barbara (2007a, 2010a).

As demarcations among tissues were shown not to rely on particular techniques, it was possible to accept that they reflected real partitions among the objects themselves. The underlying assumption was that the objects under study were all the more real when general observations made by different approaches converged, the idea being that all means of investigation are appropriate if we are to reach real objects. In this scheme, the pursuit of generality means to abstract general facts from various disciplines, in order to draw parallels and define homologies among them, and then combine them with each other so as to define the single and real objects toward which these various general facts converge. The greater the diversity of the techniques used, the higher the probability that homologies between different observations reflect the true nature of biological objects. Such ideas were original in Bichat's time and, as we shall see, became central in the nineteenth and twentieth centuries.[63]

13.4   Ranvier's *anatomie générale* at the *Collège de France*

Louis Ranvier became the French leader in microscopic anatomy, when he was given a chair of *Anatomie générale* at the *Collège de France* in 1876. In the 1850s, he was trained as an anatomopathologist in Lyons and Paris, and opened a private course in microscopy when the use of microscopes was under discussion in medical faculties.[64]

The anatomies of Bichat and Ranvier differ in the size of their objects of enquiry and the consequent need for microscopy. Ranvier's observations focused on subcellular elements such as the myelin sheath of nerve fibers. As in Claude Bernard's texts, "anatomical elements" were not only cells, but also cell parts—"anatomical details," to use Bernard's term—, the definite and stable structures which could be explained in terms of local functions. Even with these differences in scale, Ranvier and Bichat conceived of generality in a similar way. However, the examination of Ranvier's studies is essential for understanding one of the meanings of generality in present-day biology, since Ranvier needed to extend generality to the microscopic world. Moreover, Ranvier's research illustrates how Bichat's wish to discover real objects was tenable at lower scales, if one merges together different technological procedures, as well as both anatomical and physiological observations.

In the first phase of the use of microscope in life sciences, the inadequacy of the instruments and the abundance of microscopic observations of living or dead materials favored passive descriptions of an infinite world of microscopic entities. In this context, one cannot perceive an interest in general facts from which theoretical explanations of the functions of organs could be derived.

Speculative approaches developed in this domain. Leibniz' ideas were influential from the perspective of defining life by general components. Their impact can be traced in the

---

[63] See the quotation from Renaut (1889) in note 86.
[64] Barbara (2007b).

works of Maupertuis, Buffon, and Lamarck in France, or Oken in Germany.[65] The cellular proto-theories defined by François Duchesneau[66] were developed, in a similar context, to deduce function from analytic dissections of organs into micro-systems. A fibrillar theory took shape, in which the contractile property of muscles was explained by the fibrous and geometrical aspects of its fibers.[67] However, none of these theories relied on unquestioned observations and precisely defined general microscopic elements.

The development of cell theory was a major and initial step in defining general microscopic observations. In Germany, Matthias Schleiden (1804–1881) and Theodore Schwann (1810–1882) based their explanations of the development of living structures on general mechanisms involving the cell concept, and Rudolph Virchow (1821–1902) further generalized the concept to pathology.

Microscopic anatomy rested on ideas of generality similar to those of Bichat. As a result, German histologists considered Bichat a prominent reference. Along the lines promoted by Bichat, comparison and observation were systematically used to derive general characteristics. The French histologist of the Strasbourg school, Mathias Duval (1844–1907), described how these ideas influenced microscopy:

> We must remember that, in addition to the education of the eye and the hand that lead to certifying the facts, it is even more important to observe. Any observation, like any experience, requires successive comparisons that highlight the reason of the things certified through mere contemplation. This is especially true for all microscopic observations which imply continual comparisons.[68]

This recalls Bichat's ideas on observation, which were similar to those developed later by Ranvier.[69]

However, this approach was much criticized in France, where German cell theory was regarded as highly speculative in Ranvier's time. For most French scholars, generality could not be extended to the microscopic world. They believed in a diversity of elements. In their views, this diversity paralleled that of molecules in the realm of chemistry.[70] These ideas were developed under the influence of Comte's philosophy of biology. Such stands derived from an adoption of Bichat's views in two respects: his rejection of microscopy, and his project to build generality from human senses alone.[71]

---

[65]   Canguilhem (1952: 187).

[66]   Duchesneau (1987).

[67]   Canguilhem (1952: 185–6).

[68]   "Rappelons qu'indépendamment de cette éducation de l'œil et de la main qui mène à constater le fait, il faut plus encore observer, toute observation comme toute expérience exigeant une succession de comparaisons qui donnent la raison des choses constatées par la simple contemplation. Cela est surtout vrai pour les observations microscopiques qui toutes impliquent cette comparaison d'une manière incessante" (Duval, 1878).

[69]   Ranvier (1863, 1865).

[70]   Robin often referred to the chemist Eugène Chevreul (1786–1889); see Robin and Verdeil (1853). Charles Robin (1821–1885) was the first professor of histology at the *Faculté de Médecine de Paris*. He rejected the generality of the cell, as the single anatomical unit of life, and cell theory.

[71]   See note 61 for Ranvier's opinion on Bichat's rejection of the microscope. Microscopes in Bichat's time were not reliable instruments.

In France, Ranvier's general anatomy developed thanks to the help of Claude Bernard (1813–1878) and the extension of experimental physiology, as separate from anatomical and anatomopathological studies. The medical milieu was hostile to Ranvier's personal project. Detractors of microscopy flourished,[72] and some of its prominent partisans, including the first professor of histology at the *Faculté de Médecine de Paris*, Charles Robin (1821–1885), rejected what he perceived as the German style of research, characterized in his view by a preeminence given to abstraction, the search for generality, and cell theory.

Claude Bernard helped Ranvier develop a cell-theory-based histology combining anatomy and physiology to search for general structures, in a small histological laboratory at the *Ecole Pratique des Hautes Etudes*, which was later transferred to the *Collège de France* (1867). Thus, generality at the microscopic level, as a natural extension of Bichat's work at a lower scale, was pursued despite the hostility of French scientists.

How did Ranvier manage to define new and real microscopic objects, which allowed him to found *Anatomie générale* at the *Collège de France*, along the same lines as Bichat's? Ranvier developed a physiological form of anatomy, as Bichat had done. In addition, the relationships between Bernard and Ranvier were closer than usually thought. Ranvier trained as a physiologist and practiced vivisection with Bernard. His biographer, Justin Jolly, insisted that Ranvier should be considered a physiologist, with a special interest in anatomy. I have shown elsewhere that Ranvier's program was based on Bernard's principles as laid out in his *Leçons sur la physiologie et la pathologie du système nerveux*,[73] which Ranvier attended when he was a student.[74] Bernard urged that the study of nerve fiber sheaths, ganglion cells, tactile corpuscles, that is, all structures which he considered as "anatomical details" and which constituted "anatomical elements," should be integrated into physiological explanations of tissue functions. Bernard did not doubt that these structures were cells or cell parts.[75] The cell concept allowed Bernard to consider the functioning of organisms as a whole built from the coordinated activity of elementary parts.[76] With such an integrated interpretation of life, Bernard avoided turning away a narrow physicochemical determinism[77] by adopting what he named in one of his notebooks the "true vitalism of B(ichat)."[78] Embracing Bernard's perspective, Ranvier created, through the observation of both dead and living tissues, a new field of enquiry devoted to anatomical details, which he studied with microscopy from both an anatomical and a physiological point of view.

The articulation of anatomy and physiology in Ranvier's work, in the context of cell theory and the development of histology, led to a new disciplinary approach. Ranvier contributed to the establishment of the generality of the cell. Later, he demonstrated general subcellular structures, including regularly spaced constrictions in the nerve myelin

---

[72] La Berge (2004).
[73] Bernard (1858).
[74] Barbara (2007b).
[75] Schiller (1962: 65).
[76] Canguilhem noted the complexity of relations between the parts and the whole in the views of Bernard.
[77] Canguilhem (1968b).
[78] "Le vrai vitalisme de B(ichat)" is the expression used by Bernard in one of his laboratory notebooks and which he crossed out. Bichat was alluded to by the letter "B" (Canguilhem, 1968b: 157).

sheath of a nerve fiber (the "nodes of Ranvier"). Thus, as Bichat had done, Ranvier managed to define anatomy from a new perspective, in which closer relations to physiology were made at the microscopic scale.

While Bichat usually began his studies by delineating the spatial extent of a physiological property and localizing it in anatomical elements, Ranvier first focused on describing anatomical details that he later studied physiologically in diverse experimental conditions. For example, Ranvier examined the penetration of a dye into a nerve fiber through a node in the myelin sheath, and concluded that this structure was important to the nutrition of the nerve fiber. Physiology thus met general anatomy in the search for the function of cell parts. Anatomical generality was combined with a physiological generality created by experimental histology as envisaged by Bernard and founded by Ranvier. The combination of these approaches was facilitated by Ranvier's method, which consisted of working on dissociated living elements, rather than on the fixed stained specimens used by most contemporary histologists. Thus, Ranvier's generality may be considered the counterpart of Bichat's in the field of microscopy. It contributed to the rise of physiology to the extent that physiology could then rival the status of anatomy. Scholars of the Paris science faculty and the faculty of medicine considered such an achievement impossible. It relied on great technical advances and the construction of histology as a renewed discipline both in France and Germany.

The theoretical context of Ranvier's work was also of major importance in the foundation of microscopic histology. At the moment when his chair of general anatomy was created, during his opening lecture, Ranvier recalled how Bernard's principles underpinned his methods of enquiry:

> The goal of physiology and general pathology is to study the most intimate and most essential parts of organs, the tissue elements.... Anatomical knowledge of these organic elements is not enough. One needs to study their properties and functions by means of the most delicate experiments; in a word, experimental histology is needed. That is the ultimate goal of our research. That is the basis of future medicine.[79]

Ranvier also stressed his debt to Bichat. In parallel with the way that Bichat had proceeded, Ranvier's anatomy was intimately linked to physiology:

> When a part of any one of these systems has been observed, as a histological configuration, it must be compared with the same disposition that can be recognized in other parts of the same system. Then, we must pursue the work still further to establish the generality of observed facts, comparing them with each other in the various systems of the organism. This is general anatomy, since it takes as its object not only the structure or texture of tissues (Bichat),[80] but also, and above all, their relations. When searching for a definition

---

[79] "Le problème de la physiologie et de la pathologie générale a pour objet les parties les plus intimes et les plus essentielles des organes, les éléments des tissus ... Il ne suffit pas de connaître anatomiquement les éléments organiques, il faut étudier leurs propriétés et leurs fonctions à l'aide de l'expérimentation la plus délicate ; il faut faire en un mot, l'histologie expérimentale. Tel est le but suprême de nos recherches, telle est la base de la médecine future." Ranvier's quotation of Claude Bernard (Ranvier, 1880: 1). This is the text from Ranvier's opening lecture to his course at the *Collège de France* (Ranvier, 1876: 1).

[80] Added by the author.

of this science, I already confessed that general anatomy could be viewed as comparative anatomy restricted to a single organism. It represents the science dealing with the plan of organization *par excellence*.[81]

Of prime importance was the role played by cell theory. Other schools of general anatomy, apart from that of Ranvier, flourished at the time. However, most of those in France rejected cell theory.[82] Duval reports Robin's opinion: "According to him, general anatomy is a part of anatomy. Its name indicates both its object and aim. Its aim is to determine the intimate and real nature of things, whose activity and movement is called life."[83] This statement illustrates that general anatomy could be used as a tool to classify most structures, whether or not cell theory was used as a framework for morphological comparisons and analogies. Ranvier accepted cell theory as propagated by the teaching of Bernard and used it to suggest experiments, provide hypotheses, and coordinate results.[84] While Robin's ideas also certainly played similar roles, cell theory demonstrated greater utility in providing links between anatomy and the physiology of histological elements. This is why we can consider that Ranvier's work extended to the microscopic scale the methodological generality proposed by Bichat, in which the combination of multiple techniques validated the reality of objects.

French histologist Joseph Louis Renaut (1844–1917) expressed a similar idea from the perspective of the techniques used:

> If a fact is real, all technical methods applied to the same object will concur to prove its existence: each showing one or several details not revealed by others. If, on the contrary, the fact that one believed one had observed is not real, any method other than that which created the illusion will not allow the illusion to persist, and the mistake will no longer occur as would be the case if one contented oneself of a single method of observation for the object to be analyzed.[85]

Renaut referred to this approach as the "principle of converging methods."[86]

---

[81] "Lorsque nous avons observé un département de quelqu'un de ces systèmes, une disposition histologique, il importe de le comparer avec celle que l'on peut reconnaître dans d'autres départements du même système. Il convient ensuite d'aller plus loin encore et d'établir la généralité des faits observés en les comparant entre eux dans les divers systèmes de l'organisme. C'est en cela que consiste l'anatomie générale, puisqu'elle a pour objet non seulement la structure et la texture des tissus, mais encore et surtout leurs rapports. En cherchant la définition de cette science, je vous ai déjà dit que l'anatomie générale pouvait être considérée comme l'anatomie comparée limitée à un seul organisme. C'est, par excellence, la science qui s'occupe du plan de l'organisation" (Ranvier, 1880 : 10).

[82] Robin's introduction (Cadiat, 1871: 111).

[83] "L'anatomie générale est, selon lui, une partie de l'anatomie dont le nom indique à la fois l'objet et le but. Ce dernier est la détermination de la nature intime et réelle des choses dont l'activité, le mouvement, s'appelle la vie" (Duval, 1886).

[84] Ranvier (1880: 14).

[85] "Si le fait est bien réel, toutes les méthodes techniques appliquées à un objet concourront à mettre son existence hors de conteste : chacune indiquant un ou plusieurs détails laissés dans l'ombre par les autres. Si au contraire le fait qu'on avait cru observer n'est pas réel, toute autre méthode que celle qui avait engendré l'illusion ne laissera plus subsister cette dernière, et l'erreur ne pourra plus être commise comme il arriverait si l'on se contentait, pour l'objet à analyser, d'une seule et unique méthode d'examen" (Renaut ,1889–99: ix).

[86] "L'importance du principe des méthodes convergentes" (Renaut, 1889–99: ix).

Ranvier's school of general anatomy was clearly a successful multidisciplinary effort. Even so, the histology Ranvier practiced had moved closer to physiology than that of Bichat. Accordingly, Ranvier did not share Bichat's and Bernard's doubts about anatomical deductions of physiological functions and their insistence on its indispensable cooperation with physiology. Bichat and Bernard thought that approaches other than purely structural observations were needed to understand function. Bichat had attacked Haller on an over-reliance on visual analysis of membranes. In a similar line, Bernard argued that cell types in the pancreas with similar morphologies had distinct functions. However, in his dispute with Haller, Bichat in fact deviated from his conception of anatomy and physiology as two facets of an underlying and single reality.[87]

This paradox was resolved by Ranvier, leading him to advocate in a new manner for the principle of anatomical deduction. Ranvier claimed that combining morphological and physiological evidence to define an object helped to lead to an understanding of its function. For example, the anatomical demonstration of a few muscular fibers in an organ would always indicate a local contractile function, since the intimate role of muscle fibers is to contract surrounding tissues. Therefore, Ranvier associated observations of muscle fibers with their contractile function. His anatomo-physiological correlations considered anatomy and physiology as equivalent. This was the key condition for deducing local functioning from structures and vice versa. Ranvier's confident correlations between anatomical and physiological properties were an important feature of his general anatomy. Cell types, and not tissues, became the new general categories of anatomy. And generality became possible at this smaller scale, because cell types shared common organizations in different organs. Later, in the 1890s, neurons were defined and found not only in the brain, but in the spinal cord and nervous ganglia. Thus, while Bichat gave a method for building generality and discovering possibly real objects in the realm of the common senses, Ranvier's studies showed how anatomy and physiology could converge to define cell types and their functions in all organs at a microscopic level.

Anatomy and physiology have continued an intimate association with the emergence of recent disciplines such as cellular physiology or cellular and molecular biology. The ascension of experimental physiology and belief in determinism have been crucial. Modern biology has added statistics to deal with the essential variability of complex biological systems.[88] Moreover, experimental reductionism has also improved our understanding of the determinism of such systems,[89] and made the search for general facts easier, with experiments tending to give more stable results. Similarly, the search for anatomical generality became dependent on modern histological techniques, which included complex processes of dissociation, slicing, or staining to objectivize properties of the objects under study. The methodological convergence formulated by Renaut, at play when two staining procedures reveal a similar object,[90] applies more generally to all recent biotechnologies in the definition of new objects. In Ranvier's time, microscopic biological objects were

---

[87] Béclard (1827).
[88] Hacking (1983).
[89] Rheinberger (1997).
[90] For example, methylene blue was shown to stain neurons similarly to the Golgi method.

not created by means of single theoretical concepts, but they emerged rather from closely compatible scientific practices and the coalescence of different modes of objectivization linked to specific experimental systems.[91]

13.5   Conclusion

This chapter has focused on generality through the studies of two French schools of anatomy, at the beginning and at the end of the nineteenth century. Let me now formulate my main conclusions and discuss them in a broader framework, including my current ideas on the concept of scientific object.[92]

Bichat's work cannot be solely analyzed as resulting from a simple shift in ways of creating knowledge at the turn of the nineteenth century. Some aspects of his method are representative of the early eighteenth century, while others already belonged to the nineteenth century and formed the basis of many of Auguste Comte's ideas on biology.[93]

In particular, the concept of generality in Bichat's work embodies more than an eighteenth-century principle in the ordering of things, depending on their specific aspects and distinguishing levels of generality. It is true, however, that such a principle, rooted in botany, was fundamental to him and in keeping with the spirit of analysis and eighteenth-century paradigms for teaching in anatomy and many other fields.[94] Bichat's method is consistent with such a perspective, as is evidenced by his references to mathematics, philosophy, and humanities. His method rested on the belief in a general order of things on the basis of which real objects could be discovered and defined.

What else characterizes Bichat's method? We agree with Foucault's rejection of the idea that progress in eighteenth-century natural history depended only on a novel use of analysis to provide order and enable discovery.[95] Alternatively, natural history, as well as Bichat's anatomy, can be considered as also depending on new practical procedures, while for them philosophy and mathematics represented simply parallel paradigmatic methods used as metaphors in defining new goals.

There is however one essential difference between natural history and Bichat's anatomy. In natural history, objects are conceived of as natural and concrete, whether they are rocks, animals, or plants. Classification and analysis represent the means for ordering them. In contrast, Bichat created his objects as categories, through an observation of structures and

---

[91] The birth of the neuron in the twentieth century represents one such field of investigation (Barbara, 2007a, 2010a, 2010b).

[92] Cf. the concept of scientific object developed in Barbara (2007a, 2010a).

[93] Comte (1830: Leçons 40–5). Comte considered as outdated some of Bichat's ideas, such as his opposition between life and death, rooted in eighteenth-century philosophy. However, Comte shared many of Bichat's ideas, such as the need to combine anatomy, physiology and pathology.

[94] Cooking, for example. See note 1.

[95] "La constitution de l'histoire naturelle, ... il ne faut pas y voir l'expérience forçant ... l'accès d'une connaissance qui guettait ailleurs la vérité de la nature ; ... l'histoire naturelle, c'est l'espace ouvert dans la représentation par une analyse qui anticipe sur la possibilité de nommer.... L'instauration à l'âge classique d'une science naturelle n'est pas l'effet direct ou indirect du transfert d'une rationalité formée ailleurs (à propos de la géométrie ou de la mécanique). Elle est une formation distincte" (Foucault, 1966: 142).

their variation in different physiological and pathological contexts. These observations were used as partial representations of the objects to be defined. So, unlike natural history, Bichat's anatomical classification was concerned with ordering general properties to define tissue categories as concepts, rather than as concrete objects of nature.

As a matter of fact, Bichat's objects were not visible. For example, where Haller saw one type of membranes, Bichat defined three, which were not readily discernible with the naked eye. In the *Birth of the clinic*,[96] Foucault described changes in the relations between the visible and the invisible at the end of eighteenth century. He suggested that invisible objects were progressively clarified *via* the emergence of a novel, clear, and objective language. Foucault interpreted Bichat's work by stressing his advances in deciphering the body, according to a two-dimensional order and based on similarities between surfaces.[97] He examined how a new and unitary discourse could emerge to represent and classify both organs and their diseases. Foucault insisted that Bichat's work was established at the level of visible things. Consequently, Bichat's quest for invisible aspects of things was only partially clarified by Foucault's analyses. For instance, these analyses did not place emphasis on Bichat's method as an integrated system of practices engaged in the search for invisible objects. Bichat's classifications were not analyzed with the intention of showing how they created invisible objects. These questions are nonetheless at the heart of Bichat's concept of generality, and are crucial for understanding how eighteenth-century scientific analyses of visible things moved toward an acknowledgment of invisible objects by specific practices and the definition of objects as new concepts.

With Ranvier, multiple factors such as cell theory, Bernard's physiology, the rise of experimental histology, and especially exploration at microscopic scales all changed the ways scientific objects were studied. The objects created by Bichat from physiological, anatomical, and pathological observations remained essentially anatomical. In contrast, Ranvier used anatomy to construct objects that could be studied from a physiological perspective. Placing himself at the level of cell parts allowed him to localize unitary properties to subcellular parts. His descriptions at the sub-cellular level permitted correlations between their structure and function. Such microscopic studies were possible, because all living organizational levels possess general structures.

Closer to our time, the constitution of the neuron in the twentieth century can be seen as a later multidisciplinary coalescence of distinctive generalities from physics, chemistry, pharmacology, electrophysiology, histology, histochemistry, biophysics, and immuno-histology: a coalescence which constituted a common and general scientific object.[98] In this case, a form of generality was reached by means of the convergence in the ways objects were described within particular experimental set ups. When different scientific communities realized their objects of inquiry (partially artificially created to make them suitable to experimental work, such as a "synaptic potential" or an isolated synaptic vesicle) were homologous, measures and concepts could merge in a single description.

---

[96] Foucault (1975).
[97] Foucault (1975: 130).
[98] Barbara (2007a, 2010a, 2010b).

This shift to the present permits us to conclude with two remarks on the article in which Richard Burian and coworkers consider generality in biology.[99] We agree with them that epistemology and philosophy need local studies and must escape from the search for absolute, general views. We hope that our examination of eighteenth- and nineteenth-century anatomies has shown the utility of studies on specific schools to clarify how generality was valued, built, and what it meant. Such interpretations seem likely to lead to novel, if local, conclusions whose generality can then be examined. We suggest that our analyses of generality, in the works of Bichat and Ranvier, provide new and general insights into the relations between general observations and the creation of biological objects.

Furthermore, Burian et al. (1996) discussed how the gene was constructed as a scientific concept, from different, and sometimes complex, structural and functional representations, originating in diverse disciplines. This leads to our second remark which expresses our difference with these authors with respect to general biological findings. They argued that a general understanding of the gene was hindered by the multiplicity of its representations. I would suggest instead that the essence of the biological object is a single representation highlighting and taking into account the complexity of partial representations attached to particular concrete objects, to the moment when all the bodies of knowledge, developed in different contexts, can be unified. Perhaps the gene is not yet a single concept, and, in a similar way, categories of biological objects may share the complexity of the gene. However, we are learning how to describe, model, and manipulate different related objects at any required level of generality, according to our specific needs and, in a dialectic process, these objects may appear as a single representation. This is what the history of the neuron shows. In this case, all the representations of the neuron converged when polemics died out. The search for generality is thus characterized by an unremitting attempt to assemble and manipulate what appears to be distinct objects, rather than by a passive process looking for homologies between possibly heterogeneous observations. Convergence of knowledge is the way to the essence of biological objects. The passive and archaic form of generality is almost useless for biology at the present time and it can be easily criticized. As a tool to manipulate objects however, we would argue that the active search for generality has become a route to creation, which is distinct from paths by which real objects were formerly sought for. Such a route is characterized by processes whereby distinct objects are built and then coordinated into single objects.[100] This new path involves a plurality of object phenomenologies which evolve, merge, or sometimes disappear as science progresses.

Acknowledgments

The author wishes to thank Dr. Richard Miles, Chantal Barbara, and the editors for advice and careful reading of the manuscript.

[99] Burian et al. (1996).
[100] Barbara (2008b).

REFERENCES

Barbara, J.-G. (2007a) *La Constitution d'un objet biologique au XXe siècle. Enquête épistémologique et historique des modes d'objectivation du neurone.* Ph.D. Dissertation, Université Diderot, Paris, France.

Barbara, J.-G. (2007b) 'Louis Ranvier (1835–1922): the contribution of microscopy to physiology, and the renewal of French general anatomy', *Journal of the History of the Neurosciences* 16: 413–31.

Barbara, J.-G. (2008a) 'Diversité et évolution des pratiques chirurgicales, anatomiques et physiologiques du cerveau au XVIIIe siècle', in *Querelles du cerveau à l'âge classique*, eds. C. Cherici and J.-C. Dupont. Paris: Vuibert, 19–54.

Barbara, J.-G. (2008b) 'L'étude du vivant chez Georges Canguilhem : des concepts aux objets biologiques', in *Philosophie et médecine. Hommage à Georges Canguilhem*, eds. A. Fagot-Largeault, C. Debru, and M. Morange. Paris: Vrin, 113–51.

Barbara, J.-G. (2010a) *La naissance du neurone.* Paris: Vrin.

Barbara, J.-G. (2010b) *Le paradigme neuronal.* Paris: Hermann.

Béclard, P. A. (1827) *Elemens d'anatomie générale.* Paris: Béchet.

Bernard, C. (1858) *Leçons sur la physiologie et la pathologie du système nerveux.* Paris: Baillière.

Bichat, X. (1798) *Mémoire de la société médicale d'émulation.* Paris: Maradan.

Bichat, X. (1799) *Traité des membranes en général et de diverses membranes en particulier.* Paris: Richard, Caillé et Ravier.

Bichat, X. (1801) *Anatomie générale appliquée à la physiologie et à la médecine.* Paris: Brosson, Gabon et Cie.

Bichat, X. ([1799], 1827) *Traité des membranes en général et de diverses membranes en particulier.* Paris: Méquignon-Marvis.

Bichat, X. (1829) *Anatomie descriptive.* Paris: Gabon.

Bourbon (de), L.-A. ([1741], 1999) *Le Cuisinier Gascon.* Paris: Loubatières.

Burian, R. M., Richardson, R.C., Van der Steen, W.J. (1996) 'Against generality: meaning in genetics and philosophy', *Studies in History and Philosophy of Science* 27: 1–29.

Cadiat, L. O. (1871) *Anatomie générale.* Paris: Delahaye et Cie.

Candolle (de), A. P. (1819) *Théorie élémentaire de la botanique.* Paris: Déterville.

Canguilhem, G. (1952) *La Connaissance de la Vie.* Paris: Vrin.

Canguilhem, G. (1968a) 'La constitution de la physiologie comme science', in *Etudes d'histoire et de philosophie des sciences.* Paris: Vrin, 226–71.

Canguilhem, G. (1968b) 'Claude Bernard et Bichat', in *Etudes d'histoire et de philosophie des sciences.* Paris: Vrin, 156–62.

Cherici, C. (2005) *L'anatomophysiologie du cerveau et du cervelet chez Vincenzo Malacarne (1744–1816). L'ébauche d'une médecine de l'intellect.* Ph.D. Dissertation, Université Diderot, Paris, France.

Comte, A. (1830) *Cours de philosophie positive.* Paris: Rouen frères.

Condillac, (de) E. B. (1780) *La logique, ou Les premiers développements de l'art de penser.* Paris: L'Esprit et de Bure l'aîné.

Debru, A. (1996) *Le corps respirant : la pensée physiologique chez Galien.* Leiden, New York: Brill.

Desault, P. J. (1798) *Œuvres chirurgicales de P.J. Desault.* Paris: Desault.

Duchesneau, F. (1987) *Genèse de la théorie cellulaire.* Paris: Vrin.

Duval, M. (1878) *Précis de technique microscopique et histologique.* Paris: Baillière.

Duval, M. (1886) 'L'anatomie générale et son histoire', *Revue scientifique (1st sem.)*: 107–12.

Hall, T.S. (1968) 'On biological analogs of Newtonian paradigms', *Philosophy of Science* 35: 6–27.

Flourens, P. (1858) *De la vie et de l'intelligence*. Paris: Garnier frères.

Fontenelle, B. (1708) *De l'utilité des mathématiques et de la physique*. Paris: Boudot.

Foucault, M. (1966) *Les mots et les choses*. Paris: Gallimard.

Foucault, M. (1975) *The birth of the clinic: an archaeology of medical perception*. Translated by A. M. Sheridan Smith. New York: Vintage Books.

Hacking, I. (1983) 'Nineteenth century cracks in the concept of determinism', *Journal of the History of Ideas* **44**: 455–75.

Jussieu, (de) A. L. (1798) *Genera plantarum secundum ordines naturales disposita*. Paris: Viduam Herissant and Theophilum Barrois.

Keel, O. (1979) *La généalogie de l'histopathologie*. Paris: Vrin.

Keel, O. (1982) 'Les conditions de la décomposition "analytique" de l'organisme : Haller, Hunter, Bichat', *Les Etudes philosophiques* **1**: 37–62.

La Berge, A. (2004) 'Debate as scientific practice in nineteenth-century Paris: the controversy over the microscope', *Perspectives on Science: Historical, Philosophical, Social* **12**: 424–53.

Larson, J. L. (1968) 'Linnaeus and the natural method', *Isis* **58**: 304–20.

Lorch, J. (1961) 'The natural system in biology', *Philosophy of Science* **28**: 282–95.

Malacarne V. (1780) *Encefalotomia nuova universale*. Turin: Briolo.

Ranvier, L. (1863) 'De quelques modes de préparation du tissus osseux', *Journal de Physiologie de l'Homme et des Animaux* **6**: 549–53.

Ranvier, L. (1865) *Considérations sur le développement du tissu osseux et sur les lésions élémentaires des cartilages et des os*. Ph.D. Dissertation, Paris.

Ranvier, L. (1876) *Leçon d'ouverture du cours d'anatomie générale*. Paris: Duval.

Ranvier, L. (1880) *Leçons d'anatomie générale sur le système musculaire*. Paris: Savy.

Renaut, J. (1889) *Traité d'histologie pratique*. Paris: Rueff et cie.

Rheinberger, H. J. (1997) 'Experimental complexity in biology: some epistemological and historical remarks', *Philosophy of Science* **64**(supplement): S245–54.

Robin, C. and Verdeil, F. (1853) *Traité de chimie anatomique et physiologique normale et pathologique ou des principes immédiats normaux et morbides qui constituent le corps de l'homme et des mammifères*. Paris: Baillière.

Schiller, J. (1962) 'Claude Benard and the cell', *The Physiologist* **4**: 62–8.

Pascal, B. ([1657], 1985) *De l'esprit géométrique*. Paris: Flammarion.

Vicq d'Azyr, F. (1805) *Œuvres complètes*. Paris: Duprat Duverger.

Virchow, R. (1861) *La pathologie cellulaire basée sur l'étude physiologique et pathologique des tissus*. Paris: Baillière.

14

Questions of generality as probes into nineteenth-century mathematical analysis

RENAUD CHORLAY

> *At the beginning of the century, the idea of a function was a notion both too narrow and too vague…. It has all changed today; one distinguishes between two domains, one is limitless, the other one is narrower but better-cultivated. The first one is that of functions in general, the second one that of analytic functions. In the first one, all whimsies are allowed and, every step of the way, our habits are clashed with, our associations of ideas are disrupted; thus, we learn to distrust some loose reasoning which seemed convincing to our fathers. In the second domain, those conclusions are allowable, but we know why; once a good definition had been placed at the start, rigorous logic reappeared.*[1]

In this passage from Poincaré's 1898 eulogy of Weierstrass, the French mathematician gave his version of the classical description of the rise of rigor in mathematical Analysis[2] in the nineteenth century. Though the quotation is quite straightforward, two elements raise questions. First, it is customary to associate vague ideas with limitless object-domains, and precise definitions with clearly bounded object-domains;[3] conceptual clarification walks hand in hand with domain-restriction. However, Poincaré described here the passage from a vague to a distinct notion of function as walking hand in hand with domain-extension. Second, after reading the last sentence, one would expect Poincaré to give this "good definition," which took one century to emerge and whose emergence eventually put mathematical Analysis back on the safer track of rigor. However, this definition is nowhere to be found in Poincaré's paper and, as we will see, this feature is in no way specific to Poincaré: the "function in general" is not something one defines but something one points

---

[1] Poincaré (1899: 4). Trans. RC.

[2] For the sake of clarity, we will systematically write "Analysis" with a capital A to denote mathematical Analysis (function theory). Thus, we will present an analysis of Analysis.

[3] To prevent any misunderstandings: the objects referred to here are functions; the object-domains are function classes or function-sets.

The Oxford Handbook of Generality in Mathematics and the Sciences. First Edition. Karine Chemla, Renaud Chorlay and David Rabouin. © Oxford University Press 2016. Publishing in 2016 by Oxford University Press.

to; it is not an object to be studied but the background on which objects can be studied. These topics will be discussed in the first part of this chapter.

This first part will provide the background against which we will endeavor to delineate two other historical interactions between generality issues and function theory in the nineteenth century. We will first focus on the first years of the nineteenth century and use questions of generality to attempt a comparison between two major treatises on function theory, one by Lagrange and one by Cauchy. We will attempt to show how Lagrange and Cauchy chose different strategies to take up the same threefold generality challenge: to give a general (uniform) account of the general behavior (i.e., save for isolated values of *x*) of a general (non-specified) function. On the basis of the elements gathered in the first two parts of this chapter, we will sketch a systematic comparison grid.

In the third part, we will concentrate on the end of the nineteenth century so as to show how some mathematicians used the sophisticated point-set theoretic tools provided for by the advocates of rigor to show that, in some way, Lagrange and Cauchy had been right all along: counter-revolution, as we all know, is a synthesis of pre-revolutionary and revolutionary elements. On the basis of the mathematical material covered in this third part, we will put forward a new concept, that of *embedded* generality. Though we came across it in the context of analysis, it is by no means specific to that context. We will argue that it captures an approach to generality issues that is specific to mathematics and whose mathematical treatment is a striking feature of twentieth-century mathematics.

Before we start, here are a few remarks on the nature of this chapter. To someone who usually works as a historian of mathematics, it may appear somewhat quick-paced: the various contexts are barely sketched; the collection of quotations displays a kind of imaginary dialogue between mathematicians, regardless of actual historical connections. However, it must be acknowledged that the main goal of this chapter is of an epistemological nature.[4] We aim at documenting ways of *expressing* generality and *epistemic configurations* in which generality issues became linked with other topics, be they epistemological topics such as rigor or mathematical topics such as point-set theory. In this regard, we present and try to characterize three very specific configurations: the first evolving from Abel to Weierstrass, the second in Lagrange's treatises on analytic functions,[5] and the third in Borel.

14.1   Generality, rigor, and arbitrariness

14.1.1   Abel's letter to Hansteen

We can start by reading an excerpt from one of Abel's letters to his master Hansteen, written in 1826:

> I shall devote all my strengths to shedding some light on the immense obscurity which, at present, reigns over analysis. It is so devoid of plan and system that one is astonished by the

---

[4] Here we use the adjective "epistemological" with the meaning it has in the French tradition, denoting what pertains to the theory of science and not what pertains to the theory of knowledge in general.
[5] Lagrange (1797).

fact that so many people indulge in it—and, what is even worse, it lacks rigor, absolutely so. In higher Analysis very few propositions are proved with conclusive rigor. Everywhere, we come across the sorry habit of concluding from the special to the general and, what is amazing is that, after such a procedure, one rarely finds what is called a paradox. The reason for that is indeed very interesting to think over. The reason, to my mind, lies in the fact that most of the functions dealt with by Analysis up to now can be expressed by powers. When other ones mingle with them, which, admittedly, seldom occurs, we don't do so well; were one to draw false conclusions, from them would spring an infinity of tainted propositions, all standing together.[6]

This quotation nicely parallels that of Poincaré which we gave in the introduction: Poincaré looked backward on a century-long process which Abel, among others, had kicked off. In Abel's letter, the themes of rigor and generality are beautifully intertwined, in a way which, to a large extent, will prove stable throughout the nineteenth century. We need to distinguish between two levels, the epistemic level and the object-level. On the epistemic level, this quotation is famous for its ideal image of mathematics (or its image of ideal mathematics) as a set of interrelated theorems: if one false assertion is mistakenly taken to be true, then the whole network is tainted. As a consequence, the current lack of rigor in "higher Analysis"—that is infinitesimal and integral calculus—is an outrage. But the situation described by Abel is a paradoxical one. One the one hand, logic tells us that there *might* be false assertions in analysis, since the usual mode of reasoning is in itself faulty: "concluding from the special to the general;" mathematical truth cannot rely on induction. On the other hand, it appears that there *are not* so many false assertions or pairs of contradictory assertions as one could expect. The reason for this is sought for on the object-level: the objects mathematicians usually consider—Abel writes—are functions of a special kind, namely functions which can be expressed as power series (with positive and negative integer powers and, on occasion, fractional powers as well). On this basis, two different lines of research can emerge, both of which require specific proof-methods. One can either stick to the study of this special class of functions, or try to understand how more general functions behave. To study the links between generality and rigor we need to take a closer look at the second line of research, in which, as we shall see, generality in the object-domain is expressed thanks to a notion of *arbitrariness*.

14.1.2   Investigating the generality of a theorem: Dirichlet (1829)

New developments in the history of rigor in function theory can be found in Dirichlet's 1829 article *Sur la convergence des séries trigonométriques qui servent à représenter une fonction arbitraire entre des limites données*. In the final paragraphs of this 15-page article, the flag of rigor is first waved, and then the main result summarized:

The former considerations prove in a rigorous way that, if the function $\phi(x)$, all values of which are assumed to be finished and determined, shows but a finite number of

---

[6] Abel (1992: 263). Trans. RC.

discontinuities between the limits $-\pi$ and π, and, moreover, has but a definite number of maxima and minima between these limits; series (7),[7] the coefficients of which are definite integrals depending on function $\phi(x)$, is convergent and takes on a value whose general[8] expression is

$$\tfrac{1}{2} \, [\phi(x+\varepsilon)+\phi(x-\varepsilon)],$$

where ε stands for an infinitely small number.[9]

This is what students still learn today as "Dirichlet's theorem": a 2π-periodic function which is piecewise continuous and piecewise monotonous has a converging Fourier series (series (7) in the quotation), whose limit is $\phi(x)$ if ϕ is continuous at x, and, more generally, the mean value of $\phi(x^-)$ and $\phi(x^+)$. Clearly, the theorem says nothing about the "arbitrary function," quite the contrary: the hypotheses under which the conclusion has been proved to hold are painstakingly spelled out, which is exactly what being rigorous means. This theorem took ten pages to prove, and each of the hypotheses played a part in at least one step of the proof. Carefully wording restrictions on the domain of objects about which the conclusion was proved to hold does not mean that more general cases should not to be considered, as the end of Dirichlet's paper shows:

> We still have to investigate the cases in which what we have assumed as to the number of discontinuities and that of maxima and minima ceases to be the case.[10]

Dirichlet managed to prove the conclusion under some hypotheses which emerged in the proof, but acknowledged the fact that this conclusion may hold under weaker hypotheses. Another way of saying this (but we should notice that Dirichlet never spoke of function classes or function sets) is: this theorem asserts that a given property is valid for some object-domain, but it is likely that this object-domain can be extended; it is likely—or at least worth investigating—that the conclusion is still valid for more general functions. The end of Dirichlet's paper pointed to two ways of exploring the generality of the conclusion, that is, the extent of the domain of objects for which it holds. Dirichlet first wrote that the proof could be amended for functions with an infinite number of discontinuities as long as the set of points of discontinuities is nowhere dense (in modern parlance).[11] The latter restriction came from the fact that the coefficients of the Fourier series are integrals involving a function ϕ, and that the very notion of an integral may become meaningless if restrictions are not put on the set of points of discontinuity.[12] We can see that, in order to assess the generality of a conclusion which he had established under what he felt to be

---

[7] The Fourier series.
[8] In this context, "general" means for every value of x.
[9] Dirichlet (1829: 168). Trans. RC.
[10] Dirichlet (1829: 168). Trans. RC.
[11] We need not discuss here the relevance of this integrability condition.
[12] For instance, Cauchy had proved the convergence of the rectangle-method sums for the case of functions with a finite set of discontinuities.

too strong hypotheses, Dirichlet first resorted to *proof analysis*,[13] but this proof analysis led him to the analysis of a mathematical concept, that of integrable function. In this 1829 paper, he merely pointed to this concept analysis as a research program: "But, doing things with as much clearness as one can wish for demands that one go into some details as to the fundamental principles of infinitesimal analysis; these details will be expounded in a further note ..."[14] A few lines above, Dirichlet used a different strategy to explore the generality of the property. Instead of pointing to the general concept of the integral and a possible weakening of the hypotheses under which integrability can be ascertained, he gave an example of a function that is *too* arbitrary to belong to the maximal object-domain for which the conclusion holds, more precisely, too arbitrary to be integrable:

> One would get an example of the function which doesn't fulfill this requirement if one assumed that $\phi(x)$ equaled some determined constant c when variable x takes on a rational value, and equaled some other constant d when the variable is irrational.[15]

This strange function had not until then been considered in mathematical Analysis; it had not turned up so far, whether in pure mathematics or in mathematical physics. This *display* of a specific function is an element of a new *mathematical configuration* in function theory, a configuration which encompasses epistemic values, such as "rigor," epistemic practices,[16] such as proof-analysis, and also strictly mathematical elements such as the exploration of the various properties of point-sets on the real straight line. The "Dirichlet moment" in the theory of Fourier series is perfectly characterized by Riemann in his 1854 dissertation on the same topic:

> The works on this question which we have so far mentioned endeavored to establish the Fourier series for those functions which are encountered in mathematical physics; thus, one could start the proof for completely arbitrary functions then, later, submit the course of the function to arbitrary restrictions for the purpose of the proof, so long as these restrictions don't go against the purpose.[17]

We need not remind the reader that Riemann was Dirichlet's student and that the goal of this 1854 dissertation was to fulfill the research program which Dirichlet had sketched at the end of his 1829 paper.

14.1.3   Expressing generality through arbitrariness

Before studying further the specific links between generality, rigor, and arbitrariness in this new mathematical practice, we need to pay more attention to the way general/arbitrary

---

[13] Our analysis of Dirichlet's move is, of course, very close to that of Lakatos (1976: 148–9).

[14] Dirichlet (1829: 169). Trans. RC.

[15] On aurait un exemple d'une fonction qui ne remplit pas cette condition, si l'on supposait $\phi(x)$ égale à une constante déterminée c lorsque la variable x obtient une valeur rationnelle, et égale à une autre constante d, lorsque cette variable est irrationnelle (Dirichlet, 1829: 169).

[16] We will use "practice" instead of "configuration" when we assume a more agent-based approach.

[17] Riemann (1892: 244). Trans. RC.

functional objects were referred to by mathematicians. We shall distinguish between three modes of expression.

14.1.3.1 *Referring to*

From its emergence as an autonomous mathematical concept in the eighteenth century, an element of arbitrariness had always been part of the function concept, however quickly the concept may have been sketched. As for functions which turned up in purely mathematical contexts, they could be either formed by the free juxtaposition of symbols (a freedom subject to syntactic constraints, however) or given (in the case of continuous functions) by a freely drawn plane curve. For instances, both ideas can be found in various parts of Euler's work (Youschkevitch, 1976: 68–9). Questions of generality and arbitrariness were not central in purely mathematical contexts. In mathematical physics however, the theory of vibrating strings and the subsequent development of Fourier theory placed the question of the mathematical description of the "arbitrary function" (*fonction arbitraire*)—describing a physical phenomenon—on the center stage.

In terms of vocabulary, throughout the nineteenth century, "arbitrary function" remained very much in use: in Fourier's 1822 *Théorie analytique de la chaleur*, in Dirichlet's 1829 paper. In his 1837 paper, Dirichlet used the German translation "*willkürlich*" (arbitrary) and, as if "arbitrary" would not convey the idea with sufficient strength, sometimes used "*ganz willkürlich*" (completely arbitrary); he also wrote "*ganz gesetzlos*" (completely lawless). The same words were to be found in Riemann's 1854 paper: "*willkürlich*," "*ganz willkürlich*," sometimes "*ohne besondere Voraussetzungen über die Natur der Function*"[18] (without any specific hypothesis as to the nature of the function). In his 1875 paper on the classification of functions, Paul du Bois-Reymond turned it into an adjective: "*die voraussetzungslose Function*" (the hypotheses-free function). The very same year, in France, "*arbitraires*" was replaced by "*les plus générales*" (most general) in Darboux's *Mémoire sur les fonctions discontinues*. The opening paragraph reads:

> At the risk of being too long, I was set on being rigorous, perhaps without full success. Many points which would justly be considered obvious or would be granted in the applications of science to usual functions [*fonctions usuelles*], have to undergo rigorous criticism when it comes to expounding the propositions pertaining to the most general functions.[19]

14.1.3.2 *Describing*

The easiest way to describe a general/arbitrary function is to negate a property which you feel to be specific. For instance, in his 1837 paper, Dirichlet shortly explained what his "gesetzlos" meant: "it is by no means necessary to think that the dependence [between the function and the variable] is expressible by means of mathematical operations;"[20] here, Dirichlet negated one of the two standard descriptions of what a function is. In

---

[18] The twentieth-century German spelling would be "*Funktion*."
[19] Darboux (1875: 58). Trans. RC.
[20] Dirichlet (1889: 135). Trans. RC.

the same paper, Dirichlet pointed to arbitrariness by negating another property which is usually (more or less implicitly) assumed: *usual* functions (in modern parlance: analytic functions) are completely determined by their behavior in any interval belonging to their domain of analyticity (uniqueness of analytic continuation); thus for an arbitrary function, "so long as one has determined the function for only a part of the interval, its continuation for the rest of the interval remains completely arbitrary."[21]

Alongside the negation of a property encountered in *usual* functions (using Darboux's terminology), more positive descriptions can be found. Here, a *generic mode of description* was used. The standard one, at least for continuous function, remained that of the arbitrarily drawn curve: in spite of the radical change in epistemic configurations, the description remained stable from Euler to Dirichlet and Riemann (e.g., "the arbitrary (graphically given) functions"[22] in Riemann). A significant change occurred in 1875 in du Bois-Reymond's paper on the classification of arbitrary functions; to introduce the most general function concept, that of the function "on which no hypotheses are made," he discarded the classical "arbitrary curve" image (suited for continuous functions only):

I. The hypotheses-free function.
In the case where no specific determination presents itself, the mathematical function is a table—similar to an ideal logarithmic table, thanks to which to any specified numerical value of the independent variable, one or several functional values—or one indeterminate but between limits given in the table—is associated. No horizontal row of the table has any influence on any other, that is, each and every value in the column which displays the values of the function stands by itself and may be altered; such an alteration would not prevent the column from representing a mathematical function. The mathematical concept of functions holds nothing more (nothing less either); it is thus fully exhausted.[23]

By a strange turn of events, one of the standard modes of description for the most classical and specific functions—one which pre-dates by centuries the emergence of the function concept—is used to express how a general function can be given: when dealing with usual functions such as logarithms or sines, tables of values are commonly used; a general function can be given by a similar table, but an "ideal table," in which every value is completely independent from the others (which is, of course, reminiscent of the negation of the uniqueness of continuation). With this generic description, du Bois-Reymond was close the twentieth-century concept of a map between sets, though he didn't require that two sets be declared beforehand.

14.1.3.3 *Exemplifying*

Exemplifying the general; the endeavor sounds paradoxical. Following Nelson Goodman, we shall say that to be displayed as a *sample*, an "object" has to both possess and denote

---

[21] „so lange man über eine Function nur für einen Theil des Intervalls bestimmt hat, bleibt die Art ihrer Fortsetzung für das übrige Intervall ganz der Willkür überlassen" (Dirichlet, 1889: 136).

[22] „die willkürlichen (graphisch gegebenen) Functionen" (Riemann, 1892: 227).

[23] Du Bois-Reymond (1875: 21). Trans. RC.

(or refer to) a property.[24] Thus, a *sample* of generality is a contradiction in terms, since no individual object—be it of a *usual* or of an *extraordinary* kind—can possess the property of being general. However, in the epistemic configuration which links generality, rigor, and arbitrariness in nineteenth-century function theory, mathematicians pointed to generality by displaying examples; we saw one of the first instances in Dirichlet's 1829 paper, with the example of (to rephrase) the indicatrix of the set of rational numbers within the set of real numbers. The exemplification tactics differed in a striking way from the two we have described so far. When it came to referring to or describing what a function of the most general type could be, mathematicians strove for the least specific (whether by negating common but specific properties or by describing generic *templates* to be filled arbitrarily). When examples are to be displayed in order to point to the general, the more specific the example, the more successful the denotation. We need not go into the details of the history of "pathological," "*bizarre*" (Borel), "amusing" ("*drôlatiques*" in Darboux[25]) functions; let us just mention Riemann's example of a (Riemann-)integrable function whose set of discontinuities is dense (1854), Weierstrass' continuous but nowhere differentiable function (1872), and Hankel's monster-function producing process (based on his "principle of condensation of singularities").[26]

14.1.4   Arbitrary functions: what for?

Examples of functions with extraordinary properties are sometimes used as counter-examples, but their display can fulfill other purposes. The mere displaying of the "monster" may reveal a new and unexpected feature of the world of functions. Geometric intuition is the first victim of this display:

> Indeed, the existence of the derivative in a continuous function $f(x)$ is reflected geometrically by the existence of a tangent line at any point of the continuous curve which is the geometric image of this function; and, though it is possible for us to conceive that at some singular points, even very close one to the other, the direction of the tangent line be parallel to the x-axis or to the y-axis, or even completely indeterminate, we cannot conceive that it be so in every arc of the curve, however small it may be taken. Hence the tendency to consider it unnecessary to prove the existence of the derivative in a continuous function.[27]

This quotation, by Belgian mathematician Gilbert (in 1873), is also here for the sake of irony: Gilbert is (somewhat) famous for his attempt to prove that a continuous function is piecewise differentiable, the very same year Weierstrass displayed (in the Berlin Academy of Science) his nowhere differentiable continuous function! In this passage, his goal was to remind the reader that intuition is no adequate ground for mathematical knowledge and that, consequently, this differentiability property called for a proof in spite of its

---

[24] Goodman (1976: 53).

[25] Quoted in Gispert (1983: 83).

[26] For an analysis of nineteenth-century teratology (i.e., science of monsters) in function theory, see for instance Volkert (1987).

[27] Quoted in Volkert (1988: 201). Trans. RC.

intuitive nature; he certainly didn't mean to underline the deceiving nature of geometric intuition when general (continuous) functions were being considered. It has to be noted that, contrary to the first examples such as the indicatrix of **Q**, Weierstrass' function takes a lot of mathematical machinery to describe; once the formula is written down it still takes skillful mathematical work to establish the nowhere differentiability. The know-how in the monster-making business is definitely a part of the mathematical practice which we're endeavoring to delineate.

More generally, the display of specific functions with unusual properties is a tool for the assessment of the generality of a given statement. For instance, the example of the indicatrix of **Q** showed that the Fourier-series development process is not universally valid: the theorem proved by Dirichlet in 1829 established its validity for a given (presumably not maximal) class of functions, and the example showed that the maximal class couldn't be all-encompassing. The display of the monster helped point to the task of identifying the exact contours of the right function-class. On a more general level, we endorse Klaus Volkert's interpretation of the monster-displaying business:[28] pathological functions served as milestones for the *extensional* exploring of the function-world. We can read an explicit description of this way of charting the world of functions in du Bois-Reymond's paper:

> First come a number of conditions satisfied by a function over a whole interval—however small; each of the conditions in the list restricts the function ever more, so that every former function class encompasses all the following ones—assuming all functions are finite.[29]

Fifty years after Abel's lament about the complete lack of "plan" and "system" in mathematical Analysis, a form of systematicity had emerged: functions are grouped in classes, function classes are characterized by explicit (set-theoretic) properties; logical implications between properties (on the *intentional* level) are reflected on the *extensional* level by inclusion relations between function classes. This systematic way of charting the world of functions is typical of du Bois-Reymond's work (arbitrary functions ⊃ integrable functions ⊃ continuous functions ⊃ differentiable functions) or of Camille Jordan's *Traité d'Analyse*,[30] whose second edition is a landmark in the history of "rigorous" analysis.

This interpretation also helps us understand the role of the "arbitrary function." As du Bois-Reymond strikingly put it, the "most general" function, the "arbitrary function," the "function on which no hypotheses are made" is something about which nothing can be said.[31] It is by no means an object to be studied, it is but an (intentionally) empty place in the whole epistemic configuration: not something to investigate, but a kind of background against which ever more specific function classes can be delineated; meaningless (on the intentional level) because all-encompassing (on the extensional level; see Fig. 14.1). This

---

[28] Volkert (1987).
[29] Du Bois-Reymond (1875: 21).
[30] Jordan (1893).
[31] As mentioned earlier, du Bois-Reymond's general function concept is close to the abstract map concept but differs slightly. Things can be studied in an abstract map: is it one–one, is it onto, etc.? Du Bois-Reymond fails to see these questions since the sets between which the map works are still implicit in his approach to the general function concept.

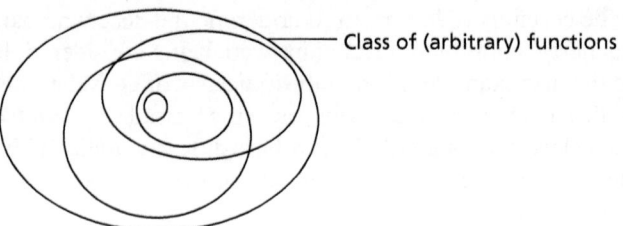

Figure 14.1 *Extensional functional world.*

interpretation also helps us clarify the relationship between the general/arbitrary function and the pathological examples. Both are necessary elements of the same epistemic configuration, which doesn't mean that pathological functions serve as samples for the class of arbitrary functions—a part which, as noted earlier, no example can play.

A few years later, a similar way of charting of the function-world was used by Hilbert in his famous 1900 Paris address. Before expounding the last series of problems (pb. 19–23), a series devoted to problems in mathematical analysis, he discussed the relevance of various function classes; let us just read the first few lines, a wonderful sample of this fine tuning of the relevant classes within the new "system" of functions:

> If we look over the development of the theory of functions in the last century, we notice above all the fundamental importance of that class of functions which we now designate as analytic functions—a class of functions which will probably stand permanently in the center of mathematical interest.
>
> There are many different standpoints from which we might choose, out of the totality of all conceivable functions, extensive classes worthy of a particularly thorough investigation. Consider, for example, the class of functions characterized by ordinary or partial algebraic differential equations. It should be observed that this class does not contain the functions that arise in number theory and whose investigation is of the highest importance. ...
>
> If, on the other hand, we are led by arithmetical or geometrical reasons to consider the class of all those functions which are continuous and indefinitely differentiable, we should be obliged in its investigation to dispense with that pliant instrument, the power series, and with the circumstance that the function is fully determined by the assignment of values in any region, however small. While, therefore, the former limitation of the field of functions was too narrow, the latter seems to me too wide.[32]

14.2 General theory as a theory of the general behavior of a function: Lagrange and Cauchy

Going backward in time and focusing on texts which do not belong to the epistemic configuration that we studied earlier, we come across another lead: other links between

---

[32] Hilbert (1902: 467).

questions of generality and the historical development of mathematical Analysis in the nineteenth century appear. We will thus follow two leads at a time: on the one hand, we will try to characterize the various ways in which Lagrange's grasp of the world of functions differs from the one we described in the former paragraph, thus delineating two *mathematical configurations*; on the other hand, we will come across a new conceptual intersection between questions of generality and the theory of functions: once a function is given, one can try to distinguish between a *general behavior* (to be studied in a uniform way) and points where the behavior is singular (to be investigated later, with specific tools). We will see that this idea of general behavior of a function is common to both Lagrange and Cauchy, but treated in very different ways by these mathematicians: we think this point of comparison is quite illuminating and helps understand some of the peculiarities of Cauchy's concept of continuity.

14.2.1   Scenes from Lagrange's *Théorie des fonctions analytiques*[33]

The introduction to Lagrange's treatise is entitled "*des fonctions en général*" (on functions in general):

> One calls *function* of one or several quantities any calculating expression in which these quantities appear in any way, along with other quantities which are considered to have given fixed values, in contrast to the quantities in the function which may take on any possible values....The word *function* was used by the first analysts to denote generally the powers of a given quantity. Since then, the meaning has been extended to any quantity formed in any way from another quantity.[34]

As mentioned earlier, an element of arbitrariness is present in Lagrange's function concept, but as a part of a very different configuration. However anachronistic, we feel the distinction between *axiomatic* and *genetic* definitions can help us to contrast Lagrange against, say, du Bois-Reymond. Du Bois-Reymond needed a definition (or, at least, a template) for the most general/arbitrary function; this concept had maximal extension and minimal intention, but was necessary for the definition of more interesting function classes in terms of characteristic properties: the most general function—this nondescript element of the class of all functions—was the starting point for the systematic exposition of mathematical Analysis. The notion of function in Lagrange is a genetic one: the basic, simple elements are known (letters standing for variable quantities) and they are to be combined at will to form any function you like; of course, the free combination of symbols has to remain within certain syntactic bounds, but for those as well only the most simple ones are known (namely, the general rules of algebra and maybe the symbols for the derivative, the partial derivative, and the integral): just as new functions can be formed,

---

[33] As the title of this paragraph indicates, we certainly do not mean to give an overview of Lagrange's work and its relationship to questions of generality in mathematics or mathematical physics. For a detailed study of the conceptual architecture of Lagrange's *Théorie des fonctions analytiques*, see Ferraro and Panza (2012).

[34] Lagrange (1797: 1).Trans. RC.

it is quite possible to add new syntactic structures. There is no need in Lagrange for a definite criterion enabling us to distinguish between functions and non-functions, no need to precisely delineate the outer rim of the function world. Quite the contrary, the function world is an open-field; generality, a mere horizon. The challenge is to find a systematic way to study these functional objects, of which the basic elements (in generic terms) but not the basic properties (in axiomatic terms) are known.

Lagrange met this challenge by resorting to a general mode of description, a general "form":

> Let us consider therefore a function $f(x)$ of any variable x. If x is replaced by $x + i$, i being any indeterminate quantity, the function becomes $f(x + i)$ and, thanks to the theory of series, it will be possible to develop it into a series of the following type
>
> $$f(x) + pi + qi^2 + ri^3 + \cdots,$$
>
> in which quantities $p, q, r \ldots$, the coefficients of the powers of i, are new functions of x, derived from the primitive function of x and independent of indeterminate i.[35]

Whatever the form of $f(x)$, seen as a formula in which x appears, it can be written in the *universal form* of a power series (with positive integer powers). The generality strategy is clear, but the claim remained to be ascertained:

> But to avoid advancing anything gratuitously, we shall examine the very form of the series which is to represent the development of any function $f(x)$ when $x + i$ is substituted for x and in which we have assumed only positive, whole powers of x appear. This requirement is indeed met by the development of the various known functions; but no one, to my knowledge, has ever tried to establish it *a priori*, what seems all the more necessary since there are particular cases in which it might fail to be met.[36]

As we shall see, Lagrange *would* attempt to provide such an *a priori* proof. It seems that, according to Lagrange, a proof for the generality of the property is all the more needed since counter-examples were known ... it could be the oddest justification for the need of a general proof ever given! This quotation reveals a feature of the epistemic configuration to which Lagrange belongs, a feature that we haven't encountered so far:

> I will first prove that, in the series which results from the development of function $f(x + i)$, no fractional power of i can appear, unless x takes on some particular values.... This proof is general and rigorous as long as x and i remain indeterminate; it would cease to be so if x took on determinate values .... We will later (n°34) deal with these particular cases and their consequences.[37]

---

[35] Lagrange (1797: 2). Trans. RC.
[36] Lagrange (1797: 7). Trans. RC.
[37] Lagrange (1797: 8). Trans. RC.

The relevant distinction is that between "indeterminate" and "determinate" values: a function, or, more generally, a variable quantity, is an object of complex nature. A variable quantity, denoted by a letter, *stands for* any possible particular value (potential level), and *may be* given any values (actual level). But, for Lagrange, the use of letters is not a mere shorthand, a way to denote any particular number; there is an *autonomous* level on which indeterminate quantities are to be dealt with, a level whose autonomy is often referred to by bringing up the "generality of algebra" (*la généralité de l'algèbre*). The theorem establishing the generality of the power-series form belongs to this theoretical level, regardless of the actualization (or specialization) of the variable quantity as a number. Yet, this *autonomous* level is by no means an *independent* level; properties proved at the "generality of algebra" level (let's call it level A) have an implicit counterpart on the "determinate value" level

(level B), as we shall see more clearly by reading paragraphs 34–6, in which Lagrange

deals with his "particular cases." Reading the first few lines is enough: commenting on

the "*forme*" $f(x) + if'(x) + \dfrac{i^2}{2} f''(x) + \cdots$

> It is thus necessary, before we proceed further, to examine when and how this form could fail to hold.
>   We showed earlier (n°2) that it may only be the case when x is given a determinate value which causes some radical to vanish in function $f(x)$ and in all its derivatives. Now, there are only two ways for a radical to vanish, either because the quantity by which the radical is multiplied vanishes, or because the radical itself vanishes.[38]

To explain what he meant, Lagrange used the example of function $f(x) = (x - a)\sqrt{x - b}$. If we draw the curve for $a = 2$ and $b = 1$ (which Lagrange didn't do), we get Fig. 14.2.

One has to remember that for Lagrange, the square root function is two-valued; for instance, 4 has two square roots, 2 and −2, which explains the symmetry of the curve. The function has two singular values, value $x = 2$ for which "the quantity by which the radical is multiplied vanishes" and value $x = 1$ for which "the radical itself vanishes." In this paragraph, Lagrange showed the reader how the development of $f(x + i)$ can be found if $x = 1$ or 2, with power series featuring negative and fractional powers of I, and not only positive integer powers. Here, a more general form (meaning: the usual form is a particular case) is needed to deal with singular values, that is to reach universality on the B-level (the extended form is valid for any values of x); the "general rule" was general because it "lived" on the A epistemic level, and,[39] because it was generally valid on the B-level, that is for all but singular (i.e., isolated on the straight line and singular for the specific function under study) values of x.

It is also worth commenting upon the generation and use of examples in Lagrange's treatise. Function $f(x) = (x - a)\sqrt{x - b}$ is clearly the simplest case showing both types

---

[38] Lagrange (1797: 32). Trans. RC.
[39] Behind this "and" lies the whole dialectic between the two levels. We can but touch here on this fascinating topic.

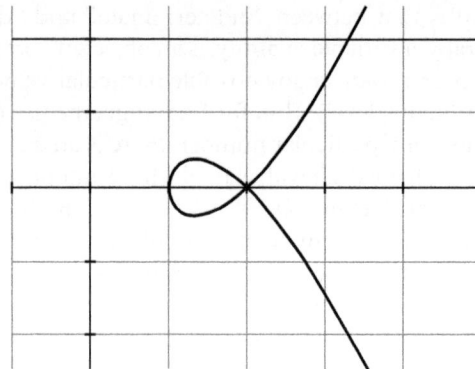

Figure 14.2 *Larange's $f(x) = (x-a)\sqrt{x-b}$.*

of singular behaviors: Lagrange didn't set out to chart a closed functional world using point-set properties, but formally generated functions from the simplest elements. In the "rigorous" configuration that we studied in the first part of this chapter, simplicity was not a central part of the picture (though not necessarily ill-considered); here, it is of the essence. The use of examples also differs. The case of $f(x) = (x-a)\sqrt{x-b}$ doesn't support any general statement, its role is pedagogical: it helps the reader spot the potentially singularity-bearing forms, and teaches her how to deal with them. Its simplicity makes it both a paradigm (to be used as a model) and a generic example: in a generically-structured open function-world, dealing with the simplest of all singular cases is the most obvious (if not the only possible) general move; anything that can be said about one of the basic building-blocks is of general interest.

14.2.2 Two mathematical configurations

Let us use a table to summarize some of the elements of comparison that we have come across so far, and add a few others as well. For the right column, I chose Camille Jordan as a representative for the "rigorous" configuration in its mature form: his 1890s *Cours d'Analyse* is a standard landmark, in which the works of Dirichlet, Riemann, Weierstrass, Heine, du Bois-Reymond, and Dini are reflected. Needless to say this table is just a rough sketch.

	Lagrange	Jordan
1	Genetic description, bottom-up journey across an open function-world	Axiomatic definition, top-down journey in a closed function-world
2	*Elementary functions* as starting point	*General/hypotheses-free* functions as starting point
3	Classification of formulae/functions according to *form*, differential diagnosis helping you make out the variety of function *types*	Classification of maps/functions according to (point-set) *properties*, delineation of function *classes*

(continued)

	Lagrange	Jordan
4	Generality derives from *simplicity* (proper identification of the most simple and elementary forms)	Generality derives from *rigor* (careful wording of hypotheses, counter-example-proof statements, detailed proofs)
5	Examples are simple cases which show how to deal with whatever you may come across out there	Examples are mainly counter-examples, they both illustrate and motivate the lengthy hypotheses and mind-boggling conceptual distinctions
6	In terms of *genre*, the book is both a *treatise* and a *textbook*.	In terms of *genre*, the book is first and foremost a *treatise*.[40]
7	*Concept image* (for the function concept)	*Concept definition*
8	Convince oneself, convince a friend	Convince an opponent

The first five points summarize things said earlier.

In points 7 and 8 we borrow concepts from the didactics of mathematics, since we feel they bring out contrasts between the two configurations quite nicely. The distinction between concept image and concept definition comes from the work of M. Tall and S. Vinner on the psychology of mathematical learning:

> We shall use the term concept image to describe the total cognitive structure that is associated with the concept, which includes all the mental pictures and associated properties and processes. It is built up over the years through experiences of all kinds, changing as the individual meets new stimuli and matures.[41]

The concept image is a much larger and piecemeal cognitive structure than the concept definition (that is, the formal definition, if there is one). To succeed in mathematics, at least in higher education, a student has to display some degree of cognitive flexibility, enabling her to switch from concept image to concept definition in some cases (say, to write down a proof) or the other way round in other cases (say, to devise an easy counter-example to a false statement); in the worst cases, the concept definition is not included

[40] The question of *genres*, their historical evolution and epistemological implication, is in itself well-worthy of study; we are not going into that here. We are simply referring to a *growing* tension between "mathematics for professional mathematicians" and mathematics for non-mathematicians (engineers, physicists, maths-teachers, or undergraduate students) through the nineteenth century. In his 1893 series of talks on the occasion of the Chicago World's fair, Klein stressed this problem: "Now, just here a practical difficulty presents itself in the teaching of mathematics, let us say of the elements of differential and integral calculus. The teacher is confronted with the problem of harmonizing two opposite and almost contradictory requirements. On the one hand, he has to consider the limited and as yet undeveloped intellectual grasp of his students and the fact that most of them study mathematics mainly with a view to practical applications; on the other, his conscientiousness as a teacher and man of science would seem to compel him to detract in nowise from perfect mathematical rigour and therefore to introduce from the beginning all the refinements and niceties of modern abstract mathematics.... The second edition of the *Cours d'analyse* of Camille Jordan may be regarded as an example of this extreme refinement in laying the foundations of infinitesimal calculus" (Klein, 1894: 49).

[41] Tall and Vinner (1981: 152).

in the concept image: the definition may be learnt by rote, still it means nothing to the student. As for the function concept, we saw that hard and fast definitions were rarely found until the second half of the nineteenth century. The few lines of explanation which can be read in the first paragraphs of Euler, Lagrange, or Lacroix's treatises are not meant as definitions on which proofs can be based, they are description of what can be called functional dependence, a description that has to be vague enough so as to be *fitted out* to anything that might come up in mathematics and mathematical physics. Lacroix put it bluntly: after his short explanation of the word "function," he added: "the use of this word will throw light on its meaning"[42] ... pure concept image.

Point 8 comes from Mason, Burton, and Stacey's *Thinking mathematically*,[43] when reflecting on the function of proof. You can try to write down a proof in order (1) to convince yourself of some mathematical fact; (2) to convince a friend or a student, someone of good will but in need of some kind of explanation; (3) to convince someone who assumes a systematically skeptical posture, someone who will look for flaws in every step of your reasoning. The proofs in Lacroix and Lagrange are, to some extent, of the first and second kinds. In the 1890s, Jordan's treatise presents a coherent system of subtle definitions and detailed proofs; those proofs and definitions are the result of 70 years of co-evolution of proofs and concepts in mathematical Analysis, an evolution triggered by fierce proof-analysis and counter-example devising.[44] This know-how in proof-design accumulated as a result of the assumption of the "opponent" role by prominent mathematicians, as is exemplified in this passage from one of Abel's letters to his friend Holmboe (in 1826):

> I doubt you will be able to put forward more than a small number of theorems dealing with infinite series, to the proof of which I can't object with good grounds. Do that, and I shall answer you.[45]

The friendly tone of this letter shows very clearly that this *epistemic posture* has nothing to do with personal enmity or scientific controversy.

14.2.3   Cauchy's concept of continuity as the answer to a generality challenge

From a historical point of view, the lectures which Cauchy[46] gave at the *Ecole Polytechnique* in the early 1820s should provide a missing link between the two configurations. On the one hand, they can be analyzed as a globally anti-Lagrangian move, and had a profound influence on the pioneers of the new epistemic style, Abel and Dirichlet. On the other hand, Cauchy shared with Lagrange some basic views as to what functions were, as to the role of singular points etc. which paint a picture of the function world in sharp contrast

---

[42] L'usage de ce mot en éclairera la signification (Lacroix, 1802: 3).
[43] Mason et al. (1982).
[44] See Volkert (1987) or the appendices in Lakatos (1976).
[45] Abel (1992: 257). Trans. RC.
[46] On Cauchy's analysis, classical references are Dugac (2003) and Grabiner (2005). For recent developments, see Sørensen (2005).

with what we described in the first part of this chapter. We think the depiction of Cauchy as an in-between figure—in between two coherent epistemic configurations—helps makes sense of his somewhat puzzling concept of continuity.

We need not expatiate on the first point (the "down with Lagrange" part); quoting the famous introduction to *l'Analyse algébrique* (1821) will suffice:

> As to the methods, I strove to give them the very rigor that is demanded in Geometry, so as to never resort to arguments based on the generality of Algebra. It seems to me that this kind of argument, though quite commonly acknowledged, most of all when passing from finite to infinite series, and from real to imaginary expressions, can be considered but mere induction; this kind of induction can help sense some truth but is not in keeping with the praised rigor of the mathematical science. It has to be noted that they led us to ascribe an indefinite scope to algebraic formulae whereas, in reality, most of these formulae only hold under certain conditions and for certain values of the quantities which appear.[47]

The arguments based of the "generality of Algebra" are deemed incompatible with mathematical rigor. Cauchy rejected the (implicit) dialectic between levels A and B: level A has at best a heuristic value (at best: it proves deceiving, more often than not), level B is the only firm ground on which to base mathematical statements; it has become "reality."

The discarding of level A created a new generality challenge for Cauchy. The didactic genre of the *traité d'analyse* called for a general treatment of all functional situations and for the laying out of a systematic exposition. Lagrange had met these requirements by resorting to a universal (enough) *form* for functions, the power series; obviously, this is not an option for level-A-skeptic Cauchy. We think that two concepts played, for Cauchy, this central role that the power-series form had played for Lagrange: the concept of limit and the concept of continuity. Both strictly operate on the B level, they refer to the numerical behavior of variable quantities. Limiting processes allowed Cauchy to tackle problems of existence for functions: the exponential function, the primitive and the derivative of a continuous function, and the solution of an ordinary differential equation (when regular initial conditions are given) are functions whose existence is proved by a limiting process. In the case of primitives, a short comparison with Lacroix's *Traité élémentaire de calcul differential et de calcul integral*[48] will help illustrate this point. After presenting formal rules for differentiating functions, Lacroix wrote that finding the primitive (or indefinite integral) $\int f(x)dx$ of a function $f(x)$ is the reverse problem to that of finding the derivative; he then presented various methods to (formally) solve this problem; the definite integral $\int_a^b f(x)dx$ was then introduced, and various numerical methods were presented to help find an approximate value for this definite integral, in case no primitive could be formally obtained. Cauchy proceeded exactly the other way round: he used the approximation methods to prove that the symbol $\int_a^b f(x)dx$ stood for a well-defined number, provided

[47] Cauchy (1989: ij). Trans. RC.
[48] Lacroix (1802).

f is continuous between *a* and *b*; then allowed quantity *b* to vary, thus defining a new numerical function which was proved to have *f*(*x*) as its derivative.

Cauchy also had to distinguish between regular values and singular values. Lagrange had done it by analyzing the syntactic expression of a function; this, again, is not an option for Cauchy. This is where, in our view, the concept of continuity comes into play. To the modern reader, Cauchy's continuity concept is a cause for puzzlement. One the one hand, his definition looks like our numerical, point-set theoretic definition. On the other hand, Cauchy used this definition in ways which the modern reader finds either inconclusive (for lack of distinction between continuity and uniform continuity, for instance) or altogether misleading; for instance, Cauchy "established" that the limit of a sequence of continuous functions is a continuous function ... a fact to which counter-examples were known by the time of Cauchy![49] Another puzzling feature is that for Cauchy, continuity was always assumed to hold in an interval (of non-null length) and discontinuity assumed to occur at isolated points, in spite of the fact that the Cauchy definition seemed to allow for functions with dense discontinuity loci (as in Dirichlet) and functions which were continuous at isolated points only (though those didn't come up until much later).

I think questions of generality help understand Cauchy's baffling "continuity" concept, in a twofold way. First, Cauchy's use of his continuity concept is understandable when one refers to its epistemic role instead of focusing on the numerical definition. Cauchy's continuity hypotheses serve him right as hypotheses of general/regular behavior: to some extent, what matters is not what Cauchy *means* when he writes "continuous," but what he *means to do*. The second element is not specific to Cauchy but plays a part in Cauchy's choice of the continuity concept as generality-bearing concept. It can also be found, for instance, in Ampère's famous proof of the "fact" that continuous functions admit a derivative. A close look at the proof shows that what Ampère meant is that continuous functions have a derivative, *save for isolated points*. In every step of the proof, Ampère acknowledged the fact that for specific values of the variable, the general behavior that is aimed for may not hold ... yet he didn't mention it when stating the final theorem. I believe this unwritten rule that "when dealing with functions, all that is stated and proved is so, *save for, maybe, isolated values of the variable*" is an implicit but essential part of a mathematical configuration which is common to Lagrange, Ampère, *and* Cauchy. Singular points may drop out of sight, and are implicitly assumed to appear only as isolated points. Formally universal statements are to be read modulo this proviso; this is the price to pay for a general statement, that is, one that deals with all functions, whatever they may be.[50]

To sum up, in spite of radically opposing views as to the legitimacy of the A level, it seems to us that Lagrange and Cauchy had at least this in common: they studied the *general* behavior (that is, except for isolated values of the variable) of what would later be termed "usual functions." The numerical setting chosen by Cauchy paved the way for the monster-making business, but it is not a business in which Cauchy engaged, or even a business which he considered; the functions which Cauchy studied were the

---

[49] This is discussed in Sørensen (2005).
[50] See chapter 4 in Chorlay (2007).

same analytic functions which Lagrange studied and the very same properties were to be attained, though in a completely different way. A systematic comparison between Lagrange and Cauchy helps identify similar *generality demands,* and stresses the *functional equivalence* between the form of the power series (in Lagrange) and the general property of continuity (in Cauchy). Both mathematicians faced two generality demands: one that pertained to the *genre* "general treatise" and called for the identification of a unifying element (be it form or property); one that was more content-specific and derives from the fact that functions operate (partially or exclusively) on the B-level, which called for a distinction between intervals of regular behavior (where general proven statements hold) and isolated irregular points.

14.3   Logical generality vs. *embedded* generality: Cauchy vindicated

We would eventually like to point to a third interaction between questions of generality and the development of function theory in the nineteenth century. On this occasion, we shall introduce the concept of *embedded* generality, in order to document both the reflexive character of mathematics and a specific form of general statement. As for the term "logical generality," against which I shall contrast "embedded generality," it is taken from Poincaré (see Section 14.3.1): it simply denotes the standard idea that a case is more general than another one if the first one extensionally encompasses the second one.

14.3.1   A "small corner" for "proper" functions?

In a 1904 paper on definitions in mathematics, Poincaré took a backward look at the development of rigor in mathematical Analysis over the nineteenth century; the mood was significantly different from that of the quotation which we gave in the introduction to this paper:

> Logic sometimes begets monsters. In the last half-century, we saw the emergence of a bunch of bizarre functions, the purpose of which seems to be to differ as much as possible from these straightforward functions [*honnêtes fonctions*] that prove useful. No more continuity, or continuity without derivability, etc. What's more, from the logical point of view, these weird functions [*fonctions étranges*] are the most general; those that we came across without looking for them now appear to be but a particular case. They are left with but a small corner. In the old days, when a new function was invented, it was for a practical purpose; nowadays, they are invented for the very purpose of finding fault in our fathers' reasoning, and nothing more will come out of it.[51]

Rigor developed, Poincaré lamented, at the cost of fruitfulness; in his view, the "straightforward functions" should have remained the main topic of study, and younger

---

[51] Poincaré (1904: 263). Trans. RC.

mathematicians spent too much time reveling in the minutiae of the general theory of functions. Yet he could but acknowledge the fact that general functions are more general from a logical viewpoint: they form the all-encompassing class and many subclasses, sub-subclasses, sub-sub-subclasses ... can be made out before that of analytic functions, a situation for which Poincaré used the metaphor of the "small corner" (*petit coin*) of the function world.

14.3.2   The return of Cauchy

At the very same time however, some mathematicians started using the sophisticated tools of general function theory, and the point-set theory it gave rise to, in order to vindicate the classical ("old days," to quote Poincaré) point of view. They would use generality arguments to show that the "small corner" is actually large *enough*. Emile Borel's work is a good example and I will focus on this case. His overall view is put in the clearest of ways in his 1912 analysis of his mathematical work; the following quotation is pretty lengthy, but we feel its skillful weaving of the various threads that we have been following makes it well worth reading:

> There were, there still are, mathematicians who choose to ignore what they deem to be refined subtleties with no practical use; this attitude is indeed legitimate since it leads to results but it seemed to me that I could not stick with it, for several reasons: one the one hand, until now, no one could draw a clear line between *straightforward* and *bizarre* functions; when studying the first, you can never be certain you will not come across the others; thus they need to be known, if only to be able to rule them out. On the other hand, one cannot decide, from the outset, to ignore the wealth of works by outstanding geometers; these works have to be studied before they can be criticized....
> To my knowledge, Cauchy never explicitly explained what he meant by "function;" a reading of his work seems to me to reveal evidence that, for him, the question didn't arise; "function" was but the general term used to denote any of the particular functions which the analysts study, each of these particular functions had its own definition based on elementary functions (by means of series, integrals, differential equations etc.); it was assumed that any argument pertaining to the general "function" would apply to all particular functions which would later be discovered, provided they meet the conditions appearing in the propositions (most of the time, these conditions are continuity for the function and its derivative).
> In the very same way, a biologist would refer to "living beings" or a chemist to "simple elements" without having had to delineate an *a priori* concept of the living being *per se*, or the simple element *per se*; they simply have in mind the living beings that they know or could know of.
> This Cauchy viewpoint was contrasted with the seemingly more general method in which one starts with a function given *a priori* as a correspondence which can be devised regardless of explicit formulation; ... here is no place to discuss whether what cannot be formulated can or cannot be an object for science; two remarks will suffice; on the one hand, this more general conception of functions led to the devising and studying of new functions, which would otherwise not have been thought of; thus it proved useful; but, on the other hand, the actual display of analytical expressions representing the newly

devised functions made the *a priori* conception useless; after a detour, one comes in fact back to Cauchy's viewpoint.

... My work on divergent series as well as those on monogenous functions can be traced directly to Cauchy's ideas; in these works just as well, I used the improvements to the rigor of analysis worked out by Cauchy's successors, while breaking free from the too narrow conceptions which they introduced along with that very rigor.[52]

Two significant examples will help understand Borel's subtle stand. To understand the first example, one must recall that in Lagrange, Lacroix, or Cauchy, singular points were always assumed to be isolated (on the straight line); in higher dimensions, the locus of singular points was always assumed to be of lower dimension than relevant the parameter space, which is why statements such as "a real-valued square matrix is, generally speaking, invertible" or "three points in a plane aren't usually collinear" could be given precise mathematical meaning (which made them not only meaningful but also true!). In the next phase, the displaying of functions whose locus of singular points was not made of isolated points was one of the most active industries in the monster-making business, and Weierstrass' everywhere singular continuous function was the monster *par excellence*. Borel went one step further, so as to ascertain the generality of the most straightforward of all functions, namely polynomial functions: y being a bounded function, defined over the 0–1 interval,

> Given two positive and arbitrarily small numbers ε and ε', one can determine a polynomial $P(x)$ such that the points at which $y - P(x)$ is, in absolute value, greater than ε make up a set of measure less than ε'.... One can also say that, by letting ε and ε' tend to zero, there is a sequence of polynomials that tend toward y, except at the points of a set of null measure.
>
> This result is essential for the theory of functions of one real variable, since it shows that the singularities of such a function *fill very little room*; it is thus possible, in many circumstances, to proceed as if they didn't exist. The in depth study of the notion of a set of null measure thereby leads to a middle stand between these geometers who are inclined to consider only "good" functions and those who could be led to think that "good" functions are but an extremely particular case. We know, in a precise way, that neither party has it completely wrong.[53]

From a logical point of view, the concept of continuous function is much more general than that of a polynomial function; but the refined tools of point-set theory help us go beyond this simple fact, they help us assess just *how much* more general they are. And it turns out that, in some way, polynomial functions are general *enough*. This kind of generality is context-dependent, in two ways. It depends on the mathematical tool with which one assesses generality (here, the measure of a set of points on the straight line); in this case the polynomial case proves general enough to found the integration theory of continuous functions (goal-dependence):

> The integral of y may be defined as the limit of the integrals of polynomials $P(x)$....
> This example shows how the notion of measure allows us to rid the theory of real

---

[52] Borel (1972: 120). Trans. RC.
[53] Borel (1972: 122). Trans. RC.

functions of much of the complication which had emerged in the logical development of analysis.[54]

The other example comes from Borel's work on power series in one complex variable. In a 1896 note to the *Comptes Rendus de l'Académie des Sciences de Paris*, Borel had given a mere heuristic argument to support the claim that a function defined by a power series cannot generally be analytically extended beyond this convergence disc. By 1912, he had devised a more rigorous and much more sophisticated argument, by applying the notion of set of zero-measure *in a function set*. Rephrasing in terms of probability, he summarized:

> I proved that for such a series, picked at random (words whose meaning I made precise), its convergence circle is generally a cut, which means that the cases in which analytic continuation is possible are to be deemed exceptional.[55]

In this example, the logical viewpoint says nothing more than: not all power series can be analytically extended; the class of analytic functions whose maximal domain of analyticity is the unit disc is strictly included in the class of analytic functions whose maximal domain of analyticity contains the unit disc ... Borel goes beyond this rather trivial statement (which could easily be proved by displaying just *one* "bizarre" power-series) and endeavors to assess *how much more general* the second class is. It turns out that, if the mathematical tool used to compare degrees of generality is a measure-theoretic tool of a function-space, the smaller of the two classes is so bulky that its complement has measure zero. Borel concluded:

> It is thus illusory to consider Taylor series *a priori*, regardless of its origin, this abstract study can only lead to negative answers.[56]

14.3.3 *Embedded* generality

In the first part of this chapter, we showed how the search for a more rigorous and systematic function theory—one that could encompass the difficult case of functions studied with Fourier series—resulted in significant changes in the function concept and in the adoption of new systematic ways of charting the function world. Essential features of this new configuration are those which Poincaré or Borel call "logical": the use of an abstract definition of a function, the delineation of function classes (or function sets) by abstract characteristic properties, and milestone standard examples. In this viewpoint, if a function class C_1 is strictly included in function class C_2 (extensional side of the logical viewpoint), then C_2 is more general than C_1 ... and that's about it.

Yet, as we saw with Borel's example, this extensional viewpoint can serve as stepping stone for a new kind of investigation: C_2 may be logically more general than C_1,

---

[54] Borel (1972: 123). Trans. RC.
[55] Borel (1972: 123). Trans. RC.
[56] Borel (1972: 124). Trans. RC.

but *how much so* ? Can't C_1 be *general enough* for some purpose ? Is C_1 so special that
it can, in some circumstances, be *neglected* altogether ? Those questions are by no
means specific to function sets or function theory; any set of objects, any parameter
space for some mathematical situation can be investigated in this way. The tools with
which the degree (or relative degree) of generality is assessed is a mathematical tool,
though, in most cases, not a number (as the word "degree" might suggest). To compare in terms of "size" two sets, one being part of the other, a wealth of methods is
available to the twentieth-century mathematician. Dimensional arguments had been
in use since classical mathematics: a doubly-infinite set is significantly larger than a
simply-infinite one, though the fact that the smaller one may disconnect the larger
one may give it a global topological importance that its "size," alone, doesn't account
for; in this respect, it is safer to neglect a subset of singular cases whose dimension
is at least two degrees lower than that of the space of all cases. With the advent of
point-set theory in the last years of the nineteenth century, a great variety of new
tools were made available.

Let us give but a few simple and context-free examples. Consider the set C_1 of positive
rational numbers less than 1 and the set C_2 of positive real numbers less than one. One of
the mathematical tools that can be used is that of density: C_1 is dense in C_2, which, loosely
speaking, means that there are elements of C_1 "everywhere" in C_2. In some cases, it makes
C_1 general enough; for instance, to check that two real-valued continuous functions f and
g are equal on C_2, it suffices to check that they are equal on C_1. Another mathematical
tool is measure theory. Let us say that the measure of an interval which is a part of C_2 is
its length, and that the probability of a denumerable infinity of pairwise disjoint intervals
is the sum of their probabilities/lengths. In this context, if C_1 has measure 0, then it seems
to be completely negligible compared to C_2; in particular, if a number is chosen at random
in C_2, the probability that a C_1 number be chosen is 0. As Borel remarked, these features
are relevant in integration theory.

We wish to coin the term *"embedded generality"* for this kind of generality assessment
which relies of the description of a mathematical structure (whether of set, ordered
set, measured space, topological space, manifold, etc.) on a set of objects or parameter
space for mathematical situations. For instance, the search for the right structure in the
case of the qualitative theory of dynamical systems is beautifully illustrated in Tatiana
Roque's chapter in this volume (see Chapter 10). One of the striking features of *embedded* generality is its *twofold context-dependence*: dependence on the purpose and on the
measuring-tool. In our simple example, C_1 can turn out to be either general enough or
completely negligible. This concept testifies to the reflexive nature of mathematics, its
ability to turn apparently (and formerly) *meta*-level questions *about* mathematics (such
as: the comparison between two theories, the degree of generality of a class of objects/
statements, the choice of the class of objects which are really worth investigating) into
mathematical questions, by designing the proper mathematical tools (e.g., group relations
to study the relationships between various geometrical theories, assessment of *embedded
generality*).

We came across this concept of *embedded* generality in our discussion of the interactions between questions of generality and the development of function theory in the

nineteenth century, but we do not mean to say it emerged in this context. For instance, Anne Robadey's chapter (Chapter 6)[57] documents Poincaré's devising of probability-theoretic arguments in celestial mechanics; her historically detailed and epistemologically informed narrative shows how Poincaré managed to turn a loosely formulated corollary to a false theorem into a full-fledged rigorous theorem about the general behavior of orbits, by describing the parameter space of orbits with tools he imported from the theory of continuous probability. He thus kicked off the theory of dynamical systems, a theory in which several types of embedded generality arguments are of the essence: Tatiana Roque's chapter on genericity (Chapter 10) documents at least two generation of such arguments since World War II. Other examples could be found in Poincaré's work, for instance in his work on the so-called "Fuchsian" functions (1881–5). Presenting the mathematical details would take us far beyond the scope of this chapter, it suffices to know that this example documents the passage from dimensional arguments to topological arguments: to show that two parameter spaces could be identified, Poincaré had to show that they not only were of the same dimension, but also topologically equivalent (homeomorphic). The proof-method he devised on this occasion, the "method of continuity" (*méthode de continuité* in French, *Kontinuitätsbeweis* in German), would stir admiration (and disbelief) until the 1920s.[58]

14.4   Conclusion

This chapter emerged from an exploration of the use of the word "general" in a well-known corpus, namely, that on the foundation of function theory in the nineteenth century. In keeping with the spirit of this handbook, we endeavored to make sense of the wealth and diversity of occurrences by focusing on topics such as the use of examples, exceptions, and singular cases; by focusing, also, on ways of expressing and assessing generality. These guiding threads led us to identify three distinct configurations which we strove to characterize. As descriptive terms, we used both *epistemic configuration* and *mathematical practice*: the first referred to closed—at least coherent—epistemic structures, with their own rules for action; the second referred to the way mathematicians actually engaged with mathematics in the context of these epistemic configurations, in accordance with or, at times, in spite of the rules.

These three configurations are by no means independent, quite the contrary; we clearly opted for a kind of dialectical narrative, in which the third phase was explicitly described—sometimes by mathematicians themselves, such as Borel—as a synthesis of the former two. In a sense, this dialectic movement relies not so much on three concepts of generality but more on a feature that is specific to the objects under study: mathematical functions. Indeed, a function is a two-faced entity: it can either be considered as an *individuum*—when a formula is written down, or when the function

---

[57] See also Robadey (2006).
[58] See, for instance, chapter 5 in Chorlay (2007).

is proved to be an element of a given class of functions—or as a *dynamic plurality*—as a correspondence between numerical values; in the latter viewpoint, the behavior of a given function can, in turn, be considered general for some numerical values of the variable and singular for some other values. This Russian doll structure accounts for much of the complexity of the story we tried to tell. On the basis of our analysis of the third phase—in which tools from point-set theory first designed to describe the singular sets of values of an arbitrary function started to be used to distinguish among functions in function sets—we eventually endeavored to define the concept of *embedded* generality, which we think is specific to the mathematical sciences but not to mathematical Analysis.

REFERENCES

Abel, H.N. (1992) *Œuvres complètes*, tome II. Paris: Gabay. Reprint of the second edition, Christiania: Grondahl & Son, 1881.

Borel, E. (1972) [1912] 'Notice sur les travaux scientifiques', in *Œuvres d'Emile Borel*, tome I. Paris: Editions du CNRS, 119–90.

Cauchy, A.-L. (1989) [1821] *Cours d'Analyse de l'Ecole Royale Polytechnique, Première partie : Analyse Algébrique*. Paris: Gabay.

Chorlay, R. (2007) *L'émergence du couple local/global dans les théories géométriques, de Bernhard Riemann à la théorie des faisceaux (1851–1953)*. Dissertation, Université Paris Diderot, France.

Darboux, G. (1875) 'Mémoire sur les fonctions discontinues', *Annales scientifiques de l'E.N.S.* 5 (2ème série): 57–112.

Dirichlet, G.L. (1829) 'Sur la convergence des séries trigonométriques qui servent à représenter une fonction arbitraire entre des limites données', *Journal für die reine und angewandte Mathematik* 4: 157–69.

Dirichlet, G.L. (1889–97) [1837] 'Ueber die Darstellung ganz willkürlicher Functionen durch sinus- und cosinusreihen', in *G. Lejeune Dirichlet's Werke*, Bd. I, ed. L. Kronecker and I.L. Fuchs. Berlin: G. Reimer, 133–60.

Du Bois-Reymond, P. (1875) 'Versuch einer Classification der willkürlichen Functionen reeller Argumente nach ihren Aenderungen in den kleinsten Intervallen', *Journal für die reine und angewandte Mathematik* 79: 21–37.

Dugac, P. (2003) *Histoire de l'analyse: autour de la notion de limite et de ses voisinages*. Paris: Vuibert.

Ferraro, G. and Panza, M. (2012) 'Lagrange's theory of analytical functions and his ideal of purity of methods (1797–1813)', *Archive for History of Exact Sciences* 66: 95–197.

Gispert, H. (1983) 'Sur les fondements de l'analyse en France', *Archive for History of Exact Sciences* 28: 37–106.

Goodman, N. (1976) *Languages of art* (2nd ed.). Indianapolis: Hackett Publishing.

Grabiner, J. (2005) *The origins of Cauchy's rigorous calculus*. Mineola, NY: Dover. Reprint of the 1981 edition.

Hilbert, D. (1902) 'Mathematical problems (lecture delivered before the International Congress of Mathematicians at Paris in 1900)'. Translated by W. Newson. *Bulletin of the American Mathematical Society* 8: 437–79.

Jordan, C. (1893) *Cours d'analyse de l'Ecole Polytechnique, tome premier : calcul différentiel (deuxième édition entièrement refondue)*. Paris: Gauthier-Villars.

Klein, F. (1894) *Lectures on mathematics (before members of the Congress of Mathematics held in connection with the World's Fair in Chicago)*. New York: MacMillan & co.

Lacroix, S. (1802) *Traité élémentaire de calcul différentiel et de calcul intégral*. Paris: Duprat.

Lagrange, J.L. (1797) *Théorie des fonctions analytiques*. Paris: Imprimerie de la République.

Lakatos, I. (1976) *Proofs and refutations: the logic of mathematical discovery*, ed. J. Worrall and E. Zahar. Cambridge: Cambridge University Press.

Mason, J., Burton, L., and Stacey K. (1982) *Thinking mathematically*. Reading, MA: Addison-Wesley.

Poincaré, H. (1899) 'L'œuvre mathématique de Weierstrass', *Acta Mathematica* 22: 1–18.

Poincaré, H. (1904) 'Les définitions générales en mathématiques', *L'enseignement mathématique* 6: 257–83.

Riemann, B. (1892) *Gesammelte mathematische Werke und Wissenschaftlicher Nachlass*. ed. R. Dedekind and H. Weber. Leipzig: Teubner.

Robadey, A. (2006) *Différentes modalités de travail sur le général dans les recherches de Poincaré sur les systèmes dynamiques*. Thèse de doctorat, Université Paris Diderot, France.

Sørensen, H.K. (2005) 'Exceptions and counterexamples: understanding Abel's comment on Cauchy's theorem', *Historia Mathematica* 32(4): 453–80.

Tall, D. and Vinner, S. (1981) 'Concept image and concept definition in mathematics with particular reference to limits and continuity', *Educational Studies in Mathematics* 12(2): 151–69.

Volkert, K. (1987) 'Die Geschichte der pathologischen Funktionen—Ein Beitrag zur Entstehung der mathematischen Methodologie', *Archive for History of Exact Sciences* 37(3): 193–232.

Volkert, K. (1988) *Geschichte der Analysis*. Manheim: Wissenschatfsverlag.

Youschkevitch, A.P. (1976) 'The concept of function up to the middle of the 19th century', *Archive for History of Exact Sciences* 16(1): 37–85.

Section III.3

A synchronic approach: controversies

15

Universality versus generality: an interpretation of the dispute over tangents between Descartes and Fermat

EVELYNE BARBIN

15.1 Introduction

During the 1630s, geometers introduced several methods of discovery for solving problems about curves, such as the method of tangents or the method of indivisibles for finding quadratures. These methods of discovery aim to alleviate what, in the eyes of the seventeenth-century geometer, is a major shortcoming of the proofs of the Ancients about curves, namely that they do not permit one to know how the statements of the propositions were obtained. While the Greek geometers demonstrate propositions about curves by contradiction, the methods of discovery must determine quadratures or tangents in a direct fashion. These methods therefore appear undistorted and are considered as processes of discovery. But they pose a question of legitimacy, which we summarize as follows: May a result obtained by a method of discovery be considered proven? Must one then prove it in the manner of the Ancients? This question is inescapable because the methods of discovery pass through new considerations, in particular the infinite and movement.[1] The diversity of responses given by geometers to this question reveals an important historic shift in the meaning granted to mathematical proof.[2]

The "dispute over tangents" between Descartes and Fermat, which began in 1638, should be recast in context if we wish to understand it, not as a dispute between two men,

---

[1] Barbin (2006).
[2] Barbin (1992: 29–49).

The Oxford Handbook of Generality in Mathematics and the Sciences. First Edition. Karine Chemla, Renaud Chorlay and David Rabouin. © Oxford University Press 2016. Publishing in 2016 by Oxford University Press.

but as a controversy between two mathematicians.[3] Now it is in terms of claiming universality and generality that the two geometers present and defend their respective methods of discovery. Prior to the title "Discourse on the method of rightly guiding reason, and seeking truth in the sciences," Descartes had thought of calling his work "The design of a universal science that could elevate our nature to its highest degree of perfection," the three essays bound to "render proof for the proposed universal science."[4] From the time of his letter of January 1638, Descartes uses the term "universal rule" to distinguish his method from that of Fermat, while Fermat speaks of his "faultless general rule" or "singular and general rule" in his tracts. We will examine the methods of discovery of tangents of Descartes and Fermat, but also the discourse of the geometers at the time of the dispute, in order to seek that which identifies, according to them, the universality and generality of a method, and therefore that which distinguishes them.

15.2   Descartes' method of tangents: an application of a universal method

In 1637, Descartes presented an essay following the method of *Discourse on the method,* titled *Geometry.* The purpose was not to write of the "elements of geometry," but to present a method of constructing "all the problems of geometry."[5] This method is summarized in several lines at the beginning of book I of the essay:

> Thus wishing to resolve any problem, one must first consider it as already done, and give names to all the lines that appear necessary for its construction, just as much to those that are unknown as to the others. Then without considering any difference between known and unknown lines, one must traverse the difficulty, according to the order that shows most naturally in which manner the lines depend upon each other, until one has found a means of expressing a same quantity in two manners, which is called an equation. [6]

The method thus consists of five stages. First, one must suppose the problem solved, that is, proceed by analysis, then give names to all known and unknown lines and designate

---

[3] Baillet relates this dispute, which Fermat called "his little war versus M. Descartes" and Descartes "his little mathematical suit versus M. de Fermat" (Baillet, 1992: 486–93). We find traces of this dispute in comments of historians. For instance, Boyer wrote: "In criticising Fermat's method of tangents, Descartes attempted to correct the method by interpreting it in terms of equal roots and coincident points, a procedure which was practically equivalent to defining the tangent as the limit of a secant" (Boyer, 1959: 167). While Kline wrote: "Though Fermat's method was general, Descartes thought his own method was better; he criticized Fermat's, which admittedly was not clear as presented then, and tried to interpret it in terms of his own ideas" (Kline, 1972: 345–6). But Whiteside did not mention the dispute: he interpreted the methods of Descartes and Fermat in modern terms of limits and pointed "slight differences of treatment required in the two approaches" (Whiteside, 1960–2: 179–387).

[4] Letter to Mersenne of March 1636 (Descartes, 1897, vol. I: 339).

[5] Descartes (1987: 333).

[6] "Ainsi voulant résoudre quelque problème, on doit d'abord le considérer comme déjà fait, et donner des noms à toutes les lignes, qui semblent nécessaires pour le construire, aussi bien à celles qui sont inconnues qu'aux autres. Puis sans considérer aucune différence entre ces lignes connues et inconnues, on doit parcourir la difficulté, selon l'ordre qui montre le plus naturellement de tous en quelle sorte elles dépendent mutuellement

them by letters. After this one should interpret the problem with the aid of equations, leading to one or more equations, and finally resolve these. This method appears to be universal, in the sense that it enables resolution of all the problems of geometry.

However, putting this to work for problems about curves requires delimitation of a field of application. At the beginning of book II on "the nature of curved lines," Descartes names as geometric the curved lines to which his method applies, namely those to which one can associate an algebraic equation:

> I could give here several other means of drawing and conceiving of curved lines, which would compound more and more by degree to infinity. But to understand together all those which occur in nature, and to distinguish them by their order into certain types, I know nothing better than to say that all the points of these that one can call Geometric, that is, that fall under precise and exact measurement, necessarily have some correspondence to all the points on a straight line, which can be expressed by some equation, altogether by a single one.[7]

He does not consider that there is therefore a restriction. On the contrary, he believes he is enlarging the universe of curves of the Ancients, which corresponds to the classification of geometric problems into plane, solid, and linear, according to whether their solution required straight lines and circles, conics, or other lines.

The question therefore is less about a restriction than a notional incorporation, that is, the necessity of exhibiting a notion of curve for which the method is universal. It is rightly to this universal meaning that Descartes aspires when he writes that the knowledge of their equations allows one to find all the properties of curves : "in order to find all the properties of curved lines, it suffices to know the relationship that all their points have to those of straight lines, and the manner of drawing other lines that cut them at right angles in all their points."[8]

The method of finding normals, that is, perpendiculars to tangents, is presented in book II of *The Geometry*, by applying the method of book I. One must suppose the problem solved and give names to the lines, known and unknown. Given C a point of a curve, AG a straight line to which the points of the curve are related, and CP the normal to the curve (Fig. 15.1). One sets

$$MA = y, \ CM = x, \ PC = s, \ \text{and} \ PA = v,$$

where s and v are thus the unknowns that must be determined.

les unes des autres, jusqu'à ce qu'on ait trouvé moyen d'exprimer une même quantité en deux façons ce qui se nomme une équation" (Descartes, 1987: 335). Original translations of Descartes' texts by David Pengelley.

[7] "Je pourrais mettre ici plusieurs autres moyens pour tracer et concevoir des lignes courbes, qui seraient de plus en plus composées par degré à l'infini. Mais pour comprendre ensemble toutes celles qui sont en la nature, et les distinguer par ordre en certains genres, je ne sache rien de meilleur que de dire que tous les points de celles qu'on peut nommer Géométriques, c'est-à-dire qui tombent sous quelque mesure précise et exacte, ont nécessairement quelque rapport à tous les points d'une ligne droite, qui peut être exprimé par quelque équation en tous par une même" (Descartes, 1987: 351).

[8] Descartes (1987: 369).

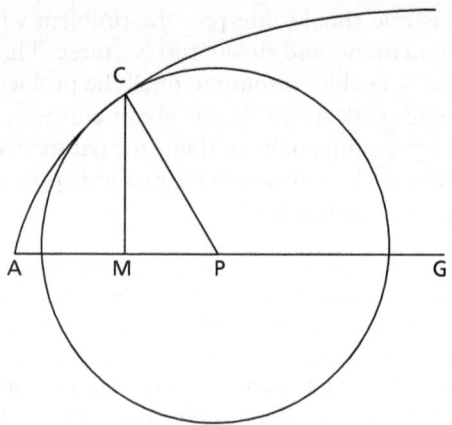

Figure 15.1 *Circle tangent to a curve in Descartes' method.*

Next one must translate the problem. For this, Descartes writes

$$s^2 = x^2 + v^2 - 2vy + y^2,$$

which is the equation of a circle with center P and radius CP, and supposes, as given the equation of the curve,

$$x^2 = ry - \frac{r}{q} y^2.$$

This leads to the equation that must be satisfied by the ordinate of a point lying simultaneously on the circle and on the curve, by using the second equation to eliminate x from the first equation:

$$y^2 - \frac{r}{q} y^2 + ry - 2vy + v^2 - s^2 = 0.$$

This last equation must have a single solution because the circle must be tangent to the curve (Fig. 15.2). Descartes therefore identifies the last equation with the equation:

$$y^2 - 2ey + e^2 = 0.$$

He then obtains the solution

$$v = y - \frac{r}{q} y + \frac{1}{2} r.$$

Applying the method thus demands not only a notional incorporation, but also a procedural incorporation, namely an algebraic identification. This allows algebraic translation of

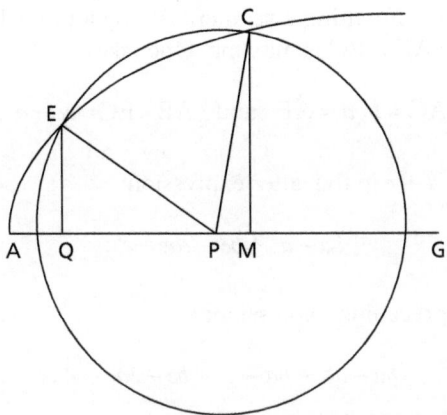

Figure 15.2 *Circle tangent to a curve in Descartes' method.*

the assumption that the curve and the circle meet in one point. From these two incorpora-
tions, Descartes concludes that the method applies to all geometric curves: "I see nothing
preventing one from extending this problem in the same manner to all curved lines, which
fall under some Geometric calculation … It is always easy to find [normals]: although
often one needs some adroitness to render it brief and simple."[9]

15.3 Fermat's method of tangents: an extension of a general method

The method of tangents of Fermat is presented in a tract titled *Method of seeking the
maximum and minimum*, where the geometer enunciates a rule for finding a maximum
or minimum:

> Let *a* be any unknown under consideration…. One expresses the maxima or minima
> quantity in terms of *a*, using terms which may be of arbitrary degrees. One then substitutes
> *a* + *e* for the primitive unknown *a*, and expresses next the maxima or minima quantity in
> terms where *a* and *e* appear to arbitrary degrees. One adequates, to speak like Diophantus,
> the two expressions for the maxima or minima quantity, and subtracts the terms common
> to both sides…. One divides all the terms by *e*, or by *a* power of *e* of higher degree, in
> such a way that *e* disappears completely in at least one of the terms of one member of
> the equation. One then suppresses all the terms where *e* or one of its powers still appears,
> and one equates the rest.[10]

[9] "Je ne vois rien qui empêche qu'on étende ce problème en même façon à toutes les lignes courbes, qui
tombent sous quelque calcul Géométrique … Il est toujours aisé de les trouver [les normales]: bien que souvent on ait besoin d'un peu d'adresse, pour les rendre courtes et simples" (Descartes, 1987: 377–8).

[10] "Soit a une inconnue quelconque de la question …. On exprimera la quantité maxima ou minima en a, au
moyen de termes qui pourront être de degrés quelconques. On substituera ensuite a + e à l'inconnue primitive

Fermat takes the example of cutting a segment AC by a point E in such a way that the product of the segments AE × EC is maxima. One takes

$$AC = b, a = AE \quad \text{and} \quad AE \times EC = ba - a^2.$$

Then one replaces a by $a + e$ in the latter expression:

$$ba - a^2 + be - 2ae - e^2;$$

one "adequates" to the preceding expression:

$$ba - a^2 \sim ba - a^2 + be - 2ae - e^2;$$

by suppressing the common terms and dividing all the terms by e:

$$b \sim 2a + e;$$

by suppressing e, one in conclusion obtains the solution: $b = 2a$.

Fermat explains in the memoir how "we bring back to the preceding method the construction of tangents at given points on arbitrary curves." He takes first the example of "the construction" of the tangent BE at a point B of a parabola with axis CE and vertex D (Fig. 15.3).

One considers an arbitrary point O on this tangent, so one has:

$$\frac{CD}{DI} > \frac{BC^2}{OI^2}.$$

One supposes CD = d given, CE = a, and CI = e:

$$\frac{d}{d-e} > \frac{a^2}{a^2 + e^2 - 2ae},$$

or further:

$$da^2 + de^2 - 2dae > da^2 - a^2e.$$

a, et on exprimera ainsi la quantité maxima ou minima en termes où entreront a et e à des degrés quelconques. On adégalera, pour parler comme Diophante, les deux expressions de la quantité maxima ou minima, et on retranchera les termes communs de part et d'autre. [...] On divisera tous les termes par e, ou par une puissance de e d'un degré plus élevé, de façon que dans l'un au moins des termes de l'un quelconque des membres e disparaisse entièrement. On supprimera ensuite tous les termes où entrera encore e ou l'une de ses puissances et l'on égalera les autres" (Fermat, 1891, vol. III: 121). Original translations of Fermat's texts by David Pengelley.

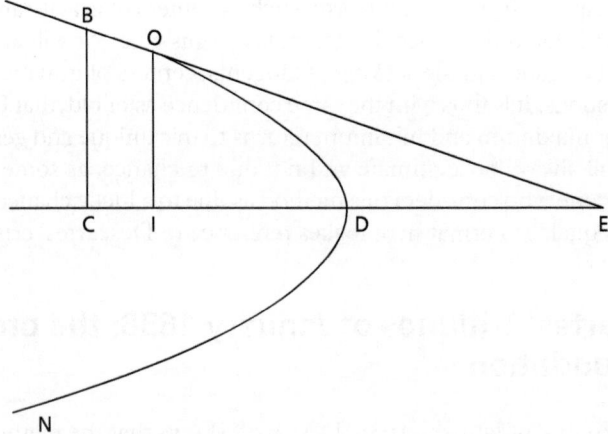

Figure 15.3 *Tangent to a parabola with Fermat's method.*[11]

Following the method of maxima and minima, one "adequates" the two expressions:

$$da^2 + de^2 - 2dae \sim da^2 - a^2 e;$$

by simplifying and dividing by e:

$$de + a^2 \sim 2da;$$

by suppressing e, one obtains the solution: $a^2 = 2da$.

Fermat insists on the generality of the method of maximum and minimum, it is general in the sense that it extends without prescribed condition to all types of problems. He writes: "this method never deludes us, and can be extended to a number of very beautiful questions; thanks to it, we have found the centers of gravity of shapes bounded by straight and curved lines, as well as those of solids and numerous other things that we will be able to treat elsewhere, if we have the time."[12] In subsequent tracts, he shows the generality of the method by extending it to new problems. He writes in a later tract that "the rule is general and faultless": "I could add a number of other examples, as many of the first as of the second case of my method, but those here suffice, and prove sufficiently that it is general and never at fault. I do not add the demonstration of the rule, nor the numerous other

---

[11] Fermat (1891, vol. III: 122).

[12] "Cette méthode ne trompe jamais, et peut s'étendre à nombre de questions très belles ; grâce à elle, nous avons trouvé les centres de gravité de figures terminées par des lignes droites et courbes, aussi bien que ceux de solides et nombre d'autres choses dont nous pourrons traiter ailleurs, si nous en avons le loisir" (Fermat, 1891, vol. III: 123).

applications that can confirm its great worth, such as centers of gravity and asymptotes."[13] In another tract, he again describes the generality of his rule: "it will apply, with a most superior ease and elegance, to the seeking of tangents, centers of gravity, asymptotes, and other similar questions. It is thus with the same confidence as of old, that I affirm still today that the search for maximum and minimum returns to this unique and general rule, whose happy success will always be legitimate and not due to chance, as some have thought. If there remains anyone who considers this method as due to a lucky chance, he may well try to encounter an equal."[14] Fermat here makes reference to Descartes' criticisms.

15.4   Descartes' critiques of January 1638: the problem of foundation

In a letter to Mersenne of January 1638, Descartes shows that the method of tangents of Fermat, applied to the ellipse and the hyperbola, leads to false conclusions. He arrives at this conclusion by taking literally the assertion that the search for tangents is a problem of maximum and minimum, and by keeping the relation that characterizes a point of a parabola for a point of the ellipse or of the hyperbola.[15] In this way, Descartes wants to put forward two large objections to Fermat's rule. The first is factual: the rule is invalid since when applied to the ellipse and the hyperbola one obtains erroneous results. The second is constitutive: it does not have the virtues of a method of construction because it is not a priori, it does not depend on a general conception of curves and it does not arise from a universal method of solution. Descartes expresses himself by exhibiting the algebraic foundation of his method, which assures its universality. He uses the French word *fondement* (we translate it by "foundation"). He compares the two methods in these terms.

> First, his own (that is to say, that which he desired to find) is such that, without skill and by chance, one can easily fall into the path that one must take to encounter it, which is nothing other than a false position, founded on the method of demonstration that reduces to the impossible, and which is the least estimable and the least ingenious of all those that one uses in Mathematics. Mine, rather, is drawn from a knowledge of the nature of Equations, which has to my knowledge never been sufficiently explained elsewhere than in the third book of my Geometry. So that it could not be invented by someone who will have ignored the foundations of Algebra; and it follows the most noble means of demonstration

---

[13] "Je pourrais ajouter nombre d'autres exemples, tant du premier que du second cas de ma méthode, mais ceux-ci suffisent, et prouvent assez qu'elle est générale et ne tombe jamais en défaut. Je n'ajoute pas la démonstration de la règle, ni les nombreuses autres applications qui pourraient en confirmer la haute valeur, comme centres de gravité et des asymptotes" (Fermat, 1891: 130).

[14] "Elle s'appliquera, avec une aisance et une élégance bien supérieure, à la recherche des tangentes, des centres de gravité, des asymptotes et d'autres questions pareilles. C'est donc avec la même confiance que jadis, que j'affirme toujours aujourd'hui que la recherche du maximum est du minimum se ramène à cette règle unique et générale, dont l'heureux succès sera toujours légitime et non pas dû au hasard, comme certains l'ont pensé. S'il reste encore quelqu'un qui considère cette méthode comme due à un heureux hasard, il peut bien essayer d'en rencontrer un pareil" (Fermat, 1891: 135).

[15] Descartes (1897, vol. I: 481–93).

possible, namely that which one calls a priori. Then beyond this, his claimed rule is not universal as he makes it seem, and it cannot reach to any questions that are a little difficult, but only to the very easy ones.[16]

Indeed, Fermat does not claim that his method should be universal, but that it should be general, that is to say, that it should be able to reach to numerous problems, with a field of extent that is not closed but open. Whereas Descartes lays claim to a universality, because he has furnished a foundation of his method, namely the knowledge of algebraic equations, at the same time he has enclosed the field of application of his method. In his critique, Descartes exhorts Fermat to provide a foundation to his method, when the latter is content with a character of generality. For Fermat, defending the character of generality of his method consists in finding new applications, always enlarging the field of possibilities. Quite the contrary, for Descartes, defending the universal character of his method consists in producing the algebraic foundation that specifies its field of application, its universe of validity.

This is manifest in the letter from Descartes to Mersenne of August 1638. Mersenne has announced that Roberval, who had taken the side of Fermat in the dispute, did not know how to find the tangent to the cycloid. Descartes explains to Mersenne that he cannot apply his method of tangents to this curve, because it is not geometric: "one must also notice that the curves described by wheels are entirely mechanical lines, and are amongst those that I have rejected in my Geometry; this is why it is no surprise that their tangents cannot be found at all by the rules that I have introduced for them."[17] He finds these tangents nonetheless by "a very short and very simple demonstration" adapted to the generation of the cycloid. Then he continues the letter by solving the problem of the jasmine flower, which he had posed and which Roberval did not know how to solve either. Descartes systematically applies the method of book I of *The Geometry*.

The problem is to find a point F of the jasmine flower with axis AK such that the tangent at F is parallel to AK (Fig. 15.4). One must suppose the problem solved and give names to all the lines, known and unknown:

$$AG = x, \ GF = y \ \text{ and } \ 2AH = n, \ AE = v.$$

---

[16] "Premièrement, la sienne (c'est-à-dire celle qu'il a eu envie de trouver) est telle que, sans industrie et par hasard, on peut aisément tomber dans le chemin qu'il faut tenir pour la rencontrer, lequel n'est autre chose qu'une fausse position, fondée sur la façon de démontrer qui réduit à l'impossible, et qui est la moins estimée et la moins ingénieuse de toutes celles dont on se sert en Mathématique. Au lieu que la mienne est tirée d'une connaissance de la nature des Equations, qui n'a jamais été que je sache, assez expliquée ailleurs que dans le troisième livre de ma Géométrie. De sorte qu'elle ne saurait être inventée par une personne qui aurait ignoré le fonds de l'Algèbre; et elle suit la plus noble façon de démontrer qui puisse être, savoir celle qu'on nomme a priori. Puis outre cela, sa règle prétendue n'est pas universelle comme il lui semble, et elle ne se peut étendre à aucune des questions qui sont un peu difficiles, mais seulement aux plus aisées" (Descartes, 1897, vol. I: 490).

[17] "Il faut aussi remarquer que les courbes décrites par des roulettes sont des lignes entièrement mécaniques, et du nombre de celles que j'ai rejetées de ma Géométrie ; c'est pourquoi ce n'est pas merveille que leurs tangentes ne se trouvent point par les règles que j'y ai mises" (Descartes, 1897, vol. II: 313).

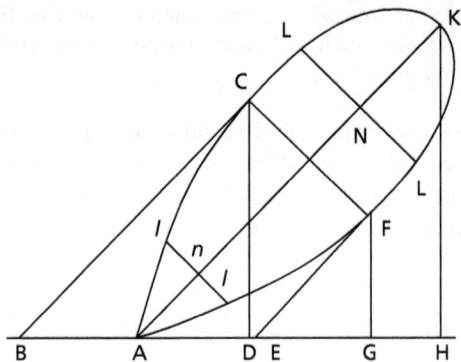

Figure 15.4 *Tangent to a jasmine flower with Descartes' method.*[18]

In order to translate the problem, one much writes the equation of the curve:

$$x^3 + y^3 = xyn,$$

and that EF is parallel to AK:

$$x - v = y.$$

Substituting $x - v$ for y results in:

$$2x^3 - 3vx^2 + 3v^2x - v^3 = nx^2 - nvx.$$

Finally, this equation is identified with

$$(x^2 - 2ex + e^2)\,(2x - 2f) = 0,$$

which permits determination of the unknown v.

We see well here how the application of the geometric method of book I to problems of tangents to curves is fully subject to a notional incorporation, namely that the curves must be associated to algebraic equations, and subordinate to a procedural incorporation, namely the algebraic identification that translates the property of tangency. It is interesting to remark that it is on the occasion of objections to the method of Fermat that Descartes brings forward the question of the foundation of his method. We can summarize the universal character of the geometric method in its application to the method of tangents with the following diagram (Fig. 15.5):

---

[18] Descartes (1897, vol. II: 313).

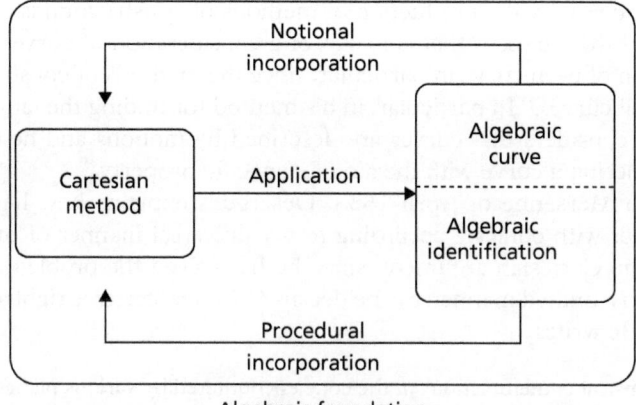

Figure 15.5 *Descartes' universal algebraic method.*

15.5   The problem of the specificity of the curve in April 1638

We have said that Descartes claimed that the method of Fermat was erroneous, when he applied it to the ellipse and to the hyperbola while keeping the relation that characterizes a point of a parabola. We shall see that this critique will compel Fermat to state more precisely the manner in which a curve is specified in the search for its tangents. On the one hand, the method of Descartes relies explicitly on the equation of the curve. Indeed, book II of *The Geometry* opens upon the equational nature of geometric curves and the method of tangents specifies fully the place of this equation in the line of argument. On the other hand, the method of Fermat does not bring forward the relation that characterizes the parabola, even if it uses it. It draws support from an inequality that becomes an "adequality" from which the rule of the maximum and minimum is practiced.

Roberval, who takes the side of Fermat in the letter "Roberval versus Descartes" of April 1638, reproaches his adversary for not knowing that a method applied to a curve must take into account what he calls "the specific property" of the curve. But he seems to recognize that Fermat has not sufficiently singled out this property. He writes:

> To conclude a specific property of any subject whatever, one must, in the propositions of which the arguments are composed, employ at least one other specific property of the same subject, that is to say, that is drawn from its own nature, and that it [the property] fits only with it [the subject] ... and M. de Fermat has not ignored this, since in his tract he has nothing not conforming to this, and he employs in his reasoning specific properties of his subject, which are adroitly mixed with generic and universal properties, serving to conclude the other specific properties that he needs.[19]

---

[19] "Pour conclure une propriété spécifique de quelque sujet que ce soit, il faut dans les propositions, desquelles les arguments sont composés, employer au moins une autre propriété spécifique du même sujet,

   This reply shows how the production of methods of construction is constitutive of a "construction of the curve,"[20] that is to say, of a consideration of curves in general and of a specification of each curve in particular, since the methods of construction must be applicable to "all curves." In particular, in his method for finding the tangents of curved lines, Roberval considers that curves are described by motions and he determines the motions engendering a curve with the aid of a specific property."

   In a letter to Mersenne of April 1638, Desargues explains how he proceeds in a universal manner with conics: "according to my universal manner of proceeding." He is sensitive to the Cartesian argument, since he has solved the problem of tangents on various conics in a unified manner. So, he declares "M. Descartes is right and Mr. Fermat is not wrong." He writes:

> Via my whimsical contemplations of the cone encountered by various planes in all manners, and of the lines and figures generated by this encounter, I have found by a single and common enunciation, construction and preparation, or to say better with a single and common discourse and with the same words, one declares a means of constructing, or properly, one declares the means of making a construction [of another order].... And one sees a similar simultaneous generation of all their tangents.[21]

Fermat arrives at the question of the "specific property" of a curve in a treatise titled *On the same method*, entirely devoted to the "theory of tangents." He responds to Descartes' first objection by insisting in each instance on the specific property of the curve for which one seeks the tangents: "the curved lines for which we seek the tangents have their specific properties, expressible either solely by straight lines, or additionally by curves as complicated as one wishes with lines or other curves."[22] This means that the method is not limited to geometric curves in the sense of Descartes. Fermat adds: "we have already satisfied the first case by our rule, which, too concise, could have appeared difficult, but has however been recognized as legitimate."[23] He considers therefore that he has emerged the victor of the « dispute » with Descartes. In what follows, he constructs the tangents to non-geometric curves, in the sense of Descartes, showing thus the superiority of his method. Indeed, he determines the tangent to the cissoid, the conchoid of Nicomedes,

---

c'est-à-dire qu'elle soit tirée de sa nature propre, et qu'elle ne convienne qu'à lui ... et laquelle M. de Fermat n'a pas ignorée, puisque dans son traité il n'y a rien qui ne lui soit conforme, et qu'il emploie dans son raisonnement des propriétés spécifiques de son sujet, lesquelles étant dextrement mêlées avec des propriétés génériques et universelles, servent pour conclure les autres propriétés spécifiques desquelles il a besoin" (Descartes, 1897, vol. II: 111).

[20] Barbin (2006).

[21] "Par mes contemplations capricieuses du cône rencontré par divers plans en toutes façons, et des lignes et des figures qui s'engendrent en cette rencontre, j'ai trouvé que par une seule et même énonciation, construction et préparation ou pour dire mieux par un seul et même discours et sous de mêmes paroles, on déclare un moyen de construire ou bien on déclare les moyens de faire une construction [d'un autre ordre].... Et l'on voit une pareille génération à même temps de toutes leurs touchantes" (Fermat, 1891, vol. IV: 44).

[22] "Les lignes courbes dont nous cherchons les tangentes ont leurs propriétés spécifiques exprimables, soit par des lignes droites seulement, soit encore par des courbes compliquées comme on voudra avec des droites ou d'autres courbes" (Fermat,1891, vol. III: 140–41).

[23] "Nous avons déjà satisfait au premier cas par notre règle, qui, trop concise, a pu paraître difficile, mais cependant a été reconnue légitime" (Fermat, 1891, vol. III: 141).

the cycloid, and the quadratrix of Dinostratus. When he attacks the cycloid, he remarks that "for the second case, which M. Descartes judged difficult, for whom nothing is, one satisfies it with a very elegant and rather subtle method."[24]

15.6   Fermat's method of tangents: from generality to universality

In the tract *On the same method*, Fermat modifies the method of tangents in making explicit the role of adequality and in responding thus to Descartes' second objection.

> We consider in fact in the plane of an arbitrary curve two lines of given position, of which one can call one the diameter, the other the ordinate. We suppose the tangent already found at a given point on the curve, and we consider by adequality the specific property of the curve, not only on the curve itself, but on the tangent to be found. Upon eliminating, according to our theory of maxima and minima, the terms required to be, we arrive at an equality that determines the meeting point of the tangent with the diameter, and consequently the tangent itself.[25]

Although Fermat does not do it, let us examine what becomes of the search for the tangent to the parabola according to this explanation. Firstly, one takes an arbitrary point O on the tangent BE and one considers by adequality the specific property of the parabola, for the point B on the curve but also for the point O on the tangent (Fig. 15.6):

$$\frac{CD}{DI} \approx \frac{BC^2}{OI^2}.$$

If one supposes given CD = d, CE = a, and CI = e, one obtains:

$$\frac{d}{d-e} \sim \frac{a^2}{a^2 + e^2 - 2ae},$$

or further:

$$da^2 + de^2 - 2dae \sim da^2 - a^2 e.$$

---

[24] "Pour le second cas, que jugeait difficile M. Descartes, à qui rien ne l'est, on y satisfait par une méthode très élégante et assez subtile" (Fermat, 1891, vol. III: 143).

[25] "Nous considérons en fait dans le plan d'une courbe quelconque deux droites données de position, dont on peut appeler l'une diamètre, l'autre ordonnée. Nous supposons la tangente déjà trouvée en un point donné sur la courbe, et nous considérons par adégalité la propriété spécifique de la courbe, non plus sur la courbe même, mais sur la tangente à trouver. En éliminant selon notre théorie des maxima et minima, les termes qui doivent l'être, nous arrivons à une égalité qui détermine le point de rencontre de la tangente avec le diamètre, par la suite la tangente elle-même" (Fermat, 1891, vol. III: 141).

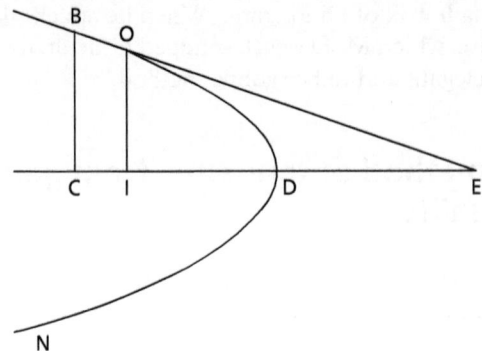

Figure 15.6 *Tangent to a parabola with Fermat's method.*[26]

Secondly, one follows the rule of the maximum and minimum. Simplifying and dividing by *e*, one obtains:

$$de + a^2 \sim 2da.$$

Then suppressing *e*: $a^2 = 2da$.

Thus Fermat delimits well the part in common to a solution of a problem of the maximum and the minimum and of a problem of tangents, namely adequality. In seeking the tangent, one considers a point of the tangent that is almost a point of the curve. Therefore, one may almost write the specific property of the curve at this point. Bringing forth the specific property of the curve has thus permitted the specification of both that which allows the use of adequality and that which limits its use. The method of tangents applies to curves for which a specific property can be given.

We have said that Descartes, in his letter of 1638, takes literally the assertion of Fermat that the search for tangents should be a problem of maximum and minimum. In effect, this poses the question of the tie between the two problems. Fermat busied himself with this in an *Appendix to the method of the maximum and minimum*, where he returns to his method. He shows there how a problem of the maximum and the minimum can be solved by means of the construction of a tangent. He writes then: "one can likewise in general bring every search for the maximum or minimum back to the geometric construction of a tangent; but this does not diminish at all the importance of the general method, because the construction of tangents depends on it, as well as the determination of maxima and minima."[27] But, it is only when he specifies the role of adequality in the search for the tangent that he also specifies more generally to which problems this adequality applies.

---

[26] Fermat (1891, vol. III: 140).

[27] "On peut de même ramener en général toute recherche de maximum ou de minimum à la construction géométrique d'une tangente ; mais cela ne diminue en rien l'importance de la méthode générale, puisque la construction des tangentes en dépend, aussi bien que la détermination des maxima et des minima" (Fermat, 189,1 vol. III: 140).

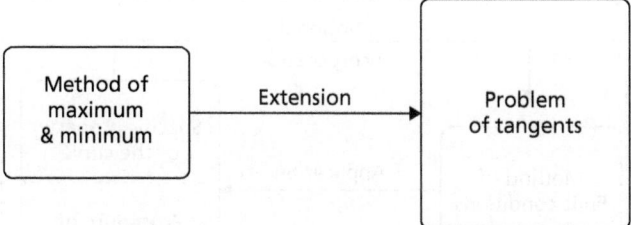

Figure 15.7 *From method to problem of tangents.*

Indeed, Fermat writes at the beginning of his tract *On the same method* that "the theory of tangents is a consequence of the method, published long ago for the construction of the maximum and the minimum, which allows the very easy solution of all questions of limits, and particularly those famous problems for which the limit conditions were shown to be difficult by Pappus."[28] He specifies thus the field of applicability of adequality, namely the problems having limit conditions. The method of tangents is no longer an extension of a general method, namely the method of the maximum and minimum, but an application of a universal method for which the foundation is adequality.

Fermat responds to the question of foundation by passage from the conception of an always extensible general method (Fig. 15.7) to the conception of a universal method, use of which is limited to curves having a specific property and to limit problems. The passage to the universal method requires a notional incorporation and a procedural incorporation (Fig. 15.8), which indicate how adequality is applied to problems of tangents. We have therefore a diagram similar to that of the Cartesian method.

15.7 The two methods of Fermat according to Descartes

Descartes wrote in his letter of January, 1638, without further commentary, that "if one changes several words of the rule which he [Fermat] proposes, for finding maximum and minimum, one can render it true and good enough."[29] He returns to this affirmation in a letter to Hardy, who had taken his side at the time of the dispute: "but for that which I have set, since my early writing, that one could render it good by correcting it, and that I always maintained the same thing, I am sure that you will not be sorry that I tell you here of the foundation, as well as I am persuaded that these gentlemen, who prize it so

---

[28] "La théorie des tangentes est une suite de la méthode, dès longtemps publiée pour l'invention du maximum et du minimum, qui permet de résoudre très aisément toutes les questions de limitation, et notamment ces fameux problèmes dont les conditions limites sont indiquées comme difficiles par Pappus" (Fermat, 1891, vol. III: 140).

[29] "Si on change quelques mots de la règle qu'il [Fermat] propose, pour trouver maxima et minima, on la peut rendre vraie et est assez bonne" (Descartes, 1897, vol. I: 489).

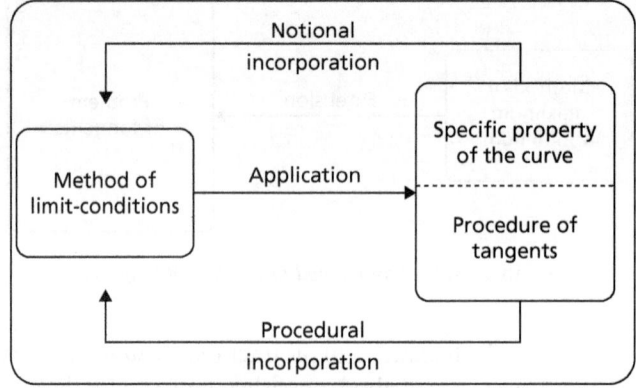

Figure 15.8 *Fermat's universal method.*

much, do not understand it, perhaps not even he who himself is its author."[30] Descartes thus intends to give a foundation for Fermat's rule.

For that, Descartes considers a point E of the diameter of a curve, from which one draws a secant to it that cuts the curve in B and D (Fig. 15.9) and he sets

$$BC = b, \ AC = c, \ EC = a \text{ and } CF = e.$$

The triangles EBC and EDF are similar, thus:

$$\frac{CE}{BC} = \frac{EF}{DF},$$

from which

$$DF = \frac{ba + be}{a}.$$

Descartes assumes that the curve is a cubic parabola. Upon writing the relation that connects the ordinates BC and DF to segments of the diameter, one has

$$\frac{AC}{FA} = \frac{BC^3}{DF^3}.$$

[30] "Mais pour ce que j'ai mis, dès mon premier écrit, qu'on la pouvait rendre bonne en la corrigeant, et que j'ai toujours soutenu la même chose, je m'assure que vous ne serez pas marri que je vous en dise ici le fondement, aussi bien je me persuade que ces Messieurs qui l'estiment tant, ne l'entendent pas, ni peut être même celui qui en est l'auteur" (Descartes, 1897, vol. II: 170).

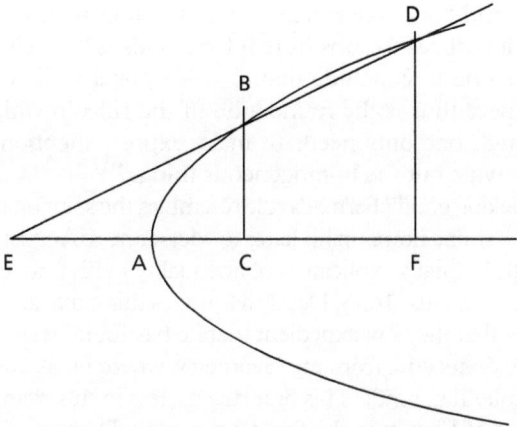

Figure 15.9 *Tangent to a parabola by Descartes in the letter to Hardy.*[31]

This yields

$$cb^3 + eb^3 = \frac{cb^3a^3 + 3b^3ca^2e + 3b^3ace^2 + cb^3e^3}{a^3}.$$

He explains that on multiplying the middles and the extremes of this proportion, simplifying by b^3, multiplying by a^3, and simplifying he obtains:

$$a^3 e = 3ca^2 e + 3\ cae^2 + \ ce^3.$$

Then, he divides by e:

$$a^3 = 3ca^2 + 3\ cae + ce^2.$$

This equation has two unknowns a and e, but these are related by another equation bringing in the relation g upon h of BC to DF:

$$ha = ga + ge.$$

When substituting, with the help of this latter equation, a or e in the preceding, one obtains an equation with a single unknown.

Descartes notes that, up to now, the calculations have followed "the ordinary path of analysis." He remarks then, that to apply all this to the construction of the tangent, "one need only consider that, when EB is the tangent, the line DF is none other than one with BC, and nevertheless that it must be sought by the same calculation that I have just made."

[31] Descartes (1897, vol. II: 171).

Consequently, $h = g$ and "one has merely $a = a + e$, that is to say e equals nothing."[32] Therefore to find a, it suffices to substitute 0 for e. This is "the elision of homogeneous things," which allows one to conclude that $a^3 = 3ca^2$, or $a = 3c$. Descartes finishes the letter by writing: "There thus is the foundation of the rule, in which there are virtually two equations, although one only needs to make express mention of one, because the other serves solely to wipe out the homogeneous things."[33]

This manner of "making good" Fermat's rule resembles the solution that Descartes will give for the problem of the jasmine flower in his letter to Mersenne of August 1638. It reconciles the Cartesian method with Fermat's explication of adequality in his tract *On the same method*. In the letter to Mersenne of August 1638, Descartes judges this explication as follows: "I do not pause at all here to say that the new expedient that he has found was very easy to encounter and that he could have deduced it from my Geometry, where I make use of a similar manner for avoiding the difficulty that renders his first rule useless in this example."[34] He thinks that the method of tangents of Fermat in the first tract is very different of the method of *On the same method*. He reproaches Fermat for wanting to persuade the opposite:

> Furthermore, I am extremely surprised at that which he wishes to try in order to convince that the fashion in which he finds this tangent is the same that he proposed at the beginning, and that he furnishes as proof of this that he makes use of the same diagram for it, as if he had to deal with people who did not know merely to read; because one needs only to read the one and the other writing, to understand that they are very different.[35]

It is true that the title of Fermat's tract is *On the same method*, and that the diagram and the calculations are almost identical. But it is true, that from the point of view of foundation, they are different, thus very different to the eyes of Descartes.

15.8   Conclusion: universality versus generality

Descartes' and Fermat's methods of constructing tangents can be compared in different ways.[36] It is from the point of view of the expectations of the method that we oppose them here. The method of Descartes aims at a universality, and that of Fermat to a generality. The distinction is particularly obvious with Fermat's passage from a general method to a universal method. In the setting of the methods that interest us here, two

---

[32] "On a seulement $a = a + e$, c'est-à-dire e égal à rien" (Descartes, 1897, vol. II: 172).

[33] "Voilà donc le fondement de la règle, en laquelle il y a virtuellement deux équations, bien qu'il ne soit besoin d'y faire mention expresse que d'une, car l'autre sert seulement à faire effacer ces homogènes" (Descartes, 1897, vol. II: 173).

[34] Descartes (1897, vol. II: 322).

[35] "Au reste, je m'étonne extrêmement de ce qu'il veut tacher de persuader que la façon dont il trouve cette tangente est la même qu'il avait proposée au commencement, et qu'il apporte pour preuve de cela qu'il s'y sert de la même figure, comme s'il avait affaire à des personnes qui ne sachent pas seulement lire ; car il n'est besoin que de lire l'un et l'autre écrit, pour connaître qu'ils sont très différents" (Descartes, 1897, vol. II: 323).

[36] Brunschvicg (1981: 177–82).

features distinguish universality from generality. The first trait is the domain of validity of the method. The universal method applies under restrictive conditions while the general method does not involve a priori restriction. The second trait is the type of validation of the method. The universal method is justified by a foundation while the general method is justified by its efficacy.

From this latter point of view, "the kinds of discourse"[37] of the two geometers are different and correspond to the expectations of the methods. Descartes writes a work of distinctive character, addressing himself largely to contemporaries and to descendants who will be able to exercise themselves to invent methodically, according to the last words of his work, whereas Fermat writes tracts of provisional character, always hoping for new extensions of the method. Each tract ends with a new promise. Here, for example, are the last phrases of *On the same method*: "But as a crowning achievement, one can further find the asymptotes of a given curve, a research that leads to remarkable properties for unlimited curves. We will one day be able to display and demonstrate these at great length."[38] The tracts are addressed to other geometers, who will know how to judge the value of the challenge, but who will not necessarily know how the method can take it up.

The opposition that we set up between universality and generality can serve to analyze other methods, in particular the methods of construction of the seventeenth century. For instance, the method of tangents of Roberval is displayed as a universal method with a foundation, a principle of construction and a condition of application. On the other hand, the *New method* of Leibniz displayed in 1684 presents itself as a general method, and the following tracts will not cease displaying extensions of this method. The passage from the general open method to the universal method occurs at the moment of closure that the latter imposes. This moment can be that at which a concept (*conceptio*) of curve is laid down, that contains and that confines a delimited universe of curves.[39]

Acknowledgment

I thank David Pengelley for the careful translation of this paper.

....

REFERENCES

Bakhtine, M. (1979) *Esthétique de la création verbale* (trad. Aucouturier). Paris: Gallimard.
Baillet, A. (1992) *Vie de Mr Descartes*. Paris: La Table Ronde.
Barbin, E. (1992) 'Démontrer : convaincre ou éclairer ? Signification de la démonstration mathématique au XVIIe siècle ', *Cahiers d'histoire et de philosophie des sciences* 40: 29–49.

---

[37] Bakhtine (1979).

[38] "Mais comme couronnement, on peut encore trouver les asymptotes d'une courbe donnée, recherche qui conduit à de remarquables propriétés pour les courbes indéfinies. Nous pourrons un jour les développer et les démontrer plus au long" (Fermat, 1891, vol. III: 147).

[39] Barbin (2006: 193–6, 262–4).

Barbin, E. (2006) *La révolution mathématique du XVIIe siècle*. Paris: Ellipses.

Boyer, C. B. (1959) *The history of the calculus and its conceptual development*. New York: Dover Publications.

Brunschvicg, L. (1981) [1912] *Les étapes de la philosophie mathématique* (reprint). Paris: Blanchard.

Descartes, R. (1897), *Œuvres*, eds. Adam and Tannery. Paris: Cerf.

Descartes, R. (1987) [1637] *Discours de la méthode* (reprint). Paris: Fayard.

Fermat, P. (1891) *Œuvres*, eds. Tannery and Henry. Paris: Gauthier-Villars.

Kline, M. (1972) *Mathematical thought from ancient to modern times*. New York: Oxford University Press.

Whiteside, D. T. (1960–2) 'Patterns of mathematical thought in the later seventeenth century', *Archive for History of Exact Sciences* 1: 179–387.

16

Algebraic generality versus arithmetic generality in the 1874 controversy between C. Jordan and L. Kronecker

FRÉDÉRIC BRECHENMACHER[1]

16.1   Introduction[2]

Throughout 1874, Camille Jordan and Leopold Kronecker quarreled over two theorems, namely Jordan's canonical form and Karl Weierstrass' elementary divisors. Although these theorems would later be considered equivalent from the perspective of modern matrix theory, not only had they been stated independently between 1868 and 1870, but they had also been devised within the distinct frameworks of separate two theories. Only later would some connections come to light. This context explains why the 1874 controversy developed over the issue of which of the two approaches was the most fundamental.[3] This controversy would eventually turn into an opposition over the algebraic or arithmetic nature of the "theory of forms." As we shall see in this chapter, this opposition sheds light on two conflicting perspectives on "generality."

---

[1] This work was supported by the French National Research Agency (CaaFÉ, "ANR-10-JCJC 0101").

[2] The author is grateful to R. Chorlay and K. Chemla for their helpful advice and comments.

[3] Jordan stated a result for linear substitutions (i.e., linear applications operating on finite fields in observer's terms), whereas Weierstrass stated a theorem in the framework of the theory of bilinear and quadratic forms. Kronecker reworked Weierstrass' theorem, introducing in it what are today known as invariant factors. See Appendix 1 for more details about the mathematics involved in the two theorems.

From the standpoint of linear algebra, the main notions we will consider in this paper, that is, substitutions, bilinear forms, and quadratic forms can all be represented by matrices. However depending on the mathematical object to which the matrix is attached, the related classes of equivalences differ (linear substitutions are represented by classes of similar matrices while bilinear forms are related to classes of equivalent matrices and quadratic forms to classes of congruent matrices). Connections between these notions were one of the issues at stake in the 1874 controversy. At the time, quadratic forms were already considered as particular bilinear forms with symmetric coefficients. However, connections between bilinear forms and linear groups were more problematic even though both theories used the characteristic equations of the linear systems involved.

The Oxford Handbook of Generality in Mathematics and the Sciences. First Edition. Karine Chemla, Renaud Chorlay and David Rabouin. © Oxford University Press 2016. Publishing in 2016 by Oxford University Press.

The 1874 debate was based on what a "truly general" approach to the theory of forms should be, and thus related to disciplinary ideals. Consequently, on a broader scale, this episode raises the issue of the mathematical disciplines that developed in the nineteenth century. In a series of essays devoted to the history of matrices, Thomas Hawkins laid special emphasis on the "generality" of Weierstrass' theorem, which he presents as a "keystone" in the history of algebra. The main reason for this, Hawkins argues, is that "Weierstrass demonstrated more than theorems. He also demonstrated the possibility and desirability of a more rigorous approach to algebraic analysis that did not rest content with the prevailing tendency to reason vaguely in terms of the 'general case'." According to Hawkins, the elementary divisors theorem thus served as a paradigm of generality with the consequence that "he [Weierstrass], more than anyone, was responsible for the emergence of the theory of matrices as a coherent, substantial branch of twentieth-century mathematics."[4]

In this chapter, we shall take the opportunity of the controversial 1874 debate to observe the various forms and meanings taken by "generality" when it is attached to an organization of knowledge of the type that was used before the emergence of object-oriented disciplines. Our aim is not to give a definition of algebraic generality, nor of arithmetic generality, but to pay attention to the ways in which the various actors used these categories. Methodologically speaking, we shall examine the two protagonists' perspectives on generality without considering issues like the origins of abstract notions, theories, ways of reasoning and, more generally, of structures. In this respect, our approach differs from that of most of the historians who have studied the history of linear algebra. A retrospective disciplinary identity has indeed often served as a lens for looking into the past and for selecting the texts and authors to take into consideration. The prevailing disciplinary identity has thereby given a structure to its own history while other identities that did not fit with this retrospective glance have remained out of sight. The question especially arises with respect to the identities and meanings taken by algebraic procedures developed within various fields—such as mechanics, arithmetic, or geometry—and which circulated from one context to another before being considered as methods within unified algebraic theories—in our case, the post-1930s theory of matrices.

As a matter of fact, during the nineteenth century, algebraic procedures such as the manipulation of the "forms" of linear systems considered in this paper were not usually identified as methods or notions within a theoretical framework. Nevertheless they did not only exist as technical tools. As will be seen by looking through the prism of the 1874 controversy, procedures such as those of Jordan's canonical reduction and Kronecker's invariant computations embodied a number of conflicting epistemic values related to generality, such as "simplicity" and "abstraction" *vs* "effectivity" and "homogeneity" *vs* "formalism" and "genericity." We shall therefore refer to these procedures as "practices"

---

[4] Hawkins (1977: 119).

not only to highlight that they were not only technical tools but to emphasize the fact that their individual natures must be considered as a problem.[5]

One of the aims of this paper is to investigate how the "generality" of such practices reflected both the individual and the cultural aspects. While the 1874 controversy was triggered by the opposition between two individual practices, it also involved two perspectives on generality carried by a specific mathematical culture. This culture had been shaped through both a long-term shared history of investigations on a *special* equation, the secular equation, and two diverging lines of developments on forms of treatments of *general* equations.

16.2 A "general" theory object of controversy

The controversy started in the winter of 1873–4 when two papers were read to the Academies of Science in Paris and Berlin.[6] The quarrel was originally caused by Jordan's ambition to reorganize the theory of bilinear forms through what he designated the "simple" notion of "canonical form." Jordan's December 1873 paper was actually the first contribution to the theory of bilinear forms to be published outside Berlin. One of the issues at stake was therefore the organization of a theory which used to be a local field of research limited to a few mathematicians based in Berlin.

In the 1870s, many new applications heralded the major role the theory of bilinear forms would play in the following decades. This theory also gave a new "homogeneity" and "generality" to the treatment to the numerous problems raised in the context of the various theories developed throughout the nineteenth century.[7] In the quotation below, Jordan alludes to geometry, with Augustin Louis Cauchy's results on the principal axis of conics and quadrics (first question), to the arithmetic of quadratic forms in relation especially to the works of Carl Gauss and Charles Hermite (second question), as well as to analytical mechanics and the solution given by Lagrange to $PY'' + QY = 0$ systems of linear differential equations with constant coefficients (third question):

> It is known that there are an infinite number of ways to reduce any bilinear polynomial,
>
> $$P = \Sigma A_{\alpha\beta} x_\alpha y_\beta \, (\alpha = 1, 2, \ldots, n, \beta = 1, 2, \ldots, n),$$

---

[5] On the methodological issues raised by questions of identities in the history of mathematics, see Sinaceur (1991: 16) and Goldstein (1995: 21).

[6] For a complete study of the 1874 controversy, see Brechenmacher (2007a).

[7] In modern parlance, bilinear forms for a long time played a role similar to the one matrices would play in twentieth-century linear algebra. After the 1870s, the theory of bilinear forms would play a key role on a global level thanks to its numerous applications in geometry (Klein, 1868), the theory of quadratic forms (Kronecker, 1874; Darboux, 1874), and various problems related to systems of differential equations (Jordan, 1871–2), such as Fuchs equations (Hamburger, 1873; Jordan, 1874) or Pfaff's problem (Frobenius and Darboux, 1875–80). See Hawkins (1977) and Brechenmacher (2006a).

to the canonical form $x_1y_1 + \ldots + x_my_m \ldots$ using linear transformations applied to the two sets of variables $x_1, \ldots, x_n, y_1, \ldots, y_n \ldots$ Among the various questions of this kind that can be raised, we consider the following:

1. Reducing a bilinear polynomial P to a simple canonical form using *orthogonal* substitutions applied, some on the $x_1, \ldots, x_n$, and others on the $y_1, \ldots, y_n$.

2. Reducing P to a simple canonical form using linear substitutions operating simultaneously on the x's and on the y's.

3. Reducing two polynomials P and Q *simultaneously* to a canonical form using any linear substitutions applied on each of the two sets of variables *separately*.[8]

As we shall see in greater detail later, the fact that Weierstrass and Jordan both claimed to have given *general* solutions to problems tackled in the past brought to light the connection between Jordan's two theorems.[9] In order to investigate the issues of generality raised by the ensuing controversy, it will prove useful to change the scale for both the contexts and for the time periods.[10]

Let first consider the local context in which the theory of bilinear forms developed in Berlin in the late 1860s. In 1866, two papers published by Elwin Christoffel and by Kronecker lay the foundations of a theory on the characterization of bilinear forms—given two forms $P = \Sigma A_{\alpha\beta}x_\alpha y_\beta$ and $P' = \Sigma B_{\alpha\beta}x_\alpha y_\beta$, find the necessary and sufficient conditions under which P can be transformed into P' by using linear substitution. The main method was to look for invariants computed from coefficient forms which would be unaltered by linear transformation. The "general" resolution to the problem of the simultaneous transformation of two forms P and Q would shortly become the main issue of the theory (i.e., the third problem in the list established by Jordan in the above quotation).

This problem challenged the "generality" of traditional polynomial methods. Pairs of forms (P, Q) can indeed be investigated by using a polynomial of forms $P + sQ$, whose determinant $|P + sQ|$ is therefore a numerical polynomial in s. This determinant is invariant up to the linear transformation of the forms: all equivalent pairs of form (P, Q) have the same determinant $|P + sQ|$. The equation $|P + sQ| = 0$, that is, what is known today as the characteristic equation, thus plays a key role in the theory of forms. Yet, a pair of forms is characterized exactly by its characteristic equation only if this equation has no multiple roots. In 1874, the whole theory of bilinear forms, that is, both its internal content and

---

[8] Jordan (1873: 1487, translation F.B.). "On sait qu'il existe une infinité de manières de ramener un polynôme bilinéaire $P = \Sigma A_{\alpha\beta}x_\alpha y_\beta$ ($\alpha = 1, 2, \ldots, n, \beta = 1, 2, \ldots, n$) à la forme canonique $x_1y_1 + \cdots + x_my_m \ldots$ par des transformations linéaires opérées sur les deux systèmes de variables $x_1, \ldots, x_n, y_1, \ldots, y_n \ldots$ Parmi les diverses questions de ce genre que l'on peut se proposer, nous considérons les suivantes : 1. Ramener un polynôme bilinéaire P à une forme canonique simple par des substitutions *orthogonales* opérées les unes sur $x_1, \ldots, x_n$, les autres sur $y_1, \ldots, y_n$; 2. Ramener P à une forme canonique simple par des substitutions linéaires quelconques, mais opérées simultanément sur les x et les y ; 3. Ramener *simultanément* à une forme canonique deux polynômes P et Q par des substitutions linéaires quelconques, opérées *isolément* sur chacune des deux séries de variables."

[9] For a comment from the standpoint of modern linear algebra, see Appendix 1.

[10] See Revel (1996).

its usefulness in applications, revolved around the polynomial invariants Weierstrass had introduced in 1868 to characterize the equivalence between forms when multiple roots occurred. On the one hand, this characterization was considered to be *general* because it worked whatever the multiplicity of the roots. On the other hand, this solution resorted to a *specific* method that called on the comparison of the algebraic factorization of the equation $|P + sQ| = 0$ and the decomposition of the determinant $|P + sQ|$ in sequences of minors.[11] In a sense, Weierstrass' elementary divisors gave the theory both its generality and its specificity.

Jordan thus struck at the core of the theory when, in 1873, he proposed reorganizing the whole theory of bilinear forms by using a method of canonical reduction which, he claimed, was "more general" than Weierstrass' invariant computation. Moreover, Jordan argued that "the problem of the simultaneous reduction of two functions *P* and *Q* is identical to the problem of the reduction of a linear substitution to its canonical form," and to the theorem he had stated in his 1870 *Traité des substitutions et des équations algébriques*:[12]

> ... the third [problem has already been dealt with] by M. Weierstrass; but the solutions given by the eminent Berlin geometers are incomplete, in so far as they left out some exceptional cases which are nevertheless not without interest. Their analysis is moreover quite difficult to follow—especially that of Mr Weierstrass. On the contrary, the new methods that we propose are extremely simple and hold no exceptions... The simultaneous reduction of two functions *P* and *Q* is a problem identical to the reduction of a linear substitution to its canonical form.[13]

In a paper he submitted to the Academy of Berlin in January 1874, Kronecker rejected the whole theoretical organization Jordan had put to the fore. Recalling that as early as 1868 Weierstrass and he had organized the theory solely around the characterization of pairs of forms, he questioned the relevance of Jordan's distinction of three canonical reduction problems:

> In Mr *Jordan*'s memoir ..., the solution to the first problem is not truly new, the solution to the second problem is false, and that of the third one is not sufficiently well established. We should add that actually the third problem includes the two others as particular cases, and that its complete solution stems from Mr *Weierstrass*' work of 1868, and can also be derived from my additional contribution to this work. Unless I am mistaken, there are serious grounds to question M. *Jordan*'s priority in the invention of his results, should they even be correct ...[14]

---

[11] Weierstrass' solution was actually limited to the non-singular case when $|P + sQ|$ did not vanish identically, another paper published by Kronecker in 1868 was devoted to the singular case $|P + sQ| = 0$.

[12] Jordan (1870).

[13] Jordan (1873: 1491, translation F.B.) "... le troisième [problème a déjà été traité] par M. Weierstrass ; mais les solutions données par les éminents géomètres de Berlin sont incomplètes, en ce qu'ils ont laissé de côté certains cas exceptionnels qui, pourtant, ne manquent pas d'intérêt. Leur analyse est en outre assez difficile à suivre, surtout celle de M. Weierstrass. Les méthodes nouvelles que nous proposons sont, au contraire, extrêmement simples et ne comportent aucune exception... La réduction simultanée de deux fonctions *P* et *Q* est un problème identique à celui de la réduction d'une substitution linéaire à sa forme canonique."

[14] Kronecker (1874b: 417, translation F.B.) "Dans le Mémoire de M. *Jordan*..., la solution du premier problème n'est pas véritablement nouvelle ; la solution du deuxième est manquée, et celle du troisième n'est

During the winter, Kronecker developed his views on the organization of the theory of bilinear forms in monthly papers sent to the Academy of Berlin. Meanwhile, he and Jordan engaged in private correspondence with the aim of settling the dispute over priority caused by the connections which had been suggested between the canonical form and the elementary divisors theorems.[15]

Although it would lead Jordan to recognize the partial anteriority of Kronecker and Weierstrass for some of his results, as well as to grasp some of the tacit knowledge specific to the Berliners,[16] the correspondence failed to allow them to reach agreement on the mathematics. Kronecker failed to bring Jordan round to his own ideas on the structure of the theory of bilinear forms and Jordan did not succeed in convincing Kronecker of his "natural right" to claim the genuine originality of the distinction he had made between three types of canonical forms related to what he designated "groups of substitutions." This distinction, Jordan added, "has more brought to light than disparaged Weierstrass' result in highlighting the resolution it implicitly gave to a fundamental problem in linear substitutions theory which, to my opinion, is a much more fertile theory than the algebraic theory of forms of the second order."[17]

In the spring, the controversy would go public again, and it would reach its climax with the publications of a series of notes and papers in the journals of the Academies of Paris and Berlin as well as in the *Journal de Mathématiques Pures et Appliquées*. The quarrel over priority would then turn into an opposition over two theories (group theory *vs* the theory of forms) and two disciplines (Algebra vs Arithmetic) as well as over two practices (canonical reduction vs invariant computation) relating to conflicting philosophies of generality.

16.3   Weierstrass' theorem as marking a rupture in the history of general/generic reasoning

Kronecker associated Weierstrass' theorem with an ideal of generality. Kronecker's criticisms of the "formal" nature he imputed to Jordan's canonical reduction carried with

---

pas suffisamment établie. Ajoutons qu'en réalité ce troisième problème embrasse les deux autres comme cas particuliers, et que sa solution complète résulte du travail de M. *Weierstrass* de 1868 et se déduit aussi de mes additions à ce travail. Il y a donc, si je ne me trompe, de sérieux motifs pour contester à M. *Jordan* l'invention première de ses résultats, en tant qu'ils sont corrects... ."

[15] Jordan's correspondence is in the archives of the École Polytechnique. The part related to the correspondence has been edited in Brechenmacher (2006a).

[16] Until Jordan received some detailed explanations from Kronecker, he had some difficulties in understanding the properties of determinants Weierstrass and Kronecker had implicitly used for computing invariants. Moreover, before the correspondence, Jordan did not understand how some papers published during the 1860s were connected with one another and with the emerging theory of bilinear forms. For instance, when he published his first note in December 1873, Jordan was unaware of the two papers in which Christoffel and Kronecker works had founded in 1866 the theory of bilinear forms (Christoffel, 1866; Kronecker, 1866). Moreover, he did not realize the relation between Kronecker's 1868 memoir on "quadratic forms" and Weierstrass' 1868 paper despite the fact that these two papers had been conceived as the "two parts" of a same development and had thus been published one after the other in *Crelle's* Journal. See Brechenmacher (2012).

[17] Jordan to Kronecker, January 1874, translation F.B.

it the historical perspective that the generality of Weierstrass' theorem was (seen as) a turning point in the history of Algebra.

Taking as a starting point the "mistake" of using an algebraic expression that may vanish as a denominator, he had picked out in Jordan's December paper, Kronecker developed his views along the lines of an opposition between the "uniform" and the "formal" and the "general" and the "homogeneous."[18] He was especially critical of the uniform formal expressions which lose their meaning in some singular cases:

> We are indeed used to discovering essentially new difficulties—especially in algebraic questions—as soon as we free ourselves from the restriction of such cases one is accustomed to term general. As soon as one forces one's way through the surface of this so called generality—which excludes any particularity-, one penetrates the true generality—which encompasses all singularities-, only then, in general, does one find the real difficulties of the study, but at the same time one finds the wealth of new viewpoints and phenomena which lie in its depths.[19]

As shall be seen in greater detail when we return to Kronecker's arithmetical agenda, he had a complex attitude toward Weierstrass' approach. On the one hand, Kronecker completely transformed the content of the elementary divisors theorem. But, on the other hand, he constantly presented Weierstrass' original statement as the model of "true generality." According to Kronecker, the elementary divisors theorem contrasted with the "inadequate results" of the "so called general" methods that had been developed sporadically "for over a century" in the particular (symmetric) case of pairs of quadratic forms. Kronecker indeed blamed these methods for the little attention that had been given to the difficulties that might be caused by singularities such as multiple roots (or equal factors) occurring in the polynomial determinant $S = |P + sQ|$.

Kronecker was implicitly referring here to the long tradition of what in the nineteenth century was usually referred to as the "equation to the secular inequalities in planetary theory"—the secular equation for short—because of its association with the problem of the stability of the solar system. As will be described in greater detail later, this equation was at the core of a specific algebraic culture that played a key role in the nineteenth century. In short, and in observer's terms, the secular equation corresponds to the characteristic equation of a pair of quadratic forms, that is, to the special case in which the forms P and Q are symmetric.

Kronecker's allusion to the "well-known problem" of the consideration of pairs of quadratic forms, "which has been dealt with so often (although sporadically) over the

---

[18] Jordan promptly corrected this mistake which was of no consequence for the organization of his theory; see Brechenmacher (2006a: 689).

[19] Kronecker (1874a: 405, translation F.B.). "Denn man ist es gewohnt – zumal in algebraischen Fragen – wesentlich neue Schwierigkeiten anzutreffen, wenn man sich von der Beschränkung auf diejenigen Fälle losmachen will, welche man als die allgemeinen zu bezeichnen pflegt. Sobald man von der Oberfläche der sogenannten, jede Besonderheit ausschliessenden Allgemeinheit in das Innere der wahren Allgemeinheit eindringt, welche alle Singularitäten mit umfasst, findet man in der Regel erst die eigentlichen Schwierigkeiten der Untersuchung, zugleich aber auch die Fülle neuer Gesichtspunkte und Erscheinungen, welche sie in ihren Tiefen enthält."

last century," was a typical way of referring to a specific and widely-read group of texts by authors such as Joseph-Louis Lagrange, Pierre-Simon Laplace, Augustin Cauchy, and Carl Gustav Jacobi. Although these authors dealt with a variety of problems, from mechanics to number theory, they all used symmetric systems of linear equations. The secular equation played a key role for solving such systems: their "general" solutions were indeed expressed by the quotient of two polynomial expressions involving factors of the secular equation. As Kronecker highlighted, assigning specific values to the symbols involved in such general algebraic expressions raised difficulties as it may lead to $\frac{0}{0}$ expressions in the case of multiple roots.[20] He therefore criticized these works from the past for their false "generality," that is, for their focus on the *generic* case in which $S = 0$ had no multiple root.

On the contrary, Kronecker argued, when Weierstrass had dealt with such issues for the first time in 1858, he had developed a "truly general" approach in shifting the focus from the nature of the roots to the decomposition of the determinant $|P + sQ|$ into sequences of sub-determinants. In Kronecker's opinion, the approach Weierstrass had developed in 1858 for the particular (quadratic) case of the secular equation heralded the complete generality of the 1868 theorem on bilinear forms:

> This holds in the few algebraic questions which have been tackled completely and in the smallest detail, such as the theory of networks of quadratic forms whose main features have been developed above. As long as one did not dare to dispense with the hypothesis that the determinant has only distinct factors, one could only reach inadequate results in the well-known problem of the simultaneous transformation of two quadratic forms—a problem which has been dealt with so often (although only sporadically) over the last century—; under this hypothesis, the true viewpoints on the investigation remained completely unacknowledged. Weierstrass' 1858 work dropped this hypothesis; and resulted in a higher insight and to a complete treatment of the case when only simple elementary divisors occur. But the general introduction of the notion of elementary divisors—of which only the first step was taken—occurred for the first time in Weierstrass' 1868 work, and it shed new light on the theory of networks for the case of an arbitrary, yet non-vanishing, determinant. As I dropped this last restriction, and from this notion of elementary divisors, developed the more general notion of elementary networks, a brighter light was shed on the richness of the newly revealed algebraic picture, and at the same time by this complete treatment of the subject, the

---

[20] These algebraic formulas were rational expressions with the determinant S in their numerator and its principal minors in their denominator. As a multiple root of S may be a common root between S and its minors, it may then lead potentially to $\frac{0}{0}$ expressions. In 1858, Weierstrass carefully investigated how the polynomial factorization of the equation $S = 0$ paralleled the decomposition of the sub-determinants extracted from $|P + sQ|$. He discovered that $\frac{0}{0}$ expressions actually never occurred in the symmetric case of the equation to the secular inequalities. The generalization of this approach from pairs of quadratic forms to pairs of bilinear forms led to the statement of the elementary divisors theorem in 1868. See Brechenmacher (2007b) for more details.

most valuable insights were reached on the theory of the higher invariants, as conceived in their true generality.[21]

Combining mathematical and historical arguments, Kronecker was the first to emphasize a history of what Thomas Hawkins would, in the 1970s, refer to as "generic reasoning" in algebra. For Hawkins, such a form of reasoning had played a key role since the development of symbolical algebra in Viète's works in the sixteenth century:

> The generality of the method of analysis had been viewed as its great virtue since its inception. Thus Viète stressed that the new method of analysis "does not employ its logic on numbers—which was the tediousness of the ancient analysts—but uses its logic through a logistic which in a new way has to do with species." ... Analysis became a method for reasoning with, manipulating, expressions involving symbols with "general" values and a tendency developed to think almost exclusively in terms of the "general" case with little, if any, attention given to potential difficulties or inaccuracies that might be caused by assigning certain specific values to the symbols. Such reasoning with "general" expressions I shall refer to for the sake of brevity as *generic reasoning*. [22]

The lines of developments Hawkins discussed however were more restricted than the entire domain of algebra. They actually coincided with the long history Kronecker had alluded to in 1874. For both Kronecker and Hawkins, the tension between the generic and the general was instrumental to the selection of a genealogy starting with the mechanical investigations by Lagrange and Laplace in the eighteenth century, involving Cauchy's 1829 memoir on the classification of conics and quadrics,[23] and ending with Weierstrass' theorem. It was by appealing to the turning-point Kronecker associated with Weierstrass' theorem that Hawkins argued that, although "historians writing on this subject have tended to emphasize the role [of] Arthur Cayley," who had developed a

---

[21] Kronecker (1874a: 405, translation F.B.). "Diess bewährt sich durchweg in den wenigen algebraischen Fragen, welche bis in alle ihre Einzelheiten vollständig durchgeführt sind, namentlich aber in der Theorie der Schaaren von quadratischen Formen, die oben in ihren Hauptzügen entwickelt worden ist. Denn so lange man es nicht wagte, die Voraussetzung fallen zu lassen, dass die Determinante nur ungleiche Factoren enthalten, gelangte man bei jener bekannten Frage der gleichzeitigen Transformation von zwei quadratischen Formen, welche seit einem Jahrhundert so vielfach, wenn auch meist blos gelegentlich, behandelt worden ist, nur zu höchst dürftigen Resultaten, und die wahren Gesichtspunkte der Untersuchung blieben gänzlich unerkannt. Mit dem Aufgeben jener Voraussetzung führte die *Weierstrass*'sche Arbeit vom Jahre 1858 schon zu einer höheren Einsicht und namentlich zu einer vollständigen Erledigung des Falles, in welchem nur einfache Elementartheiler vorhanden sind. Aber die allgemeine Einführung dieses Begriffes der Elementartheiler, zu welcher dort nur ein vorläufiger Schritt gethan war, erfolgte erst in der *Weierstrass*'schen Abhandlung vom Jahre 1868, und es kam damit ganz neues Licht in die Theorie der Schaaren für den Fall beliebiger, doch von Null verschiedener Determinanten. Als ich darauf auch diese letzte Beschränkung abstreifte und aus jenem Begriffe der Elementartheiler den allgemeineren der elementaren Schaaren entwickelte, verbreitete sich die vollste Klarheit über die Fülle der neu auftretenden algebraischen Gebilde, und bei dieser vollständigen Behandlung des Gegenstandes wurden zugleich die wertvollsten Einblicke in die Theorie der höheren, in ihrer wahren Allgemeinheit aufzufassenden Invarianten gewonnen."

[22] Hawkins (1977: 122).

[23] From the viewpoint of modern algebra, Cauchy's 1829 memoir provided the first "general" proof that the eigenvalues of a symmetric matrix are real and that the corresponding quadratic form can be transformed into a sum of square terms (i.e., diagonalized) by means of an orthogonal transformation.

symbolical theory of matrices in 1858,[24] "there is much more to the theory of matrices—and to its history—than the formal aspect, i.e. the symbolical algebra of matrices. There is also a content ... the concept of an eigenvalue, the classification of matrices into types (symmetric, orthogonal, Hermitian, unitary, etc.), the theorems on the nature of the eigenvalues of the various types and, above all, those on the canonical (or normal) forms for matrices."[25]

Moreover, the long term development of the theory of matrices was characterized as a progressive increase of the standards of rigor which eventually resulted in the rejection of the legitimacy of generic reasoning. As Hawkins highlighted:[26]

> ... neither of them [Lagrange and Laplace] had pursued the study of the solutions of systems of linear differential equations with sufficient care to justify their claim [that the characteristic roots λ must be real]. They had no difficulty treating such a system when the characteristic roots are distinct, but their analysis of the case of multiple roots was inadequate. Given the generic tendency of their analytical methods, it is noteworthy that they considered the case at all.... Weierstrass' recognition of the questionable nature of their claims formed the starting point of the investigations that culminated in his theory of elementary divisors.[27]

In a word, to both Kronecker and Hawkins, Weierstrass' theorem marked a turning point in the history of algebra. It presented a rigorous development in stark contrast to the generic nature of reasoning in the past. It also resulted in a homogeneous solution as opposed to the specific arguments which had been developed to deal with the singular cases which restricted the range of validity of general algebraic expressions. For instance, in a paper he devoted to theta functions in 1866, Kronecker himself was still looking upon the occurrence of multiple roots as a "singular" case in which the "general" algebraic approach failed and for which it was customary to use arguments specific to the context of theta functions. Arguments of the like usually were employed when (falsely) claiming that no multiple roots could occur in the context under consideration and that the algebraic expressions involved were therefore probably fully general. Later on, Kronecker would nevertheless recognize that his reasoning was circular because he had used the result he was actually aiming to prove.

In demonstrating the possibility of developing a genuinely general and homogeneous approach to the classification of pairs of quadratic and bilinear forms, the invariants introduced by Weierstrass thus provoked a change of approach in Kronecker's mathematical work. This fact calls for a reconsideration of some classic categories in the historiography of mathematics. Epistemic values such as "rigor" or "generality" have indeed often played the role of structuring categories in the historiography of mathematical theories and

---

[24] Compare with the formal nature Kronecker attributed to Jordan's canonical form. We return to this issue in Section 16.5. On the various lines of developments in the history of matrices, see Hawkins (1977) and Brechenmacher (2010).

[25] Hawkins (1977: 1).

[26] See also R. Chorlay's chapter (Chapter 14) in this volume for issues about generality, genericity, arbitrariness, and rigor in the history of mathematical analysis in the nineteenth century.

[27] Hawkins (1977: 122–4).

disciplines. As has already been alluded to, the tension between generic and general reasoning was instrumental to the role Hawkins attributed to Weierstrass, whose work marked a major step in the history of the theory of matrices and, more generally, in the history of algebra:

> I would suggest that, insofar as anyone deserves the title of founder of the theory of matrices, it is Weierstrass.... His theory of elementary divisors provided a theoretical core, a substantial foundation, upon which to build. His work demonstrated the possibility of dealing by the methods of analysis with the non-generic case, thereby opening up a whole new world to mathematical investigation, a world that his colleagues and students proceeded to explore.... One motivational force common to the entire century was a concern for a more rigorous level of reasoning in mathematics.... A concern for higher standards of reasoning was a driving force behind Weierstrass' work and also behind that of Cauchy and Dirichlet which preceded it and behind that of Kronecker and Frobenius which succeeded it.... The rise of the theory of matrices was directly related to the fall of the generic approach to algebraic analysis. A concern for rigor did not mark the end of the creative development of the theory but its beginning.[28]

It was, however, in the very special context of this controversy that Kronecker came to highlight the generic nature of most of the algebraic reasoning of the past century.[29] The question therefore arises as to the potentially different views that were being held by other actors, such as Jordan. By investigating how the latter came to develop some connections between his research on substitution groups and the theory of forms, we shall bring to light his quite different view on the issue of generality. This will therefore lead us to develop an approach complementary to that of the work of Thomas Hawkins on the generic nature of algebraic reasoning from the eighteenth to the nineteenth century.

16.4   An algebraic practice and some mechanical interpretations dating back to the time of Lagrange

Jordan's intervention in the theory of bilinear forms in 1873 was the consequence of a note the astronomer Antoine Yvon-Villarceau had submitted three years earlier to the geometers at the Academy of Paris. The latter had pointed out a mistake in a classic method dating back to Lagrange, which used to be considered as emblematic of the general mathematical treatment of numerous mechanical problems.

More precisely, Yvon-Villarceau questioned the method "for integrating the equations of a rotating solid body under the action of gravity" which had been "introduced by the illustrious author of the *Mécanique analytique* for the special case of the small oscillations of

---

[28] Hawkins (1977: 157–9).
[29] On the construction of history by mathematical texts, see Cifoletti (1995), Goldstein (1995), Dhombres (1998), and Brechenmacher (2006b).

a loaded string whose equilibrium is slightly disturbed while one of its ends remains in position." In 1766 Lagrange devised a "general method" for the "general case" involving an arbitrary (finite) number of masses—as opposed to the particular case of a string loaded with two or three masses that had already been tackled by Jean d'Alembert. Generality here meant considering a system of n linear differential equations with constant coefficients.

Villarceau illustrated Lagrange's method in the case of a system of two differential equations as follows (with x_i functions of t and a_i constant coefficients):[30]

$$\frac{d^2 x_1}{dt} = a_1 x_1 + a_2 x_2,$$

$$\frac{d^2 x_2}{dt} = a_2 x_1 + a_1 x_2.$$

The key point was to associate to the linear system its secular equation, here an equation of the second degree which has, in "general" (in the generic sense), two roots s and s'. The initial system can then be "reduced" to the two following independent equations:[31]

$$\frac{d^2 y_1}{dt} + s y_1 = 0,$$

$$\frac{d^2 y_2}{dt} + s' y_2 = 0.$$

The initial problem can thus be solved by considering each of the above equations separately. The solutions x_1 and x_2 are given by linear combinations of expressions such as $y_1 = \sin(st + \varepsilon)$ and $y_2 = \sin(s't + \varepsilon')$. In the "general" case Lagrange had considered, that is, for the case of a system of n equations with no symmetric property, a necessary condition for reducing the system to n independent equations was that the associated characteristic equation of the nth degree had n single roots.

As alluded to before, Villarceau's 1870 note had the precise aim of criticizing a mechanical interpretation dating back to Lagrange for legitimating the algebraic resolution of the systems of differential linear equations. The presupposed mechanical stability (the oscillation had to remain small) had indeed usually been considered as having the consequence that only single roots could occur. Multiple roots were assumed to cause

---

[30] Note that in Villarceau's system, the pair (a_2, a_1) of coefficients in the second row is the mirror image of the coefficients (a_1, a_2) in the first row. In Brechenmacher (2007b), we have shown that the symmetry property of mechanical systems originated in the specific practice Lagrange had devised in 1766 for the problems of small oscillations.

[31] Yvon-Villarceau (1870: 763, translation F.B.).

unbounded oscillations as the "time t would get out of the sine" and solutions would take the form $y = t \sin(st + \varepsilon)$. But, Villarceau argued,[32]

> I claim that this condition is not necessary for the oscillations to remain small.... Here is a very simple case where equal roots occur in the characteristic equation: a homogeneous solid revolving body oscillating around a point of its axis. It is plain to see, without resorting to any computation, that the smallness of the oscillations is ascertained, provided that the initial oscillatory movement is small enough, and that, at the origin of the movement, the solid's center of gravity is below its center of suspension and not too far from the vertical axis passing through this point.[33]

In 1870, Yvon-Villarceau therefore pointed to some serious "deficiencies" in the "general" resolution of problems of small oscillations as in 1874 Kronecker would similarly criticize the "so called generality" of algebraic expressions. The astronomer nevertheless did not aim to criticize a type of generic reasoning. His purpose was to question a practice which consisted in combining a mechanical interpretation with the algebraic nature of the roots of a specific equation.

Although Villarceau's intervention stemmed from mechanical concerns such as the application of Lagrange's method to long-term perturbations of the parameters determining the planetary orbits, it eventually brought a theoretical question to the attention of the Academy's geometers. Because the occurrence of multiple roots was no contradiction to mechanical stability, and hence to the possibility of "reducing" a system of n equations to n independent single equations, a question arose concerning the characterization of systems which could be reduced to separate equations.

This question prompted the publication by Jordan of two notes in 1871 and 1872. In 1871 Jordan applied the canonical form he had introduced in 1870 for linear substitutions to the reduction of a "general" system of differential equations with constant coefficients:

$$\frac{dx_1}{dt} = a_{11}x_1 + \cdots + a_{1n}x_n,$$

$$\frac{dx_2}{dt} = a_{21}x_1 + \cdots + a_{2n}x_n,$$

$$\dots\dots$$

[32] From the standpoint of modern algebra, the stability of a system depends upon whether its matrix is diagonalizable or not. The non-equality of the system's eigenvalues is a sufficient but not necessary condition. The mechanical systems studied by Lagrange are diagonalizable because they are symmetric.

[33] Yvon-Villarceau (1870: 763). "Je dis qu'il n'est pas nécessaire que cette condition soit remplie, pour que les petites oscillations se maintiennent.... Voici un cas très simple, auquel correspondent des racines égales de l'équation caractéristique : c'est celui d'un corps solide, homogène et de révolution, oscillant autour d'un point pris sur son axe de figure. Chacun comprendra, sans recourir au calcul, que la petitesse des oscillations est assurée dans ce cas, si le centre de gravité est, à l'origine du mouvement, au-dessous du centre de suspension, à une petite distance de la verticale passant par ce point, et si le mouvement oscillatoire initial est suffisamment faible."

$$\frac{dx_n}{dt} = a_{n1}x_1 + \cdots + a_{nn}x_n.$$

He therefore gave a "form" to which such a system could be reduced, whatever the multiplicity of the roots. When only single roots occur, this form is identical to that given by Lagrange. But if a multiple root s occurs, the reduction of the system may involve the following kinds of expressions:

$$\frac{dy_1}{dt} = sy_1,$$

$$\frac{dz_1}{dt} = sz_1 + y_1,$$

$$\frac{du_1}{dt} = su_1 + z_1,$$

$$\cdots$$

Such a reduced form can then be integrated directly. It yields solutions of the form $y_1 = e^{st}\psi(t)$, where $\psi(t)$ is a polynomial.[34]

In proving that the multiplicity of In 1872, Jordan devoted an additional paper to address the questions raised by Villarceau more specifically. Here, he proved that linear systems stemming from mechanical concerns can always be reduced to n separate equations $\frac{dy_i}{dt} = sy_i$ because of their symmetric nature. In doing so, Jordan inscribed the initial mechanical question into the "more general" framework of the theory of quadratic forms. Indeed, the coefficients of a symmetrical system define a quadratic form Q and the reduction of the system to its canonical form consists in considering simultaneously the forms Q and the form identity I, that is, the polynomial form $Q + sI$.[35]

roots was of no relevance to the subject of mechanical stability, Jordan reached the same conclusion that Weierstrass had already given in 1858 (symmetric case) and in 1868.[36] It was thus thanks to a hundred-year old mechanical problem that a first connection between Jordan's and Weierstrass' theorems came to light. This connection was pointed

---

[34] Moreover, Jordan gave a characterization of the systems that can be reduced to a diagonal form by the necessary and sufficient condition that each root of the characteristic polynomial of multiplicity had to be a root of order $\mu - 1$ of each of the minor of the determinant leading to the characteristic polynomial. From the standpoint of Weierstrass' 1868 theorem, Jordan's condition is tantamount to stating that all elementary divisors are linear.

[35] Jordan (1872).

[36] Although Weierstrass had already stated such a condition for the symmetric case in 1858, in 1875 he devoted a communication at the Berlin academy to the general result stated by Jordan in 1872. He applied his theorem about elementary divisors and made no reference to Jordan.

out in 1873 by Meyer Hamburger who called Jordan's attention to the Weierstrass' work on bilinear forms.[37] In 1873 Jordan showed that Weierstrass' "transformation of forms" could be seen as the "composition of linear substitutions" and eventually proved that his canonical reduction could be used to derive Weierstrass' theorem, and as a result prompting the ensuing controversy with Kronecker.

The 1874 controversy can therefore be considered from the perspective of the opposition between two different outcomes arising from a shared history. Jordan's and Weierstrass' theorems shed new light on the past as they made the results of authors such as Lagrange and Cauchy appear incomplete because they were limited to the special case in which only single roots occurred.

For the purpose of a deeper understanding of the role played by this shared history in the quarrel, bibliographic research has been carried out, starting with the authors and texts Jordan and Kronecker referred to and by systematically working out the succession of the references that appeared. This methodology produced a network of texts covering the period 1766–1874.[38] A simplified representation of this collection of texts is shown in Fig. 16.1. The main nodes in the entanglement of bibliographic references point to the mechanical work of Lagrange as well as to Cauchy's analytical geometry. However, our network can neither be identified with a theory nor to a discipline. What then gives coherence to this collection of texts?

One of the main shared characteristics in this network of texts is the role of point of origin systematically attributed to Lagrange's solution for mechanical problems of "small oscillations." As has been already mentioned in relation to Villarceau's 1870 note, this reference pointed especially to the association Lagrange had developed in 1766 between the stability of a mechanical system and the algebraic nature of the roots of the associated algebraic equation: the roots had to be "real, unequal, and negative" for the oscillations to remain small. A few years later, Lagrange transferred his approach to the study of the small oscillations of planets, called "secular inequalities." The nature of the roots of the equation that gave the solution thus concerned the crucial issue of the stability of the solar system. For this reason, this equation was by then usually called the "equation to the secular inequalities in planetary theory," that is, the secular equation. Yet, the validity of Lagrange's criterion of stability had remained unquestioned until Weierstrass, and later Jordan, proved in 1858 and 1872, respectively, that the multiplicity of the roots had nothing to do with the stability of the system. Even though the network of texts stemmed from a specific problem, this problem was considered to have been solved by Lagrange until this resolution was questioned by Weierstrass and Jordan. This collection of texts cannot therefore be considered to be sharing the common goal of finding a solution to a given problem.[39]

---

[37] Hamburger (1873).

[38] On the use of networks of texts for investigating collective organizations of knowledge, see Goldstein (1999), Goldstein and Schappacher (2007), and Brechenmacher (2006a, 2007a, 2007b).

[39] From the viewpoint of 1930s modern algebra, the different problems appearing in the discussion would be considered as depending on the theory of matrices and consisting in the reduction of a pair (A,B) of matrices in (D,I), where A is symmetric, B is definite symmetric, D is diagonal, and I the identity matrix. For more details, see Gantmacher (1959: 311).

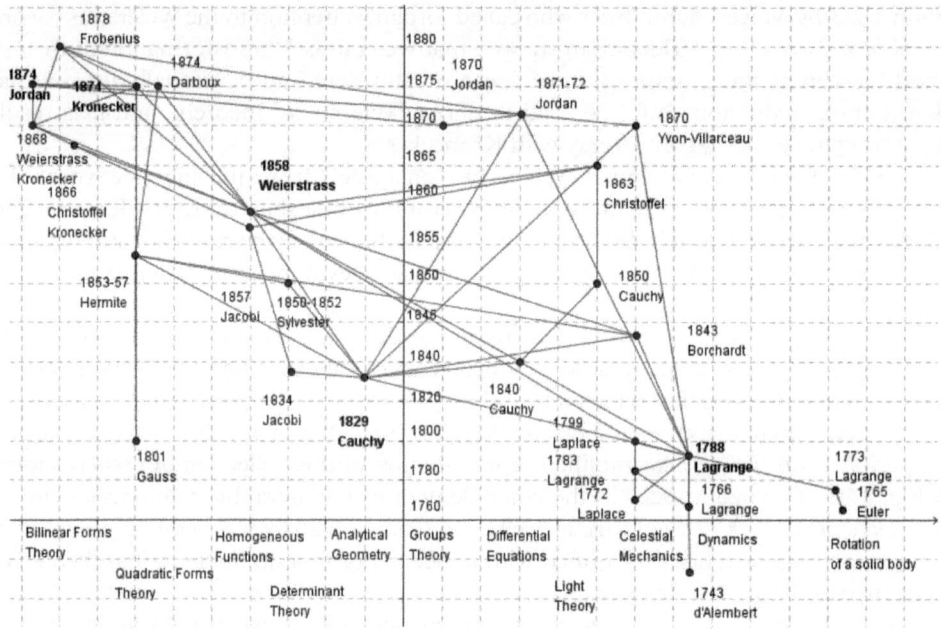

Figure 16.1 *The network of the equations to the secular inequalities on planetary theory.*

When, in the 1770s, Lagrange's method had been applied to the planetary orbits,[40] the algebraic nature of the roots of the associated equation was linked to the issue of the stability of the solar system. In 1787 Laplace[41] attempted to give a general proof of the stability of the solar system and highlighted that the nature of being real of the roots occurring in Lagrange's approach was related to the property of symmetry of the coefficients of the system. After this episode, it was the *special* nature of the secular equation which would give the network its coherence. Amongst the many authors whose interest in celestial mechanics went no further than the identification of this equation, we may cite Cauchy's 1829 "*Sur l'équation à l'aide de laquelle on détermine les inégalités séculaires des planètes*," James Joseph Sylvester's 1851 "On the equation to the secular inequalities in the planetary theory," or Hermite's 1857 "*Mémoire sur l'équation à l'aide de laquelle, etc.*"[42]

During the first half of the nineteenth century, the expression "secular equation" came to be progressively employed to identify a specific algebraic culture, shared on a European level and based on the practice developed by Lagrange for solving problems of small oscillations.[43] This practice used a polynomial procedure which offered a base supporting

[40] Lagrange (1781: 125).
[41] Laplace (1787). See also Laplace (1799).
[42] See Cauchy (1829), Sylvester (1851), and Hermite (1857).
[43] See Brechenmacher (2007b) and Brechenmacher (2014).

analogies through which meanings were extended from one domain to another.[44] It was, for instance, thanks to the secular equation that in 1829 Cauchy developed a formal analogy between various problems such as mechanical oscillations, the rotation of a solid body and the classification of conics and quadrics.[45] Even though the practice attached to the secular equation was never explicitly identified as having a method *per se*, it was both developed and spread through the network using various methods belonging to different theoretical frameworks. It especially played a key role in several developments in algebra and number theory.[46]

Although related to a special equation, the practice of the secular equation had nevertheless progressively become detached from it and eventually gave rise to a shared algebraic culture. Reflecting on the connections between the theories required by the secular equation was especially popular for the emerging communities of teachers of mathematics which revolved around journals such as the *Nouvelles annales de mathématiques* or the *Cambridge Mathematical Journal* in the 1840s–1850s. This shared algebraic culture nevertheless remained limited to a periodical form until monographs and textbooks were devoted to the theory of bilinear forms in the 1880s. Until the development of theoretical frameworks such as those Jordan and Kronecker were quarrelling about in 1874, it was therefore above all a historical identity that characterized the algebraic nature of the practice of the secular equation. It was by referring systematically to a corpus of earlier texts in periodical publications that authors pointed to its specific nature.

As we shall see in greater detail in the next paragraph, in the absence of any monograph or theoretical synthesis, referring to the secular equation discussion was a mode of legitimating some "extensions to the general" by relying on analogies carried out using a common polynomial procedure. It is therefore meaningful to look more closely at how such a way of referring to a shared algebraic culture which had at its core an issue of generality would later on be torn apart by two disciplines: algebra and arithmetic.

16.5   On generality and the algebraic status of a polynomial practice

From the earliest texts in our corpus to its two final contributions from Weierstrass and Jordan, all the authors were driven by a quest for generality.

---

[44] Supporting analogies through common operatory procedures was a common way to legitimate extensions of meanings in the nineteenth century. See Durand-Richard (1996, 2008).

[45] See Cauchy (1829: 173). From the viewpoint of modern algebra, Cauchy was interested in the transformation of a quadratic form in three variables into a sum of squares. This problem also arose in the mathematical analysis of the rotational motion of a rigid body as studied by Lagrange in the eighteenth century. For a description of Cauchy's work on the problem of the rotation of a solid body in connection to Lagrange's analytic reformulation of the solution given by Euler, see Hawkins (1975: 18).

[46] Such as the algebraic proof Sylvester and Hermite gave to Sturm's theorem, and which questioned the relationships between the functions of analysis, the equations of algebra and the quadratic forms of number theory. See Sinaceur (1991).

It was already with the aim of generalizing d'Alembert's investigations into the question of the vibrating string loaded with three masses to the "oscillations of an unspecified system of bodies" that Lagrange had been working in 1766 on the polynomial procedure at the origin of the discussion. Because of the generality he attributed to his description of a motion that Daniel Bernoulli had regarded as too irregular to be treated by analytic methods,[47] Lagrange would promote the problem of small oscillations among the first examples of applications he gave of the "general principles" in his *Mécanique analytique*.[48]

Generality was then the main issue driving the development of a discussion on the qualitative nature of characteristic roots. It nevertheless took on changing meanings between 1766 and 1874. To Lagrange's mind, the fact that his method would fail if multiple roots should occur did not restrict its generality. The method brought into play implicit mechanical representations. It was because the oscillations of one vibrating string loaded with n masses could be represented mechanically as a combination of independent oscillations of n strings loaded with a single mass, that linear differential systems were thought to be representable as combinations of independent equations. In Lagrange's method, algebraic roots could not be dissociated from their mechanical representations as periods of oscillations and the occurrence of multiple roots was therefore (wrongly) believed to be contradictory to the existence of n independent oscillations.

The issue of Lagrange's stability criteria—the roots had to be real and distinct because the oscillations had to remain bounded—changed when the method was generalized to the secular inequalities in planetary theory. The stability of the solar system could not be taken for granted and Lagrange therefore pointed out that "it would be difficult, perhaps impossible, to determine the roots of the equation in general" as it would mean demonstrating the real nature and non-equality of the roots of a very general nth degree polynomial while these roots cannot be expressed by radicals in general as soon as n is greater than five.[49] Lagrange worked out an effective computation for a system of four planets. He determined that the roots of the associated fourth degree secular equation are real, negative and distinct. Lagrange thus came to the conclusion that:

> One may wonder whether, by changing the values [of the masses of the planets], equal or imaginary roots may occur. For removing all doubt, it would be necessary to prove, in general, that the roots of the equation are always real and distinct, whatever the values of the masses. This is easy when the mutual action of only two planets is considered simultaneously, since the equation is only of the second degree, but this equation becomes more and more complicated and higher [in degree] as the number of planets increases.[50]

Laplace was not content with this numerical computation and his aim of devising a "fully general" demonstration that would not depend upon the approximate values assigned to the masses of the planets brought him to engage in the discussion about the secular equation.

---

[47] Truesdell (1960: 156).
[48] Lagrange (1788).
[49] Lagrange (1766: 538).
[50] Lagrange (1784: 316), translation T. Hawkins.

As has already been pointed out, Cauchy's 1829 intervention was motivated by his ambition to generalize a method he had devised for two or three variables in a geometric framework to n variables. The polynomial practice peculiar to the systems of n equations related to the secular equation was then translated in terms of determinants.[51] This general perspective raised new issues as some algebraic expressions involved in Cauchy's method may, in the case of multiple roots, take a $\frac{0}{0}$ value. The occurrence of multiple roots appeared as a singular case limiting the range of validity of an algebraic expression. It thus seemed necessary to introduce some particular methods for this singular case. Following d'Alembert and Lagrange, Cauchy initially moved outside algebra by appealing to the specific argument that consisted in making the roots unequal by the use of infinitesimal quantities. But Cauchy was not happy with this situation. He criticized polynomial methods for being overburdened with singular cases. With the aim of developing a fully homogeneous resolution, he then turned to the calculus of residues and to complex analysis.[52]

The change of perspective on generality induced by this ideal of homogeneity encouraged further developments involving Jacobi, Sylvester, Hermite, Borchardt, and eventually Weierstrass.[53] In 1858, Weierstrass gave a general, homogeneous, and algebraic solution to the problem in arguing that:

> However, it does not appear that special attention has been given to the strange circumstances that arise when the roots of the equations $f(s) = 0$ are not all different from each other; and the difficulties which they present then—of which I was made aware by a question to be discussed more fully later—do not seem to have been properly cleared up. At first, I also believed that this would not be possible without extensive discussion in view of the large number of different cases that can occur. It seemed all the more desirable to me to show that the solution to the problem given by the above-named mathematicians could be modified in such a way that it does not at all matter whether some of the quantities $s_1, s_2, ..., s_n$ are equal.... After Lagrange had given the form of the integrals, and shown how their arbitrary constants are determined by the initial values of $x_1, x_1, \frac{dx_1}{dt}$, etc., he introduced among the conditions that must hold—so that $x_1, \frac{dx_1}{dt}$ always remain infinitely small if they initially were—, the one (among the conditions) saying that the characteristic equation must not have multiple roots, otherwise, inside the integral, there would appear a component that could become arbitrarily large over time. The same claim happens to be repeated by Laplace, when he dealt with the secular variations of the planets in the

---

[51] Hawkins (1977: 125).

[52] Since its origin in 1826, the calculus of residues had been introduced by Cauchy as a way to deal with problems caused by multiple roots in generic algebraic expressions. See Dahan Dalmedico (1992: 197) and Brechenmacher (2007b).

[53] See Hawkins (1977: 128–33) for descriptions of the work by Jacobi, Borchardt, and Dirichlet. Sylvester's work of 1850–2 would lead him to introduce the notions of "matrix" and "minors," which would be invested by Hermite in an arithmetical framework (quadratic forms, decomposition in four squares). See Brechenmacher (2006a).

Mécanique céleste; and, as far as I know, this statement has been repeated again by all the other authors treating this subject, if they ever mentioned the case of equal roots (which, for instance, Poisson does not). But this is not justified [...] and the [same condition] can hold without all the roots of the equation *f(s)* = 0 being different from one another, if only the function Ψ remains negative and if its determinant does not vanish; since, in fact, one has repeatedly dealt with particular cases of the above equations (in which this condition is not satisfied), but without having found any components of the sort described above.[54]

We have seen that some ambitions of generality had been strongly linked to the development of the network of the secular equation since its origin. But we have seen also that these ambitions of generality originally related to the algebraic identity of a practice attached to a special equation. We should bear in mind that Algebra was not considered at the time to be an autonomous research discipline. It was actually because it was considered as algebraic that the practice attached to the secular equation was not formalized further than something that could be done with a special equation. The same kind of attitude can be illustrated with other *special* types of equations such as the binomial equation, the modular equation, etc.[55] It was precisely because of this non-formal algebraic identity that the practice attached to the secular equation circulated between theories and supported analogies and "generalizations."[56]

After the 1830s, this equation appeared as an archetype of algebraic generality. Cauchy referred to it to contrast to the homogenous methods of complex analysis with the genericity of algebra. This algebraic generality especially appealed to authors such as Jacobi and Sylvester. When Sylvester tried to state a purely algebraic proof of Sturm's theorem, he had indeed started by investigating the secular equation.[57] In this context, the algebraic

---

[54] Weierstrass (1858: 234, 243–4), translation F.B. "Dagegen scheint es nicht, als ob den eigentümlichen Umständen, die eintreten, wenn die Wurzeln der Gleichung *f(s)* = 0 nicht alle von einander verschieden sind, besondere Beachtung geschenkt, und die Schwierigkeit, die sich alsdann darbieten, und auf die ich bei einer nachher näher zu besprechenden Frage aufmerksam geworden bin, schon gehörig aufgeklärt sein. Auch glaubte ich anfangs, es würde dies bei der großen Zahl verschiedener Fälle, die vorkommen können, nicht ohne weitläufige Erörterungen möglich sein. Umso erwünschter war es mir, zu finden, dass sich die von den genannten Mathematikern gegebene Lösung der Aufgabe in einer Weise modifizieren lässt, bei der ein ganz gleichgültig ist, ob unter den Größen $s_1, s_2, ..., s_n$ gleiche vorkommen oder nicht.... Nachdem Lagrange die Form der Integral angegeben und gezeigt hat, wie die willkürlichen Constanten derselben durch die Anfangswerthe von $x_1, \dfrac{dx_1}{dt}$ u.s.w. bestimmt werden, führt er unter den Bedingungen die erfüllt sein müssen, damit $x_1, \dfrac{dx_1}{dt}$ stets unendlich klein bleiben, wenn sie es ursprünglich sind, auch die an, dass die genannte Gleichung keine gleiche Wurzeln haben dürfe, weil sonst in den Integralen Glieder vorkommen würden, die mit der Zeit beliebig gross werden könnten. Dieselbe Behauptung findet sich bei Laplace wiederholt, da wo er in der *Mécanique céleste* die Säcular-Störungen der Planeten behandelt, und ebenso, so viel mir bekannt ist, bei allen übrigen diesen Gegenstand behandelnden Autoren, wenn sie überhaupt den Fall der gleichen Wurzeln erwähnen, was z.B. bei Poisson nicht geschieht. Aber sie ist nicht begründet. ... wenn nur die Function Ψ stets negativ bleibt, und ihre Determinante nicht Null ist, was stattfinden kann, ohne dass die Wurzeln der Gleichung *f(s)* = 0 alle von einander verschieden sind ; wie man denn auch wirklich besondere Fälle der obige Gleichungen, bein denen diese Bedingung nicht erfüllt ist, mehrfach behandelt und doch keine Glieder von der angegebene Beschaffenheit gefunden hat."

[55] See Goldstein and Schappacher (2007) and Brechenmacher (2011).
[56] For more details on the circulation of algebraic practices, see Brechenmacher (2010).
[57] Sinaceur (1991).

practice attached to the secular equation became more and more connected to the theory of quadratic forms as it had stemmed from Gauss' "higher arithmetic." More precisely, it was discussed in relation to the inertia law of quadratic forms. In the context of his works on Sturm's theorem, Hermite introduced a distinction between the traditional "arithmetical theory of forms" and the "algebraic theory of forms" related to the secular equation. The first concerns single quadratic forms with integer coefficients. These forms are characterized by the sum of squares they can be reduced to (the inertia law). The second concerns pairs of quadratic forms with real coefficients whose characterization can thus be understood as a generalization of the inertia law.

As a matter of fact, Hermite, and later Weierstrass, had turned Lagrange's initial treatment into a theorem about "transformations" and "forms." These two terms had thus been given explicit mathematical definitions in the "higher arithmetic" of quadratic forms. In contrast, the uses of the term "form" in connection with the secular equation pointed to mostly implicit meanings. For Lagrange and Laplace, the existence of "integrable forms" for the differential systems of small oscillations was inferred from mechanical interpretations. It was therefore not by referring to "transformations" that independent equations were associated to the initial linear system but by the computation of the systems' mechanical parameters through the secular equation.[58] In Cauchy's 1829 paper, the "transformations of homogeneous functions" were related to geometrical meanings which led to reasoning in terms of some procedures of changes of rectangular systems of coordinates.

From the different meanings and representations the term "form" had been given in relation to the ambitions of generality of Lagrange, Laplace, and Cauchy, a succession of mathematical theories were formulated whose subject was a "fully general" characterization of "forms." Should such an issue belong to arithmetic or algebra? In 1874, Jordan and Kronecker were referring to a shared history related to a specific practice they had in common. We have seen that this practice consisted in investigating pairs of bilinear or quadratic forms (P, Q) by making use of the polynomial decomposition of $S = |P + sQ|$. As we shall see in the following section, some disciplinary ideals on algebra and arithmetic nevertheless induced conflicting perspectives on the generality in the theory of "forms."

16.6 Arithmetic generality vs algebraic generality

In 1874, Kronecker contrasted the true generality of the arithmetical nature of the theory of forms with the formal nature he attributed to Jordan's algebraic group theoretical approach. We have to bear in mind that Jordan had designated as canonical forms three different algebraic expressions in the context of the operations of three kinds of groups of substitutions. For this reason, Kronecker accused him of using a notion without any

---

[58] Even though it might seem natural nowadays to wonder about these matrices which, because of their multiple eigenvalues, might not be transformed into some diagonal forms, this question was actually irrelevant to the ways the terms "forms" and "transformations" were being considered at the times of Lagrange or Cauchy. On the "mathematical interpretation of the essential mechanical concepts" in Lagrange's analytical mechanic, see Panza (1992: 205).

"general relevance" or "objective content." According to Kronecker, Jordan had mixed up the "formal aspects" of certain "means of action" (canonical forms) with the "true subject of investigation" and its "content," that is, the characterization of equivalence classes of bilinear forms.

Although Kronecker used "normal forms" similar to Jordan's canonical form; in his opinion, using such algebraic expressions was legitimate provided that they remained bound to their status of "methods." In Kronecker's view, methods from algebra had to be distinguished from the notions specific to the "other disciplines"—such as arithmetic—it was algebra's duty to serve.[59] In his view, the reification of algebraic methods would lead to mistaking a mere "formal" development for a "general" and "uniform" presentation. Kronecker thus mocked Jordan's claims for the greater simplicity and generality of his canonical reduction as a naïve simplism which contented itself with the illusionary generality of the uniformity of a formal development.

> Should such general expressions be found, their designation as canonical forms would, at best, be motivated by their simplicity and generality; but if one does not want to stick to the purely formal viewpoint which is often put to the fore in the more recent Algebra—certainly not to the greatest advantage of the true knowledge -, one shall not omit to derive the justification of the canonical forms put forward from inner reasons. Truly, these so called canonical or normal forms are essentially determined only by the orientation of the study, but they should not be seen as the aim of the research but only as a means ... One should not be at all disconcerted if for an exposition both uniform and completely general such as is found in the above mentioned work [by Jordan], some new principles turn out to be necessary; and on the contrary it would be amazing if in accordance with Jordan's claims ("The new methods we are proposing are, on the contrary, extremely simple ...." "A very simple discussion shows that one can transform ..."), the simplest means were sufficient.[60]

The purpose of the reorganization Kronecker devised for the theory of bilinear forms in 1874 was to give a truly arithmetical foundation to various results that had been

---

[59] For instance, in order to prove that two non-singular pairs of bilinear forms can be transformed one into the other, Weierstrass proved in 1868 that both forms can be linearly transformed into what Kronecker designated as a "normal form." His normal forms were similar to Jordan's canonical form. But Kronecker would not state any theorem about such normal forms which were not the purpose of his investigations. On Weierstrass' 1868 proof, see Hawkins (1977) and Brechenmacher (2006a).

[60] Kronecker (1874a: 405–8), translation F.B. "Nachträglich, wenn dergleichen allgemeine Ausdrücke gefunden sind, dürfte die Bezeichnung derselben als canonische Formen allenfalls durch ihre Allgemeinheit und Einfachheit motiviert werden können ; aber wenn man nicht bei den bloss formalen Gesichtspunkten stehen bleiben will, welche –gewiss nicht zum Vortheil der wahren Erkenntnis– in der neueren Algebra vielfach in den Vordergrund getreten sind, so darf man nicht unterlassen, die Berechtigung der aufgestellten canonischen Formen aus inneren Gründen herzuleiten. In Wahrheit sind überhaupt die so genannten canonischen oder Normalformen lediglich durch die Tendenz der Untersuchung bestimmt und daher nur als Mittel, nicht aber als Zweck der Forschung anzusehen.... Dass sich aber für eine zugleich einheitliche und ganz allgemeine Entwickelung, wie sie in der oben erwähten Arbeit gegeben ist, gewisse neue Principien als nöthig erwiesen, kann durchaus nicht befremden, und es wäre im Gegentheil zu verwundern, wenn wirklich den *Jordan*'schen Behauptungen gemäss ("Les méthodes nouvelles que nous proposons sont, au contraire extrêmement simples ..." "On voit par une discussion très simple, que l'on peut transformer ...") die allereinfachsten Mittel dazu ausreichen sollten."

obtained in the 1860s.[61] Kronecker had already implicitly referred to the legacy of the works of Gauss and Hermite on the arithmetic of quadratic forms in 1866—as when he had preferred to make use of the term "form" to call what others would designate as a function (Weierstrass, 1858) or as a "polynomial" (Jordan, 1873). However, in fact, his monthly communications to the Academy of Berlin during the winter of 1874 were aimed at an explicit generalization of the arithmetic notion of "equivalence classes" from forms to networks of forms. "As an application of arithmetic notions to algebra," two bilinear forms or two networks of bilinear forms were designated as "equivalent" and as belonging to a same "class" when one could be linearly transformed into another.[62]

Disciplinary ideals accompanied this arithmetic orientation of the theory. These ideals expressed themselves in the criticisms Kronecker made of Jordan's statement that a sufficient condition for two forms to be equivalent was the identity of their canonical forms. According to Kronecker, despite being true, this proposition had to be rejected because it did not state any effective procedure for deciding the equivalence. Jordan's reduction indeed relied on an algebraic decomposition of the characteristic determinant for which no effective procedure could be given "in general" as soon as the polynomial degree exceeds five. It thus had to be distinguished from the "immediate possibility afforded by the theoretical criteria of equivalence to set a complete system of invariants" effectively computed from the form's coefficients. Kronecker supported his claims by introducing a new system of invariants to replace those introduced by Weierstrass. The elementary divisors indeed appealed to the resolution of general polynomial equations exactly as Jordan's canonical form did. But in 1874, Kronecker assigned the name of "elementary divisors" to the new system of invariants he introduced. These new invariants were introduced as the result of the arithmetical procedure for computing the greatest common divisors (GCDs) of the successive minors extracted from the polynomial determinant $|A + sB|$.[63] Drawing on Gauss' legacy,[64] Kronecker then contrasted the effectiveness of such a procedure to the formal nature of algebraic formulas such as Jordan's canonical form:

> In the arithmetical theory of forms, one must certainly be satisfied when one is given the indication of a procedure for deciding on the question of the equivalence, and the problem in question is thus also formulated explicitly in this way too (cf. Gauss: *Disquitiones arithmeticae, Sectio* V...). The procedure itself is here also based on the transformation to reduced forms: but it must not be forgotten that, in arithmetic theory, these [reduced forms] have a completely different meaning from the one they have in Algebra. Indeed, given that there, the invariants of the equivalent forms are, by their very nature, only number-theoretic functions of the coefficients; it cannot thus be disconcerting that such

---

[61] According to Kronecker, this arithmetization ambition stemmed from discussions with E. Kummer. On Kummer's ideal numbers, see J. Boniface's paper in this volume (Chapter 18).

[62] Two families of bilinear forms $s\Phi - \Psi$ and $s\Phi' - \Psi'$ are equivalent if one can be transformed into the other by (possibly different) non-singular linear transformations of the x and y variables (where $\Phi = \sum_{i,j=1}^{n} A_{ij} x_i x_j$ and $\Psi = \sum_{i,j=1}^{n} B_{ij} x_i x_j$).

[63] See Appendix 1 for more details about the invariants Kronecker introduced in 1874.

[64] Gauss (1801).

[invariants] can be directly defined, but not represented explicitly and only as the final result of arithmetic operations; for much the same is true with most concepts of arithmetic, e.g. already the simplest notion of greatest common divisor.[65]

The ideal of effectiveness, which the historiography has usually associated with Kronecker's 1882 arithmetic theory of algebraic magnitudes,[66] had thus already been strongly expressed on the occasion of the 1874 controversy when Kronecker criticized "literal expressions" such as Jordan's canonical forms.

Throughout the 1874 controversy, Jordan retorted to Kronecker's assaults by claiming the greater generality and simplicity of his method. Far from the naïve simplism caricatured by Kronecker, Jordan's ideal of simplicity was linked to a practice of "reduction" of "general problems" into chains of sub problems. It supported Jordan's criticism of Kronecker's characterization of singular pairs of bilinear forms as having failed to find the "true reduced forms" which had to be simplest links in the chain of reductions without leaving any room for further simplification.[67]

Jordan's practice of reduction originated in his research on groups of substitutions in the 1860s. His main goal in this research was a *general* investigation of the *special* types of equations that could be solved by radicals. In order to handle the generality of this problem, Jordan developed a "machinery" to reduce the types of groups of substitutions attached to the equations, reducing them from the general to the special.[68] Among others, the linear group—and its properties such as the theorem stating the reduction of linear substitutions to their "simplest" or "canonical" forms—, "originated" from the practice of reduction Jordan had made use of in his 1860's investigations.[69] Jordan's canonical

---

[65] Kronecker (1874c: 383, translation F.B.). "In der arithmetischen Theorie der Formen muss man sich freilich mit der Angabe eines Verfahrens zur Entscheidung der Frage der Aequivalenz begnügen und das betreffende Problem wird deshalb auch ausdrücklich in dieser Weise formuliert (cf. Gauss: Disquitiones arithmeticae, Sectio V...) Das Verfahren selbst beruht auch dort auf dem Uebergange zu reducirten Formen: doch ist dabei nicht zu übersehen, dass denselben in den arithmetischen Theorien eine ganz andere Bedeutung zukommt als in der Algebra. Da nämlich die Invarianten äquivalenter Formen dort ihrer Natur nach nur zahlentheoretische Functionen der Coëfficienten sind, so kann es nicht befremden, wenn dieselben zwar direct definiert aber nicht explicite sondern nur als Endresultate arithmetischer Operationen dargestellt werden können ; denn ganz ähnlich verhält es sich mit den meisten arithmetischer Begriffen, z.B. schon mit jenem einfachsten Begriffe des grössten gemeinsamen Theilers."

[66] As shall be seen in greater detail later, the arithmetical properties of polynomials played an important role in Kronecker's 1850–70 work on the solvability of equations. These properties would later be essential in Kronecker's 1882 arithmetic theory of algebraic magnitudes and his concept of a "Rationalsbereich." This theory was indeed based on polynomial forms as an alternative to Dedekind's fields.

[67] Jordan (1874b: 614). See also Jordan (1874a).

[68] See Dieudonné (1962) and Brechenmacher (2011). General "solvable groups" were reduced into a sequence of particular groups ("transitive," "primitive," "linear," and "symplectic" groups) corresponding to the "simplest" links in Jordan's practice of reduction.

[69] In order to characterize which equations are solvable by radicals, Galois had asserted that the degree of such equations is of the form p^n, p prime, and that the corresponding group G of permutations has to be a solvable subgroup of the linear group. In contrast with his predecessors, Jordan made the consideration of general n-ary linear substitutions fundamental in his investigations on the determination of all the irreducible equations of a given degree which were solvable by radicals. It was in this context that Jordan stated his canonical form theorem. Moreover, in 1870 this theorem played a key role in Jordan's method for building up solvable groups from their composition series. For a detailed analysis of the role played by canonical forms in Jordan's

forms were therefore part of a broader practice of reduction of *general* problems to a chain of *special* ones.[70] This specific tension between the general and the special can be analyzed in the framework of the legacies of both the "generality" of Poinsot's "theory of order"—in the sense of something transcending disciplinary borders—and of Galois' "general" approach to model cases special equations—in the sense of a conceptual abstract oriented treatment of equations.[71]

When Jordan responded to Yvon-Villarceau in 1871, he used his practice of reduction for bringing general systems of linear equations down to the sequence of "simplest forms" corresponding to the decomposition of the secular equation into its simplest (linear) factors. Because it required the resolution of a general algebraic equation, Jordan's canonical reduction was nevertheless formal as it did not actually give astronomers any practical resolution of the problem.

In the 1870s, it was nevertheless thanks to the practices—such as the canonical reduction—he had originally devised for group theory that Jordan succeeded in extending the range of his investigations to subjects such as differential equations (1871–8), the theory of forms (1872–5), as well as arithmetic and number theory (1874–81). The application of these techniques was not limited to their technical character and Kronecker's criticisms highlight some of the algebraic ideals—such as simplicity and abstraction—attached to them.

16.7 Conclusion

We shall now come to some conclusions that may be drawn from the conflicting perspectives on generality relating to the two practices put forward by Jordan and Kronecker in 1874. In a nutshell, while Jordan criticized the lack of generality of Kronecker's invariant computations because they did not reduce pairs of forms to their simplest expression, Kronecker considered Jordan's canonical form as a "formal notion" with no "objective meaning" which therefore failed to achieve true generality. What made one generality true was exactly what made the other generality false.

It was, in the first place, the "general" solution they claimed to have achieved for various problems that had been addressed in the past by authors such as Lagrange, Laplace, Cauchy, and Hermite that had prompted some connections between Jordan's canonical reduction and Weierstrass' elementary divisors. The reference to a common history therefore played a key role in the controversy. Not only did the secular equation play a major role in identifying a specific shared practice that consisted in expressing the solutions of linear equations as polynomial factors of their characteristic equation. But Kronecker also stressed a history of what T. Hawkins would later refer to as the "generic

---

investigations on solvable groups, as well as on the evolution of the role played by linear groups between 1870 and 1900, see Brechenmacher (2011).

[70] Jordan especially used the same practice for the reduction of compound groups to simple groups, that is, what is now known as the Jordan–Hölder theorem.

[71] Brechenmacher (2011).

reasoning" in the algebra of the eighteenth and nineteenth centuries when he criticized traditional algebraic practices for their tendency to focus on the generic case with little attention given to the difficulties that may be caused by assigning specific values to algebraic symbols.

Jordan's canonical forms could nevertheless not be charged with such an indictment of "alleged generality." We have actually seen that both Jordan and Kronecker criticized the form of treatment of generality attached to the traditional practice mentioned above. Both aimed to ground the "theory of forms" on new forms of generality. The issue of generality could thus not be dissociated from that of the organization of knowledge. In contrast to a traditional way to legitimate generalizations from one domain of knowledge to another by relying on the analogies supported by operative procedures, Jordan and Kronecker both aimed to insert what used to be considered as different problems within a single "general" problem of "transformations" of pairs of "forms." But the two mathematicians did not only disagree on the nature of this theoretical organization, but also on the types of generality and on the treatments of the general they were advocating.

In the algebraic organization Jordan gave to the theory, transformations resulted from the action of certain linear groups of substitutions. In order to achieve "general results" on forms, underlying substitutions had to be reduced to their "simplest canonical forms" depending on the nature of the linear group the substitutions belonged to. For Jordan, this procedure of reduction from the general to the special was the very "essence" of his "method" and was an important component of the ontological nature he attributed to the concept of group as a notion underlying the various branches of the mathematical sciences. The concept of group, in particular, had an intimate relation to the Galois theory of algebraic equations, and therefore to the concept of algebraic numbers. In his 1870 *Traité des substitutions et des équations algébriques,* Jordan claimed that his method of reduction of groups provided a "higher point of view on the classification and the transformation of irrationals."[72] This claim may already have been meant as a challenge to Kronecker who had been very much concerned with the concept of irrational numbers since the 1850s. Yet, Kronecker had developed a specific approach to equations he claimed to be faithful to Gauss' legacy and had little interest for Galois' conceptual considerations and was much more interested in explicit expressions, such as the ones Abel had given to the roots of the quintic when proving the impossibility of solving it with radicals.[73] It was therefore the question of the legitimacy of the *general* methods and notions for dealing with the characterization of the "irrationals" that were defined by *special* equations which was at the core of the dispute between Jordan and Kronecker.

For Kronecker, the algebraic nature of Jordan's approach prevented it from reaching any general theoretical level. As Kronecker insisted, methods from algebra had to be

---

[72] Jordan (1870: VIII). For Jordan, the Galois theory did not only concern algebraic numbers, which are solutions to algebraic equations, but any kind of irrational numbers, including transcendental ones. See Brechenmacher (2011).

[73] The statement that the roots of Abelian equations with integer coefficients can be expressed as rational functions of the roots of unity of cyclotomic equations, that is, what is nowadays called the Kronecker–Weber theorem, was explicitly considered by Kronecker as aiming to separate the domains of algebra and of the theory of numbers in the investigation of the "essence" of the quantities associated to algebraic equations.

distinguished from the general notions specific to arithmetic. Methods such as groups of substitutions, he argued, were relative to the approach one would follow in his research and had therefore no inherent meaning. Confusing such methods for notions would only lead to formal and falsely general developments. Moreover, Kronecker criticized the way Jordan had dealt with the general and the special by "reducing" a general problem to a chain of simpler ones. For Kronecker, Jordan had applied this approach in his reduction of linear substitutions to their canonical forms in a way that amounted to the determination of the roots of general algebraic equations. Kronecker thus condemned the "false generality" and the "formal" nature of Jordan's explicit formula of canonical reduction because of its non-effectivity. On the contrary, Kronecker appealed directly to Gauss' legacy for his claim that the theory of forms belonged to arithmetic and should consequently focus on the characterization of equivalence classes by establishing arithmetical invariants using effective procedures such as GCD computations. Kronecker also presented his work as a natural outgrowth of life long concern for the arithmetical basis of algebraic methods. He especially stressed the notion of "domain of rationality" which limited the field of legitimate quantities to those that may be expressed effectively as rational functions of a given list of quantities. In this framework, expressions such as Jordan's canonical forms could not be expressed in general as opposed to the invariants Kronecker had introduced by rational operations based on GCD computations. A few years later, in 1882, this ideal of effectivity was at the core of the notion of "rationality" that laid the ground to Kronecker's famous arithmetic theory of algebraic magnitudes. For Kronecker, irrational numbers defined by algebraic equations posed "one the most interesting problems" in number theory, whose true "arithmetical nature" Kronecker strived to reveal.[74] He explicitly contrasted his arithmetical approach to the use of the algebraic notion of Galois. Kronecker also criticized Galois with a criticism very close to the one he had addressed to Jordan in 1874: for Kronecker, while Abel had stayed within the boundaries of a given rational domain by considering "concrete" rational functions of the roots of a given special equation, Galois had "escaped freely" by "abstracting" from the "problem of special equations" the notion of group whose importance" was only theoretical."

The collective dimensions of Jordan's and Kronecker's values for generality were quite complex. Even though, on the eve of World War I, nationalistic discourses would oppose the "French style of thinking"—as exemplified by the theory of order in the legacy of Jordan and Galois—to the German algebra,[75] it is not possible to analyze the 1874 quarrel as a direct echo of the war of 1870 or in the frame of antagonism between France and Germany or Paris and Berlin. As a matter of fact, when the controversy reached its climax in the spring of 1874, Gaston Darboux published a paper on the "algebraic theory of forms" which aimed to put the legacy of Hermite to the fore and whose orientation was much closer to Kronecker's than to Jordan's.[76]

---

[74] Kronecker (1874c: 383).

[75] Brechenmacher (2011).

[76] On Hermite's theory of forms and on Hermite's program for characterizing irrational quantities, see Goldstein (2007).

The fact that Jordan's and Kronecker's values for generality cannot be inscribed in any obvious social, institutional, or national category does not mean that these values were purely individual. We have seen that both mathematicians could be situated within what used to be a shared algebraic culture at a time when Algebra was not an autonomous discipline. This shared history took the form of a network of mainly periodical publications. For this reason, it put into play a way of addressing generality which was neither reflexively identified to a method nor to a theory. What we have called a "practice" took on an identity which could not be dissociated from the network of texts in which it circulated. But the controversy was not only the consequence of the two different conclusions. Jordan and Weierstrass had given to this shared history. It actually opposed two more local lines of development on the *general* theory of the irrationals defined by *special* equations. We shall conclude this paper in highlighting that the 1874 quarrel actually proceeded more from the divergence of these two lines of thought than from their meeting. In this sense, in parallel to the broad phenomenon of the universalization of new ideals of rigors which would pay attention to singularities and would separate homogeneity from genericity, various other forms of treatment of the general emerged and circulated on more local scales.

In two influential papers he published between 1878 and 1880, Georg Frobenius structured the organization of the theory of forms for the next fifty years. In doing so, he proposed a synthesis between the notions put to the fore by Jordan and Kronecker. But at the deeper level of the procedures he favored, Frobenius and followed Kronecker in aiming to build a rational theory of forms based on the effective computation of invariants. In this context, Jordan's canonical form lost its status of a theorem and appeared as a mere consequence of a more "general approach."[77]

Despite the influence of Frobenius' theory, Jordan's approach kept circulating for decades on a model close to the one we have seen in the case of the network of the secular equation. While Frobenius' theory characterized forms by the computation of polynomial invariants, the representation Jordan had given to his reduction allowed the steps in the transformation of a given form into its simplest expression to be "seen" dynamically.[78] By contrast to the static nature of invariant computations, the representation Jordan advocated for reducing general problems to chains of simpler problems induced some dynamic ways of thinking about "transformations," "reductions," or "decompositions." This representation depicted how a general problem could be reduced into a chain of simpler problems. It would later be the basis of the 1930s method of matrix decomposition, such as that shown in Fig. 16.2.

Between 1874 and 1930, the tension between canonical forms and invariants would play a major role in the complex history of the practices that eventually gave a

---

[77] Frobenius (1879: 483), translation T. Hawkins (1977: 153). For almost half a century, Jordan's canonical form was only considered as a theorem in a specific network of texts involving especially American and French works on linear groups and Galois fields. See Brechenmacher (2011).

[78] This representation especially allows the simultaneous decomposition into subgroups of the indices on which a substitution operated and of the substitution itself to be seen. From the viewpoint of linear algebra, this corresponds to a decomposition of a vector space under the action of an operator into a sum of stable subspaces. See Appendix 1 and Brechenmacher (2006a: 167–87).

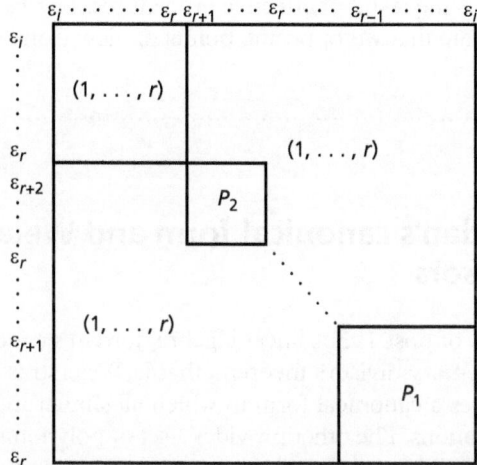

Figure 16.2 *A representation of matrix decomposition.*

universal dimension to the operatory procedures attached to the pictorial representation of matrices.[79]

In the 1930s, "Jordan's canonical form theorem" was considered as a central result in most of the Treatises on the "theory of matrices." But this "general" theorem would actually be connected with two kinds of canonical forms:[80]

$$
A = \begin{Vmatrix}
\lambda_1 & 1 & 0 & & & & & 0 \\
0 & \lambda_1 & 0 & & & & & \\
& & \lambda_2 & 1 & 0 & & & \\
& & 0 & \lambda_2 & 1 & & & \\
& & 0 & & \lambda_2 & 0 & 0 & 0 \\
& & & & 0 & \lambda_3 & 0 & 0 \\
& & & & & & \lambda_4 & 1 \\
& & & & & & & \lambda_4
\end{Vmatrix}, \quad
B = \begin{Vmatrix}
0 & 0 & \cdots & & \cdots & 0 & -\alpha_8 \\
1 & 0 & \cdots & & \cdots & & -\alpha_7 \\
0 & 1 & 0 & \cdots & \cdots & & -\alpha_6 \\
\cdots & 0 & 1 & 0 & 0 & \cdots & \\
& \cdots & 0 & 1 & 0 & 0 & \cdots & \cdots \\
& & \cdots & 0 & 1 & 0 & 0 & \cdots \\
& & & & 0 & 1 & 0 & -\alpha_2 \\
0 & 0 & \cdots & & \cdots & 0 & 1 & -\alpha_1
\end{Vmatrix}
$$

On the one hand, the Jordan canonical matrix A was considered as the "simplest" form for the maximal decomposition of a matrix. On the other hand, the "rational canonical

---

[79] Brechenmacher (2010).

[80] The two matrices given in this example both relate to the minimal polynomial $\lambda^8 + \alpha_1\lambda^7 + \cdots + \alpha_7\lambda + \alpha_8 = (\lambda - \lambda_1)^2(\lambda - \lambda_2)^3(\lambda - \lambda_3)(\lambda - \lambda_4)$

form" B was obtained as the result of effective procedures. The 1930s "general" theorem would therefore articulate the two opposing points of view from 1874.[81]

....

APPENDIX 1

More about Jordan's canonical form and Weierstrass' elementary divisors

From the point of view of post 1930s linear algebra, Jordan's canonical form theorem is equivalent to the elementary divisors theorem, that is, Weierstrass theorem as reworked by Kronecker. One gives a canonical form to which all similar matrices can be reduced by similarity transformations. The other provides a set of polynomials which are invariant for similarity transformations and therefore characterize similarity classes.

Similarity of square matrices

Both theorems provide a characterization of classes of similarities of square matrices with coefficients belonging to $\mathbb{C}$ (or, more generally, to any algebraically closed field): two square matrices A and B are similar if and only if there is an invertible matrix U such that

$$U^{-1}AU = B.$$

Similar matrices always have the same characteristic determinants $|A - sI|$, which is therefore an invariant for classes of similarity. Yet, the reciprocal property (i.e., two matrices with the same characteristic determinants are similar matrices) is true only if all the roots of the characteristic equation $|A - sI| = 0$ are distinct (which imply that the matrix A can be reduced to a diagonal matrix). Jordan's canonical form and Weierstrass' elementary divisor both provide a complete characterization of classes of similarity, whatever the multiplicity of roots. Both go beyond the characteristic equation. The first gives a canonical form to which all similar matrices can be reduced; the second provides a set of invariant polynomials. The example below shows how different types of canonical forms and elementary divisors can be associated to the same characteristic determinant:[82]

$$|A - sI| = (s - 1)^2 (s - 2)^3 (s - 3).$$

---

[81] See, for instance, Aitken and Turnbull (1932: 1).
[82] See Gantmacher (1959) for more details.

Jordan's
canonical
forms

$$\begin{bmatrix} 1 & & & & \\ & 1 & & & \\ & & 2 & & \\ & & & 2 & \\ & & & & 2 \\ & & & & & 3 \end{bmatrix} \quad \begin{bmatrix} 1 & & & & \\ & 1 & & & \\ & & 2 & 1 & \\ & & & 2 & 1 \\ & & & & 2 \\ & & & & & 3 \end{bmatrix} \quad \begin{bmatrix} 1 & & & & \\ & 1 & & & \\ & & 2 & 0 & \\ & & & 2 & 1 \\ & & & & 2 \\ & & & & & 3 \end{bmatrix}$$

Elementary
divisors

$(s-1)$, $(s-1)$, $(s-1)$, $(s-1)$, $(s-1)$, $(s-1)$,

$(s-2)$, $(s-2)$, $(s-2)$, $(s-2)^3$, $(s-2)$, $(s-2)^2$,

$(s-3)$ $(s-3)$ $(s-3)$

Jordan's list of problems on bilinear forms

Similarity is not the only type of equivalence which can be considered for matrices, as is illustrated by the three problems of reduction to canonical forms Jordan gave in his December 1873 paper of on bilinear forms. From the viewpoint of modern algebra, for matrices are now associated to bilinear forms, the canonical form $x_1 y_1 + \cdots + x_m y_m$ concerns the equivalence relation on square matrices:

$$ARB \Leftrightarrow \exists U, V \in GL_n(\mathbb{C}), UAV = B.$$

Jordan's first problem concerns the similarity relation of symmetric matrices using orthogonal matrices:

$$ARB \Leftrightarrow \exists U \in O(\mathbb{C}), U^{-1} AV = B.$$

The second problem relates to the congruence relation for square matrices:

$$ARB \Leftrightarrow \exists U \in GL_n(\mathbb{C}), {}^t UAV = B.$$

The third problem focuses on the equivalence of pairs of square matrices (A, B):

$$(P, Q) R (P', Q') \Leftrightarrow \exists U, V \in GL_n(\mathbb{C}), UPV = P'; UQV = Q'.$$

In the case of non-singular pairs of matrices, the third problem is equivalent to that of the similarity of a single matrix and can therefore be solved by the use of Jordan's canonical form theorem or of Weierstrass' elementary divisors theorem. Indeed, considering a pair of matrices (P, Q) is equivalent to considering $P - sQ$, that is, a polynomial of matrices in a parameter s, or, equivalently, $PQ^{-1} - sI$ (Q is invertible because it is non-singular), which can be written $A - sI$ with $A = PQ^{-1}$. The problem of characterizing the equivalence class of the pair (P, Q) is eventually the same as that of characterizing the equivalence class of the pair (A, I), which boils down to characterizing the similarity classes of the matrix A by analyzing the characteristic equation $|A - sI| = 0$.

Kronecker's invariant factors

In modern parlance, the "invariant factors" Kronecker introduced in 1874 give a method for deciding of the equivalence of (non-singular) pairs of matrices on a principal ring (such as the ring of integers or of polynomials with integer coefficients) whereas Jordan's canonical form and Weierstrass' elementary divisors are only valid for algebraic closed fields such as the field of complex numbers.

Consider a pair of bilinear forms $P - sQ$ and let $S(s)$ be the characteristic determinant $|P - sQ|$. Let $S_1(s)$ be the greatest common divisor of all the first minors of $S(s)$ (which are polynomials in s). Similarly, $S_2(s)$ is defined as the greatest common divisor of all the second minors of $S(s)$ and so on. Then $S_i(s)$ divides $S_{i-1}(s)$ and if $E_i(s)$ denotes the polynomial $S_{i-1}(s)/S_i(s)$ then $E_i(s)$ divides $E_{i-1}(s)$. Thus, $S(s)$ differs from the product of the $E_i(s)$ by a constant.

Let now consider the situation in an algebraic closed field so that each polynomial expression can be split into linear factors. Let $s_1, s_2, ..., s_k$ bet the distinct roots of $S(s)$, then

$$E_i(s) = c_i \prod_{j=1}^{n} (s - s_j)^{m_{ij}},$$

where c_i is constant and the m_{ij} are positive integers or zero. Each factor $e_{ij} = (s - S_j)^{m_{ij}}$ with $m_{ij} > 0$ is what Weierstrass had called an elementary divisor of $S(s)$. In the case of an algebraic closed field, Kronecker's invariants are therefore equivalent to Weierstrass' elementary divisors.

....

REFERENCES

Aitken, A. G. and Turnbull, H. W. (1932) *An introduction to the theory of canonical matrices.* London, Glasgow: Blame & Son.
Brechenmacher, F. (2006a) *Histoire du théorème de Jordan de la décomposition matricielle (1870–1930),* Thèse de doctorat, Ecole des Hautes Etudes en Sciences sociales, Paris, France.
Brechenmacher, F. (2006b) 'A controversy and the writing of a history: the discussion of "small oscillations" (1760–1860) from the standpoint of the controversy between Jordan and Kronecker (1874)', *Bulletin of the Belgian Mathematical Society* 13: 941–4.
Brechenmacher, F. (2007a) 'La controverse de 1874 entre Camille Jordan et Leopold Kronecker', *Revue d'histoire des mathématiques* tome 13, fasc.2: 187–257.

Brechenmacher, F. (2007b) 'L'identité algébrique d'une pratique portée par la discussion sur l'équation à l'aide de laquelle on détermine les inégalités séculaires des planètes (1766–1874)', *Sciences et Techniques en Perspective* IIe série, fasc. 1: 5–85.

Brechenmacher, F. (2010) 'Une histoire de l'universalité des matrices mathématiques', *Revue de Synthèse* 131: 569–603.

Brechenmacher, F. (2011) 'Self-portraits with Évariste Galois (and the shadow of Camille Jordan)', *Revue d'histoire des mathématiques* 17: 271–369.

Brechenmacher, F. (2012) 'Linear groups in Galois fields. A case study of tacit circulation of explicit knowledge', *Oberwolfach reports* 4-2012: 48–54.

Brechenmacher, F. (2014) 'Lagrange and the secular equation', Lettera Matematica int. edition. 2: 79–91.

Cauchy, A. L. (1829) 'Sur l'équation à l'aide de laquelle on détermine les inégalités séculaires du mouvement des planètes', *Exercices de mathématiques* 4, in *Œuvres complètes d'Augustin Cauchy*, Paris: Gauthier-Villars et fils, (2) 9: 174–95.

Christoffel, E. B. (1866) 'Theorie der Bilinearen Formen', *Journal für die reine und angewandte Mathematik* 68: 253–72.

Cifoletti, G. (1995) 'The creation of the history of algebra in the sixteenth century', in Goldstein, C., Gray, J., and Ritter, J. (eds.), *L'Europe mathématique: Mythes, histoires, identités—Mathematical Europe: Myth, History, Identity*. Paris: Editions de la Maison des sciences de l'homme, 123–44.

Dahan dalmedico, A. (1992) *Mathématisations. Augustin-Louis Cauchy et l'Ecole Française*. Paris: Blanchard.

Darboux, G. (1874) 'Mémoire sur la théorie algébrique des formes quadratiques', *Journal de mathématiques pures et appliqués* **XIX**: 347–96.

Dhombres, J. (1998) 'Une histoire de l'objectivité scientifique et le concept de postérité', in Guesnerie, R. and Hartog, F. (eds.), *Des sciences et des techniques: un débat*, Paris: Editions de l'EHESS, Armand Colin, 127–48.

Dieudonné, J. (1962) 'Notes sur les travaux de Camille Jordan relatifs à l'algèbre linéaire et multi-linéaire et la théorie des nombres', in (Jordan, *Œuvres*, 3: V–XX).

Durand-Richard, M. J. (1996) 'L'École algébrique anglaise: les conditions conceptuelles et insti-tutionnelles d'un calcul symbolique comme fondement de la connaissance', in Goldstein, C., Gray, J., and Ritter, J. (eds.), *L'Europe mathématique: Mythes, histoires, identités—Mathematical Europe: Myth, History, Identity*. Paris: Editions de la Maison des sciences de l'homme, 445–77.

Durand-Richard, M. J. (2008) *L'analogie dans la démarche scientifique*. Paris: l'Harmattan

Frobenius, F.G. (1878) 'Ueber lineare Substitutionen und bilineare Formen', *Journal für die reine und angewandte Mathematik* 84: 1–63.

Frobenius, F.G. (1879–80) 'Theorie der linearen Formen mit ganzen Coefficienten', *Journal für die reine und angewandte Mathematik* 86: 146–208; 88: 96–117.

Gantmacher, F. (1959) *The theory of matrices*. New York: Chelsea.

Gauss, C. F. (1801) *Disquitiones arithmeticae*. Leipzig: Fleischer.

Goldstein, C. (1995) *Un théorème de Fermat et ses lecteurs*. Saint-Denis: PUV.

Goldstein, C. (2007) 'The Hermitian Form of Reading the *Disquisitiones*', in Goldstein, C., Schappacher, N., and Schwermer, J. (eds.), *The shaping of arithmetics after C. F. Gauss's Disquisitiones Arithmeticae*. Berlin: Springer, 377–410.

Goldstein, C. and Schappacher N. (2007) 'Several disciplines and a book (1860–1901)', in Goldstein, C., Schappacher, N., and Schwermer, J. (eds.), *The shaping of arithmetics after C. F. Gauss's Disquisitiones Arithmeticae*. Berlin: Springer, 67–104.

Hamburger, M. (1873) 'Bemerkung über die Form der Integrale der linearen Differentialgleichungen mit veränderlicher Coefficienten', *Journal für die reine und angewandte Mathematik* 76: 113–25.

Hawkins, T. (1975) 'Cauchy and the spectral theory of matrices', *Historia Mathematica* **2**: 1–20.

(1977) 'Weierstrass and the theory of matrices', *Archive for History of Exact Sciences* **17**: 119–63.

Hermite, C. (1855) 'Remarque sur un théorème de M. Cauchy', *Comptes rendus de l'Académie des sciences de Paris* **41**: 181–83.

Hermite, C. (1857) 'Sur l'invariabilité du nombre des carrés positifs et des carrés négatifs dans la transformation des polynômes homogènes du second degré', *Journal für die reine und angewandte Mathematik* **53**: 271–74.

Jordan, C. (1870) *Traité des substitutions et des équations algébriques*. Paris: Gauthier-Villars.

Jordan, C. (1871) 'Sur la résolution des équations différentielles linéaires', *Comptes rendus de l'Académie des sciences de Paris* **73**: 787–91.

Jordan, C. (1872) 'Sur les oscillations infiniment petites des systèmes matériels', *Comptes rendus de l'Académie des sciences de Paris* **74**: 1395–9.

Jordan, C. (1873) 'Sur les polynômes bilinéaires', *Comptes rendus de l'Académie des sciences de Paris* **77**: 1487–91.

Jordan, C. (1874a) 'Mémoire sur les formes bilinéaires', *Journal de mathématiques pures et appliquées* **19**: 35–54.

Jordan, C. (1874b) 'Sur la réduction des formes bilinéaires', *Comptes rendus de l'Académie des sciences de Paris* **78**: 614–17.

Klein, F. (1868) *Ueber die Transformation der allgemeinen Gleichung des zweiten Grades zwischen Linien Coordinaten auf eine canonische Form*, Bonn. Inaugural Dissertation, reprinted in *Mathematische Annalen* **23**(1884): 539–78.

Kronecker, L. (Werke) *Leopold Kronecker's Werke*, ed. by K. Hensel, Leipzig: Teubner, 1895–1931.

Kronecker, L. (1866) 'Ueber bilineare Formen', *Monatsberichte der Königlich Preussischen Akademie der Wissenschaften zu Berlin*: 597–612; *Journal für die reine und angewandte Mathematik*, **68**: 273–285. (Kronecker, *Werke*, 1, pp. 145–62).

Kronecker, L. (1868) 'Ueber Schaaren quadratischer Formen', *Monatsberichte der Königlich Preussischen Akademie der Wissenschaften zu Berlin*: 339–46. (Kronecker, *Werke*, 1, pp. 163–74).

Kronecker, L. (1874a) 'Ueber Schaaren von quadratischen und bilinearen Formen', *Monatsberichte der Königlich Preussischen Akademie der Wissenschaften zu Berlin*: 59–76. (Kronecker, *Werke*, 1, pp. 349–72).

Kronecker, L. (1874b) 'Sur les faisceaux de formes quadratiques et bilinéaires', *Comptes rendus de l'Académie des sciences de Paris* **78**: 1181–2. (Kronecker, *Werke*, 1, pp. 415–19).

Kronecker, L. (1874c) 'Ueber Schaaren von quadratischen und bilinearen Formen', Nachtrag *Monatsberichte der Königlich Preussischen Akademie der Wissenschaften zu Berlin*: 149–56, 206–32. (Kronecker, *Werke*, 1, pp. 373–413).

Lagrange, J. L. (1766) 'Solution de différents problèmes de calcul intégral', *Miscellanea Taurinensia* **3**: 471–668.

Lagrange, J. L. (1774)'Recherches sur les équations séculaires des mouvements des nœuds, et des inclinaisons des orbites des planètes', *Mémoires de l'Académie royale des sciences de Paris* **177**: 97–174

Lagrange, J. L. (1784) 'Théorie des variations séculaires des éléments des planètes ; Seconde partie', *Nouveaux mémoires de l'académie des sciences de Berlin*: 169–292.

Lagrange, J. L. (1788) *Méchanique analytique* (2de éd.). Paris.

Laplace, P. S. (1787) 'Mémoire sur les variations séculaires des orbites des planètes', *Mémoires de l'Académie des sciences de Paris*: 267–79.

Laplace, P. S. (1799) *Traité de mécanique céleste*, Vol.1, Paris, 1799, [*Œuvres*, 1].

Panza, M. (1992) 'The evolution of Lagrange's research programme in the analytical foundations of mechanics with particular respect to its realisation in the 'Théorie des Fonctions analytiques", *Historia Scientarum* 1–2 (1991) and 1–3 (1992).

Revel, J. (ed.) (1996) *Jeux d'échelles: La micro-analyse* à *l'expérience*. Paris: Gallimard.

Sinaceur, H. (1991) *Corps et modèles*. Paris: Vrin.

Sylvester, J.J. (1852) 'Sur une propriété nouvelle de l'équation qui sert à déterminer les inégalités séculaires des planètes', *Nouvelles Annales de Mathematiques* (first series) **11**: 434–40.

Weierstrass, K. (Werke) *Mathematische Werke von Karl Weierstrass*. Berlin: Mayer & Müller, 1894–1927.

Weierstrass, K. (1858) 'Ueber ein die homogenen Functionen zweiten Grades betreffendes Theorem', *Monatsberichte der Königlich Preussischen Akademie der Wissenschaften zu Berlin*: 207–20. (Weierstrass, *Werke*, 1, pp. 233–46).

Weierstrass, K. (1868) 'Zur Theorie der bilinearen und quadratischen Formen', *Monatsberichte der Königlich Preussischen Akademie der Wissenschaften zu Berlin*: 310–38. (Weierstrass, *Werke*, 2, pp. 19–43).

Yvon-Villarceau, A. (1870) 'Note sur les conditions des petites oscillations d'un corps solide de figure quelconque et la théorie des équations différentielles linéaires', *Comptes rendus de l'Académie des sciences de Paris* **71**: 762–66.

17

Practices of generalization in mathematical physics, in biology, and in evolutionary strategies

EVELYN FOX KELLER

On first impression, the value of generality seems self-evident, and an appreciation of this value is manifest across many if not all scientific cultures. But on closer inspection, the meaning of generality turns out not to be at all self-evident; indeed, it varies considerably across scientific cultures. Recognition of that variation brings with it a corresponding recognition of the variety of values, all going under the name of generality, that are appreciated in different scientific practices. In this paper, I explore the different meanings, and values, associated with the mathematical and the biological sciences, and I compare these meanings with the kinds of generality prized by the practices of biological evolution.

17.1   Introducing spherical cows

I do not know when the term "spherical cow" first appeared, but this much is clear: from the start, it was a joke. Or rather, it was the punch line of a joke. The joke is this: a farmer, in his desperation over the fall in milk prices and his consequent need to increase milk production, turns, as a last resort, to a theoretical physicist. A few weeks later, the physicist phones the farmer, "I've got the answer. The solution turned out to be a bit more complicated than I thought and I'm presenting it at this afternoon's theory seminar." At the seminar the farmer finds a handful of people drinking tea and munching on cookies—none of whom looks like a farmer. As the talk begins the physicist approaches the blackboard and draws a big circle. "First, we assume a spherical cow."[1]

---

[1] http://en.wikipedia.org/wiki/Spherical_cow#_note-0

The Oxford Handbook of Generality in Mathematics and the Sciences. First Edition. Karine Chemla, Renaud Chorlay and David Rabouin. © Oxford University Press 2016. Publishing in 2016 by Oxford University Press.

Importantly, this is a joke told by physicists, ostensibly about themselves, but of which, in the end, the farmer turns out to be the butt. What makes it a joke is the manifest falsity of the assumption underlying one of the oldest and most basic strategies used by physicists to solve complex problems. Cows are obviously not spherical, and the assumption of a spherical cow is therefore a lie; it introduces a model that is a fiction, that one will never find instantiated in the real world. For the farmer, such a starting point is not only absurd, but patently useless. What finally turns the joke against the farmer is the recognition shared among the listeners but not by the farmer of the extraordinary usefulness this strategy has had in the history of physics, despite its reliance on fiction.

But the same strategy has also seemed absurd to generations of experimental biologists, and the spherical cow might also be a joke told about them—rephrased, perhaps, as a joke about the spherical cell. Only now it is less funny, and not because of the shift from cow to cell.[2] Biologists warrant more respect than do farmers; they can be expected to recognize the usefulness of the physicist's strategy in physics. So why should they not appreciate its value for their own discipline? The fact is, they have not. Where the spherical cow may be a joke to physicists, to biologists, it is an abomination. I have argued elsewhere[3] that this difference in perceptions reflects a difference in epistemological cultures that, in turn, reflects (at least in part) differences between the subjects. Here, I want to focus on what these differences can tell us about practices of generalization in mathematical physics, in biology, and even in strategies of biological evolution.

17.2   Generality in the physical sciences

To physicists, the spherical cow (or cell) is clearly an idealization, a simplification, an abstraction away from the specificities of real cows (or cells). It is a model rather than an instance, a representation of the object found in the real world that cannot itself be found in that world, created with the hope, or "chief aim," as A. S. Eddington put it, "to obtain 'insight'—to see which of the numerous factors are particularly concerned in any effect and how they work together to give it." Generality, here, is to be found in the dynamical process by which particular factors yield particular kinds of effects. Eddington went on to explain:

> For this purpose a legitimate approximation is not just an unavoidable evil; it is a discernment that certain factors—certain complications of the problem—do not contribute appreciably to the result. We satisfy ourselves that they may be left aside; and the mechanism stands out more clearly freed from these irrelevancies. This discernment is only a continuation of a task begun by the physicist before the mathematical premises of the

---

[2] Rather famously, however, Alan Turing did in fact assume a spherical cow in his mathematical model for morphogenesis (1952). And like Rashevsky (discussed below), he too was severely criticized by biologists for his neglect of the complexities of biological reality. Not only is no real cow a sphere, but so too, no real biological cell is a homogeneous distribution of enzymes (as Turing had also assumed).

[3] Fox Keller (2002).

problem could even be stated; for in any natural problem the actual conditions are of extreme complexity and the first step is to select those which have an essential influence on the result—in short, to get hold of the right end of the stick.[4]

In a similar vein, D'Arcy W. Thompson—the man sometimes claimed as the father of mathematical biology—referred to this strategy as "the principle of negligence," and he railed against his colleagues in biology for what he described as their

> ingrained and deep-seated belief that even when we seem to discern a regular mathematical figure in an organism, ... [it is] the details in which the figure differs from its mathematical prototype [that] are more important and more interesting than the features in which it agrees; and even that the peculiar aesthetic pleasure with which we regard a living thing is somehow bound up with the departure from mathematical regularity which it manifests as a peculiar attribute of life.

"This view," he claimed, "involve[s] a misapprehension: ... We may be dismayed too easily by contingencies which are nothing short of irrelevant compared to the main issue; there is a *principle of negligibility*. Someone has said that if Tycho Brahé's instruments had been ten times as exact there would have been no Kepler, no Newton, and no astronomy."[5] But what is it that undergirds Thompson's *principle of negligibility*? Or, in Eddington's terms, what implies that the representation of the cow as a sphere is a way to "get hold of the right end of the stick"? What guarantees that the spherical structure is "the main issue," and that the details that have been neglected are less relevant? To be sure, the convention of starting with a sphere has a powerful pragmatic justification by virtue of the mathematical simplicity of the sphere. But what justifies the slide between simple and fundamental? What guarantees that the simplest representation captures "the main issue," that the discarded complications have not made an appreciable contribution to the effect? For Thompson, and perhaps for Eddington as well, I suggest that the answer lies in a residual Platonism grounding reality in ideal forms. Indeed, one might say that, to the extent that we group together the "ideal," the "simple," and the "abstract," and automatically equate this collective of properties with "fundamental," we all betray our residual Platonism. Also, a more fully blown Platonism could also account for yet another way in which the conjunction between ideal, simple, and general works, and that is in the notion of model as prototypical or generic—in the sense, that is, of "a generic, idealized model of a person, object, or concept from which similar instances are derived, copied, patterned, or emulated."[6]

---

[4] Eddington (1926: 101–2).

[5] Thompson (1942: 1028–89). One might think that an insistence on the importance of detail, and the deviation from mathematical regularity, would imply a rejection of generality altogether, but in fact, biologists often employ a "principle of negligibility" of their own. Here, it is not departure from mathematical regularity but departures from what are taken as biological norms that tend to be neglected. The notion of biological norm inheres in the use of model organisms to represent particular species of organisms, and even, on occasion, organisms "in general." In this practice, it is an actual organism rather than idealized representation of an organism that serves as a model—a model that, once standardized, replicated, and distributed, provides the basis for a practice of generalization in biology of a different kind.

[6] http://en.wikipedia.org/wiki/Archetype.

17.3 Generality in the life sciences

As I've already indicated, modern experimental biologists come from very different traditions. They have for the most part eschewed the conjunctions taken for granted in the physical sciences, and, accordingly, have developed rather explanatory styles, and different approaches to generality. To illustrate these differences, particularly as they bear on practices of generalization, I want to focus on a critical episode in the history of the relations between mathematical and experimental biology in the twentieth century—an episode I have presented before, but which I now want to place under closer scrutiny. It concerns Nicolas Rashevsky's attempt in 1934 to present to practicing biologists some preliminary results of his efforts, as a theoretical physicist, to understand cell division as a consequence of the physical forces that would be acting on an idealized (spherical) cell were it a real cell. This effort was part of a much larger project, inspired by D'Arcy W. Thompson, to develop the "*Physico-Mathematical Foundations of Biology*" —that is, to "To bring mathematical biology to the same level [as mathematical physics]." The particular occasion was the second meeting of the Cold Spring Harbor (CSH) Symposia on Quantitative Biology, on the topic, "Aspects of Growth," and I offer excerpts of the recorded text of this encounter as a text we might probe for a better understanding of the different understandings of generic, general, and generalization employed in the two communities.

Addressing an audience that included Charles Davenport, E. B. Wilson, and Eric Ponder, Rashevsky began by asking: "Do we need to assume some special independent mechanisms, which produce at a certain stage of the cellular life a division, or are those mechanisms merely the consequences of more general phenomena, which we know occur in all cells?"[7] He thought not. Cell division, he argued, can be explained as a direct consequence of the forces arising from nothing more than cell metabolism, and he illustrated this claim by analyzing the mechanics of a spherical model of a cell undergoing metabolic growth. The discomfort among the biologists was palpable. As Davenport explained:

> I think the biologist might find that whereas the explanation of the division of the spherical cell is very satisfactory, yet it doesn't help as a general solution because a spherical cell isn't the commonest form of cell. The biologist knows all the possible conditions of the cell form before division.... In the special cases of egg cells and cleavage spheres, this analysis may prove very valuable. But after all, these are only special cases.[8]

In attempting to explain his view of things, Rashevsky not only defends his modeling strategy (the spherical cell), but links that defense to a strong claim for the applicability of mathematical physics to biology; his counter-response could hardly have been reassuring.

---

[7] Rashevsky (1934: 188).
[8] Discussion following Rashevsky's paper, Rashevsky (1934: 197–8).

It restated the traditional faith of the mathematician, though in a way that would only have fueled Davenport's discontent:

> It would mean a misunderstanding of the spirit and methods of mathematical sciences should we attempt to investigate more complex cases without a preliminary study of the simple ones. The generalization of the theory, to include non-spherical cells, is indeed needed, and this will be the subject of research after the simpler cases are thoroughly and exhaustively studied…. To my mind it is already quite a progress that a general physico-mathematical approach to the fundamental phenomena of cellular growth and division … has been shown to be possible. Judging by the development of other mathematical sciences, I would say that it will take at least twenty-five years of work by scores of mathematicians to bring mathematical biology to a stage of development comparable to that of mathematical physics.[9]

In the final discussion, Eric Ponder, Director of the CSH Laboratories, summed up the mood of the biologists as follows:

> One point upon which there seems to be pretty general agreement is that there is little relation between the amount of which has been done on the mathematics of growth and the clarification of the subject which has resulted. As [James] Gray said some six years ago: "It is intrinsically improbable that the behavior of a growing system should conform to that of a simple chemical system, and the conception of growth as a simple chemical process should not be accepted in the absence of rigid and direct proof." Both Dr. Wilson and Dr. Davenport seem to be of the same opinion, and I question if the conclusion ought not to be put even more strongly. Work on the mathematics of growth as opposed to the statistical description and comparison of growth, seems to me to have developed along … unprofitable lines…. [I]t is futile to conjure up in the imagination a system of differential equations for the purpose of accounting for facts which are not only very complex, but largely unknown…. It is said that if one asks the right question of Nature, she will always give you an answer, but if your question is not sufficiently specific, you can scarcely expect her to waste her time on you…. [W]hat we require at the present time is more measurement and less theory.[10]

17.4   A failure of communication

Perhaps the first thing to be noted from this exchange is the difference in meanings the interlocutors give to the terms generic, general, and generalization, and to the relation between them. To Davenport, the spherical cell doesn't help as a general solution because it is a particular kind of cell—furthermore, a kind that is not commonly encountered in the world of real cells; it is only a special case. The implicit meaning of general here is that

---

[9] Rashevsky (1934: 198). Note that Rashevsky uses the word *case* in this text to refer to kinds of models, rather than to kinds of cells. In the world of models, as in the world of mathematics more generally, the sphere (or circle) can be regarded as a particular case.

[10] *Ibid.*, p. 201.

of common, usual, widespread, and frequently encountered—corresponding to what the OED gives as its definition *1.b.*: "Pertaining in common to various persons or things;" by this reading, general (i.e., generally applicable) solutions cannot, by definition, be obtained from the study of special cases.

For Rashevsky, however, the spherical cell is not an example of a cell but a model of a cell (or perhaps, one should say, an example of a model of a cell). As such, it might be considered as a special case in the class of models, but it does not refer to a special case in the class of cells. In relation to models, a spherical cell, while specific or special in the world of biological cells, may in fact be quite general in the sense of being a model commonly employed—indeed, it might be said to be generic in the sense that the OED provides as its first definition of that term: "Belonging to a genus or class; applied to a large group or class of objects; general (opposed to *specific* or *special*); ... representing a class or genus of objects, whether formed (as is usually supposed) by blending images of several particular members of that class or by preventing an image from becoming fully determinate" (OED, A. *adj.* a.). The relevant point here is that the value of such a model (or image) has nothing to do with whether or not it is found in the real world as such. It admits of "a general physico-mathematical approach" in the sense referred to above (i.e., an approach that is commonly employed in mathematical physics), and it acquires this sense of generality by virtue of being simple enough to be tractable. More important however is that the scientific value of such a model is assured by the underlying (and unarticulated) assumption of Rashevsky's tradition that it is precisely in the simplest, stripped down, version of the cell that one finds the fundamental mechanism defining the cell's true essence. Of course, the real cells one sees under the microscope will not be expected to conform exactly to such an idealized form; indeed, it is by virtue of their departure from that ideal that the real cells, rather than the sphere, are what are to be regarded as special cases. Yet even in its stripped down form, the theory is expected to provide results that are *roughly* applicable to many special cases—that is, that are *general* (in the sense of the definition of *general* given by the OED as *8. a.*, namely, "Comprising, dealing with, or directed to the main elements, features, purposes, etc., with neglect of unimportant details or exceptions," or, as Eddington put it, in the sense that the complications of the problem that have been ignored do not contribute appreciably to the result). Better fitting applications of the theory to such special cases will however require elaboration (or extension) of the theory to include the complications or details that make these cases special (what Rashevsky calls its "generalization").

17.5   Generality, universality, and laws

One often hears it said that the aims of modern (at least physical) science are to discover properties of nature that are, at least within a particular domain, of unrestricted generality, and here we encounter yet another meaning of our main term. Such properties are products of the so-called "laws of nature"—laws that are, within the domain in which they apply, supposed to be exception-free. Whether or not there are in fact any such laws of nature is a subject of intense debate in the philosophy of science, but certainly, there has long been a strong commitment in theoretical physics to the importance of such a pursuit. Not so, however, in biology. Indeed, the question of whether or not there are "laws of biology" in the same sense in which, for example, Newton's laws are presumed to be laws of physics, whether or not there is anything that can be called lawful in biology that is specifically biological—that is, left over after all the implications of the laws of physics and chemistry have been fully pursued—is also a matter of long-standing discussion among philosophers of science.

The majority view among philosophers of biology today is that that biology does not have laws of its own, and for reasons that are closely related to biologists' distrust of the habits of idealization in theoretical physics. But to these philosophers, it does not therefore follow either that biology is any less of a science, or that biologists do not seek generality. They simply reject the view that laws are the *sine qua non* of a proper science, as well as the conviction that such laws provide the meaning of what a "fundamental explanation" is. Which gets us to the question of the proper relation between the life and physical sciences.

D'Arcy Thompson's hope for biology was that it would become a subset of the physical sciences, its own questions "reduced" (or reframed) as special cases, or applications, of the more general laws of physics and chemistry. As he wrote:

> Cell and tissue, shell and bone, leaf and flower, are so many portions of matter, and it is in obedience to the laws of physics that their particles have been moved, molded, and conformed... Their problems of form are in the first instance mathematical problems, their problems of growth are essentially physical problems, and the morphologist is, ipso facto, a student of physical science.[11]

And indeed, this seems to be the hope of more than a few physical scientists today. But as most other scientists have learned, even in physics, there are many ways of being scientific besides that of searching for laws.

Biologists, for example, may not be committed to the search for laws in the sense that we speak of Newton's laws, they may not pursue formulations of unrestricted generality, but they are deeply committed to the search for formulations that we might describe as being of "restricted" generality. Indeed, their science abounds in claims about general properties, and even in statements that are often referred to as "laws" (think, e.g., of "Mendel's laws," the central dogma, or even the "law" of natural selection). But there is no domain in which these "laws" are presumed to be exception-free. They are generalities, but not unrestricted generalities. It is evident that generality is valued in biology, but exceptions are neither a cause for alarm, nor do they necessarily send researchers back to the drawing board in search of better—exception-free—laws. Rather, they are reminders of how complex biological reality is. And as the product of evolution by natural selection, of how contingent, reminding us once again that biology may fundamentally be a different kind of science from that of physics. Living beings are produced not simply by the laws of physics and chemistry, but by the cumulative effects of evolution, acting over time. Every organism we see today has been shaped and formed by a particular trajectory

---

[11] Thompson (1942: 16).

through evolutionary space time; it is the product of eons of tinkering; of building on what had accumulated over the course of a particular evolutionary trajectory; of a history of transforming chance innovations into future requirements for survival. Because of evolution, life is both contingent and particular (in the sense, that is, of depending on the dynamics of very particular arrangements of its biochemical and physical constituents). Of course, the laws of physics and chemistry are crucial to the functioning of biological systems, but if, beyond such laws, so little of logical necessity seems to be left to biology, if biological generalizations can never be more than provisional it is because of the historical contingencies on which the emergence, and future elaboration, of "life itself" have depended. Indeed, history is the canonical trump card of biologists in these discussions. It not only accounts for the lack of interest of biologists in universal laws, but may also provide us with the key to understanding the importance of restricted (as distinct from unrestricted) and provisional generality.

17.6   What generality might mean in biological evolution

While the experience of biological evolution offers a cautionary warning against the expectation of universal laws, it also, and at the same time, offers biologists a model for a practice of generalization based on a meaning of generality that is both restricted and provisional. The property of generality to which I want to draw attention refers not to laws but to procedures: borrowing from Karine Chemla's discussion of the mathematical practices of ancient China,[12] I want to suggest that a procedure that can be employed to solve a more general class of problems is, from an evolutionary perspective, by definition, a superior procedure. In other words, I want to build on the analogical resemblance that I first suggested in 2007[13] between the value placed on generality in the mathematical practices of ancient China and its role in the dynamics of biological evolution.

Chemla has shown that while procedures are initially developed to solve a particular problem (albeit, one arising in several situations), the overriding aim in the development of new procedures to solve the same problem was to extend the range of situations or problems to which the new procedure could be applied. This search for ever greater generality was carried out iteratively, in a manner somewhat suggestive of modern computer programs: sub-procedures that had been demonstrated to be successful with respect to one class of problems were absorbed into more general procedures that could then be shown to be applicable with respect to a larger class of problems.

Much the same can be said for evolution. To the extent that one might describe the task of evolution as that of solving problems—in the sense, that is, of forging mechanisms that enable individual organisms to cope with an ever-larger range of challenges, its aim is similarly one of increasing generality, and the procedure correspondingly iterative.

[12] See, for example, Chemla (2003, 2005).
[13] Fox Keller (2007).

Mechanisms that succeed in coping with those situations encountered by an organism are selected for, and once established their components are used over and over again, differently combined, and often modified, with the net effect that, over time, the range of situations that can be effectively coped with is vastly extended. Here too is a process reaching toward every greater generality.[14]

However, a moment's pause is warranted. I have argued for the critical importance of both generality and contingency in biological evolution, but these two terms are often invoked as opposing concepts. Is there in fact a conflict between the two? My answer to this question is no—or, more specifically, not if the term is understood in the sense in which I am using it here, that is, as restricted generality. For it is not generality itself that suggests a problem, but rather, the way in which it is commonly interpreted. More specifically, the problem lies in the dual implication, first, of universality (or *unrestricted generality*), and second, of necessity.[15]

In fact, the opposite of contingency is not generality at all; in the first instance, it is necessity that is contingency's antonym. Thus, the contradiction that is so frequently perceived (or invoked) between general and contingent depends on establishing two accessory relations (or conjunctions) with one's starting concept of general: In the first such conjunction, general is associated with natural law (i.e., as prevailing everywhere in nature), and in the second, lawful is associated with necessary (as, e.g., when we speak of the necessity with which natural laws must be obeyed). I claim that it is only with this dual set of conjunctions that the appearance of contradiction arises. Both the universality and the necessity of laws of nature are now hotly contested in the philosophy of science (especially in the philosophy of physics[16]), but they have behind them a long and influential history that makes them exceedingly hard to shake. Even so, if we can keep the distinction between restricted and unrestricted generality clearly in mind, we can readily recognize a longstanding proclivity in the philosophy of science for reading generality, first, as *unrestricted generality*, and second, as implying natural necessity (i.e., on reading laws of nature as binding). By contrast, the far less demanding notion of restricted generality entails no such contradiction with contingency for the simple reason that it does not lend itself to either of the precursor conjunctions with legal, physical, or logical necessity.

17.7   The spherical cow *redux*

Let us return to the spherical cow, and more specifically, to Rashevsky's dispute with the cell biologists at Cold Spring Harbor. Four years later, in the preface to the first

[14] One might say that the value at work in evolution is ontological rather than epistemological, as it is in the mathematical practices of ancient China, but this is a human distinction, and it is entirely unclear what it might mean from an evolutionary perspective.

[15] The distinction between universality and necessity was especially important to Popper who argued that only laws characterized by "natural or physical" necessity qualified as true laws of nature (see, e.g., Popper, 1968a, 1968b).

[16] See, for example, Cartwright (1980, 1983) and Lange (2000).

edition of his book, *Mathematical Biophysics,* Rashevsky (clearly having this encounter in mind) wrote:

> Objection has been raised against the use of the word "cell" to describe a highly oversimplified conceptual system, which, true enough, possesses some properties of a living cell but lacks a much larger number of other properties. To this our answer is as follows: In the early days of the kinetic theory of gases the whole theory was built on the concept of a molecule as an elastic billiard ball. The present development shows us a molecule as a system of tremendous complexity, having scarcely any analogy to an elastic rigid ball. Of course, even in the older days, the creators of the kinetic theory were perfectly well aware that the concept of a molecule as an elastic ball was far from the actual truth. Yet they used it up to a point with great success.[17]

Once again, Rashevsky links his defense of the choice to start with a highly idealized and over-simplified model with his conviction of the ubiquitous applicability of the methods of theoretical physics. Furthermore, he assumes that the reason for the success of the billiard ball model—a spherical cow par excellence—is that it captures the features of gas molecules that are of greatest importance, and perhaps he is right. But what does this assumption have to do with the search for generality, either of the restricted or unrestricted sort, and why does it seem not to apply in biology?

In a meditation on these questions, delivered at the Thousandth meeting of the Connecticut Academy of Arts and Sciences, Max Delbrück thought that, here too, the answers (especially to why there might be limits to the applicability of physics) should be sought in the facts of evolution. The physicist learning about the problems of biology, he wrote, may be "puzzled by the circumstance there are no 'absolute phenomena' in biology." The problem, as he went on to explain, is that:

> The animal or plant or micro-organism he is working with is but a link in an evolutionary chain of changing forms, none of which has any permanent validity.... The organism ... is not a particular expression of an ideal organism, but one thread in the infinite web of all living forms. The physicist has been reared in a different atmosphere. The materials and the phenomena he works with are the same here and now as they were at all times and as they are on the most distant stars.[18]

Delbrück also thought that the dependence of living forms on evolution similarly accounts for biologists' disinterest in both abstraction and idealization. He wrote:

> On the whole, the successful theories of biology always have been and are still today simple and concrete. Presumably, this is not accidental, but is bound up with the fact that every biological phenomenon is essentially an historical one, one unique situation in the infinite total complex of life.
>
> Such a situation from the outset diminishes the hope of understanding any one living thing by itself and the hope of discovering universal laws, the pride and ambition of

---

[17] Rashevsky (1938: ix–x).
[18] Delbrück (1949: 174).

physicists. The curiosity remains, though, to grasp more clearly how the same matter, which in physics and in chemistry displays orderly and reproducible and relatively simple properties, arranges itself in the most astounding fashions as soon as it is drawn into the orbit of the living organism. The closer one looks at these performances of matter in living organisms the more impressive the show becomes. The meanest living cell becomes a magic puzzle box full of elaborate and changing molecules, and far outstrips all chemical laboratories of man in the skill of organic synthesis performed with ease, expedition, and good judgment of balance. The complex accomplishment of any one living cell is part and parcel of the first-mentioned feature, that any one cell represents more an historical than a physical event. These complex things do not arise every day by spontaneous generation from the nonliving matter—if they did, they would really be reproducible and timeless phenomena, comparable to the crystallization of a solution, and would belong to the subject matter of physics proper. No, any living cell carries with it the experiences of a billion years of experimentation by its ancestors. You cannot expect to explain so wise an old bird in a few simple words.[19]

In the 60 years since Delbrück wrote these words, biological research has more than vindicated his vision of the living cell as "a magic puzzle box," powered by a variety of elaborate and intricate molecular processes that continue to resist the search for universal laws, and, at the same time, to validate and indeed exacerbate distrust of the spherical cell as a conceptual starting point. Such a starting point has undoubtedly proved productive for physical scientists in the past, but whether or not it continues to be productive as physical scientists turn their attention to more and more complex systems remains to be seen. But thus far, it has not proven very useful in biology, and the more successful efforts of mathematical and physical scientists turning their attention to biology today have incorporated this lesson. Like Delbrück, these scientists have learned that evolution seems to have built living systems not as departures or elaborations upon idealized simple forms, but as cumulative accretions of particularity and specificity. The end result may exhibit generality, but as a historical outcome rather than as a starting point—as the consequence of the fact that when a particular structure or mechanism has proven useful, it is used over and over, elaborated and built upon by subsequent accretions, until it is firmly embedded in the structure of the cell. Think, for example, of DNA. For the vast majority of living cells, DNA is the primary carrier of heredity. As such, it has enormous generality. But, as we have learned, even this most fundamental and most widely used molecule need not be a universal feature of the living cell.

If there is a moral to this story, it is not only that generality has different meanings in different epistemological cultures, but also that the kind of generality pursued by different scientific cultures is shaped by their cultural history, as well as by the objects on which they focus. If biologists have pursued a different kind of generality than have theoretical physicists, it is in large part because their subjects of study, and hence the problems they face, are in themselves of a different kind. Today, as a new discipline that seeks to meld the experience of physics with the problems of biology comes into being, a new generation of mathematical scientists is learning the lesson that Max Delbrück

---

[19] *Ibid.*, 174–5.

was so quick to learn when he turned his attention from physics to biology. At the same time, a new generation of biologists is learning both the power of mathematical models and the appeal of the traditional values of physics. Not surprisingly, the net result is simultaneously a breaking down of traditional cultural barriers and the emergence of a new epistemological culture. Perhaps this new culture will bring us yet further variations in our understanding of generality.

REFERENCES

Cartwright, N. (1980) 'Do the laws of physics state the facts', *Pacific Philosophical Quarterly* **61**: 75–84.

Cartwright, N. (1983) *How the Laws of Physics Lie*. Oxford: Oxford University Press.

Chemla, K. (2003) 'Generality above abstraction: the general expressed in terms of the paradigmatic in mathematics in ancient China', *Science in Context* **16**: 413–58.

Chemla, K. (2005) 'The interplay between proof and algorithm in 3rd century China: the operation as prescription of computation and the operation as argument', in *Visualization, Explanation and Reasoning Styles in Mathematics*, eds. P. Mancosu, K. F. Jorgensen, and S. A. Pedersen, Synthese Library Series, vol. 327. New York/Berlin: Springer, 123–45.

Delbrück, M. (1949) 'A physicist looks at biology', *Transactions of The Connecticut Academy of Arts and Sciences* **38**: 173–90.

Eddington, A. S. (1926) *The Internal Constitution of the Stars*. Cambridge: Cambridge University Press.

Fox Keller, E. (2002) *Making Sense of Life*. Cambridge, MA: Harvard University Press.

Fox Keller, E. (2007) '"Continents of Meaning": Donner [du] sens aux pratiques que font sens dans les communautés scientifiques.' Chaire Blaise Pascal Conference, 15 June 2007, http://www.rehseis.cnrs.fr/pdf/EFK.pdf.

Lange, M. (2000) *Natural Laws in Scientific Practice*. Oxford: Oxford University Press.

Popper, K. R. (1968a) *The Logic of Scientific Discovery*, 2nd (English) ed. New York: Harper & Row.

Popper, K. R. (1968b) 'A revised definition of natural necessity', *British Journal of Philosophical Science* **18**: 316–21.

Rashevsky, N. (1934) 'Physico-mathematical aspects of cellular multiplication and development', *Cold Spring Harbor Symposium for Quantitative Biology* **II**: 188–98.

Rashevsky, N. (1938) *Mathematical Biophysics: Physico-Mathematical Foundations of Biology*. Chicago: University of Chicago Press.

Thompson, D. W. (1942) *On Growth and Form*, 2nd ed. Cambridge: Cambridge University Press.

Turing, A. M. (1952) 'The chemical basis of morphogenesis', *Philosophical Transactions of the Royal Society of London* **B237**: 37–72.

Section III.4

Circulation between epistemological cultures

18

A process of generalization: Kummer's creation of ideal numbers

JACQUELINE BONIFACE

18.1   Introduction

An interesting example of generalization within the set of complex numbers is found in Kummer's creation of ideal factors. By the creation of such ideal factors, Kummer's goal was to generalize arithmetical properties of natural numbers by extending these properties to certain complex numbers. We will see that the idea of the creation of these ideal factors came from the desire to make complex numbers analogous to natural ones. One may thus consider that Kummer's results are an important stage in the trend of the refoundation of mathematics on an arithmetical basis. We will have to clarify Kummer's place in this trend, in particular by situating Kummer's investigations on complex numbers with respect to Gauss's investigations, and by comparing Kummer's theory of ideal factors with Dedekind's ideals theory. We will show that Kummer's method of generalization rests on the distinction he established between "permanent" and "accidental" properties of complex numbers, and that this distinction was premised on his conception of mathematics, which was essentially different from those of Gauss and Dedekind.

18.2   Arithmetic and complex numbers

In the course of the nineteenth century, arithmetic gained first place among mathematical disciplines to the detriment of geometry, which had been considered, since Euclid, as the model of mathematical rigor. Such a change has been attributed to the development of function theory in the eighteenth century and at the beginning of the nineteenth, as well as to the emergence of non-Euclidean geometries. Indeed, new functions, which were not representable geometrically, and new geometries deprived, each in a different way,

The Oxford Handbook of Generality in Mathematics and the Sciences. First Edition. Karine Chemla, Renaud Chorlay and David Rabouin. © Oxford University Press 2016. Publishing in 2016 by Oxford University Press.

the Euclidean geometry of its status of model for the other mathematical disciplines, and geometrical proofs no longer seemed sufficient to secure the validity of a result. Therefore, mathematicians turned toward arithmetic, which had become the most secure discipline, "the queen of mathematics," as Gauss said.

But what did "arithmetic" mean? Gauss had noted in his *Disquisitiones Arithmeticae*, published in 1801, that the term "arithmetic" meant the study, in particular, of integer numbers, sometimes of fractions, but never of irrationals or what were called "imaginaries," that is, complex numbers.[1] Irrationals were considered to be geometrical magnitudes and complex numbers were thought of as fictitious algebraic expressions or magnitudes—both belonging to analysis. Thus, neither of them had the status of numbers, and arithmetic was the science of the only magnitudes considered as numbers, that is, rational integers and, to a lesser extent, fractions. However, by the end of the nineteenth century, both irrationals and "imaginaries" had gained this status, and beside the traditional arithmetic a new science of numbers appeared. How did this change happen?

During the nineteenth century, there was a refoundation of mathematics on the basis of arithmetic, known later as the "arithmetization of mathematics," which aimed to define the concepts of analysis solely by means of arithmetic. This arithmetization of mathematics was often considered to go together with an extension of the domain of rational numbers through successive creations of irrational and complex numbers. Thus, Dedekind explained in his essay *Was sind und was sollen die Zahlen?*, published in 1888, "how subsequently the step-by-step extension of the number-concept—the creation of zero, of the negative, rational, irrational, and complex numbers—is to be carried out always by a reduction to earlier concepts, and indeed without introducing conceptions outside of the discipline."[2] From the narrow conception of arithmetic, conceived as the study of rational integers, to the "arithmetized" mathematics and to a wider concept of number, there were many steps. One of the most important of these steps was the extension of the domain of arithmetic to "Gaussian integers."

18.2.1 The extension of arithmetic to Gaussian integers

Gauss, in a well-known letter to Bessel written in 1811, stated as his "basic proposition" (*Grundsatz*) that: "within the realm of magnitudes, the imaginaries $a + b\sqrt{-1} = a + bi$ must enjoy the same rights as the real [magnitudes]."

---

[1] We should note that Gauss' conception of arithmetic as an autonomous discipline was not shared by all mathematicians at the beginning of the nineteenth century. Some of them, such as Legendre, considered arithmetic as a part of analysis. See Goldstein and Schappacher (2007: 26–7).

[2] "In welcher Art später die Schrittweise Erweiterung des Zahlbegriffes, die Schöpfung der Null, der negativen, gebrochenen, irrationalen und komplexen Zahlen stets durch Zurückführung auf die früheren Begriffe herzustellen ist, und zwar ohne jede Einmischung fremdartiger Vorstellungen ..." (Dedekind, 1930–2, vol. 3: 338)

Before Dedekind, Gauss had formulated the same idea in nearly the same words; he wrote: "Our general arithmetic, whose scope so greatly outstrips the geometry of the ancients, is entirely the creation of modern time. It started from the concept of absolute integers, and has gradually extended its territory; the fractions have been added to the integers, the irrationals to the rationals, the negatives to the positives, and the imaginaries to the reals" (Gauss, 1831, in Ewald, 1996, vol. 1: 311).

And he added:

> The matter is not of practical usefulness, but analysis is for me an independent science, which by rejecting these fictitious magnitudes would lose enormously in beauty and roundedness, and at every instant would be forced to add quite cumbersome restrictions to truths that otherwise would be generally valid.[3]

The "beauty and roundedness" of analysis, which dealt with complex "magnitudes,"[4] were, according to Gauss, the reasons for accepting some "imaginaries" as having the same status as rational numbers. And the beauty and the roundedness of this science would be given by the conformity of its objects to general arithmetical "truths." Gauss expressed clearly his ideas about extending the domain of arithmetic to complex numbers at the time of his investigations on biquadratic residues. He wrote in a communication to the Academy of Sciences in Göttingen in 1825:

> [...] for the true foundation of the theory of biquadratic residues, one must widen the field of higher arithmetic, which had otherwise been extended only to the real integers, into the imaginary ones, and must grant to the latter exactly the same rights as the former. As soon as one has seen this once, that theory appears in an entirely new light, and its results acquire a most astonishing simplicity.[5]

To explain his conception of an "extended arithmetic" (*erweiterten Arithmetik*, p. 172), Gauss considered then what are now called "Gaussian integers," that is the complex numbers of the form $a + ib$ (with $i^2 = -1$ and rational integers a and b). He showed, in particular, that these integer complex numbers admit, just like the rational integers, a decomposition into prime complex numbers. And he added that these prime numbers play a leading role in the enlarged field of arithmetic, just as prime numbers do in the higher arithmetic of rational numbers.

18.2.2 Kummer's conception of complex numbers

Kummer's investigations on complex numbers were another step toward the development of number theory. However, we will show that this development was not conceived of by

---

[3] "... in dem Reiche der Grössen die imaginären $a + b\sqrt{-1} = a + bi$ als gleiche Rechte mit den reellen geniessend müsse. Es ist hier nicht von praktischem Nutzen die Rede, sondern die Analyse ist mir eine selbständige Wissenschaft, die durch Zurücksetzung jener fingierten Grössen ausserordentlich an Schönheit und Rundung verlieren und alle Augenblick Wahrheiten, die sonst allgemein gelten, höchst lästige Beschränkungen beizufügen genöthigt sein würde." Gauss to Bessel, December 1811, in Gauss and Bessel (1880: 156).

[4] In 1811, Gauss named "magnitudes" reals and imaginaries; in 1825, he would name them "numbers." See, for example, Gauss (1828).

[5] "... für die wahre Begründung der Theorie der biquadratischen Reste das Feld der höhern Arithmetik, welches man sonst nur auf die reellen ganzen Zahlen ausdehnte, auch über die imaginären erstreckt werden, und diesen das völlig gleiche Bürgerrecht mit jenen eingeräumt werden muss. Sobald man dies einmal eingesehen hat, erscheint jene Theorie in einem ganz neuen Lichte, und ihre Resultate gewinnen eine höchst überraschende Einfachheit" (Gauss, 1828, in Gauss, 1863/1876, vol. 2: 171). This lecture to the Academy of Sciences was published in 1828 in the *Göttingische gelehrte Anzeigen* and was followed by another one on the same subject matter, which was given in 1831 and published in 1832. The latter is translated into English in Ewald (1996, vol. 1: 306–13).

Kummer as it had been for Gauss. But let us first make clear what arithmetic and complex numbers mean for Kummer.

According to Kummer, the theory of complex numbers[6] is part of a branch of higher arithmetic, as he said in his memoir of 1851:

> The theory of complex numbers is fundamentally the same as the theory of these forms [i.e., of the homogeneous forms, which are decomposable in linear factors], and in this respect, it is part of one of the most beautiful branches of higher arithmetic.[7]

The image of a branch evokes the development of number theory from a common trunk—the classical arithmetic—by ramification. It is quite different from that of a gradually extending territory, as used by Gauss. In a development by ramification, each branch keeps an autonomy relative to the trunk and to the other branches. By contrast, in development by extension, the annexed territories and the initial one have to become a single homogeneous territory. We will see later the significance of these different views.

Although Kummer noticed the similarity between the theory of the complex numbers he introduced and the theory of decomposable forms,[8] he preferred to consider the former as the theory of the decomposition of numbers into irrational factors,[9] in the following sense. As we have seen above, a complex number was for Kummer "an entire function with integer coefficients of irrational roots of one or many algebraic equations, whose coefficients are also integer numbers." This definition is quite different from that of Gauss. For the latter, a complex number is an expression of the form $a + ib$, with $i^2 = -1$ and a and b real numbers, that is, a complex number is obtained from real ones, by adding an imaginary component. For Kummer, a complex number is obtained from the rational ones by adding irrational components of a specific kind. In other words, for Kummer, the basic domain is that of rationals, whereas for Gauss it is that of reals.[10] We will see

---

[6] A "complex number" was for Kummer "an entire [i.e., polynomial] function with integer coefficients of irrational roots of one or several algebraic equations, whose coefficients are also integer numbers" (Kummer, 1851: 377, trans. J.B.). See below for the difference between Kummer and Gauss on this point.

[7] "La théorie des nombres complexes revient, au fond, à la théorie des formes, et, à cet égard, elle fait partie d'une des plus belles branches de l'Arithmétique supérieure" (Kummer, 1851: 377, in Kummer, 1975: 363).

[8] Regarding this analogy between complex numbers and decomposable homogeneous forms, Kummer referred to Kronecker's investigations. "Ich kann ... auch auf eine Arbeit von *Herrn Kronecker* verweisen, welche näschtens erscheinen wird, in welcher die Theorie der allgemeinsten complexen Zahlen, in ihrer Verbindung mit der Theorie der zerlegbaren Formen aller Grade, vollständig und in grossartiger Einfachheit entwickelt wird." Translation J.B.: "I may also refer to a work by *Herr Kronecker* which will appear soon, in which the theory of the most general complex numbers, in its connection with the theory of decomposable forms of all degrees, is developed completely and in a great simplicity" (Kummer, 1975, vol.1: 737).

[9] See Kummer (1851: 378, in Kummer, 1975: 364) : "D'un autre côté, la théorie des nombres complexes peut être considérée comme la théorie de la décomposition des nombres en facteurs irrationnels, et c'est sous ce point de vue qu'elle a un grand intérêt, aussi bien en elle-même que pour les applications nombreuses et importantes qu'on en a faites dans plusieurs questions relatives à l'Arithmétique et à l'Algèbre supérieures." Translation J.B.: "On another side, the theory of complex numbers can be considered as the theory of the decomposition of numbers into irrational factors, and it is from this angle that it is of great interest, as well in itself as for the many important applications that have been done in several questions concerning higher arithmetic and higher algebra."

[10] In Kronecker's work, one will also find the same derivation of complex numbers from rational ones as in Kummer's.

that Dedekind followed Gauss's view and Kronecker followed Kummer's. Later we will try to explain these differences.

After giving his definition of complex numbers, Kummer added a remark which would guide his later investigations. He noted that the product of all complex numbers obtained by replacing the roots in them by conjugated roots "will always be delivered from all irrationality,"[11] that is, this product is a rational integer. Thus, it is as factors of rational integers that Kummer considered complex numbers; consequently, decomposition into prime factors would be the point of his investigations.

18.2.3   The fundamental law of arithmetic

In order to clarify Kummer's investigations, we should first review some arithmetical notions and explain the failure of the fundamental law of arithmetic for certain complex numbers. This fundamental law asserts that all rational integers admit a decomposition into irreducible factors and that this decomposition is unique. Thus 6 decomposes into 2 and 3 (since $6 = 2 \times 3$), and this decomposition is unique (except for the order of factors). The integers 2 and 3 are irreducible, because one cannot decompose them into rational integers, and they are also prime numbers, because of the uniqueness of the decomposition. The fundamental law still holds for Gauss's integers, but fails for some other complex numbers. For example, when we consider the complex numbers of the form $a + b\sqrt{-5}$ (i.e., the numbers of the ring $\mathbf{Z}[\sqrt{-5}]$), 6 admits two different decompositions into irreducible factors : $6 = 2 \times 3$ and $6 = (1 + \sqrt{-5})(1 - \sqrt{-5})$. In other words, for these complex numbers (i.e., in the ring $\mathbf{Z}[\sqrt{-5}]$), 2, 3, $1 - \sqrt{-5}$, $1 + \sqrt{-5}$, are irreducible factors, but not prime factors; the ring $\mathbf{Z}[\sqrt{-5}]$ is said to be not *factorial* (the fundamental law does not hold for all the numbers of this ring).

18.3   Kummer's ideal factors

Kummer studied complex integers that were a little more general than the Gaussian ones: expressions of the form $a_0 + a_1r + a_2r^2 + \ldots + a_nr^n$, with r a root of the unit, and a_i rational integers, called "cyclotomic integer numbers" today. For Kummer, as for Gauss, classical arithmetic was the model for building this new theory, and Kummer's aim was to prove that these complex integers had the same properties as rational ones. But, Kummer, in contrast to Gauss, did not conceive these complex integers as extending the domain of rational ones. For Kummer, arithmetic was a well delimited domain, and complex numbers belonged to another one. His purpose was not to extend the domain of rational numbers (unlike Gauss and most of mathematicians of the second half of the century), but to bring to light the analogy between complex numbers and rational ones. More precisely Kummer wanted to show that the fundamental arithmetical property, which states that every rational integer can be decomposed in a unique way into a product of irreducible factors also held for the complex numbers he was studying.

---

[11] "sera toujours délivré de toute irrationalité" (Kummer, 1851: 377, in Kummer, 1975: 363).

18.3.1 Kummer's context for creating ideal factors

A lot of mathematicians of the nineteenth century, among them the French mathematicians Gabriel Lamé and Augustin Cauchy, thought that the fundamental law of arithmetic was still valid for complex numbers composed of integer numbers and roots of unity. This law had been the basis of Lamé's and Cauchy's researches into a general proof of Fermat's conjecture, namely, the impossibility of solving the equation $x^n + y^n = z^n$, for an integer n greater than 2, and three integers x, y, and z not equal to zero. These researches had been the subject of a competition within the Paris Academy of Sciences which had occupied both mathematicians for several months in 1847. The editor of the *Journal de Mathématiques pures et appliquées*, Liouville, who participated in the debates of the Academy, brought an end to the competition by publishing, in the *Comptes rendus* of the Academy, a letter by Kummer which asserted that, for some complex numbers, the factorization into irreducible factors was not unique. This assertion had been proved in a paper written three years earlier, and a copy of this paper was published with the letter.[12]

18.3.2 Kummer's generalization of the fundamental law of arithmetic by the creation of ideal numbers

Kummer's results made it irrefutable that some complex numbers[13] did not have the fundamental property of rational integers. However, like most of his contemporaries, Kummer considered his results unsatisfying, even if irrefutable. Indeed, for all of them, the development of complex number theory had to be carried out according to the model of rational integer number theory. Thus, Kummer considered the failure of the fundamental property as an anomaly, and consequently, all of Kummer's investigations for a number of years became devoted to removing this anomaly and were concentrated on giving this property to complex numbers composed of integer numbers and roots of unity. This problem appears clearly in the letter Kummer wrote to Liouville:

> With regard to the elementary proposition, for these complex numbers, that *a composed complex number can be decomposed into prime factors but in only one way*, [...] I can assure you that this proposition *does not hold in general*, as long as we deal with complex numbers that are in the form:
>
> $$a_0 + a_1 r + a_2 r + \cdots + a_{n-1} r^{n-1}.$$
>
> However, the proposition can be saved by introducing a new kind of complex numbers that I have called *an ideal complex number*.[14]

---

[12] The paper was published under the title "De numeris complexis, qui radicibus unitatis et numeris integris realibus constant," in the *Journal de mathématiques pures et appliquées*, XII, in 1847. See Kummer (1847a).

[13] Kummer dealt with complex numbers composed of integer numbers and roots of unity.

[14] "Quant à la proposition élémentaire pour ces nombres complexes, *qu'un nombre complexe composé ne peut être décomposé en facteurs premiers que d'une seule manière,* ... je puis vous assurer *qu'elle n'a pas lieu généralement*

In his paper written in 1844 and published by Liouville, Kummer introduced his work as a continuation of Gauss's theory of complex numbers, and above all of Jacobi's research into the decomposition of prime numbers of the form $m\lambda + 1$ into cyclotomic integer factors. He noted that Jacobi had looked at the cases of factors comprising fifth, eighth, and twelfth roots of unity. Kummer, for his part, undertook to prove that every prime number of the form $m\lambda + 1$, where λ is an odd prime number, can be decomposed into $\lambda - 1$ prime factors belonging to what we call today the ring $\mathbf{Z}[\alpha]$, where α is a λth complex root of unity. With this aim in mind, he used the notion of the norm of a complex number $f(\alpha)$ of the cyclotomic ring, which he took from Lejeune–Dirichlet—as above, f is "an entire function with integer coefficients of irrational roots," here a root of the equation $x^\lambda = 1$:

$$Nf(\alpha) = f(\alpha)f(\alpha^2)f(\alpha^3) \cdots f(\alpha^{\lambda-1}).$$

For reasons of symmetry, the norm of such a complex number is a rational integer. Thus, every complex number composed of integer coefficients and roots of unity can be considered as a complex factor of a rational integer number, its norm. Conversely, if one can show that a rational integer number is the norm of a complex number, one can immediately obtain a decomposition of the rational integer number into complex factors.

Kummer also showed that for $\lambda = 5, 7, 11, 13, 17$, and 19, every prime number of the form $m\lambda + 1$ and smaller than a thousand is the norm of a complex number of the ring $\mathbf{Z}[\alpha]$, and thus can be decomposed into $\lambda - 1$ prime cyclotomic integer factors for the given value of λ. However, for $\lambda = 23$, Kummer observed that only three of the eight prime numbers of the form $m\lambda + 1$ and smaller than a thousand have this property. And he also showed that the five other prime numbers are not the norm of any cyclotomic integer, and thus that they cannot be decomposed into 22 conjugated factors.[15]

Kummer showed that this impossibility of a decomposition of prime numbers of the form $m\lambda + 1$ into $\lambda - 1$ complex factors had as a consequence that their decomposition into irreducible factors is not unique. In other words, the possibility of a decomposition for a given λ of all prime numbers of the form $m\lambda + 1$ into $\lambda - 1$ prime complex factors is a necessary condition of the factoriality of the ring.

tant qu'il s'agit des nombres complexes de la forme $a_0 + a_1 r + a_2 r^2 + \ldots + a_{n-1} r^{n-1}$, mais qu'on peut la sauver en introduisant un nouveau genre de nombres complexes que j'ai appelé *nombre complexe idéal*." Kummer to M. Liouville (Kummer, 1847b: 136, in Kummer, 1975: 298).

[15] His proof made use of the Gaussian theory of quadratic forms, which allowed him to establish the fact that, for $\lambda = 23$, the norm of a cyclotomic integer can be written in the form $\dfrac{a^2 + 23b^2}{4}$, where a and b are rational integers, and that five of the prime numbers of the form $m\lambda + 1$ cannot be written in this quadratic form and thus cannot be decomposed into a product of $\lambda - 1$ complex factors. For example, one cannot find rational integers a and b such that :

$$4 \times 47 = 188 = a^2 + 23b^2.$$

Kummer's idea, which he would develop in two later papers,[16] was to artificially generalize, for all values of λ, the decomposition of prime numbers of the form $m\lambda + 1$ into $\lambda - 1$ prime factors by creating *ideal* factors in the case where this decomposition does not exist. For $\lambda = 23$, for example, he had to obtain 22 ideal factors of 47. As he wrote:

> If p is a prime number of the form $m\lambda + 1$, then it can, in many cases, be represented as a product of $\lambda - 1$ complex factors $p = f(\alpha)f(\alpha^2)...f(\alpha^{\lambda-1})$. Where, however, a decomposition into existing complex factors is not possible, one has to introduce ideal prime factors in order to obtain it.[17]

To clarify Kummer's idea, we can take again the example of the ring $\mathbf{Z}[\sqrt{-5}]$. We saw that, in this ring, 6 admits two different decompositions into irreducible factors: $6 = 2 \times 3 = (1 + \sqrt{-5})(1 - \sqrt{-5})$. In order to restore the uniqueness of the decomposition, one can imagine that the irreducible factors 2, 3, $1 + \sqrt{-5}$, and $1 - \sqrt{-5}$ contain themselves ideal factors, and that one can write them as products of these ideal factors. The case of the number 2 is rather particular, because it behaves as if it was a square number.[18] Thus, we can write it in the form $2 = \alpha^2$. The other rational integers can be written in the form of a product of two different ideal factors, for example, $3 = \beta\gamma$, $1 + \sqrt{-5} = \alpha\beta$, $1 - \sqrt{-5} = \alpha\gamma$. Thus, we have $6 = \alpha^2\beta\gamma$ and the decomposition of 6 into the ideal factors α, β, and γ is unique.[19]

In order to determine the ideal factors which are contained in his complex numbers, Kummer used congruence calculations. This entails an analysis of the situation in the case where p can be decomposed into existing prime factors. Without going too deeply into the technical details,[20] we can say that the importance of this congruence condition is that it can be applied to ideal factors. Kummer used it as a definition, not of the ideal factors — he remarked that it would not be sufficient for such a definition — but of the divisibility of a cyclotomic complex number by an ideal factor of p.

---

[16] The first of the two later papers on the ideal factors was published in German in 1847, in the *Journal für die reine und angewandte Mathematik*, under the title "Zur Theorie der complexen Zahlen" (Kummer, 1847a). The other paper was published in French in 1851 under the title "Mémoire sur la théorie des nombres complexes composés de racines de l'unité et de nombres entiers," in the *Journal de mathématiques pures et appliquées* (Kummer, 1851).

[17] "Ist p eine Primzahl von der Form $m\lambda + 1$, so lässt sie sich in vielen Fällen als Product von folgenden $\lambda - 1$ complexen Factoren darstellen: $p = f(\alpha)f(\alpha^2) \ldots f(\alpha^{\lambda-1})$; wo aber eine Zerlegung in wirkliche complexe Primfactoren nicht möglich ist: dann sollen die idealen Primfactoren eintreten, um dieselbe zu leisten" (Kummer, 1847a, in Kummer, 1975: 204).

[18] Indeed, if we consider $w = x + y\sqrt{-5}$ and $w' = x' + y'\sqrt{-5}$, we will have $w^2 \equiv N(w) \pmod{2}$ and $w'^2 \equiv N(w')$ $\pmod{2}$. It follows that $w^2w'^2 \equiv N(w)N(w') \pmod{2}$. Now, in order to divide the product $w^2w'^2$ and thus also the product of the two rational integers $N(w)$ and $N(w')$, the number 2 has to divide one at least of the two norms, and thus also one of the two square numbers w^2, w'^2. If, in addition, we choose for x and y two odd integer numbers, we obtain a number $x + y\sqrt{-5}$ which is not divisible by 2 and whose square is divisible by 2. By analogy with the rational integers, we can consider that the number 2 behaves as if it was a square number.

[19] This example is given by Dedekind in Dedekind (1877: 278–81).

[20] For a development of that point and a modern interpretation of Kummer's method, see the Appendix 1.

18.4   Kummer's use of analogy

We have seen that Kummer's aim, by creating ideal factors of complex numbers, was to eliminate from the theory of complex numbers anomalies with regard to rational integers. And the use of analogy gave him both a motivation and a justification for his creation; it testifies to Kummer's conception of complex number theory as being separate from arithmetic. Indeed, if Kummer had conceived the domain of complex numbers as extending that of rational integers, he would not have used analogies, which require two different domains. Gauss, for example, unlike Kummer, did not speak of analogy between rationals and complex numbers, since he conceived rationals, and even reals, as particular cases of complex numbers.[21]

In fact, Kummer made use of several analogies, primarily with arithmetic, but also with other mathematical domains, as well as with chemistry. Each of these analogies had a particular role, as will be shown below.

18.4.1   The analogy with arithmetic

Arithmetic, as we saw, was the model of Kummer's theory. The following text clearly shows that Kummer's principal reason for creating ideal factors was indeed to preserve an analogy between complex numbers and rational ones. The text reveals that the analogy with arithmetic was more important for Kummer than the complex numbers themselves, since Kummer seemed to find it preferable to change these numbers rather than to lose the analogy with rational numbers.

> It is largely to be deplored that this property of real numbers [i.e., the rational integers] of being decomposable into prime factors, always in the same way for a given number, does not belong to complex numbers. If it were the case, all of the theory, which can only be elaborated with great difficulty, could easily be brought to its conclusion. For this reason, the complex numbers we consider seem imperfect, and this may generate doubt whether one should prefer to these numbers other complex numbers that could be found, and should not these other numbers be looked for, which would preserve the *analogy* [my emphasis] with real integer numbers for this fundamental property.[22]

It is interesting to remark that Kummer called "real integer numbers" what we call today rationals (or even rational integers). We saw that, unlike Kummer, Gauss distinguished already between rationals and reals, and that he contrasted complex numbers (in his terms, imaginaries) with reals, whereas Kummer contrasted them with rationals.

---

[21] See, for example, Gauss (1831, § 7, in Ewald, 1969: 309): "So the complex quantities [$a + ib$] are not opposed to the real, but contain them as a special case where $b = 0$."

[22] "Maxime dolendum videtur, quod haec numerorum realium virtus, ut in factores primos dissolvi possint, qui pro eodem numero semper iidem sint, non eadem est numerorum complexorum, quae si esset, tota haec doctrina, quae magnis adhuc difficultatibus laborat, facile absolvi et ad finem perduci posset. Eam ipsam ob causam numeri complexi, quos hic tractamus, imperfecti esse videntur, et dubium inde oriri posset, utrum hi numeri complexis ceteris qui fingi possint praeferendi, an alii quaerendi essent, qui in hac re fundamentali analogiam cum numeris integris realibus servarent" (Kummer, 1847a: 202, in Kummer, 1975: 182).

This remark underlines the fact that, for Kummer, the basis of number theory was the arithmetic of rational integer numbers. Rational integer numbers were for him the only true (or existing) numbers, and we will see later that he viewed these numbers as natural objects. For Gauss, there are no "true" numbers; a number is *"just* a product of our mind."[23] With these words, Gauss will appear as the pioneer of a major conceptual trend in mathematics at the end of the nineteenth century,[24] whereas Kummer, as we will see, marked the beginning of another less dominant trend in this period.

Thus, for Kummer, as for Gauss, the arithmetic of rational numbers was the basis of number theory, but for the former this basic discipline was a model to imitate—and that by means of analogy—whereas for the latter it had been a domain to be extended.

18.4.2   The analogy with geometry

Kummer also looked into algebra and geometry, which used methods analogous to his own, as shown in the following text.

> Algebra, arithmetic and geometry offer a lot of *analogies* [my underlining] to our theory. One decomposes, for example, rational and entire functions of only one variable into linear factors, even if these isolated factors only exist in some particular cases; it is to this aim that imaginary quantities have been created. In geometry, one speaks of a line going through the points of intersection of two circles, even if the points of intersection do not exist.... Lastly, the idea of considering ideal factors of complex numbers is, in fact, the same as the idea which has created complex numbers themselves. Indeed, one knows that, by observing that, in the search for the laws of reciprocity between biquadratic residues, the prime numbers of the form $4n+1$ behaved like composed numbers, M. Gauss decomposed them into imaginary factors of the form $a + b\sqrt{-1}$, and that he consequently established the foundations of the general theory of complex numbers.[25]

It clearly appears that the analogies with algebra, geometry, and again arithmetic, allowed Kummer to justify his creation of ideal factors. However, one notices that the analogy with geometry plays a different role than that with the other disciplines. Indeed, ideal factors had been created in algebra and arithmetic in the same context of the decomposition into prime factors as in Kummer's theory. In each of these cases, it was the same idea which was aimed at by the mathematicians: the decomposition into prime factors. On the

---

[23] "... die Zahl *bloss* unseres Geistes Product ist ..." (Gauss, 1880: 497).

[24] For more details about this point, see Boniface (2007, in Goldstein et al., 2007: 321–30).

[25] "L'Algèbre, l'Arithmétique et la Géométrie offrent des analogies nombreuses à notre théorie. On décompose, par exemple, les fonctions rationnelles et entières d'une seule variable en facteurs linéaires, quoique ces facteurs isolés n'existent qu'en des cas particuliers ; c'est pour ce but qu'on a créé les quantités imaginaires. En Géométrie, on parle d'une droite passant par les points d'intersection de deux cercles, quand même les points d'intersection n'existent pas. [...] Enfin, l'idée de considérer des facteurs idéaux des nombres complexes est, au fond, la même que celle qui a procréé les nombres complexes eux-mêmes. En effet, on sait que M. Gauss, en observant que, dans la recherche des lois de réciprocité entre les résidus biquadratiques, les nombres premiers de la forme $4n + 1$ se comportaient comme nombres composés, les a décomposés en facteurs imaginaires de la forme $a + b\sqrt{-1}$, et qu'il a jeté par là les fondements de la théorie générale des nombres complexes" (Kummer, 1851: 430, in Kummer, 1975: 416).

contrary, the geometrical example of ideal elements no longer had any relation with the fundamental property of arithmetic. It illustrates more clearly than the other examples the status of ideal entities, according to Kummer. These entities were seen by him, like ideal geometrical points, as ideal[26] entities, as opposed to real ones, introduced in order to simplify the theory. We will be in a position later to judge the importance of the analogy with geometry with regard to Kummer's conception of mathematical objects.

18.4.3 The analogy with chemistry

Kummer concluded the paragraph quoted above with a remark which underlines "the analogy of the theory of the composition of ideal complex numbers with the fundamental principles of chemistry." He wrote:

> The composition of complex numbers can be seen as *analogous* [my underlining] to chemical combination; the prime factors correspond to elements, or rather they are equivalent to these elements. Ideal complex numbers can be compared to hypothetical radicals which do not exist by themselves, but only in combinations; the fluorine, in particular, as an element which we cannot represent in isolation, can be compared to an ideal prime factor.[27]

This analogy with chemistry, again justifying Kummer's ideal factors, showed, even more clearly than the geometrical example, the status Kummer gave to mathematical objects. Indeed, Kummer conceived mathematics as a natural science and mathematical objects as natural ones. Numbers have to be seen, he said, as chemical compounds and ideal factors as chemical elements. This last consideration also explains Kummer's use of the properties of complex numbers, as we will see in the next section. Thus, one can say that Kummer used geometry and chemistry as ontological (or intuitive) models—they allowed him to express the ontological status of mathematical objects—whereas arithmetic and algebra are rather theoretical models—they allowed him to discover hidden properties of these objects.

18.5 Permanent properties/accidental properties of complex numbers

Kummer's conception of numbers as natural objects led him to turn his attention to their properties, which he divided into permanent properties and accidental ones. According to him, the properties of complex numbers which did not correspond to those of rational integers were to be considered as accidental properties, or as irregularities, and, hence,

---

[26] We will clarify below the sense of this term for Kummer.

[27] "La composition des nombres complexes peut être envisagée comme l'analogue de la combinaison chimique; les facteurs premiers correspondent aux éléments, ou plutôt aux équivalents de ces éléments. Les nombres complexes idéaux sont comparables aux radicaux hypothétiques qui n'existent pas par eux-mêmes, mais seulement dans les combinaisons; le fluor, en particulier, comme élément qu'on ne sait pas représenter isolément, peut être comparé à un facteur premier idéal" (Kummer, 1851: 433, in Kummer, 1975: 447).

as imperfections which hid permanent properties. Thus, mathematicians should focus on the permanent properties and eliminate the accidental ones. Such was the approach followed by Kummer in creating ideal numbers. Furthermore, the analogy with chemistry allowed him to justify his introduction of ideal factors by remarking that these factors "are making visible, so to speak, the internal constitution of numbers, so that their essential properties would be brought to light."[28]

On this point again, Kummer did not follow Gauss. Indeed, Gauss had considered, in opposition to Kummer, that relations between objects are more important than the objects themselves and, consequently, he did not give particular attention to the properties of mathematical objects. It would seem more relevant to compare Kummer with geometers working on projective geometry, who had analogous views on the properties of geometrical objects.[29] Thus, Kummer illustrated his distinction between permanent properties and accidental ones by having recourse to the analogy stated within geometry. He called the property of tangents drawn from an arbitrary point of the line going through the (*extant or ideal*) points of intersection of two circles to both circles to be equal to each other a "*permanent property*," and he stated that this property is analogous to a "permanent" property for complex numbers to be decomposed into (*extant or ideal*) prime factors. Then, he called the property of this line of going through *extant* points of intersection of the two circles an "*accidental property*," and he stated that this property is analogous to the "accidental" property of a complex number to have an *extant* prime factor.[30]

Thus, Kummer's interpretation of arithmetical phenomena in terms of properties of numbers brings him closer to some geometers than to an arithmetician (or algebraist) like Gauss. Gauss's focus on the relations between objects rather than on the objects themselves showed a new way of conceiving mathematics, which was to develop at the end of the nineteenth century. The new trend initiated by Gauss, which attached less importance to the nature of mathematical objects, would lead mathematicians to speak about laws of a domain rather than properties of objects. Dedekind's commentary on Kummer's theory testifies to this trend:

> This geometer [Dedekind wrote about Kummer] succeeded in reducing all apparent irregularities to rigorous laws. By considering numbers which were non decomposable and yet devoid of the character of genuine prime numbers, as products of *ideal* prime factors appearing and revealing their effects only when combined together, and not isolated, he obtained the surprising result that the laws of divisibility in the domains he studied coincided now completely with the ones which govern the domain of rational integer numbers.[31]

---

[28] "... les facteurs idéaux rendent visible, pour ainsi dire, la constitution intérieure des nombres, en sorte que leurs propriétés essentielles soient mises dans leur jour, ..." (Kummer, 1851, in Kummer, 1975: 430).

[29] See Karine Chemla, "The value of generality in Michel Chasles's historiography of geometry," in Chapter 2.

[30] Kummer (1851: 430, in Kummer 1975: 416).

[31] "Ce géomètre est parvenu à ramener toutes les irrégularités apparentes à des lois rigoureuses, et en considérant les nombres indécomposables, mais dépourvus du caractère de véritables nombres premiers, comme des produits de facteurs premiers *idéaux*, qui n'apparaissent et ne manifestent leur effet que combinés ensemble, et non pas isolés, il a obtenu ce résultat surprenant, que les lois de la divisibilité dans les domaines étudiés

Dedekind's text showed clearly that it was no longer a question of properties of the objects, but of laws in particular domains. Thus, the decomposition into prime factors, seen by Kummer as a property of numbers, was alternatively seen by Dedekind only as a law of arithmetic. Both Kummer and Dedekind considered as irregularities those complex numbers that were not decomposable in a single way into a product of irreducible factors. For Kummer, therefore, these irregularities were inherent in numbers themselves, whereas for Dedekind they were inherent in the particular domain. The consequence of this, as we will show, was that Kummer's creations were numbers, whereas Dedekind's were sets of numbers or concepts.

18.6 Ideal numbers after Kummer: Dedekind, Kronecker, and Hilbert

Kummer's theory of complex numbers was generalized by Dedekind and Kronecker in two different ways. Dedekind chose a more conceptual way, whereas Kronecker followed Kummer more closely. However, both Kronecker and Dedekind tried to avoid the use of ideal factors. Continuing from Dedekind's and Kronecker's works but unlike them, Hilbert was also interested in Kummer's theory, in particular, in his creation of ideal numbers, which he took as an example of his own method of "ideal elements."

18.6.1 Dedekind's ideals theory

Although Dedekind did not share Kummer's conception of mathematics as a natural science, the theory of ideal factors allowed him to establish his own creations. It was also the opinion of J. Cavaillès, who, in his 1938 "Remarques sur la formation de la théorie abstraite des ensembles," saw in Kummer's theory the model of Dedekind's creations.

> The model — and doubtless the first idea — of similar creations lies in Kummer's ideals, whose introduction "conducted with enough rigor, has to be considered as entirely legitimate."[32]

Dedekind himself explained how Kummer's theory of ideal factors had been the "germ of the ideals theory"[33] and he chose the term "ideal" to pay tribute to his instigator. However, Dedekind thought that Kummer's theory was not completely satisfying. Indeed,

---

par lui coïncident maintenant complètement avec celles qui régissent le domaine des nombres entiers rationnels" (Dedekind, 1877, in Dedekind 1930–2, vol. 3: 267).

[32] "Le modèle — et sans doute l'idée première — de semblables créations se trouve dans les idéaux de KUMMER dont l'introduction, 'conduite avec assez de rigueur, doit être considérée comme absolument légitime'" (Cavaillès, 1962: 38). The Dedekind quote is from a letter to Weber, dated 24 January 1888 (Dedekind, 1888: 490).

[33] This is the title of section 2 of Dedekind's theory of algebraic numbers, in which Dedekind explained Kummer's theory of ideal factors. See Dedekind (1930–2: 224).

Kummer's theory could not easily be generalized, and in particular, Kummer's method of calculation did not fit in with Dedekind's conception of mathematics, as he explained:

> Such a theory, based on computation, would not offer the highest degree of perfection. It is preferable, as in modern function theory, to derive the proofs, no longer from computation, but immediately from characteristic fundamental concepts, and to build the theory in such a way that it will be able to predict the results of the computation.[34]

Moreover, Kummer's creation of ideal numbers lacked, according to Dedekind, a commonly shared definition of these numbers—Dedekind criticized Kummer for defining only divisibility by ideal numbers, not ideal numbers themselves.[35] Only such a definition, which could generate all the ideal numbers of a given domain, could testify to a true creation, that is, could allow one to consider these fictitious entities as objects of the theory by integrating them in the considered realm.[36] Thus, it appears clear that it was existing objects, and not ideal or fictitious entities, that Dedekind aimed for.

Like Kummer, Dedekind first considered the divisibility of a number in the domain he was studying, in this case, the ring $\mathbf{Z}[\sqrt{-5}]$. A complex number $x + y\sqrt{-5}$ of this domain will be said to be divisible by the ideal factor α[37] if and only if its square, and consequently its norm, is divisible by α^2, that is, by 2, if we take up again the example developed in Section 18.2.2. It follows that $x + y\sqrt{-5}$ will be divisible by α if and only if $x \equiv y \pmod 2$. But Dedekind wanted to go further and give a characterization of all the numbers in the domain in question which are divisible by α. With this in view, he noted as a consequence of the previous congruence that $x = y + 2z$, where z is any rational integer. All the numbers of the ring which are divisible by α will be of the form $(y + 2z) + y\sqrt{-5}$, that is, $2z + (1 + \sqrt{-5})y$. Dedekind's absolutely

---

[34] "... une telle théorie, fondée sur le calcul, n'offrirait pas encore, ce me semble, le plus haut degré de perfection ; il est préférable, comme dans la théorie moderne des fonctions, de chercher à tirer les résultats, non plus du calcul, mais immédiatement des concepts fondamentaux caractéristiques, et d'édifier la théorie de manière qu'elle soit, au contraire, en état de prédire les résultats du calcul ..." (Dedekind, 1877, in Dedekind, 1930–2: 296).

[35] See Dedekind (1877, in Dedekind, 1930–2: 268): "Kummer n'a pas défini les nombres idéaux eux-mêmes, mais seulement la divisibilité par ces nombres. ... D'autre part, une définition exacte et qui soit commune à tous les nombres idéaux qu'il s'agit d'introduire dans un domaine déterminé O, et en même temps une définition générale de leur multiplication paraissent d'autant plus nécessaires, que ces nombres idéaux n'existent nullement dans le domaine numérique considéré O. Pour satisfaire à ces exigences, il sera nécessaire et suffisant d'établir une fois pour toutes le caractère commun de toutes les propriétés A, B, C, ..., qui toujours et elles seules, servent à l'introduction de nombres idéaux déterminés, et ensuite d'indiquer généralement comment de deux de ces propriétés A, B, auxquelles correspondent deux nombres idéaux déterminés, on pourra déduire la propriété C qui doit correspondre au produit de ces deux nombres idéaux." ("Kummer did not define ideal numbers themselves, but only divisibility by such numbers. [...] On the other hand, an exact definition which would be common to all ideal numbers that are introduced in a given numerical domain O and, at the same time, a general definition of their multiplication, would seem to be all the more necessary in view of the fact that these ideal numbers do not exist within the given numerical domain O. To satisfy these requirements, it will be necessary and sufficient to establish once and for all the characteristics common to all the properties A, B, C, ..., which always and only serve to introduce specific ideal numbers, and then to indicate generally how, from two such properties A, B, to which there correspond two specific ideal numbers, one can deduce the property C which must correspond to the product of these ideal numbers.")

[36] See Dedekind (1877, in Dedekind, 1930–2: 268).

[37] See the example given in Section 18.2.2.

new idea was to consider the set of all the numbers of the domain which are divisible by a determined ideal number, and he called such a set (or system) an *ideal*. He also showed that every ideal number corresponds to an ideal, and that conversely, every ideal corresponds to an ideal number.

18.6.2   Kronecker's theory of divisors

Concurrent to Dedekind's theory of ideals, Kronecker elaborated a theory of divisors. Both theories aimed to replace and to generalize Kummer's ideal factors. But instead of *creating* new abstract entities, the ideals, and a new calculus with these ideals, Kronecker *built* algebraic expressions composed of algebraic integers of a given "domain of rationality"[38] and indeterminates. His idea was to interpret Kummer's ideal number as the greatest common divisor (GCD) of a finite number of algebraic integers, that is, of numbers of the form $x^n + a_1x^{n-1} + a_2x^{n-2} + \cdots + a_n$, and to find for this GCD a concrete algebraic expression. Now, this GCD does not belong to the domain in question. Kronecker was therefore forced to extend this domain, in order to obtain (by an adequate definition of the divisibility in the new domain) the existence of the GCD of any finite number of algebraic magnitudes of the new domain. In this new domain, the GCD of any set of algebraic integers $x, x', x'', \ldots$, that is, the Kummer ideal number which is contained in them, is expressed by Kronecker as divisor $[x + u'x' + u''x'' \ldots]$ (or module $[x + u'x' + u''x'' \ldots]$), where $u', u'', \ldots$, are indeterminates.[39] Thus, the generalization of Kummer's ideal numbers, which had been obtained by Dedekind by the use of sets, was obtained by Kronecker in a less abstract way by means of indeterminates.

18.6.3   Hilbert's method of ideal elements

Later, in the 1920s, Kummer's creation of ideal factors was further extended by Hilbert into the "method of ideal elements," which consisted of the possibility of adding new formal objects within any determined theory. The extended theory obtained was then uniquely justified by the proof of its consistency. Hilbert praised this method for solving mathematical problems and he qualified it as "a method of genius" since it often saved mathematicians, he said, from some considerable difficulties. He himself applied this method to logical theory in order to justify transfinite assertions.[40]

Kummer can thus be considered as a precursor of the algebraists of the turn of the nineteenth century. However, his creation of ideal factors differed from Dedekind's creations of ideals and even more from Hilbert's method of ideal elements. Indeed, Kummer's ideal factors were only implicit components of existing numbers and neither new concepts (such as Dedekind's ideals), nor formal objects (such as Hilbert's ideal elements). They

---

[38] Kronecker used the term "domain of rationality" to designate an extension (algebraic or transcendent) of the domain of the rational numbers. He refused to use the term "field" (*Körper*) introduced by Dedekind and still used today, in order to avoid introducing a new designation to only mean the gathering of magnitudes.

[39] Kronecker (1881, in Kronecker, 1895–1930, vol. 2).

[40] Hilbert (1926).

differed also from Kronecker's divisors, which are concrete algebraic expressions, even if Kronecker's constructive method was closer to Kummer's own work than the more conceptual and formal methods of Dedekind and Hilbert.

18.7   Conclusion

We have seen that Kummer's theory of ideal numbers aimed at a generalization of the properties of integer numbers to complex numbers, more precisely as a generalization of the arithmetical fundamental property to cyclotomic numbers. Kummer's method was to create ideal factors, which were not, strictly speaking, numbers, but only components of numbers and which revealed their effects only when combined together. We have also seen that analogies with other theories, especially with the arithmetical theory of integer numbers, guided and justified Kummer's method. Finally, Kummer's conception of numbers as natural objects led him to consider arithmetical laws as properties of these objects. Arithmetical laws were also seen as permanent properties of all numbers. Conversely, all those properties which infringed these laws were considered irregularities and, hence, accidental properties. Consequently, Kummer's only recourse was to cancel out these irregularities and to bring to light the implicit permanent properties.

Kummer's theory offers in effect an example of generalization obtained through the notion of "ideality," but this notion was still connected to a conception of mathematics in line with the natural sciences. "Ideal" in Kummer's sense did not mean fictitious, as it would later become, but rather implicit, that is, not yet realized. Ideal factors, thus, were not fictions or pure inventions of a mathematician, but real and implicit components of existing numbers, which any mathematician who was working in this area would necessarily have to discover.

....

APPENDIX 1

In this appendix, we will try to clarify Kummer's use of congruence calculations for determining ideal factors.

Let us recall Kummer's formulation in which f designates an entire function with integer coefficients of roots of unity:

> If p is a prime number of the form $m\lambda + 1$, then it can, in many cases, be represented as a product of $\lambda - 1$ complex factors $p = f(a)f(a^2)\ldots f(a^{\lambda-1})$. Where, however, a decomposition into existing complex factors is not possible, one has to introduce ideal prime factors in order to obtain it.

Relying on the article published in 1847, Kummer's method can be illustrated in the simplest case where p is of the form $m\lambda + 1$ and can be decomposed into a product of $\lambda - 1$ complex factors. In this case, if $f(\alpha)$ is an existing factor of p, it has the property that, if one substitutes a determined root of the congruence $\xi^\lambda \equiv 1 \pmod{p}$ for the equation $\alpha^\lambda = 1$, it follows that $f(\xi) \equiv 0 \pmod{p}$.

As an example, let us consider $\lambda = 5$; $p = 11$ is of the form $2\lambda + 1$, and if α is a primitive fifth root of unity, 11 can be written as the product of four conjugated factors of this number: $11 = (2 + \alpha)(2 + \alpha^2)(2 + \alpha^3)(2 + \alpha^4)$.

In the congruence calculation, 9 is the determined root of the congruence $\xi^\lambda \equiv 1 \pmod p$ [$9^5 \equiv 1 \pmod{11}$] such that $f(\xi) \equiv 0 \pmod p$ [$2 + 9 \equiv 0 \pmod{11}$].

In the cases where p could not be decomposed into a product of $\lambda - 1$ complex factors, Kummer used the previous process in order to determine the ideal factors. Indeed, the congruence equation $\xi^\lambda \equiv 1 \pmod p$ always contains $\lambda - 1$ distinct solutions. Kummer's idea was to connect an ideal factor to every solution of that congruence. Kummer first noted that, when a complex number $\varphi(\alpha)$ contains (i.e., is divisible by) a prime ideal factor $f(\alpha)$, we have also the congruence $\varphi(\xi) \equiv 0 \pmod p$. Conversely, if we have $\varphi(\xi) \equiv 0 \pmod p$, and if p is decomposable into $\lambda - 1$ prime factors, then $\varphi(\alpha)$ contains the prime ideal factor $f(\alpha)$. Kummer generalized this property to ideal factors and used it to determine them. When an irreducible factor is not a prime factor, it contains an ideal factor connected to a solution of the congruence equation $\xi^\lambda \equiv 1 \pmod p$. Thus, this ideal factor will be determined by this congruence relation: a complex number $\varphi(\alpha)$ will be said to be divisible by an ideal factor defined by ξ, if we have the congruence relation $\varphi(\xi) \equiv 0 \pmod p$. One can see that it is the divisibility by the ideal factor which is defined, not the ideal factor itself.[41]

When $p = \lambda$, Kummer recalled that p could always be decomposed into $\lambda - 1$ prime factors: $p = \lambda = (1 - \alpha)(1 - \alpha^2)\ldots(1 - \alpha^{\lambda-1})$. Thus, in this case, p is always decomposable into prime factors.

When p is not of the form $m\lambda + 1$ and distinct from λ, the definition of its prime factors is more difficult and Kummer's proofs are less convincing. We will not deal with this case here.

....

REFERENCES

Boniface, J. (2004) *Hilbert et la notion d'existence en mathématiques*. Paris: Vrin Mathesis.

Boniface, J. (2007) 'The concept of number from Gauss to Kronecker', in Goldstein, C., Schappacher, N., and Schwermer, J. (eds.) *The shaping of arithmetic*. Berlin, Heidelberg, New York: Springer, 315–42.

Cavaillès, J. (1962) *Philosophie mathématique*. Paris: Hermann.

Dedekind, R. (1877) 'Sur la Théorie des Nombres algébriques entiers', *Bulletin des Sciences mathématiques et astronomiques* 1ère série, t. XI: 278–88, 2ème série t. 1: 17–41, in Dedekind (1930–2, vol. 3: 262–96).

Dedekind, R. (1888) *Was sind und was sollen die Zalhen?* Brunswick: Vieweg, in Dedekind (1930–2, vol. 3: 335–91).

Dedekind, R. (1930–2) *Gesammelte mathematische Werke*, vol. 1 1rst publ., Braunschweig, 1930, vol. 2, 1rst publ., Braunschweig, 1931, vol. 3, 1rst publ., New York, Braunschweig, 1932; reed. in 2 vol., 1969.

Ewald, W. (1996) *From Kant to Hilbert, a source book in the foundations of mathematics* (2 vols.). Oxford: Clarendon Press.

Gauss, C.F. (1801) *Disquisitiones Arithmeticae*, in Gauss (1863–1927, vol. 1). English trans. New Haven, 1966.

---

[41] For an interpretation of Kummer's method in modern terms of finite fields and cyclotomic polynomials, see Boniface (2004: 84).

Gauss, C.F. (1828) 'Theoria residuorum biquadraticorum. Commentatio prima.' *Commentationes societatis regiae scientiarum Gottingensis recentiores* Vol. **6**. Presented 5 April 1825. Reprinted in Gauss (1863–1927, vol. 2: 65–92).

Gauss, C.F. (1831) 'Anzeige der Theoria residuorum biquadraticorum. Commentatio secunda.' *Göttingische gelehrte Anzeigen,* 23 April 1831. Reprinted in Gauss 1863–1927, vol. 2: 169–78). English translation by W. Ewald, in Ewald (1996, vol. 1: 306–13).

Gauss, C. F. & Bessel, F.W. (1880) *Briefwechsel zwischen Gauss und Bessel.* Leipzig: ed. A. Auwers. Repr. in Gauss, C.F. (1975) *Werke, Ergänzungsreihe* I. Hildesheim: G. Olms.

Gauss, C.F. (1863–1927) *Werke* (**12** vols.). Göttingen: Königlichen Gesellschaft der Wissenschaften.

Goldstein, C. and Schappacher, N. (2007) 'A Book in search of a Discipline', in Goldstein, C., Schappacher, N., and Schwermer, J. (eds.) (2007) *The shaping of arithmetic.* Berlin, Heidelberg, New York: Springer, 3–65.

Goldstein, C., Schappacher, N., and Schwermer, J. (eds.) (2007) *The shaping of arithmetic.* Berlin, Heidelberg, New York: Springer.

Hilbert, D. (1926) 'Über das Unendliche', *Mathematische Annalen* **95**: 161–90.

Kronecker, L. (1881) *Grundzüge einer arithmetischen Theorie der algebraischen Grössen.* Berlin: G. Reimer, 1882. In Kronecker (1895–1930, vol. 3: 245–387).

Kronecker, L. (1895–1930) *Werke* (K. Hensel (ed.), **5** vols.). Leipzig, Berlin: Teubner. Repr. New York, 1968.

Kummer, E.E. (1844) 'De numeris complexis, qui radicibus unitatis et numeris integris realibus constant', *Journal de mathématiques pures et appliquées* 12 (1[rst] print, Breslau, 1844): 185–212. In Kummer (1975: 165–212).

Kummer, E.E. (1847a) 'Zur Theorie der complexen Zahlen', *Journal für die reine und angewandte Mathematik* 35: 319–26. In Kummer (1975: 203–210).

Kummer, E.E. (1847b) 'Extrait d'une lettre de M. Kummer à M. Liouville', *Journal de mathématiques pures et appliquées* **XII**: 136. In Kummer (1975 : 298).

Kummer, E.E. (1851) 'Mémoire sur la théorie des nombres complexes composés de racines de l'unité et de nombres entiers', *Journal de mathématiques pures et appliquées* **XVI**: 377–498. In Kummer (1975: 363–484).

Kummer, E.E. (1975) *Collected papers vol. I. Contributions to number theory* (A. Weil (ed.)). Berlin, Heidelberg, New York: Springer.

Index